SIXTH EDITION

Fundamentals of Information Systems
Sixth Edition

Ralph M. Stair
Professor Emeritus, Florida State University

George W. Reynolds
Instructor, Strayer University

COURSE TECHNOLOGY
CENGAGE Learning

Australia • Canada • Mexico • Singapore • Spain • United Kingdom • United States

COURSE TECHNOLOGY
CENGAGE Learning™

Fundamentals of Information Systems,
Sixth Edition
Ralph M. Stair & George W. Reynolds

Executive Vice President and Publisher:
Jonathan Hulbert

Executive Vice President of Editorial, Business:
Jack W. Calhoun

Publisher: Joe Sabatino

Sr. Acquisitions Editor: Charles McCormick, Jr.

Sr. Product Manager: Kate Mason

Marketing Manager: Adam Marsh

Sr. Marketing Communications Manager:
Libby Shipp

Marketing Coordinator: Suellen Ruttkay

Editorial Assistant: Nora Heink

Sr. Content Project Manager: Jill Braiewa

Media Editor: Chris Valentine

Sr. Art Director: Stacy Jenkins Shirley

Print Buyer: Julio Esperas

Cover Designer: cmiller design

Cover Photos: © Getty Images

Compositor: Value Chain

Proofreader: Green Pen Quality Assurance

Indexer: Alexandra Nickerson

For product information and technology assistance, contact us at
Cengage Learning Customer & Sales Support, 1-800-354-9706

For permission to use material from this text or product,
submit all requests online at **www.cengage.com/permissions**
Further permissions questions can be emailed to
permissionrequest@cengage.com

Student Edition:
ISBN-13: 978-0-8400-6218-5
ISBN-10: 0-8400-6218-4

Instructor's Edition:
ISBN-13: 978-1-111-53165-2
ISBN-10: 1-111-53165-X

Course Technology
20 Channel Center Street
Boston, MA 02210
USA

Some of the product names and company names used in this book have been used for identification purposes only and may be trademarks or registered trademarks of their respective manufacturers and sellers.

Any fictional data related to persons or companies or URLs used throughout this book is intended for instructional purposes only. At the time this book was printed, any such data was fictional and not belonging to any real persons or companies.

Course Technology, a part of Cengage Learning, reserves the right to revise this publication and make changes from time to time in its content without notice.

Cengage Learning is a leading provider of customized learning solutions with office locations around the globe, including Singapore, the United Kingdom, Australia, Mexico, Brazil and Japan. Locate your local office at:
www.cengage.com/global

Cengage Learning products are represented in Canada by Nelson Education, Ltd.

To learn more about Course Technology, visit **www.cengage.com/coursetechnology**

Purchase any of our products at your local college store or at our preferred online store **www.cengagebrain.com**

Printed in China
2 3 4 5 6 7 16 15 14 13 12 11

For Lila and Leslie
—RMS

To my grandchildren: Michael, Jacob, Jared, Fievel, Aubrey, Elijah, Abrielle, Sofia, Elliot
—GWR

BRIEF CONTENTS

CONTENTS

We are proud to publish the sixth edition of *Fundamentals of Information Systems*. This new edition builds on the success of the previous editions in meeting the need for a concise introductory information systems text. We have listened to feedback from the previous edition's adopters and manuscript reviewers and incorporated many suggestions to refine this new edition. We hope you are pleased with the results.

Like the previous editions, the overall goal of the sixth edition is to develop an outstanding text that follows the pedagogy and approach of our flagship text, *Principles of Information Systems*, with less detail and content. The approach in developing *Fundamentals of Information Systems* is to take the best material from *Principles of Information Systems* and condense it into a text containing nine chapters. So, our most recent edition of *Principles of Information Systems* is the foundation from which we built this new edition of *Fundamentals of Information Systems*.

We have always advocated that education in information systems is critical for employment in almost any field. Today, information systems are used for business processes from communications to order processing to number crunching and in business functions ranging from marketing to human resources to accounting and finance. Regardless of your future occupation, even if you are an entrepreneur, you need to understand what information systems can and cannot do and be able to use them to help you accomplish your work. You will be expected to suggest new uses of information systems and participate in the design of solutions to business problems employing information systems. You will be challenged to identify and evaluate IS options. To be successful, you must be able to view information systems from the perspective of business and organizational needs. For your solutions to be accepted, you must identify and address their impact on coworkers. For these reasons, a course in information systems is essential for students in today's high-tech world.

Fundamentals of Information Systems, Sixth Edition, continues the tradition and approach of the previous editions of this text and *Principles of Information Systems*. Our primary objective is to develop the best IS text and accompanying materials for the first information systems course required of all business students. Using surveys, questionnaires, focus groups, and feedback that we have received from adopters and others who teach in the field, we have been able to develop the highest-quality teaching materials available.

Fundamentals of Information Systems stands proudly at the beginning of the IS curriculum, offering the basic IS concepts that every business student must learn to be successful. This text has been written specifically for the first course in the IS curriculum, and it discusses computer and IS concepts in a business context with a strong managerial emphasis.

APPROACH OF THE TEXT

The overall vision, framework, and pedagogy that make *Principles of Information Systems* so popular are retained in this text. *Fundamentals of Information Systems, Sixth Edition*, offers the traditional coverage of computer concepts, but it places the material within the context of meeting business and organizational needs. Placing IS concepts in this context and taking a general management perspective has always set the text apart from general computer books, thus making it appealing not only to IS majors but to students from other fields of study. The text isn't overly technical, but rather deals with the role that information systems play in an organization and the key principles a manager needs to grasp to be successful. These principles of IS are brought together and presented in a way that is both understandable and relevant. In addition, this book offers an overview of the entire IS discipline, while giving students a solid foundation for further study in advanced IS courses such as the Internet, systems analysis and design, programming, project management, database management, data communications, electronic commerce and mobile commerce applications, decision support, and knowledge management. As such, it serves the needs of both general business students and those who will become IS professionals.

IS Principles First, Where They Belong

Exposing students to fundamental IS principles provides a service to students who do not later return to the discipline for advanced courses. Because most functional areas in business rely on information systems, an understanding of IS principles helps students in other course work. In addition, introducing students to the principles of information systems helps future managers and entrepreneurs avoid mishaps that often result in unfortunate and sometimes costly consequences. Furthermore, presenting IS principles at the introductory level creates interest among general business students who will later choose information systems as a field of concentration.

Author Team

Ralph Stair and George Reynolds have teamed up again for the sixth edition. Together, they have more than sixty years of academic and industrial experience. Ralph Stair brings years of writing, teaching, and academic experience to this text. He has written numerous books and a large number of articles while at Florida State University. George Reynolds brings a wealth of computer and industrial experience to the project, with more than 30 years of experience working in government, institutional, and commercial IS organizations. He has also authored numerous texts and has taught the introductory IS course at the University of Cincinnati, the College of Mount St. Joseph, and Strayer University. The Stair and Reynolds team brings a solid conceptual foundation and practical IS experience to students.

GOALS OF THIS TEXT

Fundamentals of Information Systems has four main goals:

1. To present a core of IS principles with which every business student should be familiar
2. To offer a survey of the IS discipline that will enable all business students to understand the relationship of advanced courses to the curriculum as a whole
3. To present the changing role of the IS professional
4. To show the value of the discipline as an attractive field of specialization

Because *Fundamentals of Information Systems* is written for all business majors, we believe it is important not only to present a realistic perspective of information systems in business but also to provide students with the skills they can use to be effective leaders in their companies.

IS Principles

Fundamentals of Information Systems, Sixth Edition, although comprehensive, cannot cover every aspect of the rapidly changing IS discipline. The authors, having recognized this, provide students an essential core of guiding IS principles to use as they face the career challenges ahead. Think of principles as basic truths, rules, or assumptions that remain constant regardless of the situation. As such, they provide strong guidance in the face of tough decisions. A set of IS principles is highlighted at the beginning of each chapter. The application of these principles to solve real-world problems is driven home from the opening vignettes to the end-of-chapter material. The ultimate goal of *Fundamentals of Information Systems* is to develop effective, thinking, action-oriented employees by instilling them with principles to help guide their decision making and actions.

Survey of the IS Discipline

This text not only offers the traditional coverage of computer concepts, but stresses the broad framework to provide students with solid grounding in business uses of technology. In addition to serving general business students, this book offers an overview of the entire IS discipline and solidly prepares future IS professionals for advanced IS courses and their careers in the rapidly changing IS discipline.

Changing Role of the IS Professional

As business and the IS discipline have changed, so too has the role of the IS professional. Once considered a technical specialist, today the IS professional operates as an internal consultant to all functional areas of the organization, being knowledgeable about their needs and competent in bringing the power of information systems to bear throughout the organization. The IS professional views issues through a global perspective that encompasses the entire organization and the broader industry and business environment in which it operates, including the entire interconnected network of suppliers, customers, competitors, regulatory agencies, and other entities—no matter where they are located.

The scope of responsibilities of an IS professional today is not confined to just his/her employer but encompasses the entire interconnected network of employees, suppliers, customers, competitors, regulatory agencies, and other entities, no matter where they are located. This broad scope of responsibilities creates a new challenge: how to help an organization survive in a highly interconnected, highly competitive global environment. In accepting that challenge, the IS professional plays a pivotal role in shaping the business itself and ensuring its success. To survive, businesses must now strive for the highest level of customer satisfaction and loyalty through innovative products and services, competitive prices, and ever-improving product and service quality. The IS professional assumes the critical responsibility of determining the organization's approach to both overall cost and quality performance and therefore plays an important role in the ongoing survival of the organization. This new duality in the role of the IS employee—a professional who exercises a specialist's skills with a generalist's perspective—is reflected throughout the book.

IS as a Field for Further Study

Despite the deep recession and effects of outsourcing, a survey of Human Resources professionals still puts technology and health care as the top fields of study. Business administration and computer science remain among the most sought after majors by employers. Indeed the long-term job prospects for skilled information systems professionals are optimistic. Employment of such workers is expected to grow faster than the average for all occupations through the year 2018.

A career in information systems can be exciting, challenging, and rewarding! This text shows the value of the discipline as an appealing field of study and the IS graduate as an integral part of today's organizations.

CHANGES IN THE SIXTH EDITION

We have implemented a number of exciting changes to the text based on user feedback on how the text can be aligned even more closely with how the IS principles and concepts course is now being taught. The following list summarizes these changes:

All new opening vignettes. All of the chapter-opening vignettes are new, and continue to raise actual issues from foreign-based or multinational companies.

All new Information Systems @ Work special interest boxes. Highlighting current topics and trends in today's headlines, these boxes show how information systems are used in a variety of business career areas.

All new Ethical and Societal Issues special interest boxes. Focusing on ethical issues today's professionals face, these boxes illustrate how information systems professionals confront and react to ethical dilemmas.

All new case studies. Two new cases at the end of every chapter provide a wealth of practical information for students and instructors. Each case explores a chapter concept or problem that a real-world company or organization has faced. The cases can be assigned as homework exercises or serve as a basis for class discussion. In addition, a new integrated, comprehensive, Web case builds from chapter to chapter, featuring a single fictional company facing issues and decisions introduced in the chapter.

Each chapter has been completely updated with the latest topics and examples. The following list summarizes these changes.

Chapter 1, An Introduction to Information Systems in Organizations includes over 130 new references, examples, and material on how organizations use information to deliver the right information to the right person at the right time and how they convert raw data into useful information. Topics include high-frequency trading, knowledge workers and knowledge management systems, and organizational efficiency and effectiveness. We updated the material on computer-based information systems, EDR used by police officers, operating systems such as Android by Google and Mobile 6.5 by Microsoft, Nvidia's GeForce 3D software, database systems, and social networking sites. We also updated coverage of supply chain management and customer relationship management, employee empowerment, organizational change, reengineering, outsourcing, utility computing, and return on investment. Other updates cover quality control, careers in information systems, the H-1B program, the best places to work in the information systems field, the U.S. federal government's CIO position, IS entrepreneurs, professional organizations and users groups, and guidelines for finding a good IS job.

Chapter 2, Hardware and Software has been completely updated to cover the latest hardware and software developments. New hardware topics include a discussion of smaller and less energy-consuming processors, netbook and nettop computers, smartphones, and the relative power usage of various types of monitors. We provide the latest information on the world's most powerful supercomputers. A new section titled "Green Computing" examines the efficient and environmentally responsible design, manufacture, operation, and disposal of IS-related products. We also added new material, figures, examples, and photos to the software section to cover the latest software developments including rich Internet applications, software for mobile devices and smartphones, touch display interfaces and spoken commands, Windows 7, application programming interfaces (APIs), software development kits (SDKs), Mac OS X Snow Leopard, Linux-based Chrome OS from Google, and enterprise-scale systems such as z/OS and HP-UX. We also explore Ford Sync, utility programs, virtualization software, application service providers (ASPs), software as a Service (SaaS), cloud computing, and software suites from Microsoft and others. This chapter concludes by exploring important software issues and trends.

Chapter 3, Database Systems, Data Centers, and Business Intelligence has a new title, new material, and updated examples on data centers in today's organizations, including modular data centers built in shipping containers packed with racks of prewired servers. We highlight new database applications, the effort to collect data on the damage caused by the 2010 oil spill disaster, the importance of database security in database administration, relational database management techniques and vendors, the Unified Database of Places, popular database management systems, open source database systems, databases as a service (DaaS), and a variety of database applications for mobile devices, PCs, workgroup systems, and mainframe computers. We provide new material on middleware, data warehouses, data marts, data mining, predictive analysis, online transaction processing, distributed databases, Online Analytical Processing (OLAP), and Object-Relational Database Management Systems.

Chapter 4, Telecommunications, the Internet, Intranets, and Extranets provides new material on circuit switching and packet switching networks, Intel's Centrino strategy to deliver innovative mobile platforms "anytime, anywhere," and the availability of smartphones and hundreds of smartphone applications. We included new examples on the practical use of all popular guided and unguided communications methods such as broadband over power lines, Bluetooth communications, Wi-Fi, satellite communications, 3G, 4G, and Wi-Max. We also offer new examples for common telecommunications applications such as remote network monitoring, virtual private networks, voice over Internet protocol (VoIP), and call centers. A new table shows the evolution of 3G and 4G networks. Additional new examples illustrate the use of personal, local area, metropolitan area, and wide area networks to achieve business benefits. We expanded the information on global positioning satellites, and added new examples and material on the Internet, intranets, and extranets, the transformation of the Web to a platform for computing and community, how the Internet works, and accessing the Internet. We revised many sections to reflect today's developments, including new material on cloud computing, the World Wide Web, how the Web works, Web programming languages, developing Web content and applications, and Internet and Web applications.

Chapter 5, Electronic and Mobile Commerce and Enterprise Systems has been thoroughly updated with new material, tables, and examples. We updated and expanded the coverage of mobile commerce to include mobile price comparison services, new information about mobile advertising networks, and the controversy behind Google's purchase of Admob. We provide new examples of phishing, click fraud, and online fraud. We also discuss the p-card as a form of credit card used to streamline the traditional purchase order and invoice payment processes. We updated the section on the use of mobile phones for payment by discussing new modes of payments and new services such as Obopay and Boku. New tables summarize the primary differences between B2B, B2C, and C2C and list the top rated m-commerce retail and B2C Web sites. New examples show how organizations that implemented transaction processing systems or ERP systems improved their operations. Other new examples highlight organizations that had significant problems implementing their ERP systems. A new section identifies and provides a brief description of the leading ERP systems.

Chapter 6, Information and Decision Support Systems includes more than 80 new references and new material, including a potential data security breakthrough in the section on the intelligence stage of decision making. We also discuss the importance of cost reduction, saving money, and preserving cash reserves. Among the many new examples of management information systems are a children's hospital that monitored and then reduced the death rate for its patients by improving its information systems and a company that used optimization to assign medical personnel to home healthcare patients in Sweden. Other examples include how many drug manufacturing companies use JIT to produce flu vaccinations just before the flu season and how Chrysler uses FMS to quickly change from manufacturing diesel minivans with right-hand drive to gasoline minivans with left-hand drive. We completely updated the DSS section, providing new material and examples, such as a California software company executive that used his cell phone to get rapid feedback on corporate financial performance and timely reports on the performance of his key executives; Organic's use of a team of economists and statisticians to develop models that predicted the effectiveness of advertising alternatives; and a new pilot program from IBM that allows companies to analyze data and make strategic decisions on the fly.

Chapter 7, Knowledge Management and Specialized Information Systems includes about 80 new references, examples, and material on knowledge management and specialized business information systems, artificial intelligence, expert systems, multimedia, virtual reality, and many other specialized systems. A study of a large information systems consulting firm found a return of $18.60 on every dollar invested in its knowledge management system, representing over 1,000 percent return on investment (ROI). We have also updated the material on communities of practice, the brain computer interface (BCI), robotics, voice recognition, neural networks, expert systems, and virtual reality. The section on virtual reality now includes a new section on multimedia.

Chapter 8, Systems Development includes about 110 new references, examples, and material on systems development. We included a new section on individual systems developers and users that covers developing applications for Apple's applications store (App Store), BlackBerry's App World, and Google's Android Market store. The section on long-range planning in systems development projects has new material and examples. We also have new examples in the section on IS planning and aligning corporate and IS goals and systems development failure. We discuss the failover approach, including SteelEye's LifeKeeper and Continuous Protection by NeverFail. We have new material and examples on environmental design (green design) including systems development efforts that slash power consumption, take less physical space, and result in systems that can be disposed of without harming the environment. We updated the sections on systems implementation and maintenance and review to include new references, material, or examples.

Chapter 9, The Personal and Social Impact of Computers includes all new examples and updated statistics. We quote CIA Director Leon Panetta and FBI Director Robert Mueller to illustrate the increasing concern about potential cyberterrorist attacks. We present new information about the potential of Internet gambling as an untapped source of income for state and federal governments. We added a new section called "The Computer as a Tool to Fight Crime," which provides examples of systems used by law enforcement. We reorganized the material in the "Computer as the Object of Crime" section to consolidate information about types of computer crimes and criminals, and now discuss all the tools and measures to prevent crime in the "Preventing Computer- Related Crime" section. A single table now summarizes the common types of malware. We also discuss vishing (a variation on phishing), explain that Google is involved in the largest copyright infringement case in history, and cover HTC and Apple as examples of competing organizations that sue each other over patent infringement. We also discuss money laundering with an example of one person arrested for laundering over $500 million.

WHAT WE HAVE RETAINED FROM THE FIFTH EDITION

The sixth edition builds on what has worked well in the past; it retains the focus on IS principles and strives to be the most current text on the market.

Overall principle. This book continues to stress a single, all-encompassing theme: The right information, if it is delivered to the right person, in the right fashion, and at the right time, can improve and ensure organizational effectiveness and efficiency.

Information systems principles. Information system principles summarize key concepts that every student should know. These principles are highlighted at the start of each chapter and covered thoroughly in the text.

Global perspective. The global aspects of information systems is a major theme.

Learning objectives linked to principles. Carefully crafted learning objectives are included with every chapter. The learning objectives are linked to the Information Systems principles and reflect what a student should be able to accomplish after completing a chapter.

Opening vignettes emphasize international aspects. All of the chapter-opening vignettes raise actual issues from foreign-based or multinational companies.

Why Learn About features. Each chapter has a "Why Learn About" section at the beginning of the chapter to pique student interest. The section sets the stage for students by briefly describing the importance of the chapter's material to the students-whatever their chosen field.

Information Systems @ Work special interest boxes. Each chapter has an entirely new Information Systems @ Work box that shows how information systems are used in a variety of business career areas.

Ethical and Societal Issues special interest boxes. Each chapter includes a new "Ethical and Societal Issues" box that presents a timely look at the ethical challenges and the societal impact of information systems

Current examples, boxes, cases, and references. As we have in each edition, we take great pride in presenting the most recent examples, boxes, cases, and references throughout the text. Some of these were developed at the last possible moment, literally weeks before the book went into publication. Information on new hardware and software, the latest operating systems, mobile commerce, the Internet, electronic commerce, ethical and societal issues, and many other current developments can be found throughout the text. Our adopters have come to expect the best and most recent material. We have done everything we can to meet or exceed these expectations.

Summary linked to principles. Each chapter includes a detailed summary with each section of the summary tied to an associated information system principle.

Self-assessment tests. This popular feature helps students review and test their understanding of key chapter concepts.

Career exercises. End-of-chapter career exercises ask students to research how a topic discussed in the chapter relates to a business area of their choice. Students are encouraged to use the Internet, the college library, or interviews to collect information about business careers.

End-of-chapter cases. Two new end-of-chapter cases provide students with an opportunity to apply the principles covered to real-world problems from actual organizations. The cases can be assigned as individual homework exercises or serve as a basis for class discussion.

Integrated, comprehensive, Web case. The Altitude Online cases at the end of each chapter provide an integrated and comprehensive case that runs throughout the text. The cases follow the activities of the Altitude Online firm as it is challenged to complete various IS- related projects. The cases provide a realistic fictional work environment in which students may imagine themselves in the role of systems analyst. Information systems problems are addressed using the state-of-the-art techniques discussed in the chapters.

STUDENT RESOURCES

CourseMate

The more you study, the better the results. Make the most of your study time by accessing everything you need to succeed in one place. Read your textbook, take notes, review flashcards, and take practice quizzes—online with CourseMate. *Fundamentals of Information Systems, Sixth Edition* CourseMate includes:

- An interactive eBook with highlighting, note taking, and an interactive glossary
- Interactive learning tools, including:
  - Quizzes
  - Flashcards
  - PowerPoint presentations
  - Classic cases
  - Links to useful Web sites
  - and more!

PowerPoint Slides

Direct access is offered to the book's PowerPoint presentations that cover the key points from each chapter. These presentations are a useful study tool.

Classic Cases

A frequent request from adopters is that they'd like a broader selection of cases to choose from. To meet this need, a set of over 200 cases from previous editions of the text are included here. These are the authors' choices of the "best cases" and span a broad range of companies and industries.

Links to Useful Web Sites

Chapters in *Fundamentals of Information Systems, Sixth Edition* reference many interesting Web sites. This resource takes you to links you can follow directly to the home pages of those sites so that you can explore them. There are additional links to Web sites that the authors think you would be interested in checking out.

Hands-On Activities

Use these hands-on activities to test your comprehension of IS topics and enhance your skills using Microsoft® Office applications and the Internet. Using these links, you can access three critical-thinking exercises per chapter; each activity asks you to work with an Office tool or do some research on the Internet.

Quizzes

This tool allows you to access 20 multiple-choice questions for each chapter; test yourself and then submit your answers. You will immediately find out what questions you got right and what you got wrong. For each question that you answer incorrectly, you are given the correct answer and the page in your text where that information is covered.

Glossary of Key Terms

The glossary of key terms from the text is available to search.

Online Readings

This feature provides you access to a computer database that contains articles relating to hot topics in Information Systems.

INSTRUCTOR RESOURCES

The teaching tools that accompany this text offer many options for enhancing a course. And, as always, we are committed to providing one of the best teaching resource packages available in this market.

CourseMate

Cengage Learning's *Fundamentals of Information Systems, Sixth Edition's* CourseMate brings course concepts to life with interactive learning, study, and exam preparation tools that support the printed textbook. Watch student comprehension soar as your class works with the printed textbook and the textbook-specific Web site. CourseMate goes beyond the book to deliver what you need! Learn more at *cengage.com/coursemate*.

- **Engagement Tracker**
  How do you assess your students' engagement in your course? How do you know your students have read the material or viewed the resources you've assigned? How can you tell if your students are struggling with a concept? With CourseMate, you can use the included Engagement Tracker to assess student preparation and engagement. Use the tracking tools to see progress for the class as a whole or for individual students. Identify students at risk early in the course. Uncover which concepts are most difficult for your class. Monitor time on task. Keep your students engaged.

- **Interactive Teaching and Learning Tools**
  CourseMate includes interactive teaching and learning tools:
  - Quizzes
  - Flashcards

- Games
- and more

These assets enable students to review for tests, prepare for class, and address the needs of students' varied learning styles.

- **Interactive eBook**

  In addition to interactive teaching and learning tools, CourseMate includes an interactive eBook. Students can take notes, highlight, search for, and interact with embedded media specific to their book. Use it as a supplement to the printed text, or as a substitute—the choice is your students' with CourseMate.

Instructor's Manual

An all-new *Instructor's Manual* provides valuable chapter overviews; highlights key principles and critical concepts; offers sample syllabi, learning objectives, and discussion topics; and features possible essay topics, further readings and cases, and solutions to all of the end-of-chapter questions and problems, as well as suggestions for conducting the team activities. Additional end-of-chapter questions are also included. As always, we are committed to providing the best teaching resource packages available in this market.

Sample Syllabus

A sample syllabus for both a quarter and semester-length course are provided with sample course outlines to make planning your course that much easier.

Solutions

Solutions to all end-of-chapter material are provided in a separate document for your convenience.

Test Bank and Test Generator

ExamView® is a powerful objective-based test generator that enables instructors to create paper-, LAN- or Web-based tests from test banks designed specifically for their Course Technology text. Instructors can utilize the ultra-efficient QuickTest Wizard to create tests in less than five minutes by taking advantage of Course Technology's question banks or customizing their own exams from scratch. Page references for all questions are provided so you can cross-reference test results with the book.

PowerPoint Presentations

A set of impressive Microsoft PowerPoint slides is available for each chapter. These slides are included to serve as a teaching aid for classroom presentation, to make available to students on the network for chapter review, or to be printed for classroom distribution. Our presentations help students focus on the main topics of each chapter, take better notes, and prepare for examinations. Instructors can also add their own slides for additional topics they introduce to the class.

Figure Files

Figure files allow instructors to create their own presentations using figures taken directly from the text.

ACKNOWLEDGMENTS

A book of this scope and undertaking requires a strong team effort. We would like to thank all of our fellow teammates at Cengage for their dedication and hard work. We would like to thank Charles McCormick, our Sr. Acquisitions Editor, for his overall leadership and guidance on this effort. Special thanks to Kate Mason, our Product Manager. Our appreciation goes out to all the many people who worked behind the scenes to bring this effort to fruition including Abigail Reip, our photo researcher. We would like to acknowledge and thank Lisa Ruffolo, our development editor, who deserves special recognition for her tireless effort and help in all stages of this project. Thanks also to Jill Braiewa, our Sr. Content Project Manager, who shepherded the book through the production process.

We are grateful to the salesforce at Cengage whose efforts make this all possible. You helped to get valuable feedback from current and future adopters. As Cengage product users, we know how important you are.

We would especially like to thank Ken Baldauf for his excellent help in writing the boxes and cases and revising several chapters for this edition. Ken also provided invaluable feedback on other aspects of this project.

Ralph Stair would like to thank the chairman and faculty members of the Department of Management, College of Business Administration, at Florida State University. He would also like to thank his family, Lila and Leslie, for their support.

George Reynolds would like to thank his wife, Ginnie, for her patience and support in this major project.

TO OUR PREVIOUS ADOPTERS AND POTENTIAL NEW USERS

We sincerely appreciate our loyal adopters of the previous editions and welcome new users of *Fundamentals of Information Systems, Sixth Edition.* As in the past, we truly value your needs and feedback. We can only hope this new edition continues to meet your high expectations.

OUR COMMITMENT

We are committed to listening to our adopters and readers and to developing creative solutions to meet their needs. The field of Information Systems continually evolves, and we strongly encourage your participation in helping us provide the freshest, most relevant information possible.

We welcome your input and feedback. If you have any questions or comments regarding *Fundamentals of Information Systems, Sixth Edition,* please contact us through Course Technology or your local representative, or via the Internet at *www.cengage.com/mis/ stairreynolds.com.*

Information Systems in Perspective

CHAPTER
· 1 ·

An Introduction to Information Systems in Organizations

PRINCIPLES	LEARNING OBJECTIVES
▪ The value of information is directly linked to how it helps decision makers achieve the organization's goals.	▪ Distinguish data from information and describe the characteristics used to evaluate the quality of data.
▪ Knowing the potential impact of information systems and having the ability to put this knowledge to work can result in a successful personal career, organizations that reach their goals, and a society with a higher quality of life.	▪ Identify the basic types of business information systems and discuss who uses them, how they are used, and what kinds of benefits they deliver.
▪ System users, business managers, and information systems professionals must work together to build a successful information system.	▪ Identify the major steps of the systems development process and state the goal of each.
▪ The use of information systems to add value to the organization can also give an organization a competitive advantage.	▪ Identify the value-added processes in the supply chain and describe the role of information systems within them.
	▪ Identify some of the strategies employed to lower costs or improve service.
	▪ Define the term *competitive advantage* and discuss how organizations are using information systems to gain such an advantage.
▪ IS personnel is a key to unlocking the potential of any new or modified system.	▪ Define the types of roles, functions, and careers available in information systems.

Information Systems in the Global Economy
Braskem S.A., Brazil

The Power of Information in the Petrochemical Industry

You've probably heard that "information is power." In fact, the power of information depends on how it serves a specific need at a certain time. For example, when you are deciding which automobile to buy, the fact that the Yankees won the 2009 World Series is of no value to you. Information is most powerful when it enables strategic decision making. It must be delivered to the right person at the right time with as little effort as possible. For businesses, correctly managing strategic information can mean the difference between success and failure. Consequently, today's businesses invest a large percentage of their budgets in systems designed to deliver the right information to the right people at the right time. Such is the case for Braskem S.A.

Braskem S.A. is the largest petrochemical company in Latin America, with annual revenue of $13 billion (US) and 5,500 employees. Braskem was created in 2002 out of the merger of six Brazilian companies. Its 13 chemical plants produce basic raw materials such as ethylene, propylene, and chlorine, which are used in the production of thermoplastic resins. Braskem then sells the resins to manufacturers of plastic products. Toothbrushes, baby bottles, backpacks, automotive parts, and computer parts are all made from thermoplastic resins produced by Braskem, ExxonMobile, Dow Chemical, and other petrochemical companies.

Recently, Braskem invested heavily in an information systems (IS) development effort to provide all of its 4,000 office and production staff access to information from one central source using one system. In planning and developing the new system, Braskem IS managers needed to consider many factors. The system would handle science and research information as well as production, business, and financial information. Such enterprise-wide systems are often referred to as enterprise resource planning systems (ERPs). Braskem wanted the system to be implemented within a year—a tall order for an ERP. Braskem executives also wanted the system to help the company's employees make it one of the world's top 10 petrochemical companies.

Although this may seem a lot to ask of an IS, information systems do directly influence the implementation of smart business processes. An IS can either hamper people from proper business practices or it can help them establish best practices across an organization. "Best practices" refers to insightful business practices that are proven to provide a competitive advantage. Braskem wanted its new information systems to help establish best practices and streamline its essential business processes. Braskem's chief information officer (CIO), Stefan Lanna Lepecki, investigated what type of information systems the top global petrochemical companies were using. He soon discovered that 9 of the top 10 companies used information systems developed by SAP.

SAP is a multinational software development and consulting corporation with headquarters in Waldorf, Germany. Having worked with major petrochemical companies, SAP system engineers were well acquainted with the business and with systems that guide best business practices. After gaining the approval of the steering committee, top executives, and even the workers in the plant, Braskem hired SAP to build the new system. Rather than viewing the project as a technology initiative, Braskem embraced it as a business process transformation. Systems engineers, business managers, and hourly employees would all be involved.

Braskem's CIO kept customization requests to a minimum to implement a system that, for the most part, used the same standard SAP software that other petrochemical companies used. The system required Braskem to get a new technology infrastructure including new hardware, databases, telecommunications equipment, and software. It was implemented within one year. In the final stages of development, Braskem instituted a rigorous training regimen for the 4,000 employees who would be working with the system. Using simulations, each employee was required to advance through eight skill levels before being allowed to use the real system. Although training required 63,930 people hours, it ensured that employees used the best practices and procedures that the system supported. The result was an improvement of business processes across the enterprise.

Braskem no longer suffers the frustration of working with different systems at different sites. Today, information flows freely among Braskem's plants and offices, with executives, managers, and employees accessing up-to-the-minute information from any Braskem location. They can also access the system from mobile devices when they travel. The company has reduced its maintenance, repair, and operations costs. The improved efficiency of its systems also allows Braskem to reduce the amount of inventory it keeps on hand because inventory now ships when it rolls off the production line. In general, business tasks require fewer people and take less time with the new system. The system also complies with government regulations such as the Sarbanes-Oxley Act designed to keep business practices transparent. The new IS puts Braskem in an ideal position to gain market share and reach its goals.

As you read this chapter, consider the following:

- How might the information system used at Braskem depend on the various components of a computer-based information system: hardware, software, databases, telecommunications, people, and procedures?
- How do computer-based information systems like Braskem's help businesses implement best practices?

Why Learn About Information Systems in Organizations?

Information systems are used in almost every imaginable profession. Entrepreneurs and small business owners use information systems to reach customers around the world. Sales representatives use information systems to advertise products, communicate with customers, and analyze sales trends. Managers use them to make multi-million-dollar decisions, such as whether to build a manufacturing plant or research a cancer drug. Financial advisors use information systems to advise their clients to help them save for their children's education and retirement. From a small music store to huge multinational companies, businesses of all sizes could not survive without information systems to perform accounting and finance operations. Regardless of your college major or chosen career, information systems are indispensable tools to help you achieve your career goals. Learning about information systems can help you land your first job, earn promotions, and advance your career.

Why learn about information systems in organizations? What is in it for you? Learning about information systems will help you achieve your goals. Let's get started by exploring the basics of information systems.

People and organizations use information every day. Many retail chains, for example, collect data from their stores to help them stock what customers want and to reduce costs. The components that are used are often called an information system. An **information system (IS)** is a set of interrelated components that collect, manipulate, store, and disseminate data and information and provide a feedback mechanism to meet an objective.[1] It is the feedback mechanism that helps organizations achieve their goals, such as increasing profits or improving customer service.[2] Businesses can use information systems to increase revenues and reduce costs. This book emphasizes the benefits of an information system, including speed, accuracy, increased revenues, and reduced costs.

Today we live in an information economy.[3] Information itself has value, and commerce often involves the exchange of information rather than tangible goods. Systems based on computers are increasingly being used to create, store, and transfer information. Using information systems, investors make multimillion-dollar decisions, financial institutions transfer billions of dollars around the world electronically, and manufacturers order supplies and distribute goods faster than ever before. Computers and information systems will continue to change businesses and the way we live. To prepare for these innovations, you need to be familiar with fundamental information concepts.

INFORMATION CONCEPTS

Information is a central concept of this book. The term is used in the title of the book, in this section, and in almost every chapter. To be an effective manager in any area of business, you need to understand that information is one of an organization's most valuable resources. This term, however, is often confused with *data*.

Data, Information, and Knowledge

Data consists of raw facts, such as an employee number, total hours worked in a week, inventory part numbers, or sales orders. As shown in Table 1.1, several types of data can represent these facts. When facts are arranged in a meaningful manner, they become information. **Information** is a collection of facts organized and processed so that they have additional value beyond the value of the individual facts. For example, sales managers might find that knowing the total monthly sales suits their purpose more (i.e., is more valuable) than knowing the number of sales for each sales representative. Providing information to customers can also help companies increase revenues and profits. FedEx, a worldwide leader in shipping packages and products around the world, believes that information about a package can be as important as the package itself for many of its customers.[4] Increasingly, information generated by FedEx and other organizations is being placed on the Internet. In addition, many universities are now placing course information and content on the Internet. Using the Open Course Ware program, the Massachusetts Institute of Technology (MIT) places class notes and contents on the Internet for many of its courses.[5]

data
Raw facts, such as an employee number, total hours worked in a week, inventory part numbers, or sales orders.

information
A collection of facts organized in such a way that they have additional value beyond the value of the individual facts.

Data	Represented by
Alphanumeric data	Numbers, letters, and other characters
Image data	Graphic images and pictures
Audio data	Sound, noise, or tones
Video data	Moving images or pictures

Table 1.1

Types of Data

Data represents real-world things. Hospitals and healthcare organizations, for example, maintain patient medical data, which represents actual patients with specific health situations. In many cases, hospitals and healthcare organizations are converting data to electronic form. Some have developed electronic records management (ERM) systems to store, organize, and control important data. However, data—raw facts—has little value beyond its existence. The U.S. federal stimulus plan could invest as much as $2 billion into helping healthcare organizations develop a medical records program to store and use the vast amount of medical data that is generated each year.[6] Medical records systems can be used to generate critical health-related information, saving money and lives.

Here is another example of the difference between data and information. Consider data as pieces of railroad track in a model railroad kit. Each piece of track has limited inherent value as a single object. However, if you define a relationship among the pieces of the track, they will gain value. By arranging the pieces in a certain way, a railroad layout begins to emerge (see Figure 1.1a). Data and information work the same way. Rules and relationships can be set up to organize data into useful, valuable information.

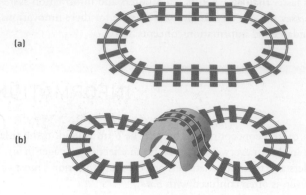

(a)

(b)

The type of information created depends on the relationships defined among existing data. For example, you could rearrange the pieces of track to form different layouts. Adding new or different data means you can redefine relationships and create new information. For instance, adding new pieces to the track can greatly increase the value—in this case, variety and fun—of the final product. You can now create a more elaborate railroad layout (see Figure 1.1b). Likewise, a sales manager could add specific product data to his or her sales data to create monthly sales information organized by product line. The manager could use this information to determine which product lines are the most popular and profitable.

Turning data into information is a **process**, or a set of logically related tasks performed to achieve a defined outcome. The process of defining relationships among data to create useful information requires knowledge. **Knowledge** is the awareness and understanding of a set of information and the ways that information can be made useful to support a specific task or reach a decision. Having knowledge means understanding relationships in information. Part of the knowledge you need to build a railroad layout, for instance, is the understanding of how much space you have for the layout, how many trains will run on the track, and how fast they will travel. Selecting or rejecting facts according to their relevancy to particular tasks is based on the knowledge used in the process of converting data into information. Therefore, you can also think of information as data made more useful through the application of knowledge. *Knowledge workers (KWs)* are people who create, use, and disseminate knowledge and are usually professionals in science, engineering, business, and other areas.[7] A *knowledge management system (KMS)* is an organized collection of people, procedures, software, databases, and devices used to create, store, and use the organization's knowledge and experience.[8] Research has shown that the success of a KMS is linked to how easy it is to use and how satisfied users are with it.[9]

In some cases, people organize or process data mentally or manually. In other cases, they use a computer. Where the data comes from or how it is processed is less important than whether the data is transformed into results that are useful and valuable. This transformation process is shown in Figure 1.2.

process
A set of logically related tasks performed to achieve a defined outcome.

knowledge
The awareness and understanding of a set of information and ways that information can be made useful to support a specific task or reach a decision.

Figure 1.2

The Process of Transforming
Data into Information

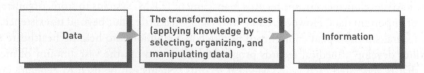

The Characteristics of Valuable Information

To be valuable to managers and decision makers, information should have the characteristics described in Table 1.2. These characteristics make the information more valuable to an organization. Many shipping companies, for example, can determine the exact location of inventory items and packages in their systems, and this information makes them responsive to their customers. In contrast, if an organization's information is not accurate or complete, people can make poor decisions, costing thousands, or even millions, of dollars. If an inaccurate forecast of future demand indicates that sales will be very high when the opposite is true, an organization can invest millions of dollars in a new plant that is not needed. Furthermore, if information is not relevant, not delivered to decision makers in a timely fashion, or too complex to understand, it can be of little value to the organization.

Table 1.2

Characteristics of Valuable Information

Characteristics	Definitions
Accessible	Information should be easily accessible by authorized users so they can obtain it in the right format and at the right time to meet their needs.
Accurate	Accurate information is error free. In some cases, inaccurate information is generated because inaccurate data is fed into the transformation process. (This is commonly called garbage in, garbage out [GIGO].)
Complete	Complete information contains all the important facts. For example, an investment report that does not include all important costs is not complete.
Economical	Information should also be relatively economical to produce. Decision makers must always balance the value of information with the cost of producing it.
Flexible	Flexible information can be used for a variety of purposes. For example, information on how much inventory is on hand for a particular part can be used by a sales representative in closing a sale, by a production manager to determine whether more inventory is needed, and by a financial executive to determine the total value the company has invested in inventory.
Relevant	Relevant information is important to the decision maker. Information showing that lumber prices might drop might not be relevant to a computer chip manufacturer.
Reliable	Reliable information can be trusted by users. In many cases, the reliability of the information depends on the reliability of the data-collection method. In other instances, reliability depends on the source of the information. A rumor from an unknown source that oil prices might go up might not be reliable.
Secure	Information should be secure from access by unauthorized users.
Simple	Information should be simple, not overly complex. Sophisticated and detailed information might not be needed. In fact, too much information can cause information overload, whereby a decision maker has too much information and is unable to determine what is really important.
Timely	Timely information is delivered when it is needed. Knowing last week's weather conditions will not help when trying to decide what coat to wear today.
Verifiable	Information should be verifiable. This means that you can check it to make sure it is correct, perhaps by checking many sources for the same information.

Depending on the type of data you need, some quality attributes become more valuable than others. For example, with market-intelligence data, some inaccuracy and incompleteness is acceptable, but timeliness is essential. Getco, a Chicago-based stock-trading company, requires the most timely market information possible so it can place profitable trades.[10] Getco uses an approach called high-frequency trading that requires powerful and very fast computers to make its trades. On some days, Getco can account for 10 to 20 percent of the total trading volume for some stocks. Market intelligence might alert you that competitors are about to make a major price cut. The exact details and timing of the price cut might not be as important as being warned far enough in advance to plan how to react. On the other hand, accuracy, verifiability, and completeness are critical for data used in accounting to manage company assets such as cash, inventory, and equipment.

The Value of Information

The value of information is directly linked to how it helps decision makers achieve their organization's goals. Valuable information can help people in their organizations perform tasks more efficiently and effectively. Consider a market forecast that predicts a high demand for a new product. If you use this information to develop the new product and your company makes an additional profit of $10,000, the value of this information to the company is $10,000 minus the cost of the information. Valuable information can also help managers decide whether to invest in additional information systems and technology. A new computerized ordering system might cost $30,000 but generate an additional $50,000 in sales. The *value added* by the new system is the additional revenue from the increased sales of $20,000. Most corporations have cost reduction as a primary goal. Using information systems, some manufacturing companies have slashed inventory costs by millions of dollars. Other companies have increased inventory levels to increase profits. Walmart, for example, uses information about certain regions of the country and specific situations to increase needed inventory levels of certain products and improve overall profitability. In other cases, the value of information can be realized in cost savings. Shermag, a Canadian furniture manufacturing company, was able to use a sophisticated computer system to achieve the company's cost reduction goal.[11] The company was able to reduce total costs by more than 20 percent by using optimization software to reduce material and manufacturing costs.

WHAT IS AN INFORMATION SYSTEM?

As mentioned previously, an information system (IS) is a set of interrelated elements or components that collect (input), manipulate (process), store, and disseminate (output) data and information and provide a corrective reaction (feedback mechanism) to meet an objective (see Figure 1.3). The feedback mechanism is the component that helps organizations achieve their goals, such as increasing profits or improving customer service.

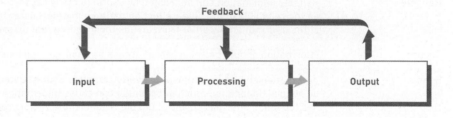

Input, Processing, Output, Feedback

Input

input
The activity of gathering and capturing raw data.

In information systems, **input** is the activity of gathering and capturing raw data. In producing paychecks, for example, the number of hours every employee works must be collected before paychecks can be calculated or printed. In a university grading system, instructors must submit student grades before a summary of grades for the semester or quarter can be compiled and sent to students.

Processing

processing
Converting or transforming data into useful outputs.

In information systems, **processing** means converting or transforming data into useful outputs. Processing can involve making calculations, comparing data and taking alternative actions, and storing data for future use. Processing data into useful information is critical in business settings.

Processing can be done manually or with computer assistance. In a payroll application, the number of hours each employee worked must be converted into net, or take-home, pay. Other inputs often include employee ID number and department. The processing can first involve multiplying the number of hours worked by the employee's hourly pay rate to get gross pay. If weekly hours worked exceed 40, overtime pay might also be included. Then deductions—for example, federal and state taxes or contributions to insurance or savings plans—are subtracted from gross pay to get net pay.

After these calculations and comparisons are performed, the results are typically stored. *Storage* involves keeping data and information available for future use, including output, discussed next.

Output

In information systems, **output** involves producing useful information, usually in the form of documents and reports. Outputs can include paychecks for employees, reports for managers, and information supplied to stockholders, banks, government agencies, and other groups. In some cases, output from one system can become input for another. For example, output from a system that processes sales orders can be used as input to a customer billing system.

output
Production of useful information, usually in the form of documents and reports.

Feedback

In information systems, **feedback** is information from the system that is used to make changes to input or processing activities. For example, errors or problems might make it necessary to correct input data or change a process. Consider a payroll example. Perhaps the number of hours an employee worked was entered as 400 instead of 40. Fortunately, most information systems check to make sure that data falls within certain ranges. For number of hours worked, the range might be from 0 to 100 because it is unlikely that an employee would work more than 100 hours in a week. The information system would determine that 400 hours is out of range and provide feedback. The feedback is used to check and correct the input on the number of hours worked to 40. If undetected, this error would result in a very high net pay on the printed paycheck!

Feedback is also important for managers and decision makers. For example, a furniture maker could use a computerized feedback system to link its suppliers and plants. The output from an information system might indicate that inventory levels for mahogany and oak are getting low—a potential problem. A manager could use this feedback to decide to order more wood from a supplier. These new inventory orders then become input to the system. In addition to this reactive approach, a computer system can also be proactive—predicting future events to avoid problems. This concept, often called **forecasting**, can be used to estimate future sales and order more inventory before a shortage occurs. According to the CIO of Coty Fragrance, which produces Jennifer Lopez and Vera Wang brands, "If we can't meet demand, it annoys the retailers, the consumers lose interest, and we lose sales."[12] Forecasting is also used to predict the strength and landfall sites of hurricanes, future stock-market values, and who will win a political election. Disappointed with existing weather forecasting systems, Robert Baron developed a more sophisticated forecasting approach that used radar data along with other meteorological data to forecast storms and weather. Today, his weather forecasting software generates about $25 million in annual revenues.[13]

feedback
Output that is used to make changes to input or processing activities.

forecasting
Predicting future events to avoid problems.

Forecasting systems can help meteorologists predict the strength and landfall sites of tropical storms.

(Source: Courtesy of AP Photo/Bullit Marquez.)

Manual and Computerized Information Systems

As discussed earlier, an information system can be manual or computerized. For example, some investment analysts manually draw charts and trend lines to assist them in making investment decisions. Tracking data on stock prices (input) over the last few months or years, these analysts develop patterns on graph paper (processing) that help them determine what stock prices are likely to do in the next few days or weeks (output). Some investors have made millions of dollars using manual stock analysis information systems. Of course, today many excellent computerized information systems follow stock indexes and markets and suggest when large blocks of stocks should be purchased or sold (called *program trading*) to take advantage of market discrepancies.

Computer-Based Information Systems

A **computer-based information system (CBIS)** is a single set of hardware, software, databases, telecommunications, people, and procedures that are configured to collect, manipulate, store, and process data into information. Lloyd's Insurance in London used a CBIS to reduce paper transactions and convert to an electronic insurance system. The CBIS allows Lloyd's to insure people and property more efficiently and effectively. Lloyd's often insures the unusual, including actress Betty Grable's legs, Rolling Stone Keith Richards's hands, and a possible appearance of the Loch Ness Monster (Nessie) in Scotland, which would result in a large payment for the person first seeing the monster.

The components of a CBIS are illustrated in Figure 1.4. *Information technology (IT)* refers to hardware, software, databases, and telecommunications. Telecommunications also includes networks and the Internet. A business's **technology infrastructure** includes all the hardware, software, databases, telecommunications, people, and procedures that are configured to collect, manipulate, store, and process data into information. The technology infrastructure is a set of shared IS resources that form the foundation of each computer-based information system.

computer-based information system (CBIS)
A single set of hardware, software, databases, telecommunications, people, and procedures that are configured to collect, manipulate, store, and process data into information.

technology infrastructure
All the hardware, software, databases, telecommunications, people, and procedures that are configured to collect, manipulate, store, and process data into information.

Figure 1.4

The Components of a Computer-Based Information System

Hardware

Hardware consists of the physical components of a computer that perform the input, processing, storage, and output activities of the computer. Input devices include keyboards, mice, and other pointing devices; automatic scanning devices; and equipment that can read magnetic ink characters. Processing devices include computer chips that contain the central processing unit and main memory. Advances in chip design allow faster speeds, less power consumption, and larger storage capacity. Some specialized computer chips will be able to monitor power consumption for companies and homeowners.[14] SanDisk and other

hardware
The physical components of a computer that perform the input, processing, storage, and output activities of the computer.

companies make small, portable chips that are used to conveniently store programs, data files, and more.[15] The publisher of this book, for example, used this type of chip storage device to send promotional material for this book to professors and instructors.

Processor speed is also important. Today's more advanced processor chips have the power of 1990s-era supercomputers that occupied a room measuring 10 feet by 40 feet. A large IBM computer used by U.S. Livermore National Laboratories to analyze nuclear explosions is one of the fastest computers in the world (up to 300 teraflops—300 trillion operations per second).[16] The super-fast computer, called Blue Gene, costs about $40 million.[17] It received the *National Medal of Technology and Innovation* award from President Barack Obama. Small, inexpensive computers and handheld devices are also becoming popular. Inexpensive netbooks are small, inexpensive laptop computers that can cost less than $500 and be used primarily to connect to the Internet.[18] In addition, the iPhone by Apple Computer can perform many functions that can be done on a desktop or laptop computer.[19] The One Laptop Per Child computer costs less than $200.[20] The Classmate PC by Intel will cost about $300 and include some educational software. Both computers are intended for regions of the world that can't afford traditional personal computers. The country of Peru, for example, has purchased about 350,000 laptops loaded with about 100 books for children, who also teach their parents how to use the inexpensive computers.[21] According to the founder of One Laptop Per Child, "If that doesn't give you goose bumps, I don't know what will."

The One Laptop Per Child Computer costs less than $200, and is designed for regions of the world that can't afford traditional personal computers.

(Source: Courtesy of AFP/Getty Images.)

The many types of output devices include printers and computer screens. Some touch-sensitive computer screens, for example, can be used to execute functions or complete programs, such as connecting to the Internet or running a new computer game or word processing program.[22] Many special-purpose hardware devices have also been developed. Computerized event data recorders (EDRs) are now being placed into vehicles. Like an airplane's black box, EDRs record vehicle speed, possible engine problems, driver performance, and more. The technology is being used to document and monitor vehicle operation, determine the cause of accidents, and investigate whether truck drivers are taking required breaks. In one case, an EDR was used to help convict a driver of vehicular homicide. In another case, an EDR in a police officer's car showed that the officer may have run a stop light and accelerated to more than 70 miles per hour on a road with a speed limit of 35 miles per hour before an accident that killed two teenagers.[23]

Software

Software consists of the computer programs that govern the operation of the computer. These programs allow a computer to process payroll, send bills to customers, and provide managers with information to increase profits, reduce costs, and provide better customer service. Fab Lab software, for example, controls tools such as cutters, milling machines, and other devices.[24] One Fab Lab system, which costs about $20,000, has been used to make radio frequency tags to track animals in Norway, engine parts to allow tractors to run on processed castor beans in India, and many other fabrication applications. SalesForce (*www.salesforce.com*) sells software to help companies manage their salesforce and help improve customer satisfaction.[25]

The two types of software are *system software*, such as Microsoft Windows Vista and Windows 7, which controls basic computer operations, including start-up and printing, and *applications software*, such as Microsoft Office 2010, which allows you to accomplish specific tasks, including word processing or tabulating numbers.[26] Software is needed for computers of all sizes, from small handheld computers to large supercomputers. The Android operating system by Google and Microsoft's Mobile 6.5, for example, are operating systems for cell phones and small portable devices.[27] Although most software can be installed from CDs, many of today's software packages can be downloaded through the Internet.

software
The computer programs that govern the operation of the computer.

Sophisticated application software, such as Adobe Creative Suite 4, can be used to design, develop, print, and place professional-quality advertising, brochures, posters, prints, and videos on the Internet.[28] Nvidia's GeForce 3D is software that can display images on a computer screen that appear three-dimensional (3D) when viewed using special glasses.[29]

Windows 7 is systems software that controls basic computer operations, including start-up and printing.

(Source: Courtesy of Microsoft Corporation.)

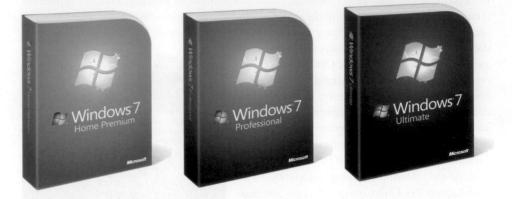

Databases

database
An organized collection of facts and information.

A **database** is an organized collection of facts and information, typically consisting of two or more related data files. An organization's database can contain facts and information on customers, employees, inventory, competitors' sales, online purchases, and much more. A database manager for a large bank, for example, has developed a patented security process that generates a random numeric code from a customer's bank card that can be verified by a computer system through a customer database.[30] Once the bank card and customer have been verified, the customer can make financial transactions.

Data can be stored in large data centers, within computers of all sizes, in the Internet, and in smart cell phones and small computing devices.[31] The New York Stock Exchange (NYSE) and other exchanges are using database systems to get better business information and intelligence to help them run successful and profitable operations.[32] The huge increase in database storage requirements, however, often requires more storage devices, more space to house the additional storage devices, and additional electricity to operate them.[33]

Telecommunications, Networks, and the Internet

telecommunications
The electronic transmission of signals for communications; enables organizations to carry out their processes and tasks through effective computer networks.

Telecommunications is the electronic transmission of signals for communications, which enables organizations to carry out their processes and tasks through effective computer networks. Telecommunications can take place through wired, wireless, and satellite transmissions.[34] The Associated Press was one of the first users of telecommunications in the 1920s, sending news over 103,000 miles of wire in the United States and almost 10,000 miles of cable across the ocean. Today, telecommunications is used by organizations of all sizes and individuals around the world. With telecommunications, people can work at home or while traveling. This approach to work, often called *telecommuting,* allows a telecommuter living in England to send his or her work to the United States, China, or any location with telecommunications capabilities.

networks
Computers and equipment that are connected in a building, around the country, or around the world to enable electronic communication.

Networks connect computers and equipment in a building, around the country, or around the world to enable electronic communication. Wireless transmission allows aircraft drones, such as Boeing's Scan Eagle, to fly using a remote control system to monitor commercial buildings or enemy positions.[35] The drones are smaller and less- expensive versions of the Predator and Global Hawk drones that the U.S. military used in the Afghanistan and Iraq conflicts. According to a Navy Rear Admiral, "There are all sorts of levels of stealthiness. Operators have been deploying it in an undetectable fashion; at a certain low altitude, you can't hear it or see it."

The **Internet** is the world's largest computer network, consisting of thousands of inter-connected networks, all freely exchanging information. Research firms, colleges, universities, high schools, hospitals, and businesses are just a few examples of organizations using the Internet. Beth Israel Deaconess Medical Center, for example, allows doctors to use its Internet site to provide better patient care and reduce costs.[36] The doctors pay a monthly service fee to use the hospital's Internet site. Increasingly, businesses and people are using the Internet to run and deliver important applications, such as accessing vast databases, performing so-phisticated business analysis, and getting a variety of reports. This concept, called *cloud computing,* allows people to get the information they need from the Internet (the cloud) instead of from desktop or corporate computers.[37] According to the CIO of Avon Products, "Today, wherever you are, you can connect to all the information you need." Some appli-cations are available to everyone (public cloud computing), while other applications are only available to corporate employees and managers (private cloud computing).[38]

Internet
The world's largest computer network, consisting of thousands of interconnected networks, all freely exchanging information.

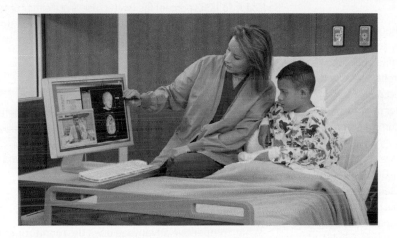

Doctors use cloud computing and other types of Web sites to provide better patient care and reduce costs.

(Source: © B Busco/Getty Images.)

People use the Internet to research information, buy and sell products and services, make travel arrangements, conduct banking, download music and videos, read books, and listen to radio programs, among other activities.[39] Bank of America allows people to check their bank balances and pay their bills on the Internet using Apple's iPhone and other handheld devices.[40] Internet sites like MySpace (*www.myspace.com*) and Facebook (*www.facebook.com*) have become popular places to connect with friends and colleagues. People can also send short messages of up to 140 characters using Twitter (*www.twitter.com*) over the Internet.[41] Some people, however, fear that this increased usage can lead to problems, including criminals hacking into the Internet and gaining access to sensitive personal information.

Large computers, personal computers, and today's cell phones, such as Apple's iPhone, can access the Internet.[42] This not only speeds communications, but also allows people to conduct business electronically. Internet users can create *Web logs (blogs)* to store and share their thoughts and ideas with others around the world. Using *podcasting*, you can download audio programs or music from the Internet to play on computers or music players. One of the authors of this book uses podcasts to obtain information on information systems and technology.

The *World Wide Web (WWW),* or the *Web,* is a network of links on the Internet to documents containing text, graphics, video, and sound. Information about the documents and access to them are controlled and provided by tens of thousands of special computers called Web servers. The Web is one of many services available over the Internet and provides access to millions of documents. New Internet technologies and increased Internet commu-nications and collaboration are collectively called *Web 2.0.* [43]

Who Is Interested in Your Social Network Updates?

More than two-thirds of the world's online population use social networks such as Facebook, MySpace, and Twitter to stay in touch with friends. It is likely that you are one of them. In 2008, social networks became more popular than e-mail, with 66.8 percent of Internet users accessing member communities. Most members of social networks use a posting feature that allows them to share their day-to-day thoughts and activities with their circle of friends. Facebook calls these postings "updates," while Twitter calls them "tweets." Most users do not realize the value of their comments, updates, or tweets to people outside their circle.

Businesses are flocking to social networks to harvest consumer sentiment for use in guiding product development. They are also watching social networks to confront negative publicity. The broad scale use of social networks and the careful analysis of billions of messages have made it possible to collect public sentiment and build customer relations in a manner never done before. But sifting through the babble to discover comments of interest is challenging.

A number of information system companies have sprung up to provide products designed to monitor social media. Companies such as Alterian, Radian6, Attensity, Visible Technologies, Conversion, and Nielsen Online provide social media monitoring systems for businesses and organizations. As a young technology, there is no standard approach to social media monitoring. Similar to a search engine, the systems typically traverse the continuous streams of comments in social networks, looking for key terms related to specified products. Artificial intelligence (AI) techniques that automate the interpretation of user comments make it possible to quickly identify comments of particular interest. Ultimately, they generate analytic and performance reports for the human expert to evaluate. Systems that monitor social media enable useful information to be drawn from billions of seemingly mundane and unrelated messages.

Monitoring social media can focus on brand reputation management, public relations, or even market research. Companies such as Comcast, a major communications company, hire full-time social media experts who interact with customers online to address problems and complaints. For example, if you complain about Comcast service on Twitter, you might be contacted by a Comcast employee offering to help you.

The social network service owners are well aware of the value of the information that flows over their networks. Most of them intend to build their business through the comments and attention of their members. Whether through targeted ads or selling access to user data, social networks can become very lucrative businesses. Why else would Twitter, a service with apparently no business model, be worth over a billion dollars? Twitter's goal is to grow to one billion members and provide interested parties with the pulse of the planet.

How do users feel about their "personal" comments being harvested to make billions for Internet companies? With social network growth rates in 2009 ranging from 228 percent for Facebook to 1,382 percent for Twitter, users are either unaware or unconcerned. Regardless of what users think, it is likely that businesses will increasingly analyze the continuous flow of data over social networks to generate insights they can use.

Discussion Questions

1. Do you think it is ethical for social networks to sell access to user information to businesses for market research and other uses? Why or why not?
2. What service does the monitoring of social media ultimately provide for consumers?

Critical Thinking Questions

1. What competitive advantage does the monitoring of social media provide to companies that invest in it?
2. Why is the monitoring of social media considered a CBIS?

SOURCES: Ostrow, Adam, "Social Networking More Popular than Email," *Mashable*, March 9, 2009, *http://mashable.com/2009/03/09/social-networking-more-popular-than- email*; Zabin, Jeff, "Finding Out What They're Saying About You Is Worth Every Penny," *E-Commerce Times*, November 12, 2009, *www.ecommercetimes.com/rsstory/68624.html*; Bensen, Connie, "Do you know what people are saying about you?" *Reuters UK*, September 14, 2009, *http://blogs.reuters.com/great-debate-uk/2009/09/14/do-you-know-what-people-are- saying-about-you*; Schonfeld, Erick, "Twitter's Internal Strategy Laid Bare: To Be "The Pulse of the Planet," *TechCrunch*, July 16, 2009, *www.techcrunch.com/2009/07/16/twitters-internal-strategy-laid-bare-to-be-the-pulse-of- the-planet*; Reisner, Rebecca, "Comcast's Twitter Man," *Business Week*, January 13, 2009, *www.businessweek.com/managing/content/jan2009/ca20090113_373506.htm*; McCarthy, Carolina, "Nielsen: Twitter's growing really, really, really, really fast," CNET, March 2009, *http://news.cnet.com/8301-13577_3-10200161-36.html*; Nielsen Staff, "Social Networking's New Global Footprint," *NielsenWire*, March 9, 2009, *http://blog.nielsen.com/nielsenwire/global/social-networking-new-global-footprint/*.

The technology used to create the Internet is also being applied within companies and organizations to create **intranets**, which allow people in an organization to exchange information and work on projects. ING DIRECT Canada (*www.ingdirect.ca/en*), for example, used its intranet to get ideas from its employees. According to one corporate executive, "Many of the ideas we've been able to implement are from front-line staff who talk to our customers every day and know what they want." [44] Companies often use intranets to connect its employees around the globe. An **extranet** is a network based on Web technologies that allows selected outsiders, such as business partners and customers, to access authorized resources of a company's intranet. Many people use extranets every day without realizing it—to track shipped goods, order products from their suppliers, or access customer assistance from other companies. Penske Truck Leasing, for example, uses an extranet (*www.MyFleetAtPenske.com*) for Penske leasing companies and its customers. [45] The extranet site allows customers to schedule maintenance, find Penske fuel stops, receive emergency roadside assistance, participate in driver training programs, and more. If you log on to the FedEx site (*www.fedex.com*) to check the status of a package, for example, you are using an extranet.

intranet
An internal network based on Web technologies that allows people within an organization to exchange information and work on projects.

extranet
A network based on Web technologies that allows selected outsiders, such as business partners and customers, to access authorized resources of a company's intranet.

When you log on to the FedEx site (*www.fedex.com*) to check the status of a package, you are using an extranet.

(Source: *www.fedex.com*.)

People

People are the most important element in most computer-based information systems. They make the difference between success and failure for most organizations. Information systems personnel include all the people who manage, run, program, and maintain the system, including the CIO, who manages the IS department. [46] Users are people who work with information systems to get results. Users include financial executives, marketing representatives, manufacturing operators, and many others. Certain computer users are also IS personnel.

The chief information officer (CIO) manages the Information Systems department, which includes all the people who manage, run, program, and maintain a computer-based information system.

(Source: © Ryan McVay/Getty Images.)

procedures
The strategies, policies, methods, and rules for using a CBIS.

Procedures

Procedures include the strategies, policies, methods, and rules for using the CBIS, including the operation, maintenance, and security of the computer. For example, some procedures describe when each program should be run. Others describe who can access facts in the database or what to do if a disaster, such as a fire, earthquake, or hurricane, renders the CBIS unusable. Good procedures can help companies take advantage of new opportunities and avoid potential disasters. Poorly developed and inadequately implemented procedures, however, can cause people to waste their time on useless rules or result in inadequate responses to disasters, such as hurricanes or tornadoes.

Now that we have looked at computer-based information systems in general, we will briefly examine the most common types used in business today. These IS types are covered in greater detail in Part 3.

BUSINESS INFORMATION SYSTEMS

The most common types of information systems used in business organizations are those designed for electronic and mobile commerce, transaction processing, management information, and decision support. In addition, some organizations employ special-purpose systems, such as virtual reality, that not every organization uses. Although these systems are discussed in separate sections in this chapter and explained in greater detail later, they are often integrated in one product and delivered by the same software package. See Figure 1.5. For example, some business information systems process transactions, deliver information, and support decisions. Figure 1.6 shows a simple overview of the development of important business information systems discussed in this section.

Figure 1.5

Business Information Systems

Business information systems are often integrated in one product and can be delivered by the same software package.

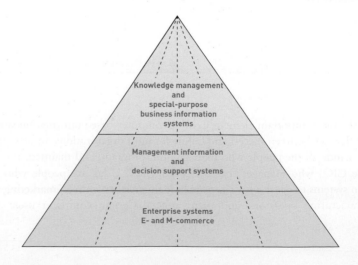

Knowledge management
and
special-purpose
business information
systems

Management information
and
decision support systems

Enterprise systems
E- and M-commerce

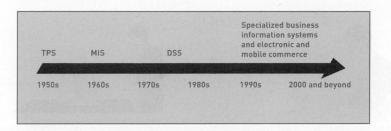

Figure 1.6

The Development of Important
Business Information Systems

Electronic and Mobile Commerce

E-commerce involves any business transaction executed electronically between companies (business-to-business, or B2B), companies and consumers (business-to-consumer, or B2C), consumers and other consumers (consumer-to-consumer, or C2C), business and the public sector, and consumers and the public sector.[47] Some of the stimulus funds in 2009, for example, were aimed at increasing electronic record keeping and electronic commerce for healthcare facilities.[48] E-commerce offers opportunities for businesses of all sizes to market and sell at a low cost worldwide, allowing them to enter the global market. **Mobile commerce (m-commerce)** is the use of mobile, wireless devices to place orders and conduct business. M-commerce relies on wireless communications that managers and corporations use to place orders and conduct business with handheld computers, portable phones, laptop computers connected to a network, and other mobile devices. Today, mobile commerce has exploded in popularity with advances in smartphones, including Apple's iPhone.[49] Customers are using their cell phones to purchase concert tickets from companies such as Ticketmaster Entertainment (*www.ticketmaster.com*) and Tickets (*www.tickets.com*).[50]

e-commerce
Any business transaction executed electronically between companies (business-to-business, or B2B), companies and consumers (business-to-consumer, or B2C), consumers and other consumers (consumer-to-consumer, or C2C), business and the public sector, and consumers and the public sector.

mobile commerce (m-commerce)
Transactions conducted anywhere, anytime.

With mobile commerce (m-commerce), people can use cell phones to pay for goods and services anywhere, anytime.

(Source: Courtesy of Davie Hinshaw/ MCT/Landov.)

E-commerce offers many advantages for streamlining work activities. Figure 1.7 provides a brief example of how e-commerce can simplify the process of purchasing new office furniture from an office supply company. In the manual system, a corporate office worker must get approval for a purchase that exceeds a certain amount. That request goes to the purchasing department, which generates a formal purchase order to procure the goods from the approved vendor. Business-to-business e-commerce automates the entire process. Employees go directly to the supplier's Web site, find the item in a catalog, and order what they need at a price set by their company. If management approval is required, the manager is notified automatically. As the use of e-commerce systems grows, companies are phasing out their traditional systems. The resulting growth of e-commerce is creating many new business opportunities.

Figure 1.7

E-Commerce Greatly Simplifies
Purchasing

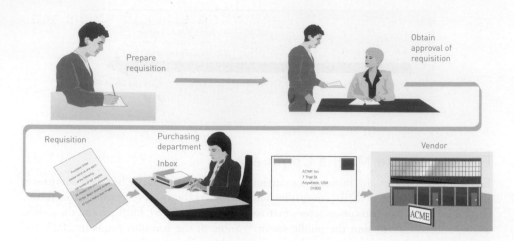

Traditional process for placing a purchase order

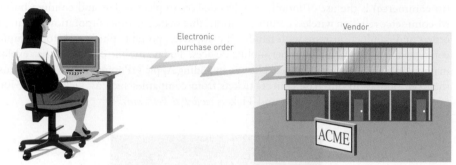

E-commerce process for placing a purchase order

E-commerce can enhance a company's stock prices and market value. Today, several e-commerce firms have teamed up with more traditional brick-and-mortar businesses to draw from each other's strengths. For example, e-commerce customers can order products on a Web site and pick them up at a nearby store.

In addition to e-commerce, business information systems use telecommunications and the Internet to perform many related tasks. *Electronic procurement (e-procurement)*, for example, involves using information systems and the Internet to acquire parts and supplies. **Electronic business (e-business)** goes beyond e-commerce and e-procurement by using information systems and the Internet to perform all business-related tasks and functions, such as accounting, finance, marketing, manufacturing, and human resource activities. E-business also includes working with customers, suppliers, strategic partners, and stakeholders. Compared to traditional business strategy, e-business strategy is flexible and adaptable. See Figure 1.8.

**electronic business
(e-business)**

Using information systems and the Internet to perform all business-related tasks and functions.

Figure 1.8

Electronic Business

E-business goes beyond e-commerce to include using information systems and the Internet to perform all business-related tasks and functions, such as accounting, finance, marketing, manufacturing, and human resources activities.

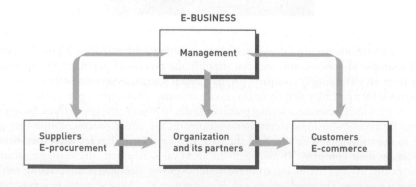

Enterprise Systems: Transaction Processing Systems and Enterprise Resource Planning

Enterprise systems that process daily transactions have evolved over the years and offer important solutions for businesses of all sizes. Traditional transaction processing systems are still being used, but increasingly, companies are turning to enterprise resource planning systems. These systems are discussed next.

Transaction Processing Systems

Since the 1950s, computers have been used to perform common business applications. Many of these early systems were designed to reduce costs by automating routine, labor-intensive business transactions. A **transaction** is any business-related exchange such as payments to employees, sales to customers, or payments to suppliers. Processing business transactions was the first computer application developed for most organizations. A **transaction processing system (TPS)** is an organized collection of people, procedures, software, databases, and devices used to perform and record business transactions. If you understand a transaction processing system, you understand basic business operations and functions.

One of the first business systems to be computerized was the payroll system (see Figure 1.9). The primary inputs for a payroll TPS are the number of employee hours worked during the week and the pay rate. The primary output consists of paychecks. Early payroll systems produced employee paychecks and related reports required by state and federal agencies, such as the Internal Revenue Service. Other routine applications include sales ordering, customer billing and customer relationship management, and inventory control.

transaction
Any business-related exchange, such as payments to employees, sales to customers, and payments to suppliers.

transaction processing system (TPS)
An organized collection of people, procedures, software, databases, and devices used to perform and record business transactions.

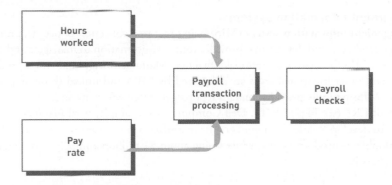

Figure 1.9

A Payroll Transaction Processing System

In a payroll TPS, the inputs (numbers of employee hours worked and pay rates) go through a transformation process to produce outputs (paychecks).

Enterprise systems help organizations perform and integrate important tasks, such as paying employees and suppliers, controlling inventory, sending invoices, and ordering supplies. In the past, companies accomplished these tasks using traditional transaction processing systems. Today, they are increasingly being performed by enterprise resource planning systems.

Enterprise Resource Planning

An **enterprise resource planning (ERP) system** is a set of integrated programs that manages the vital business operations for an entire multisite, global organization.[51] Pick n Pay, a South African (SA) food retailer, used ERP to reduce costs and the prices paid by customers. According to the chief executive officer, "We are happy to play our part in ensuring that SA's economy continues to perform well, particularly given the pressures being felt globally."[52]

enterprise resource planning (ERP) system
A set of integrated programs capable of managing a company's vital business operations for an entire multisite, global organization.

Information and Decision Support Systems

The benefits provided by an effective TPS or ERP, including reduced processing costs and reductions in needed personnel, are substantial and justify their associated costs in computing equipment, computer programs, and specialized personnel and supplies. Companies soon realized that they could use the data stored in these systems to help managers make better decisions, whether in human resource management, marketing, or administration. Satisfying the needs of managers and decision makers continues to be a major factor in developing information systems.

SAP AG, a German software company, is one of the leading suppliers of ERP software. The company employs more than 50,000 people in more than 120 countries.

(Source: *www.sap.com*.)

Management Information Systems

management information system (MIS)

An organized collection of people, procedures, software, databases, and devices that provides routine information to managers and decision makers.

A **management information system (MIS)** is an organized collection of people, procedures, software, databases, and devices that provides routine information to managers and decision makers. An MIS focuses on operational efficiency. Manufacturing, marketing, production, finance, and other functional areas are supported by MISs and linked through a common database. MISs typically provide standard reports generated with data and information from the TPS or ERP (see Figure 1.10). Dell Computer, for example, used manufacturing MIS software to develop a variety of reports on its manufacturing processes and costs.[53] Dell was able to double its product variety, while saving about $1 million annually in manufacturing costs as a result.

Figure 1.10

Management Information System

Functional management information systems draw data from the organization's transaction processing system.

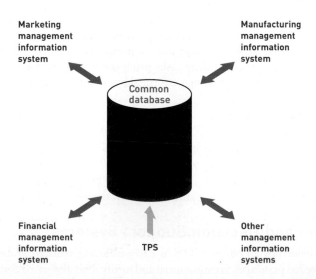

MISs were first developed in the 1960s and typically use information systems to produce managerial reports. In many cases, these early reports were produced periodically—daily, weekly, monthly, or yearly. Because of their value to managers, MISs have proliferated throughout the management ranks.

Decision Support Systems

By the 1980s, dramatic improvements in technology resulted in information systems that were less expensive but more powerful than earlier systems. People quickly recognized that computer systems could support additional decision-making activities. A **decision support system (DSS)** is an organized collection of people, procedures, software, databases, and devices that support problem-specific decision making. The focus of a DSS is on making effective decisions. Whereas an MIS helps an organization "do things right," a DSS helps a manager "do the right thing."[54]

decision support system (DSS)
An organized collection of people, procedures, software, databases, and devices used to support problem-specific decision making.

Endeca provides Discovery for Design, decision support software that helps businesspeople assess risk and analyze performance. The data shown here is for electronic component development.

(Source: Courtesy of Endeca Technologies, Inc.)

A DSS can include a collection of models used to support a decision maker or user (model base), a collection of facts and information to assist in decision making (database), and systems and procedures (user interface or dialogue manager) that help decision makers and other users interact with the DSS (see Figure 1.11). Software is often used to manage the database—the database management system (DBMS)—and the model base—the model management system (MMS). Not all DSSs have all of these components.

In addition to DSSs for managers, other systems use the same approach to support groups and executives. A *group support system* includes the DSS elements just described as well as software, called *groupware*, to help groups make effective decisions. Kraft, for example, used iPhones and other mobile devices to help managers and workers stay connected and work together on important projects.[55] An executive support system, also called an *executive information system*, helps top-level managers, including a firm's president, vice presidents, and members of the board of directors, make better decisions. Healthland and Performance Management Institute, a healthcare company, has developed an executive information system to help small community and rural hospital executives make better decisions about delivering quality health care to patients and increasing the efficient delivery of healthcare services for hospitals.[56] The American Recovery and Reinvestment Act provides funds for qualifying healthcare companies that invest in better information and decision support systems. An executive support system can assist with strategic planning, top-level organizing and staffing, strategic control, and crisis management.

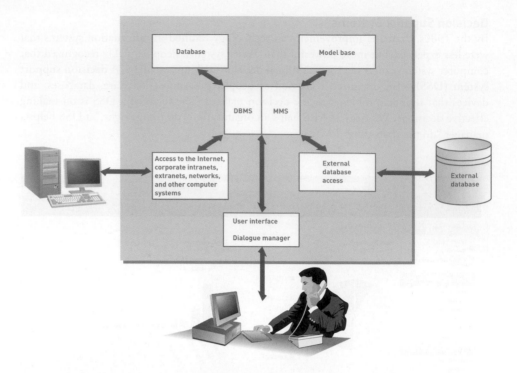

Specialized Business Information Systems: Knowledge Management, Artificial Intelligence, Expert Systems, and Virtual Reality

In addition to TPSs, MISs, and DSSs, organizations often rely on specialized systems. Many use *knowledge management systems (KMSs)*, an organized collection of people, procedures, software, databases, and devices, to create, store, share, and use the organization's knowledge and experience.[57] Advent, a San Francisco company that develops investment software for hedge funds, used a KMS to help its employees locate and use critical knowledge to help its customers.[58]

In addition to knowledge management, companies use other types of specialized systems. Experimental systems in cars can help prevent accidents. These new systems allow cars to communicate with each other using radio chips installed in their trunks. When two or more cars move too close together, the specialized systems sound alarms and brake in some cases. Some specialized systems are based on the notion of **artificial intelligence** (AI), in which the computer system takes on the characteristics of human intelligence. The field of artificial intelligence includes several subfields (see Figure 1.12). Some people predict that, in the future, we will have nanobots, small molecular-sized robots, traveling throughout our bodies and in our bloodstream, monitoring our health.[59] Other nanobots will be embedded in products and services.[60]

artificial intelligence (AI)
A field in which the computer system takes on the characteristics of human intelligence.

A Nissan Motor Company car swerves back into its lane on its own shortly after it ran off the track during a test of the Lane Departure Prevention feature, which also sounds a warning when the car veers out of its lane.

(Source: © AP Photo/Katsumi Kasahara.)

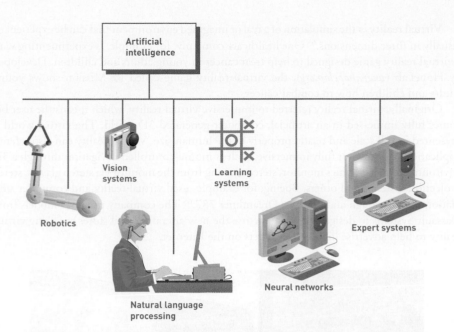

Figure 1.12

The Major Elements of Artificial Intelligence

Artificial Intelligence

Robotics is an area of artificial intelligence in which machines take over complex, dangerous, routine, or boring tasks, such as welding car frames or assembling computer systems and components. Honda Motor has spent millions of dollars on advanced robotics that allows a person to give orders to a computer using only his or her thoughts. The new system uses a special helmet that can measure and transmit brain activity to a computer. [61] A robot used by a Staples distribution center in the Denver area is able to locate items in a 100,000 square foot warehouse and pack them into containers to be shipped to other Staples stores.[62] Vision systems allow robots and other devices to "see," store, and process visual images. Natural language processing involves computers understanding and acting on verbal or written commands in English, Spanish, or other human languages. Learning systems allow computers to learn from past mistakes or experiences, such as playing games or making business decisions. Neural networks is a branch of artificial intelligence that allows computers to recognize and act on patterns or trends.[63] Some successful stock, options, and futures traders use neural networks to spot trends and improve the profitability of their investments.

Expert Systems

Expert systems give the computer the ability to make suggestions and function like an expert in a particular field, helping enhance the performance of the novice user. The unique value of expert systems is that they allow organizations to capture and use the wisdom of experts and specialists.[64] Therefore, years of experience and specific skills are not completely lost when a human expert dies, retires, or leaves for another job. The U.S. Army uses the Knowledge and Information Fusion Exchange (KnIFE) expert system to help soldiers in the field make better military decisions based on successful decisions made in previous military engagements. The collection of data, rules, procedures, and relationships that must be followed to achieve value or the proper outcome is contained in the expert system's **knowledge base**.

Virtual Reality and Multimedia

Virtual reality and multimedia are specialized systems that are valuable for many businesses and nonprofit organizations. Many imitate or act like real environments. These unique systems are discussed in this section.

expert system
A system that gives a computer the ability to make suggestions and function like an expert in a particular field.

knowledge base
The collection of data, rules, procedures, and relationships that must be followed to achieve value or the proper outcome.

virtual reality

The simulation of a real or imagined environment that can be experienced visually in three dimensions.

Virtual reality is the simulation of a real or imagined environment that can be experienced visually in three dimensions.[65] One healthcare company, for example, is experimenting with a virtual reality game designed to help treat cancer in young adults and children. Developed by HopeLab (*www.hopelab.org*), the virtual reality game called Re-Mission shows young adults and children how to combat cancer.

Originally, virtual reality referred to immersive virtual reality, which means the user becomes fully immersed in an artificial, computer-generated 3D world. The virtual world is presented in full scale and relates properly to the human size. Virtual reality can also refer to applications that are not fully immersive, such as mouse-controlled navigation through a 3D environment on a graphics monitor, stereo viewing from the monitor via stereo glasses, stereo projection systems, and others. Boeing, for example, used virtual reality and computer simulation to help design and build its Dreamliner 787.[66] The company used 3D models from Dassault Systems to design and manufacture the new aircraft. Retail stores are using virtual reality to help advertise high-end products on the Internet.

The Cave Automatic Virtual Environment (CAVE) is a virtual reality room that allows users to completely immerse themselves in a virtual car interior while operating a workstation in a factory.

(Source: © Sipa via AP Images.)

No: 584338.06 Date: 26.08.2009 Credit: MEIGNEUX/SIPA
Headline: VELIZY; The CAVE (Cave Automatic Virtual Environment)
Caption: The CAVE (Cave Automatic Virtual Environment) is a virtual reality room that enables complete immersion in a virtual car interior, workstation at factory ... This is the first immersive center based in France.
Sipa Press   Velizy, FRANCE-26/08/2009

Figure 1.13

A Head-Mounted Display

The head-mounted display (HMD) was the first device to provide the wearer with an immersive experience. A typical HMD houses two miniature display screens and an optical system that channels the images from the screens to the eyes, thereby presenting a stereo view of a virtual world. A motion tracker continuously measures the position and orientation of the user's head and allows the image-generating computer to adjust the scene representation to the current view. As a result, the viewer can look around and walk through the surrounding virtual environment.

(Source: Courtesy of 5DT, Inc. *www.5dt.com*.)

A variety of input devices, such as head-mounted displays (see Figure 1.13), data gloves, joysticks, and handheld wands, allow the user to navigate through a virtual environment and to interact with virtual objects. Directional sound, tactile and force feedback devices, voice recognition, and other technologies enrich the immersive experience. Because several people can share and interact in the same environment, virtual reality can be a powerful medium for communication, entertainment, and learning.

Multimedia is a natural extension of virtual reality. It can include photos and images, the manipulation of sound, and special 3D effects. Once used primarily in movies, 3D technology can be used by companies to design products, such as motorcycles, jet engines, bridges, and more.[67] Autodesk, for example, makes exciting 3D software that companies can use to design large skyscrapers and other buildings.[68] The software can also be used by Hollywood animators to develop action and animated movies.

SYSTEMS DEVELOPMENT

Systems development is the activity of creating or modifying information systems. Systems development projects can range from small to very large and are conducted in fields as diverse as stock analysis and video game development. Individuals from around the world are using the steps of systems development to create unique applications for the iPhone.[69] Apple has special tools for iPhone application developers, including GPS capabilities and audio streaming, to make it easier for people to craft unique applications. Apple is also allowing these systems developers to charge users in a variety of ways, including fixed prices and subscription fees. Recall that individuals and companies are increasingly developing "cloud computing" applications that can be run from the Internet.[70] These applications have additional systems development challenges, such as making sure that the data and programs on the Internet are safe and secure from hackers and corporate spies.

People inside a company can develop systems, or companies can use *outsourcing*, hiring an outside company to perform some or all of a systems development project. Outsourcing allows a company to focus on what it does best and delegate other functions to companies with expertise in systems development. The drug company Pfizer, for example, used outsourcing to allow about 4,000 of its busy employees to outsource some of their jobs functions to other individuals or companies around the globe, allowing them to concentrate on key tasks.[71] Any outsourcing decision should depend on the company and the project being considered for outsourcing.

Some systems development efforts fail to meet their cost or schedule goals. Systems development failures can be a result of poor planning and scheduling, insufficient management of risk, poor requirements determination, and lack of user involvement. One strategy for improving the results of a systems development project is to divide it into several steps, each with a well-defined goal and set of tasks to accomplish (see Figure 1.14). These steps are summarized next.

systems development
The activity of creating or modifying existing business systems.

Systems Investigation and Analysis

The first two steps of systems development are systems investigation and analysis. The goal of the *systems investigation* is to gain a clear understanding of the problem to be solved or opportunity to be addressed. After an organization understands the problem, the next question is, "Is the problem worth solving?" Given that organizations have limited resources—people and money—this question deserves careful consideration. If the decision is to continue with the solution, the next step, *systems analysis*, defines the problems and opportunities of the existing system. During systems investigation and analysis, as well as design maintenance and review, discussed next, the project must have the complete support of top-level managers and focus on developing systems that achieve business goals.

Figure 1.14

An Overview of Systems
Development

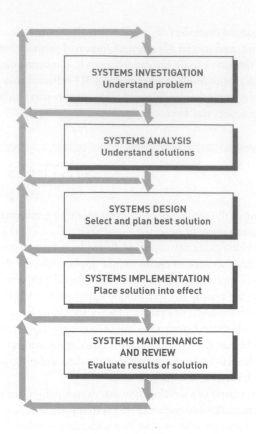

Systems Design, Implementation, and Maintenance and Review

Systems design determines how the new system should be developed to meet the business needs defined during systems analysis. For some companies, this involves environmental design that attempts to use systems development approaches that are kind to the environment and make a profit. Gazelle, for example, used systems design to develop the software and systems needed to recycle computer and electronic systems for a profit. According to the company founder, "What we're doing here is buying dollars for 80 cents."[72] *Systems implementation* involves creating or acquiring the various system components (hardware, software, databases, etc.) defined in the design step, assembling them, and putting the new system into operation. For many organizations, this includes purchasing software, hardware, databases, and other IS components. The purpose of *systems maintenance and review* is to check and modify the system so that it continues to meet changing business needs. Increasingly, companies are hiring outside companies to do their design, implementation, maintenance, and review functions.

ORGANIZATIONS AND INFORMATION SYSTEMS

organization
A formal collection of people and
other resources established to
accomplish a set of goals.

An **organization** is a formal collection of people and other resources established to accomplish a set of goals. The primary goal of a for-profit organization is to maximize shareholder value, often measured by the price of the company stock. Nonprofit organizations include social groups, religious groups, universities, and other organizations that do not have profit as their goal. As discussed in this chapter, the ability of an organization to achieve its goals is often a function of the organization's overall structure, culture, and ability to change.

An organization is a system, which means that it has inputs, processing mechanisms, outputs, and feedback. An organization constantly uses money, people, materials, machines and other equipment, data, information, and decisions. As shown in Figure 1.15, resources such as materials, people, and money serve as inputs to the organizational system from the environment, go through a transformation mechanism, and then are produced as outputs to the environment. The outputs from the transformation mechanism are usually goods or services, which are of higher relative value than the inputs alone. Through adding value or worth, organizations attempt to increase performance and achieve their goals. According to one chief information officer (CIO) for a large healthcare company, "As business executives, other than the CEO, CIOs are best positioned to help drive business outcomes ... to increase top- and bottom-line performance."[73]

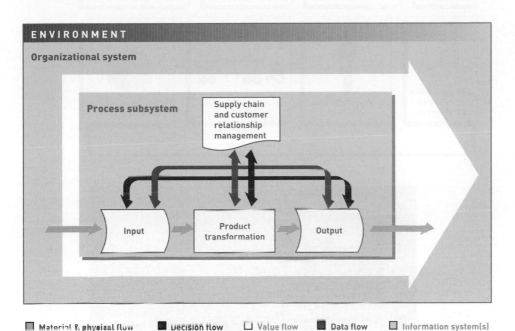

ENVIRONMENT

Organizational system

Process subsystem

Supply chain and customer relationship management

Input

Product transformation

Output

■ Material & physical flow   ■ Decision flow   □ Value flow   ■ Data flow   □ Information system(s)

Figure 1.15

A General Model of an Organization

Information systems support and work within all parts of an organizational process. Although not shown in this simple model, input to the process subsystem can come from internal and external sources. Just prior to entering the subsystem, data is external. After it enters the subsystem, it becomes internal. Likewise, goods and services can be output to either internal or external systems.

Providing value to a stakeholder—customer, supplier, manager, shareholder, or employee—is the primary goal of any organization. The value chain, first described by Michael Porter in a 1985 *Harvard Business Review* article, reveals how organizations can add value to their products and services. The **value chain** is a series (chain) of activities that includes inbound logistics, warehouse and storage, production and manufacturing, finished product storage, outbound logistics, marketing and sales, and customer service (see Figure 1.16). You investigate each activity in the chain to determine how to increase the value perceived by a customer. Depending on the customer, value might mean lower price, better service, higher quality, or uniqueness of product. The value comes from the skill, knowledge, time, and energy that the company invests in the product or activity. The value chain is just as important to companies that don't manufacture products, such as tax preparers, retail stores, legal firms, and other service providers. By adding a significant amount of value to their products and services, companies ensure success.

value chain
A series (chain) of activities that includes inbound logistics, warehouse and storage, production, finished product storage, outbound logistics, marketing and sales, and customer service.

Figure 1.16

The Value Chain of a Manufacturing Company

Managing raw materials, inbound logistics, and warehouse and storage facilities is called *upstream management*. Managing finished product storage, outbound logistics, marketing and sales, and customer service is called *downstream management*.

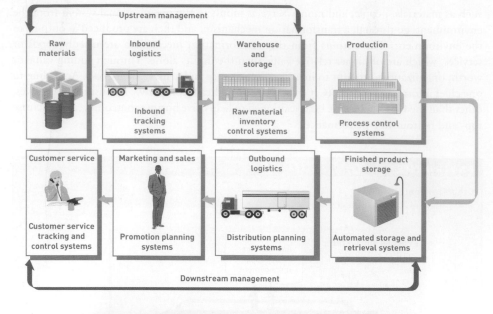

Combining a value chain with just-in-time (JIT) inventory means companies can deliver materials or parts when they are needed. Ball Aerospace uses JIT to help reduce inventory costs and enhance customer satisfaction.

(Source: AP Photo/Denver Post, R. J. Sangosti.)

Aldra Manages Workflow to Support Customization

Aldra Fenster und Türen GmbH, or Aldra for short, is a leading door and window manufacturer with over 300 dealers in Germany and Scandinavia. Aldra is well known for its precision craftsmanship in manufacturing intricate, custom-designed windows. In the early 1970s, the company developed a unique method of manufacturing windows from plastic. Combined with its customization service, this cost-saving manufacturing innovation gave Aldra a leg up on the competition.

Aldra's custom window design and manufacturing has created challenges in its corporate workflow and information processing. Mass-producing windows and doors in standard sizes is far easier than creating custom designs, where production techniques change from one item to the next. At Aldra, most orders have unique requirements in terms of size, shape, materials, function, and embedded technology. To support custom orders, Aldra must provide considerable flexibility in both its manufacturing processes and its information systems.

Providing customized manufacturing does not excuse Aldra from meeting the tight deadlines imposed by costly construction projects. Aggressive construction schedules rarely allow for the extra time required to produce custom products. Aldra found that the complexities of building its high-quality products were causing confusion in the order processing system and delays in manufacturing, leading to missed deadlines. Order specifications were sometimes incomplete or incorrect, and correcting orders is time consuming. Lack of coordination among departments resulted in additional errors that occasionally resulted in costly idle time on the production line. The lack of coordination also led to errors in calculating manufacturing costs, which reduced profits. Aldra set out to implement a new system that would assist the company in managing its value chain and corporate workflow.

Aldra purchased information systems from Infor Corporation that allowed the company to better coordinate efforts across departments. Using the software, Aldra now models its critical core processes (workflows) and then uses the models to improve communication across the value chain. The models define the specific employees involved in the various stages of the process.

The system then generates daily activities for each employee displayed in a particular area on the computer desktop. As activities approach their deadline, they are moved to the top of the list. Employees also receive e-mail notices of new or pressing actions needing attention.

Aldra's new workflow management system depends on a corporate-wide system that stores and manipulates all order details. Top managers can view orders to see how they are progressing through the value chain so that they can intervene when necessary.

Aldra implemented the new system in an unusually short amount of time. The company spent three days installing the system, another three days training managers in how to model workflow processes, and two weeks to model processes and train users. The benefits of the new system were almost immediately apparent. Within weeks, the company's adherence to delivery dates was improved by over 95 percent. Cost estimates are now reliably calculated. Employees make more productive use of their time, and customers are happy. Aldra is looking to expand the use of its new systems to other areas of its business.

Discussion Questions

1. What problems did Aldra's new information systems address, and what was the root of those problems?
2. How did Aldra's new systems assist employees in being more productive?

Critical Thinking Questions

1. What lessons can be learned from this case in terms of managing information in a value chain?
2. How does an organization determine when it is worthwhile to invest in a system such as Aldra's workflow management system?

SOURCES: Infor Staff, "Aldra Fenster und Türen GmbH," Aldra Customer Profile, accessed December 24, 2009, *www.infor.com/content/casestudies/296661*; Infor ERP systems Web site, accessed December 24, 2009; Aldra Web site (translated), accessed December 24, 2009, *www.aldra.de*.

Walmart's use of information systems is an integral part of its operation. The company gives suppliers access to its inventory system, so the suppliers can monitor the database and automatically send another shipment when stocks are low, eliminating the need for purchase orders. This speeds delivery time, lowers Walmart's inventory carrying costs, and reduces stockout costs.

(Source: *www.walmart.com.*)

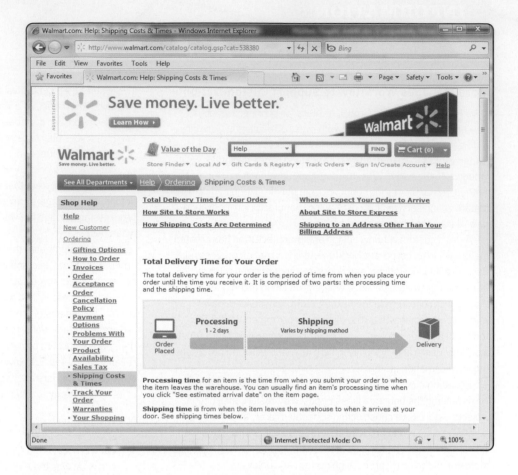

Managing the supply chain and customer relationships are two key elements of managing the value chain. *Supply chain management (SCM)* helps determine what supplies are required for the value chain, what quantities are needed to meet customer demand, how the supplies should be processed (manufactured) into finished goods and services, and how the shipment of supplies and products to customers should be scheduled, monitored, and controlled.[74] Companies use a number of approaches to manage their supply chain. Some automotive companies, for example, require that their suppliers locate close to manufacturing plants. Other companies have considered purchasing suppliers to manage their supply chain.[75] Sysco, a Texas-based food distribution company, uses a sophisticated supply chain management system that incorporates software and databases to prepare and ship over 20 million tons of meats, produce, and other food items to restaurants and other outlets every year.[76] The huge company supplies one in three cafeterias, sports stadiums, restaurants, and other food stores.

Customer relationship management (CRM) programs help companies of all sizes manage all aspects of customer encounters, including marketing and advertising, sales, customer service after the sale, and programs to retain loyal customers.[77] Often, CRM software uses a variety of information sources, including sales from retail stores, surveys, e-mail, and Internet browsing habits, to compile comprehensive customer profiles. CRM systems can also get customer feedback to help design new products and services. See Figure 1.17. To be of most benefit, CRM programs must be tailored for each company or organization. Duke Energy, an energy holding company, uses Convergys (*www.convergys.com*) to provide CRM software that is specifically configured to help the energy company manage its customer's use of energy grids and energy services.[78] Oracle, SalesForce, and other companies develop and sell CRM software.[79] CRM software can also be purchased as a service and delivered over the Internet instead of being installed on corporate computers.

Figure 1.17

SAP CRM

Companies in more than 25 industries use SAP CRM to reduce cost and increase decision-making ability in all aspects of their customer relationship management.

(Source: *www.sap.com*.)

Organizational Culture and Change

Culture is a set of major understandings and assumptions shared by a group, such as within an ethnic group or a country. **Organizational culture** consists of the major understandings and assumptions for a business, corporation, or other organization. The understandings, which can include common beliefs, values, and approaches to decision making, are often not stated or documented as goals or formal policies. For example, Procter & Gamble has an organizational culture that places an extremely high value on understanding its customers and their needs. As another example, employees might be expected to be clean-cut, wear conservative outfits, and be courteous in dealing with all customers. Sometimes organizational culture is formed over years. In other cases, top-level managers can form it rapidly by starting a "casual Friday" dress policy. Organizational culture can also have a positive effect on the successful development of new information systems that support the organization's culture. Some healthcare professionals believe that a good organizational culture can improve patient health and safety.[80]

Organizational change deals with how for-profit and nonprofit organizations plan for, implement, and handle change. Change can be caused by internal factors, such as those initiated by employees at all levels, or by external factors, such as activities wrought by competitors, stockholders, federal and state laws, community regulations, natural occurrences (such as hurricanes), and general economic conditions. Organizational change occurs when two or more organizations merge. When organizations merge, however, integrating their information systems can be critical to future success. When VeriSign, for example, acquired and merged with a number of companies, it had to integrate various information systems.[81] According to the chief information officer of VeriSign, "By being decisive and making the goals and objectives clear, we were able to fuse multiple teams into a single unit, which in the end was smaller and far more productive."

Change can be sustaining or disruptive.[82] *Sustaining change* can help an organization improve the supply of raw materials, the production process, and the products and services it offers. Developing new manufacturing equipment to make disk drives is an example of a sustaining change for a computer manufacturer. The new equipment might reduce the costs of producing the disk drives and improve overall performance. *Disruptive change,* on the other hand, can completely transform an industry or create new ones, which can harm an organization's performance or even put it out of business. In general, disruptive technologies might not originally have good performance, low cost, or even strong demand. Over time, however,

culture
A set of major understandings and assumptions shared by a group.

organizational culture
The major understandings and assumptions for a business, corporation, or other organization.

organizational change
How for-profit and nonprofit organizations plan for, implement, and handle change.

they often replace existing technologies. They can cause profitable, stable companies to fail when they don't change or adopt the new technology. On a positive note, disruptive change often results in new, successful companies and offers consumers the potential of new products and services at reduced costs and superior performance. An institute called Singularity University, located at the NASA Ames Research Center in California, offers workshops on how to deal with disruptive change.[83] The purpose of the institute is to prepare managers and executives for the fast, ever-changing nature of information systems.

When VeriSign acquired and merged with a number of companies, it had to integrate various information systems.

(Source: *www.verisign.com*.)

User Satisfaction and Technology Acceptance

To be effective, reengineering and continuous improvement efforts must result in satisfied users and be accepted and used throughout the organization. Over the years, IS researchers have studied user satisfaction and technology acceptance as they relate to IS attitudes and usage.[84] Although user satisfaction and technology acceptance started as two separate theories, some believe that they are related concepts.[85]

User satisfaction with a computer system and the information it generates often depend on the quality of the system and the value of the information it delivers to users.[86] A quality information system is usually flexible, efficient, accessible, and timely. Recall that quality information is accurate, reliable, current, complete, and delivered in the proper format.[87]

technology acceptance model (TAM)
A model that describes the factors leading to higher levels of acceptance and usage of technology.

The **technology acceptance model** (TAM) specifies the factors that can lead to better attitudes about the information system, along with higher acceptance and usage of the system in an organization.[88] These factors include the perceived usefulness of the technology, the ease of its use, the quality of the information system, and the degree to which the organization supports its use.[89] Studies have shown that user satisfaction and technology acceptance are critical in health care.[90] Doctors and other healthcare professionals need training and time to accept and use medical records technology and databases to reduce medical errors and save lives.

technology diffusion
A measure of how widely technology is spread throughout the organization.

You can determine the actual usage of an information system by the amount of technology diffusion and infusion.[91] **Technology diffusion** is a measure of how widely technology is spread throughout an organization. An organization in which computers and information systems are located in most departments and areas has a high level of technology diffusion.[92] Some online merchants such as Amazon.com have a high diffusion and use computer systems to perform most of their business functions, including marketing, purchasing, and billing. **Technology infusion**, on the other hand, is the extent to which technology permeates an area or department. In other words, it is a measure of how deeply embedded technology is in an area of the organization. Some architectural firms, for example, use computers in all aspects of designing a building from drafting to final blueprints. See Figure 1.18. The design area, thus, has a high level of infusion. Of course, a firm can have a high level of infusion in one part of its operations and a low level of diffusion overall. The architectural firm might use computers in all aspects of design (high infusion in the design area), but not to perform other business functions, including billing, purchasing, and marketing (low diffusion). Diffusion and infusion often depend on the technology available now and in the future, the size and type of the organization, and the environmental factors that include the competition, government regulations, suppliers, and so on. This is often called the technology, organization, and environment (TOE) framework.[93]

technology infusion
The extent to which technology is deeply integrated into an area or department.

Technology Infusion

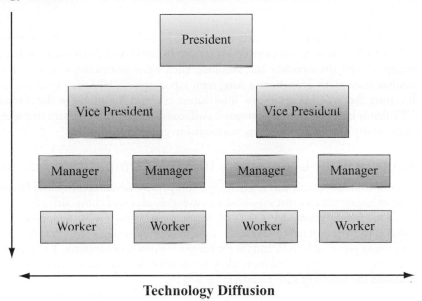

Technology Diffusion

Figure 1.18
Technology Infusion and Diffusion

Although an organization might have a high level of diffusion and infusion, with computers throughout the organization, this does not necessarily mean that information systems are being used to their full potential. In fact, the assimilation and use of expensive computer technology throughout organizations varies greatly.[94] Providing support and help to employees usually increases the use of a new information system.[95] Companies also hope that a high level of diffusion, infusion, satisfaction, and acceptance will lead to greater performance and profitability.[96] How appropriate and useful the information system is to the tasks or activities being performed, often called *Task-Technology Fit (TTF)*, can also lead to greater performance and profitability.[97]

COMPETITIVE ADVANTAGE

A **competitive advantage** is a significant and (ideally) long-term benefit to a company over its competition, and can result in higher-quality products, better customer service, and lower costs. According to the chief information officer of a large consulting company, "An efficiently run IT organization can be a significant source of competitive advantage."[98] An organization often uses its information system to help achieve a competitive advantage. A large Canadian furniture manufacturing company, for example, achieved a competitive advantage by reducing total operating costs by more than 20 percent using its information

competitive advantage
A significant and (ideally) long-term benefit to a company over its competition.

system to streamline its supply chain and reduce the cost of wood and other raw materials.[99] In his book *Good to Great,* Jim Collins outlines how technology can be used to accelerate companies to greatness.[100] Table 1.3 shows how a few companies accomplished this move. Ultimately, it is not how much a company spends on information systems but how it makes and manages investments in technology. Companies can spend less and get more value.

Company	Business	Competitive Use of Information Systems
Gillette	Shaving products	Developed advanced computerized manufacturing systems to produce high-quality products at low cost
Walgreens	Drug and convenience stores	Developed satellite communications systems to link local stores to centralized computer systems
Wells Fargo	Financial services	Developed 24-hour banking, ATMs, investments, and increased customer service using information systems

Taking advantage of the existing situation, including an economic downturn, can also help a firm achieve a competitive advantage. In 2009 and 2010, while some companies struggled with the economy and slumping sales, other companies were investing in information systems to give them a long-term advantage.[101] UPS, for example, planned on investing about $1 billion in new information systems. According to the company's CIO, "We firmly believe the strong companies will come out of this downturn stronger. This is an opportunity to get your company positioned to grow on the upturn."

Factors That Lead Firms to Seek Competitive Advantage

A number of factors can lead to attaining a competitive advantage. Michael Porter, a prominent management theorist, suggested a now widely accepted competitive forces model, also called the **five-forces model**. The five forces include (1) the rivalry among existing competitors, (2) the threat of new entrants, (3) the threat of substitute products and services, (4) the bargaining power of buyers, and (5) the bargaining power of suppliers. The more these forces combine in any instance, the more likely firms will seek competitive advantage and the more dramatic the results of such an advantage will be.

five-forces model

A widely accepted model that identifies five key factors that can lead to attainment of competitive advantage, including (1) the rivalry among existing competitors, (2) the threat of new entrants, (3) the threat of substitute products and services, (4) the bargaining power of buyers, and (5) the bargaining power of suppliers.

Rivalry Among Existing Competitors

Typically, highly competitive industries are characterized by high fixed costs of entering or leaving the industry, low degrees of product differentiation, and many competitors. Although all firms are rivals with their competitors, industries with stronger rivalries tend to have more firms seeking competitive advantage. To gain an advantage over competitors, companies constantly analyze how they use their resources and assets. This *resource-based view* is an approach to acquiring and controlling assets or resources that can help the company achieve a competitive advantage. For example, a transportation company might decide to invest in radio-frequency technology to tag and trace products as they move from one location to another.

Threat of New Entrants

A threat appears when entry and exit costs to an industry are low and the technology needed to start and maintain a business is commonly available. For example, a small restaurant is threatened by new competitors. Owners of small restaurants do not require millions of dollars to start the business, food costs do not decline substantially for large volumes, and food processing and preparation equipment is easily available. When the threat of new market entrants is high, the desire to seek and maintain competitive advantage to dissuade new entrants is also usually high.

Threat of Substitute Products and Services

Companies that offer one type of goods or services are threatened by other companies that offer similar goods or services. The more consumers can obtain similar products and services that satisfy their needs, the more likely firms are to try to establish competitive advantage. For example, consider the photographic industry. When digital cameras became popular, traditional film companies had to respond to stay competitive and profitable. Traditional film companies, such as Kodak and others, started to offer additional products and enhanced services, including digital cameras, the ability to produce digital images from traditional film cameras, and Web sites that could be used to store and view pictures.

In the restaurant industry, competition is fierce because entry costs are low. Therefore, a small restaurant that enters the market can be a threat to existing restaurants.

(Source: © 2010, Emin Kuliyev. Used under license from Shutterstock.com.)

Bargaining Power of Customers and Suppliers

Large customers tend to influence a firm, and this influence can increase significantly if the customers can threaten to switch to rival companies. When customers have a lot of bargaining power, companies increase their competitive advantage to retain their customers. Similarly, when the bargaining power of suppliers is strong, companies need to improve their competitive advantage to maintain their bargaining position. Suppliers can also help an organization gain a competitive advantage. Some suppliers enter into strategic alliances with firms and eventually act as a part of the company. Suppliers and companies can use telecommunications to link their computers and personnel to react quickly and provide parts or supplies as necessary to satisfy customers.

Strategic Planning for Competitive Advantage

To be competitive, a company must be fast, nimble, flexible, innovative, productive, economical, and customer oriented. It must also align its IS strategy with general business strategies and objectives. [102] Given the five market forces previously mentioned, Porter and others have proposed a number of strategies to attain competitive advantage, including cost leadership, differentiation, niche strategy, altering the industry structure, creating new products and services, and improving existing product lines and services. [103] In some cases, one of these strategies becomes dominant. For example, with a cost leadership strategy, cost can be the key consideration, at the expense of other factors if need be.

- **Cost leadership.** Deliver the lowest possible cost for products and services. Walmart and other discount retailers have used this strategy for years. Cost leadership is often achieved by reducing the costs of raw materials through aggressive negotiations with suppliers, becoming more efficient with production and manufacturing processes, and reducing warehousing and shipping costs. Some companies use outsourcing to cut costs when making products or completing services.

- **Differentiation.** Deliver different products and services. This strategy can involve producing a variety of products, giving customers more choices, or delivering higher-quality products and services. Many car companies make different models that use the same basic parts and components, giving customers more options. Other car companies attempt to increase perceived quality and safety to differentiate their products and appeal to consumers who are willing to pay higher prices for these features. Companies that try to differentiate their products often strive to uncover and eliminate counterfeit products produced and delivered by others.

Walmart and other discount retailers have used a cost leadership strategy to deliver the lowest possible price for products and services.

(Source: © Jeff Zelevansky/Getty Images.)

- **Niche strategy.** Deliver to only a small, niche market. Porsche, for example, doesn't produce inexpensive economy cars. It makes high-performance sports cars and SUVs. Rolex only makes high-quality, expensive watches. It doesn't make inexpensive, plastic watches that can be purchased for $20 or less.

Porsche is an example of a company with a niche strategy, producing only high-performance sports cars and SUVs.

(Source: © 2010, Max Earey. Used under license from Shutterstock.com.)

strategic alliance (strategic partnership)
An agreement between two or more companies that involves the joint production and distribution of goods and services.

- **Altering the industry structure.** Change the industry to become more favorable to the company or organization. The introduction of low-fare airline carriers, such as Southwest Airlines, has forever changed the airline industry, making it difficult for traditional airlines to make high profit margins. Creating strategic alliances can also alter the industry structure. A **strategic alliance**, also called a **strategic partnership**, is an agreement between two or more companies that involves the joint production and distribution of goods and services. The investment firm American Diversified Holdings, for example, developed a strategic alliance with Invent Pharmaceuticals to help the pharmaceutical company with investments, regulatory issues, and business operations.[104] According to the chairman of American Diversified Holdings, "This alliance with Invent Pharma will enhance our investment focus in the biotech industry."
- **Creating new products and services.** Introduce new products and services periodically or frequently. This strategy always helps a firm gain a competitive advantage, especially for the computer industry and other high-tech businesses. If an organization does not introduce new products and services every few months, the company can quickly stagnate, lose market share, and decline. Companies that stay on top are constantly developing

new products and services. Apple Computer, for example, introduced the iPod, iPhone, and iPad as new products.

- **Improving existing product lines and services.** Make real or perceived improvements to existing product lines and services. Manufacturers of household products are always advertising new and improved products. In some cases, the improvements are more perceived than actual refinements; usually, only minor changes are made to the existing product, such as to reduce the amount of sugar in breakfast cereal.
- **Other strategies.** Some companies seek strong *growth* in sales, hoping that it can increase profits in the long run due to increased sales. Being the *first to market* is another competitive strategy. Apple Computer was one of the first companies to offer complete and ready-to-use personal computers. Some companies offer *customized* products and services to achieve a competitive advantage. Dell, for example, builds custom PCs for consumers. *Hire the best people* is another example of a competitive strategy. The assumption is that the best people will determine the best products and services to deliver to the market and the best approach to deliver these products and services. Having *agile* information systems that can rapidly change with changing conditions and environments can be a key to information systems success and a competitive advantage.[105] Achieving a high level of efficiency and effectiveness is an important challenge of developing an agile information system. Other challenges included satisfying various governmental regulations, meeting customer requirements, and maintaining a good growth level. *Innovation* is another competitive strategy.[106] Vodafone relied on outside help to provide innovative solutions in its wireless business.[107] According its chief executive, "The only way to create a fertile environment for innovation is to have open platforms and leverage them." Natural Selection, a San Diego company, originally developed a computer program that attempted to analyze past inventions and suggest future ones.[108] Although the original program was not an immediate success, the approach has been used by General Electric, the U.S. Air Force, and others to cut costs and streamline delivery routes of products. According to one expert, "Successful innovations are often built on the back of failed ones." A lack of innovation can lead to a loss in competitiveness and long-term profitability.[109] Some believe that less innovation has led to lower productivity, lower profits, and lower wages and salaries for managers and workers. Companies can also combine one or more of these strategies. In addition to customization, Dell attempts to offer low-cost computers (cost leadership) and top-notch service (differentiation).

PERFORMANCE-BASED INFORMATION SYSTEMS

Businesses have passed through at least three major stages in their use of information systems. In the first stage, organizations focused on using information systems to reduce costs and improve productivity. TransUnion, a large credit reporting company, reduced computer-related costs by about $2.5 million annually by investing $50,000 in a corporate social networking Internet site.[110] According to the chief technology officer, "The savings mostly come out of teams that would have historically said 'Buy me more hardware' or 'I need a new software tool' who figured out how to solve their problems without asking for those things." In another example, the National ePrescribing Patient Safety Initiative offers powerful software to doctors to reduce medication errors and costs. Companies can also use software tools, such as Apptio's IT Cost Optimization Solutions, to cut the costs of computer upgrades, reduce the number of computers, and help determine what to charge business units for providing computer services and equipment.[111]

The second stage was defined by Porter and others. It was oriented toward gaining a competitive advantage. In many cases, companies spent large amounts on information systems and downplayed the costs.

Today, companies are shifting from strategic management to performance-based management of their information systems. In this third stage, companies carefully consider both strategic advantage and costs. They use productivity, return on investment (ROI), net present value, and other measures of performance to evaluate the contributions their information systems make to their businesses. Figure 1.19 illustrates these stages. This balanced approach attempts to reduce costs and increase revenues.

Figure 1.19

Three Stages in the Business Use of Information Systems

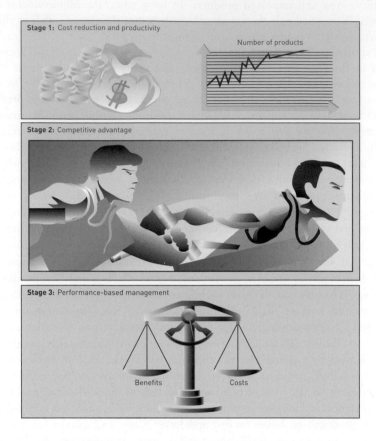

Productivity

productivity

A measure of the output achieved divided by the input required.

Developing information systems that measure and control productivity is a key element for most organizations. **Productivity** is a measure of the output achieved divided by the input required. A higher level of output for a given level of input means greater productivity; a lower level of output for a given level of input means lower productivity. The numbers assigned to productivity levels are not always based on labor hours—productivity can be based on factors such as the amount of raw materials used, resulting quality, or time to produce the goods or service. The value of the productivity number is not as significant as how it compares with other time periods, settings, and organizations. Xerox has developed an information system to increase printer productivity and reduce costs called Lean Document Production (LDP) solutions.[112] According to one researcher, "These solutions, which Xerox has implemented in approximately 100 sites to date, have provided dramatic productivity and cost improvements for both print shops and document-manufacturing facilities."

Productivity = (Output / Input) × 100%

After a basic level of productivity is measured, an information system can monitor and compare it over time to see whether productivity is increasing. Then, a company can take corrective action if productivity drops below certain levels. An automotive company, for example, might use robots in assembling new cars to increase its labor productivity and reduce costs. In addition to measuring productivity, an information system can be used within a process to significantly increase productivity. Thus, improved productivity can result in faster customer response, lower costs, and increased customer satisfaction.

Return on Investment and the Value of Information Systems

One measure of IS value is **return on investment** (**ROI**). This measure investigates the additional profits or benefits that are generated as a percentage of the investment in IS technology. A small business that generates an additional profit of $20,000 for the year as a result of an investment of $100,000 for additional computer equipment and software would have a return on investment of 20 percent ($20,000/$100,000). ROI calculations can be complex, including investment returns over multiple years and the impact of the time value of money. According to the chief technology officer for the Financial Industry Regulatory Authority, "ROI is a key metric for technology initiatives."[113] Some researchers believe that how an IS function is managed and run is one of the best indicators of the value of the system to the organization and its return on investment.[114] Because of the importance of ROI, many computer companies provide ROI calculators to potential customers. ROI calculators are typically provided on a vendor's Web site and can be used to estimate returns. Kodak, for example, has an ROI calculator for many of its products based on lifetime value to customers.[115]

return on investment (ROI)
One measure of IS value that investigates the additional profits or benefits that are generated as a percentage of the investment in IS technology.

The National ePrescribing Patient Safety Initiative offers software to doctors to reduce medication errors and costs.

(Source: *www.nationalerx.com.*)

Earnings Growth

Another measure of IS value is the increase in profit, or earnings growth, the system brings. For instance, a mail-order company might install an order-processing system that generates a seven percent earnings growth compared with the previous year.

Market Share and Speed to Market

Market share is the percentage of sales that a product or service has in relation to the total market. If installing a new online catalog increases sales, it might help a company increase its market share by 20 percent. Information systems can also help organizations bring new products and services to customers in less time. This is often called speed to market. Speed can also be a critical performance objective for many organizations. The New York Stock Exchange, for example, is building a large facility the size of several football fields to house super-fast trading systems that can be used by large hedge funds and institutional investors.[116]

Customer Awareness and Satisfaction

Although customer satisfaction can be difficult to quantify, about half of today's best global companies measure the performance of their information systems based on feedback from internal and external users. Some companies and nonprofit organizations use surveys and questionnaires to determine whether the IS investment has increased customer awareness and satisfaction.

Total Cost of Ownership

total cost of ownership (TCO)
The sum of all costs over the life of an information system, including the costs to acquire components such as the technology, technical support, administrative costs, and end-user operations.

Another way to measure the value of information systems was developed by the Gartner Group and is called the **total cost of ownership (TCO)**. TCO is the sum of all costs over the life of the information system, including the costs to acquire components such as the technology, technical support, administrative costs, and end-user operations. Hitachi uses TCO to promote its projectors to businesses and individuals.[117] TCO is also used by many other companies to rate and select hardware, software, databases, and other computer-related components.

Return on investment, earnings growth, market share, customer satisfaction, and TCO are only a few measures that companies use to plan for and maximize the value of their IS investments. Regardless of the difficulties, organizations must attempt to evaluate the contributions that information systems make to assess their progress and plan for the future. Information systems and personnel are too important to leave to chance.

Risk

In addition to the return-on-investment measures of a new or modified system discussed earlier, managers must also consider the risks of designing, developing, and implementing these systems. Information systems can sometimes be costly failures. The risks of designing, developing, and implementing new or modified systems are covered in more detail in Chapter 8, which discuss systems development.

CAREERS IN INFORMATION SYSTEMS

Realizing the benefits of any information system requires competent and motivated IS personnel, and many companies offer excellent job opportunities. As mentioned earlier, *knowledge workers (KWs)* are people who create, use, and disseminate knowledge. They are usually professionals in science, engineering, business, and other areas that specialize in information systems. Numerous schools have degree programs with such titles as information systems, computer information systems, and management information systems. These programs are typically offered by information schools, business schools, and within computer science departments. Information systems skills can also help people start their own companies.

Skills that some experts believe are important for IS workers to have include those in the following list.[118] Nontechnical skills are also important for IS personnel, including communication skills, a detailed knowledge of the organization, and how information systems can help the organization achieve its goals. All of the following skills are discussed in the chapters throughout this book.

1. Program and application development
2. Help Desk and technical support
3. Project management

4. Networking
5. Business intelligence
6. Security
7. Web 2.0
8. Data center
9. Telecommunications.

The U.S. Department of Labor's Bureau of Labor Statistics (*www.bls.gov*) publishes the fastest growing occupations and predicts that many technology jobs will increase through 2012 or beyond. Table 1.4 summarizes some of the best places to work as an IS professional.[119] Career development opportunities, training, benefits, retention, diversity, and the nature of the work itself are just a few of the qualities these top employers offer.

Table 1.4

Best Places to Work as an IS Professional

Source: "Best Places to Work in IT," *Computerworld,* June 16, 2009.

Company	Additional Benefits
General Mills	Auto service facilities and fitness center
Genentech	Relaxation, meditation, and mindfulness programs
San Diego Gas & Electric	Good retirement program
University of Pennsylvania	Many excellent campus events and activities
Monsanto	Flex schedules and telecommuting options
Securian Financial Group	Career growth opportunities
Verizon	Innovation and working with new technologies
JM Family Enterprises	Employee growth and deals on the Toyota vehicles the company represents
USAA	Flexible work schedules
University of Miami	Good compensation plan and many university benefits

Opportunities in information systems are also available to people from foreign countries, including Russia and India. The U.S. H-1B and L-1 visa programs seek to allow skilled employees from foreign lands into the United States. These programs, however, are limited and usually in high demand. The L-1 visa program is often used for intracompany transfers for multinational companies. The H-1B program can be used for new employees.

Roles, Functions, and Careers in IS

IS offers many exciting and rewarding careers. Professionals with careers in information systems can work in an IS department or outside a traditional IS department as Web developers, computer programmers, systems analysts, computer operators, and many other positions. There are also opportunities for IS professionals in the public sector. The U.S. stimulus package of 2009, for example, budgeted about $1 billion to develop better systems, including computer programs to deliver disability claims for the federal government.[120] This massive project will require a large number of IS professionals. In addition to technical skills, IS professionals need skills in written and verbal communication, an understanding of organizations and the way they operate, and the ability to work with people and in groups. Today, many good information, business, and computer science schools require these business and communications skills of their graduates. At the end of every chapter, you will find career exercises that will help you explore careers in IS and career areas that interest you.

Most medium to large organizations manage information resources through an IS department. In smaller businesses, one or more people might manage information resources, with support from outsourced services. (Recall that outsourcing is also popular with larger organizations.) As shown in Figure 1.20, the IS organization has three primary responsibilities: operations, systems development, and support.

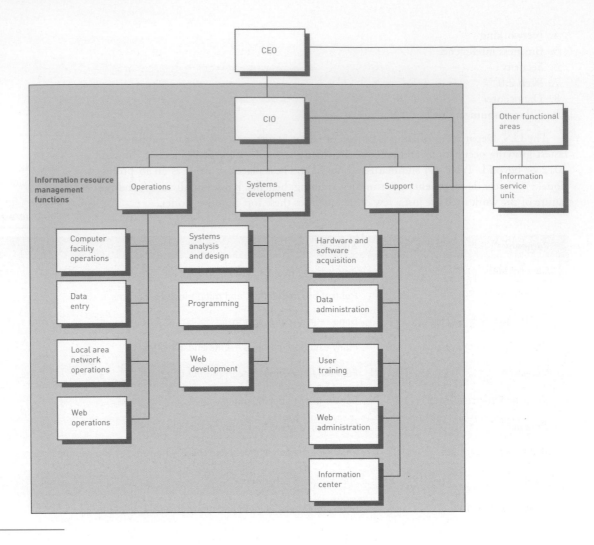

Figure 1.20

The Three Primary Responsibilities of Information Systems

Each of these elements— operations, systems development, and support—contains sub- elements that are critical to the efficient and effective performance of the organization.

Web developers create and maintain company Web sites.

(Source: © iStockphoto/David H. Lewis.)

Operations

System operators primarily run and maintain IS equipment, and are typically trained at technical schools or through on-the-job experience. They are responsible for efficiently start- ing, stopping, and correctly operating mainframe systems, networks, tape drives, disk devices, printers, and so on. Other operations include scheduling, hardware maintenance, and preparing input and output. Data-entry operators convert data into a form the computer system can use, using terminals or other devices to enter business transactions, such as sales orders and payroll data. In addition, companies might have local area network and Web operators who run the local network and any Web sites the company has.

Systems Development

The systems development component of a typical IS department focuses on specific development projects and ongoing maintenance and review. Systems analysts and programmers, for example, address these concerns to achieve and maintain IS effectiveness. The role of a systems analyst is multifaceted. *Systems analysts* help users determine what outputs they need from the system and construct plans for developing the necessary programs that produce these outputs. Systems analysts then work with one or more programmers to make sure that the appropriate programs are purchased, modified from existing programs, or developed. A *computer programmer* uses the plans created by the systems analyst to develop or adapt one or more computer programs that produce the desired outputs. A meteorologist and several part-time programmers from the University of Alabama developed weather forecasting software that used radar data along with other meteorological data to forecast storms and weather.[121] Today, the weather forecasting company employs about 100 people, including many programmers, to keep the software current. To help businesses select the best analysts and programmers, companies such as TopCoder offer tests to evaluate the proficiency and competence of current IS employees or job candidates. TopCoder Collegiate Challenge allows programming students to compete with other programmers around the world.[122] In addition, with the dramatic increase in the use of the Internet, intranets, and extranets, many companies have Web or Internet developers who create effective and attractive Web sites for customers, internal personnel, suppliers, stockholders, and others who have a business relationship with the company.

Support

The support component of a typical IS department provides user assistance in hardware and software acquisition and use, data administration, user training and assistance, and Web administration. Increasingly, training is done using the Internet. Microsoft, for example, offers free training in areas including time management, marketing, sales, and others (*office.microsoft.com/en-us/officelive/FX102119031033.aspx*). Other companies, such as Hewlett Packard (*www.hp.com/sbso*), also offer online training courses and programs. In many cases, support is delivered through an information center.

IS personnel provide assistance in hardware and software acquisition, data administration, user training and assistance, and Web administration.

(Source: © iStockphoto/Chris Schmidt.)

Because IS hardware and software are costly, a specialized support group often manages computer hardware and software acquisitions. This group sets guidelines and standards for the rest of the organization to follow in making purchases. A database administrator focuses on planning, policies, and procedures regarding the use of corporate data and information. For example, database administrators develop and disseminate information about the corporate databases for developers of IS applications. In addition, the database administrator monitors and controls database use. Web administration is another key area for support staff. With the increased use of the Internet and corporate Web sites, Web administrators are sometimes asked to regulate and monitor Internet use by employees and managers to make sure that it is authorized and appropriate. User training is a key to get the most from any information system, and the support area ensures that appropriate training is available. Training can be provided by internal staff or from external sources.

information center
A support function that provides users with assistance, training, application development, documentation, equipment selection and setup, standards, technical assistance, and troubleshooting.

The support component typically operates the information center. An **information center** provides users with assistance, training, application development, documentation, equipment selection and setup, standards, technical assistance, and troubleshooting. Although many firms have attempted to phase out information centers, others have changed their focus from technical training to helping users find ways to maximize the benefits of the information resource.

Information Service Units

information service unit
A miniature IS department.

An **information service unit** is basically a miniature IS department attached and directly reporting to a functional area in a large organization. Notice the information service unit shown in Figure 1.20. Even though this unit is usually staffed by IS professionals, the project assignments and the resources necessary to accomplish these projects are provided by the functional area to which it reports. Depending on the policies of the organization, the salaries of IS professionals staffing the information service unit might be budgeted to either the IS department or the functional area.

Typical IS Titles and Functions

The organizational chart shown in Figure 1.20 is a simplified model of an IS department in a typical medium-sized or large organization. Many organizations have even larger departments, with increasingly specialized positions such as librarian or quality assurance manager. Smaller firms often combine the roles shown in Figure 1.20 into fewer formal positions.

Chief Information Officer

The role of the chief information officer (CIO) is to employ an IS department's equipment and personnel to help the organization attain its goals.[123] The CIO is usually a vice president concerned with the overall needs of the organization, sets corporate-wide policies, and plans, manages, and acquires information systems.[124] In one survey, more than 60 percent of CIOs reported directly to the president of the company or the chief executive officer (CEO). According to another survey, almost 80 percent of CIOs are actively involved in or consulted on most major decisions. The CIO of Sunoco and President of the Society for Information Management described one of his duties as follows: "In 30 seconds, be able to describe how your company makes money. Make sure that your style and behavior are aligned with your company's culture and style."[125] The chief information officer of the Financial Industry Regulatory Authority agrees with this approach and said: "New CIOs need to understand how the business functions and build strong relationships with their business partners."[126] CIOs can also help companies avoid damaging ethical challenges by monitoring how companies are complying with a large number of laws and regulations.[127]

The high level of the CIO position reflects that information is one of the organization's most important resources. A good CIO is typically a visionary who provides leadership and direction to the IS department to help an organization achieve its goals. CIOs need both technical and business skills. In giving advice to other CIOs, the CIO of Wipro said, "Keep in close touch with the business side and focus on delivering continuous business value."[128] For federal agencies, the Clinger-Cohen Act of 1996 requires that a CIO coordinate the purchase and management of information systems.[129] The U.S. federal government has also instituted a CIO position to manage federal IS projects, including budgets and deadlines.[130] In 2009, Vivek Kundra was the first person appointed to this new position—CIO of the United States.

A company's CIO is usually a vice president who sets corporate-wide policies, and plans, manages, and acquires information systems.

(Source: © iStockphoto/Jacob Wackerhausen.)

Depending on the size of the IS department, several people might work in senior IS managerial levels. Some job titles associated with IS management are the CIO, vice president of information systems, manager of information systems, and chief technology officer (CTO). A central role of all these people is to communicate with other areas of the organization to determine changing needs. Often these employees are part of an advisory or steering committee that helps the CIO and other IS managers make decisions about the use of information systems. Together they can best decide what information systems will support corporate goals. The CTO, for example, typically works under a CIO and specializes in networks and related equipment and technology.

LAN Administrators

Local area network (LAN) administrators set up and manage the network hardware, software, and security processes. They manage the addition of new users, software, and devices to the network. They also isolate and fix operations problems. LAN administrators are in high demand and often solve both technical and nontechnical problems.

Internet Careers

The use of the Internet to conduct business continues to grow and has stimulated a steady need for skilled personnel to develop and coordinate Internet usage. As shown in Figure 1.20, these careers are in the areas of Web operations, Web development, and Web administration. As with other areas in IS, many top-level administrative jobs are related to the Internet. These career opportunities are found in both traditional companies and those that specialize in the Internet.

Internet jobs within a traditional company include Internet strategists and administrators, Internet systems developers, Internet programmers, and Internet or Web site operators. Some companies suggest a new position, chief Internet officer, with responsibilities and a salary similar to the CIO's.

In addition to traditional companies, Internet companies offer exciting career opportunities. These companies include Google, Amazon.com, Yahoo!, eBay, and many others. Systest, for example, specializes in finding and eliminating digital bugs that could halt the operation of a computer system.[131]

Often, the people filling IS roles have completed some form of certification. **Certification** is a process for testing skills and knowledge resulting in an endorsement by the certifying authority that an individual is capable of performing a particular job. Certification frequently involves specific, vendor-provided or vendor-endorsed coursework. Popular certification programs include Microsoft Certified Systems Engineer, Certified Information Systems Security Professional (CISSP), Oracle Certified Professional, Cisco Certified Security Professional (CCSP), and many others.

certification

A process for testing skills and knowledge, which results in a statement by the certifying authority that confirms an individual is capable of performing a particular kind of job.

Other IS Careers

To respond to the increase in attacks on computers, new and exciting careers have developed in security and fraud detection and prevention. Today, many companies have IS security positions, such as a chief information security officer or a chief privacy officer. Some universities offer degree programs in security or privacy. It is even possible to work from home in an IS field. Programmers, systems developers, and others are also working from home in developing new information systems.

In addition to working for an IS department in an organization, IS personnel can work for large consulting firms, such as Accenture (*www.accenture.com*), IBM (*www.ibm.com/services*), EDS (*www.eds.com*), and others.[132] Some consulting jobs can entail frequent travel because consultants are assigned to work on various projects wherever the client is. Such roles require excellent project management and people skills in addition to IS technical skills. Related career opportunities include computer training, computer and computer equipment salespersons, computer repair and maintenance, and many others.

Other IS career opportunities include being employed by technology companies, such as Microsoft (*www.microsoft.com*), Google (*www.google.com*), Dell (*www.dell.com*), and many others. Such a role enables an individual to work on the cutting edge of technology, which can be extremely challenging and exciting. As some computer companies cut their services to customers, new companies are being formed to fill the need. With names such as Speak with a Geek and Geek Squad, which is located in many Best Buy stores, these companies are helping people and organizations with their computer-related problems that computer vendors are no longer solving.

Some people start their own IS businesses from scratch, such as Craig Newmark, founder of Craig's List.[133] In the mid 1990s, Newmark was working for a large financial services firm and wanted to give something back to society by developing an e-mail list for arts and technology events in the San Francisco area. This early e-mail list turned into Craig's List. According to Newmark, to run a successful business, you should "Treat people like you want to be treated, including providing good customer service. Listening skills and effective communication are essential." Other people are becoming IS entrepreneurs, working from home writing programs, working on IS projects with larger businesses, or developing new applications for the iPhone or similar devices.

Working in Teams

Most IS careers involve working in project teams that can consist of many of the positions and roles discussed earlier. Thus, it is always good for IS professionals to have good communications skills and the ability to work with other people. Many colleges and universities have courses in information systems and related areas that require students to work in project teams. At the end of every chapter in this book, we have "team activities" that require teamwork to complete a project. You may be required to complete one or more of these team-oriented assignments.

Getting the best team of IS personnel to work on important projects is critical in successfully developing new information systems or modifying existing ones.[134] Increasingly, companies and IS departments seek teams with varying degrees of skills, ages, and approaches. According to the managing director of Accenture, a large IS consulting company, "Every project team we build has an entire spectrum of age and experience represented. Diversity guarantees the best project result and usually some layer of innovation."[135]

Finding a Job in IS

Traditional approaches to finding a job in the information systems area include on-campus visits from recruiters and referrals from professors, friends, and family members. Many colleges and universities have excellent programs to help students develop résumés and conduct job interviews. Developing an online résumé can be critical to finding a good job. Many companies accept résumés online and use software to search for keywords and skills used to screen job candidates. Thus, having the right keywords and skills can mean the difference between getting a job interview and not being considered.

Increasingly, students are using the Internet and other sources to find IS jobs. Many Web sites, such as Dice.com, CareerBuilder.com, TheLadders.com, LinkedIn.com, Computerjobs.com, and Monster.com, post job opportunities for Internet careers and more traditional careers.[136] Most large companies list job opportunities on their Web sites. These sites allow prospective job hunters to browse job opportunities, locations, salaries, benefits, and other factors. In addition, some sites allow job hunters to post their résumés. Many of the social networking sites, including MySpace and Facebook, can be used to help get job leads. Corporate recruiters also use the Internet or Web logs (blogs) to gather information on existing job candidates or to locate new job candidates. In addition, many professional organizations and user groups can be helpful in finding a job, staying current once employed, and seeking new career opportunities, including the Association for Computer Machinery (ACM – *www.acm.org*), the Association of Information Technology Professionals (AITP – *www.aitp.org*), Apple User Groups (*www.apple.com/usergroups*), and Linux users groups located around the world. Many companies, including Microsoft, Viacom, and others, use Twitter, an Internet site that allows short messages of 140 characters or less, to advertise job openings.[137] People who have quit jobs or have been laid off often use informal networks of colleagues or business acquaintances to help find new jobs.[138]

As with other areas in IS, many top-level administrative jobs are related to the Internet, such as Internet systems developers and Internet programmers.

(Source: © iStockphoto/Frances Twitty.)

GLOBAL CHALLENGES IN INFORMATION SYSTEMS

Changes in society as a result of increased international trade and cultural exchange, often called globalization, have always had a significant impact on organizations and their information systems. In his book *The World Is Flat*, Thomas Friedman describes three eras of globalization.[139] (See Table 1.5.) According to Friedman, we have progressed from the globalization of countries to the globalization of multinational corporations and individuals. Today, people in remote areas can use the Internet to compete with and contribute to other people, the largest corporations, and entire countries. These workers are empowered by high-speed Internet access, making the world flatter. In the Globalization 3.0 era, designing a new airplane or computer can be separated into smaller subtasks and then completed by a person or small group that can do the best job. These workers can be located in India, China, Russia, Europe, and other areas of the world. The subtasks can then be combined or reassembled into the complete design. This approach can be used to prepare tax returns, diagnose a patient's medical condition, fix a broken computer, and many other tasks.

Global markets have expanded. People and companies can get products and services from around the world, instead of around the corner or across town. These opportunities, however, introduce numerous obstacles and issues, including challenges involving culture, language, and many others.

Table 1.5

Eras of Globalization

Era	Dates	Characterized by
Globalization 1.0	Late 1400–1800	Countries with the power to explore and influence the world
Globalization 2.0	1800–2000	Multinational corporations that have plants, warehouses, and offices around the world
Globalization 3.0	2000–today	Individuals from around the world who can compete and influence other people, corporations, and countries by using the Internet and powerful technology tools

- **Cultural challenges.** Countries and regional areas have their own cultures and customs that can significantly affect individuals and organizations involved in global trade.
- **Language challenges.** Language differences can make it difficult to translate exact meanings from one language to another.
- **Time and distance challenges.** Time and distance issues can be difficult to overcome for individuals and organizations involved with global trade in remote locations. Large time differences make it difficult to talk to people on the other side of the world. With long distance, it can take days to get a product, a critical part, or a piece of equipment from one location to another location.
- **Infrastructure challenges.** High-quality electricity and water might not be available in certain parts of the world. Telephone services, Internet connections, and skilled employees might be expensive or not readily available.
- **Currency challenges.** The value of different currencies can vary significantly over time, making international trade more difficult and complex.
- **Product and service challenges.** Traditional products that are physical or tangible, such as an automobile or bicycle, can be difficult to deliver to the global market. However, *electronic products (e-products)* and *electronic services (e-services)* can be delivered to customers electronically, over the phone, networks, through the Internet, or other electronic means. Software, music, books, manuals, and advice can all be delivered globally and over the Internet.
- **Technology transfer issues.** Most governments don't allow certain military-related equipment and systems to be sold to some countries. Even so, some believe that foreign companies are stealing intellectual property, trade secrets, and copyrighted materials, and counterfeiting products and services.
- **State, regional, and national laws.** Each state, region, and country has a set of laws that must be obeyed by citizens and organizations operating in the country. These laws can deal with a variety of issues, including trade secrets, patents, copyrights, protection of personal or financial data, privacy, and much more. Laws restricting how data enters or exits a country are often called *transborder data-flow* laws. Keeping track of these laws and incorporating them into the procedures and computer systems of multinational and transnational organizations can be very difficult and time consuming, requiring expert legal advice.
- **Trade agreements.** Countries often enter into trade agreements with each other. The North American Free Trade Agreement (NAFTA) and the Central American Free Trade Agreement (CAFTA) are examples. The European Union (EU) is another example of a group of countries with an international trade agreement.[140] The EU is a collection of mostly European countries that have joined together for peace and prosperity. Additional trade agreements include the Australia-United States Free Trade Agreement (AUSFTA), signed into law in 2005, and the Korean-United States Free Trade Agreement (KORUS-FTA), signed into law in 2007. Free trade agreements have been established between Bolivia and Mexico, Canada and Costa Rica, Canada and Israel, Chile and Korea, Mexico and Japan, the United States and Jordan, and many others.[141]

SUMMARY

Principle:

The value of information is directly linked to how it helps decision makers achieve the organization's goals.

Data consists of raw facts; information is data transformed into a meaningful form. The process of defining relationships among data requires knowledge. Knowledge is an awareness and understanding of a set of information and the way that information can support a specific task. To be valuable, information must have several characteristics: It should be accurate, complete, economical to produce, flexible, reliable, relevant, simple to understand, timely, verifiable, accessible, and secure. The value of information is directly linked to how it helps people achieve their organization's goals.

Information systems are sets of interrelated elements that collect (input), manipulate and store (process), and disseminate (output) data and information. Input is the activity of capturing and gathering new data, processing involves converting or transforming data into useful outputs, and output involves producing useful information. Feedback is the output that is used to make adjustments or changes to input or processing activities.

Principle:

Knowing the potential impact of information systems and having the ability to put this knowledge to work can result in a successful personal career, organizations that reach their goals, and a society with a higher quality of life.

Information systems play an important role in today's businesses and society. The key to understanding the existing variety of systems begins with learning their fundamentals. The types of systems used within organizations can be classified into four basic groups: (1) e-commerce and m-commerce, (2) TPS and ERP, (3) MIS and DSS, and (4) specialized business information systems.

E-commerce involves any business transaction executed electronically between parties such as companies (business-to-business), companies and consumers (business-to-consumer), business and the public sector, and consumers and the public sector. The major volume of e-commerce and its fastest-growing segment is business-to-business transactions that make purchasing easier for big corporations. E-commerce offers opportunities for small businesses by enabling them to market and sell at a low cost worldwide, thus enabling them to enter the global market. Mobile commerce (m-commerce) are transactions conducted anywhere, anytime. M-commerce relies on the use of wireless communications to allows managers and

corporations to place orders and conduct business using handheld computers, portable phones, laptop computers connected to a network, and other mobile devices.

The most fundamental system is the transaction processing system (TPS). A transaction is any business-related exchange. The TPS handles the large volume of business transactions that occur daily within an organization. TPSs include order processing, purchasing, accounting, and related systems.

An enterprise resource planning (ERP) system is a set of integrated programs that is capable of managing a company's vital business operations for an entire multisite, global organization. Although the scope of an ERP system may vary from company to company, most ERP systems provide integrated software to support the manufacturing and finance business functions of an organization.

A management information system (MIS) uses the information from a TPS to generate information that is useful for management decision making. The focus of an MIS is primarily on operational efficiency. A decision support system (DSS) is an organized collection of people, procedures, databases, and devices used to support problem-specific decision making. The DSS differs from an MIS in the support given to users, the decision emphasis, the development and approach, and system components, speed, and output. The specialized business information systems include knowledge management systems, artificial intelligence systems, expert systems, multimedia, and virtual reality systems. Knowledge management systems are organized collections of people, procedures, software, databases and devices used to create, store, share, and use the organization's knowledge and experience.

Principle:

System users, business managers, and information systems professionals must work together to build a successful information system.

Systems development is the activity of creating or modifying existing business systems. The goal of the systems investigation is to gain a clear understanding of the problem to be solved or opportunity to be addressed. If the decision is to continue with the solution, the next step, systems analysis, defines the problems and opportunities of the existing system. Systems design determines how the new system will work to meet the business needs defined during systems analysis. Systems implementation involves creating or acquiring the various system components (hardware, software, databases, etc.) defined in the design step, assembling them, and putting the new system into operation. The purpose of systems maintenance and review is to check and modify the system so that it continues to meet changing business needs.

Principle:

The use of information systems to add value to the organization can also give an organization a competitive advantage.

An organization is a formal collection of people and various other resources established to accomplish a set of goals. The primary goal of a for-profit organization is to maximize shareholder value. Nonprofit organizations include social groups, religious groups, universities, and other organizations that do not have profit as the primary goal. Organizations are systems with inputs, transformation mechanisms, and outputs.

Value-added processes increase the relative worth of the combined inputs on their way to becoming final outputs of the organization. The value chain is a series (chain) of activities that includes (1) inbound logistics, (2) warehouse and storage, (3) production, (4) finished product storage, (5) outbound logistics, (6) marketing and sales, and (7) customer service.

Supply chain management (SCM) helps determine what supplies are required, what quantities are needed to meet customer demand, how the supplies are to be processed (manufactured) into finished goods and services, and how the shipment of supplies and products to customers is to be scheduled, monitored, and controlled. Customer relationship management (CRM) programs help a company manage all aspects of customer encounters, including marketing and advertising, sales, customer service after the sale, and programs to help keep and retain loyal customers. CRM can help a company collect customer data, contact customers, educate customers on new products, and actively sell products to existing and new customers.

Organizations use information systems to support organizational goals. Because information systems typically are designed to improve productivity, methods for measuring the system's impact on productivity should be devised. In the late 1980s and early 1990s, overall productivity did not seem to increase with increases in investments in information systems. Often called the *productivity paradox*, this situation troubled many economists who were expecting to see dramatic productivity gains. In the early 2000s, however, productivity again seemed on the rise.

Organizational culture and change are important internal issues that affect most organizations. Organizational culture consists of the major understandings and assumptions for a business, a corporation, or an organization. Organizational change deals with how for-profit and nonprofit organizations plan for, implement, and handle change. Change can be caused by internal or external factors. Many European countries, for example, adopted the euro, a single European currency, which changed how financial companies do business and how they use their information systems.

User satisfaction with a computer system and the information it generates often depends on the quality of the system and the resulting information. A quality information system is usually flexible, efficient, accessible, and timely.

The extent to which technology is used throughout an organization is a function of technology diffusion, infusion, and acceptance. Technology diffusion is a measure of how widely technology is in place throughout an organization. Technology infusion is the extent to which technology permeates an area or department. The technology acceptance model (TAM) investigates factors, such as perceived usefulness of the technology, ease of use of the technology, the quality of the information system, and the degree to which the organization supports the use of the information system, to predict IS usage and performance.

Competitive advantage is usually embodied in either a product or service that has the most added value to consumers and that is unavailable from the competition or in an internal system that delivers benefits to a firm not enjoyed by its competition. The five-forces model covers factors that lead firms to seek competitive advantage: rivalry among existing competitors, the threat of new market entrants, the threat of substitute products and services, the bargaining power of buyers, and the bargaining power of suppliers. Three strategies to address these factors and to attain competitive advantage include altering the industry structure, creating new products and services, and improving existing product lines and services.

The ability of an information system to provide or maintain competitive advantage should also be determined. Several strategies for achieving competitive advantage include enhancing existing products or services or developing new ones, as well as changing the existing industry or creating a new one.

Developing information systems that measure and control productivity is a key element for most organizations. A useful measure of the value of an IS project is return on investment (ROI). This measure investigates the additional profits or benefits that are generated as a percentage of the investment in IS technology. Total cost of ownership (TCO) can also be a useful measure.

Principle:

IS personnel is a key to unlocking the potential of any new or modified system.

Information systems personnel typically work in an IS department that employs a chief information officer, systems analysts, computer programmers, computer operators, and a number of other people. The overall role of the chief information officer (CIO) is to employ an IS department's equipment and personnel in a manner that will help the organization attain its goals. Systems analysts help users determine what outputs they need from the system and construct the plans for developing the necessary programs that produce these outputs. Systems analysts then work with one or more programmers to make sure that the appropriate programs are purchased, modified from existing programs, or developed. The major responsibility of a computer programmer is to use the plans developed by the systems analyst to develop or adapt one or more computer programs that produce the desired outputs. Computer operators are responsible for starting, stopping, and correctly operating mainframe systems, networks, tape drives, disk devices, printers, and so on. LAN administrators set up and manage the network hardware, software, and security processes. Trained personnel are also increasingly needed to set up and manage a company's Internet site, including Internet strategists, Internet systems developers, Internet programmers, and Web site operators. Information systems personnel may also work in other functional departments or areas in a support capacity. In addition to technical skills, IS personnel also need skills in written and verbal communication, an understanding of organizations and the way they operate, and the ability to work with people (users). In general, IS personnel are charged with maintaining the broadest enterprise-wide perspective.

In addition to working for an IS department in an organization, IS personnel can work for one of the large consulting firms, such as Accenture, EDS, and others. Another IS career opportunity is to be employed by a hardware or software vendor developing or selling products.

Today's information systems have led to greater globalization. High-speed Internet access and networks that can connect individuals and organizations around the world create more international opportunities. Global markets have expanded. People and companies can get products and services from around the world, instead of around the corner or across town. These opportunities, however, introduce numerous obstacles and issues, including challenges involving culture, language, and many others.

CHAPTER 1: SELF-ASSESSMENT TEST

The value of information is directly linked to how it helps decision makers achieve the organization's goals.

1. A(n) _____ is a set of interrelated components that collect, manipulate, and disseminate data and information and provide a feedback mechanism to meet an objective.

2. What consists of raw facts, such as an employee number?
   a. bytes
   b. data
   c. information
   d. knowledge

Knowing the potential impact of information systems and having the ability to put this knowledge to work can result in a successful personal career, organizations that reach their goals, and a society with a higher quality of life.

3. A(n) _____ consists of hardware, software, databases, telecommunications, people, and procedures.

4. Computer programs that govern the operation of a computer system are called _____.
   a. feedback
   b. feedforward
   c. software
   d. transaction processing system

5. What is an organized collection of people, procedures, software, databases and devices used to create, store, share, and use the organization's experience and knowledge?
   a. TPS (transaction processing system)
   b. MIS (management information system)
   c. DSS (decision support system)
   d. KMS (knowledge management system)

System users, business managers, and information systems professionals must work together to build a successful information system.

6. What involves creating or acquiring the various system components (hardware, software, databases, etc.) defined in the design step, assembling them, and putting the new system into operation?
   a. systems implementation
   b. systems review
   c. systems development
   d. systems design

7. _____ involves anytime, anywhere commerce that uses wireless communications.

8. _____ involves contracting with outside professional services to meet specific business needs.

The use of information systems to add value to the organization can also give an organization a competitive advantage.

9. _____ change can help an organization improve raw materials supply, the production process, and the products and services offered by the organization.

10. Technology infusion is a measure of how widely technology is spread throughout an organization. True or False?

IS personnel is a key to unlocking the potential of any new or modified system.

11. Who is involved in helping users determine what outputs they need and constructing the plans needed to produce these outputs?

a. the CIO
b. the applications programmer
c. the systems programmer
d. the systems analyst

12. An information center provides users with assistance, training, and application development. True or False?

13. The _____ is typically in charge of the information systems department or area in a company.

REVIEW QUESTIONS

1. What are the components of any information system?
2. Describe the different types of data.
3. Identify at least six characteristics of valuable information.
4. What is a computer-based information system? What are its components?
5. What are the most common types of computer-based information systems used in business organizations today? Give an example of each.
6. What is the difference between e-commerce and m-commerce?
7. Describe three applications of multimedia.
8. What is a knowledge management system? Give an example.
9. What is the technology acceptance model (TAM)?
10. What is user satisfaction?
11. What are some general strategies employed by organizations to achieve competitive advantage?
12. Define the term *productivity*. Why is it difficult to measure the impact that investments in information systems have on productivity?
13. What is customer relationship management?
14. What is the total cost of ownership?
15. What is the role of the systems analyst? What is the role of the programmer?
16. What is the operations component of a typical IS department?
17. What is the role of the chief information officer?

DISCUSSION QUESTIONS

1. Describe the "ideal" automated auto license plate renewal system for the drivers in your state. Describe the input, processing, output, and feedback associated with this system.
2. Describe the "ideal" automated class registration system for a college or university. Compare this "ideal" system with what is available at your college or university.
3. You have decided to open an Internet site to buy and sell used music CDs to other students. Describe the value chain for your new business.
4. How is it that useful information can vary widely from the quality attributes of valuable information?
5. What is the difference between DSS and knowledge management?
6. Discuss the potential use of virtual reality to enhance the learning experience for new automobile drivers. How might such a system operate? What are the benefits and potential drawbacks of such a system?
7. Discuss how information systems are linked to the business objectives of an organization.
8. You have been hired to work in the IS area of a manufacturing company that is starting to use the Internet to order parts from its suppliers and offer sales and support to its customers. What types of Internet positions would you expect to see at the company?
9. How would you measure technology diffusion and infusion?
10. You have been asked to participate in the preparation of your company's strategic plan. Specifically, your task is to analyze the competitive marketplace using Porter's five-forces model. Prepare your analysis, using your knowledge of a business you have worked for or have an interest in working for.
11. Based on the analysis you performed in the preceding discussion question, what possible strategies could your organization adopt to address these challenges? What role could information systems play in these strategies? Use Porter's strategies as a guide.
12. You have been hired as a sales representative for a sporting goods store. You would like the IS department to develop

new software to give you reports on which customers are spending the most at your store. Describe your role in getting the new software developed. Describe the roles of the systems analysts and the computer programmers.

13. Imagine that you are the CIO for a large, multinational company. Outline a few of your key responsibilities.

14. You have decided to open an Internet site to buy and sell used music CDs to other students. Describe the supply chain for your new business.

15. What sort of IS position would be most appealing to you—working as a member of an IS organization, being a consultant, or working for an IS hardware or software vendor? Why?

16. What are your career goals, and how can a computer-based information system be used to achieve them?

PROBLEM-SOLVING EXERCISES

1. Prepare a data disk and a backup disk for the problem-solving exercises and other computer-based assignments you will complete in this class. Create one directory for each chapter in the textbook (you should have 9 directories). As you work through the problem-solving exercises and complete other work using the computer, save your assignments for each chapter in the appropriate directory. On the label of each disk be sure to include your name, course, and section. On one disk, write "Working Copy"; on the other, write "Backup."

2. Search through several business magazines (*Business Week, Computerworld, PC Week,* etc.) or an Internet search engine for recent articles that describe potential social or ethical issues related to the use of an information system. Use word-processing software to write a one-page report summarizing what you discovered.

3. Using a word-processing program, write a detailed job description of a systems analyst for a medium-sized manufacturing company. Use a graphics program to make a presentation on the requirements for the new CIO.

TEAM ACTIVITIES

1. Before you can do a team activity, you need a team! The class members may self-select their teams, or the instructor may assign members to groups. Once your group has been formed, meet and introduce yourselves to each other. You will need to find out the first name, hometown, major, and e-mail address and phone number of each member. Find out one interesting fact about each member of your team, as well. Come up with a name for your team. Put the information on each team member into a database and print enough copies for each team member and your instructor.

2. Have your team interview a company that recently introduced new technology. Write a brief report that describes the extent of technology infusion and diffusion.

3. With your team, interview one or more instructors or professors at your college or university. Describe how they keep current with the latest teaching and research developments in their field.

WEB EXERCISES

1. Throughout this book, you will see how the Internet provides a vast amount of information to individuals and organizations. We will stress the World Wide Web, or simply the Web, which is an important part of the Internet. Most large universities and organizations have an address on the Internet, called a Web site or home page. The address of the Web site for the publisher of this text is *www.cengage.com.* You can gain access to the Internet through a browser, such as Internet Explorer or Netscape. Using an Internet browser, go to the Web site for this publisher. What did you find? Try to obtain information on this book. You may be asked to develop a report or send an e-mail message to your instructor about what you found.

2. Go to an Internet search engine, such as *www.google.com* or *www.yahoo.com*, and search for information about artificial intelligence. Write a brief report that summarizes what you found.

3. Use the Internet to search for information about a company that has excellent or poor product quality in your estima-

tion. You can use a search engine, such as Google, or a database at your college or university. Write a brief report describing what you found. What leads to higher-quality products? How can an information system help a company produce higher quality products?

CAREER EXERCISES

1. In the Career Exercises found at the end of every chapter, you will explore how material in the chapter can help you excel in your college major or chosen career. Write a brief report on the career that appeals to you the most. Do the same for two other careers that interest you.

2. Research careers in finance, management, information systems, and two other career areas that interest you. Describe

the job opportunities, job duties, and the possible starting salaries for each career area in a report.

3. Pick the five best companies for your career. Describe how each company uses information systems to help achieve a competitive advantage.

CASE STUDIES

Case One

Information System as an Effective Force Against H1N1 Pandemic

Information systems are valuable to businesses for tracking business activities in real-time, as they occur. They are also valuable to the medical community for tracking the spread of viruses such as the H1N1 virus, also known as the swine flu. New Jersey-based Emergency Medical Associates (EMA) operates 21 emergency rooms in hospitals across New Jersey, New York, and Pennsylvania. With information mined from its diverse locations, EMA is in an ideal position to spot an outbreak of the flu in its early stages. All it requires is an information system to provide valuable information in a timely manner.

EMA's CIO and information systems specialists applied proven business information management techniques to their medical information needs. They understood that tracking medical statistics across their 21 emergency rooms was similar to tracking sales statistics across retail outlets. They required the same business intelligence (BI) and reporting tools used by successful businesses. Business intelligence or BI systems are designed to extract, or mine, useful information out of the data collected by businesses or organizations into databases. That data may consist of detailed sales information collected at the time of a sale or patient symptom information collected at the time of an examination.

EMA began by installing a database management system from Oracle. The database was shared by all of its 21 emergency rooms over a high-speed private network. EMA then contracted with SAP to install its BusinessObjects XI tool set to function as the company's BI platform. BusinessObjects can sort and sift through data in the database to find patterns and exceptions. Combining the BusinessObjects system with other software including Xcelsius and Crystal Reports (powerful reporting software), and Web Intelligence (providing a Web interface to the system), EMA created a system that generates insightful reports and visualizations about medical conditions on a regular schedule and on demand.

Today, EMA physicians and nurses, depending on their needs, can access 27 dashboards, which provide statistics displayed in charts and lists that are updated as information is entered into the database. They also have access to 30 daily reports from the system informing them of the current status in all of their emergency rooms and of any changes in the status quo. The system allows users to customize their view of the data to focus on the information that is most important to their work.

Using its new information system, EMA was the first to spot the outbreak of H1N1 in the Northeast. Doctors knew that about 6 percent of patients complain of flu-like symptoms on any given day. When the EMA BI system reported that 30 percent of patients were arriving with flu symptoms, the doctors warned the country that H1N1 was on the move. This alert provided medical professionals and citizens the time needed to take action.

Discussion Questions

1. What role did business intelligence software play in catching an H1N1 outbreak in the northeastern United States.?
2. How does a system such as EMA's BI system use human intelligence and machine intelligence to support decision making?

Critical Thinking Questions

1. How do the BI needs of business professionals and medical professionals differ? How are they alike?
2. How does this case study reflect the need for standardized digital medical records systems in the U.S.? How might such standards influence the country's ability to keep its population healthy?

SOURCES: Lai, Eric, "BI helps New York-area hospitals track, fight H1N1," *Computerworld*, October 8, 2009, *www.computerworld.com/s/article/9139121/BI_helps_New_York_area_hospitals_track_fight_H1N1?source=rss_news*; EMA Web site, accessed November 12, 2009, *www.ema-ed.com*; SAP staff, "Emergency Medical Associates Stays Ahead of Swine Flu Outbreaks This Back-To-School Season with Sap® Software," SAP Press Release, September 14, 2009, *www.sap.com/about/newsroom/topic-rooms/business-objects/press.epx?pressid=11826*.

Case Two
Creativity Moves Up the Value Chain

Creativity Inc. deals in beads, baubles, and stylized paper to "bring crafters' dreams to reality by providing the materials to give life to their ideas and imagination." Creativity owns five well-known brands in the craft industry: Autumn Leaves, Blue Moon Beads, Crop in Style, DND, Hip in a Hurry, and Westrim Crafts. The company is one of the top five wholesale suppliers to national craft chains in the United States with 500 employees at four office and warehouse locations in California and one in Hong Kong.

Creativity outsources the manufacturing of its designs to production facilities across Asia. Crates of assorted beads, scrapbooking supplies, and papercrafting materials flow through Creativity's port-side warehouses to craft stores and department stores across the U.S. In this way, Creativity facilitates the value chain for craft retailers.

In 2007, Creativity found its business model challenged by growing globalization and economic hardships. To save money, some of its customers decided to "do away with the middleman," and purchase crafting materials directly from the Asian manufacturers. Creativity needed to find new ways to provide value to its customers.

Creativity's challenges are not unique. Many businesses are facing growing competition from low-cost manufacturers and service providers in developing countries. To survive, they need to find a way to move up the value chain—that is, to provide valuable services beyond upstream management of the supply chain. Many are turning to information systems to assist in that move.

Creativity turned to IBM's Cognos 8 Business Intelligence suite to identify high-value products that could not be manufactured by its low-cost overseas competitors. The company acquired data about purchase transactions from retailers in craft-related markets and added that data to its data warehouse. Using the Cognos software and Smart Software's SmartForecast program, Creativity determined a need for more "design-oriented, fashion-oriented" products—especially ones associated with popular U.S. media, such as television shows and celebrities.

By shifting its focus to fashion-based craft products, Creativity made up for the business it lost in the low-cost crafting material market. In fact, fashion-oriented products are now the dominant portion of its business, comprising more than 50 percent of its products and a much higher percentage of its profits.

Creativity also uses Cognos to determine which customer segments are most profitable. The company can then focus its efforts in those areas to boost profitability. In addition, Creativity created an "Analytical Center of Excellence" composed of representatives from all of its brands. By improving communication between its brands and sharing its research findings, Creativity elevated the corporate awareness of the entire company and created an environment where everyone is working towards common goals. To further communication, CIO Jim Mulholland used Cognos to develop a software dashboard that provides corporate news and information on the desktops of company managers across its brands. These communication improvements help safeguard against duplication of effort. Each brand is aware of what the other brands are experiencing and working on, allowing brands to learn from each other.

Creativity and other struggling businesses want to create valuable information from low cost data to learn how to work more intelligently and efficiently. Integrating data from transactions, call centers, Web logs, sales reps, external sources, and elsewhere into data warehouses for analysis allows companies to discover what products are likely to sell, what products return the highest profits, where to cut costs, where to invest for the highest return, and other key information to fuel smart decision making. Many businesses are counting on information systems to provide the knowledge to survive tough economic times.

Discussion Questions

1. Describe the global economic forces that pushed Creativity to move up the value chain.
2. What information did Creativity use to boost its profits and remain solvent?

Critical Thinking Questions

1. What role does communication play in creating savings for a multibrand company like Creative?
2. What lessons does Creative's story provide for U.S. businesses? What does this forecast for the global marketplace in general?

SOURCES: Mitchell, Robert, "Smart and cheap: Business intelligence on a budget," *Computerworld*, May 14, 2009, *www.computerworld.com*; Creativity Inc. Web site, access December 26, 2009, *www.creativityinc.com*; Cognos Web site, accessed December 26, 2009, *www-01.ibm.com/software/data/cognos*.

Questions for Web Case

See the Web site for this book to read about the Altitude Online case for this chapter. The following questions cover this Web case.

Altitude Online: Outgrowing Systems

Discussion Questions

1. Why do you think it's a problem for Altitude Online to use different information systems in its branch locations?
2. What information do you think Jon should collect from the branch offices to plan the new centralized information system?

Critical Thinking Questions

1. With Jon's education and experience, he could design and implement a new information system for Altitude Online himself. What would be the benefits and drawbacks of doing the job himself compared to contracting with an information systems contractor?

2. While Jon is visiting the branch offices, how might he prepare them for the inevitable upheaval caused by the upcoming overhaul to the information system?

Altitude Online: Addressing the Needs of the Organization

Discussion Questions

1. What are the advantages of Altitude Online adopting a new ERP system compared to simply connecting existing corporate systems?
2. Why isn't an out-of-the-box ERP system enough for Altitude Online? What additional needs does the company have? Is this the case for businesses in other industries as well?

Critical Thinking Questions

1. Why do you think Jon is taking weeks to directly communicate with stakeholders about the new system?
2. Why do you think Jon and the system administrators decided to outsource the software for this system to an ERP company rather than developing it from scratch themselves?

NOTES

Sources for the opening vignette: SAP staff, "Braskem - Pursuing Growth and Synergy Through Mergers and Acquisitions," SAP Customer References Web page, accessed November 12, 2009, *www.sap.com/usa/solutions/business-suite/customers*; Braskem Web site, accessed November 12, 2009, *www.braskem.com.br*; Accenture staff, "Braskem: SAP Solutions," Accenture Client Successes Web page, accessed November 12, 2009, *www.accenture.com/Global/Technology/Enterprise_Solutions/SAP_Solutions/Client_Su ccesses/Braskem.htm*.

1   Ramiller, N., et al, "Management Implications in Information Systems Research," *Journal of the Association for Information Systems*, Vol. 10, 2009, p. 474.
2   Lurie, N., & Swaminathan, J. M., "Is timely information always better? The effect of feedback frequency on decision making," *Organizational Behavior and Human Decision Processes*, March 2009, p. 315.
3   Pinch, Trevor, "Selling Technology: The Changing Shape of Sales in an Information Economy," *Industrial & Labor Relations Review*, Vol. 23, October 2008, p. 331.
4   Marcial, Gene, "How Expeditors Move the Freight," *Business Week*, November 16, 2009, p. 93.
5   MIT Open Course Ware home page, *http://ocw.mit.edu/OcwWeb/web/home/home/index.htm*, accessed October 6, 2009.
6   Kolbasuk McGee, Marianne, "A $20 Billion Shot In the Arm," *InformationWeek*, March 16, 2009, p. 27.
7   Moon, M, "Knowledge Worker Productivity," *Journal of Digital Asset Management*, August 2009, p. 178.

8   Hahn, J., et al, "Knowledge Management Systems and Organizational Knowledge Processing Systems," *Decision Support Systems*, November 2009, pp. 332.
9   Lai, J., "How Reward, Computer Self-efficacy, and Perceived Power Security Affect Knowledge Management Systems Success," Journal of the American Society for Information Science and Technology, February 2009, p. 332.
10 Patterson, Scott, "Meet Getco," *The Wall Street Journal*, August 27, 2009, p. C1.
11 D'Amours, M., et al, "Optimization Helps Shermag Gain Competitive Advantage," *Interfaces*, July-August, 2009, p. 329.
12 Henschen, Doug, "Predictive Analysis: A Matter of Survival," *InformationWeek*, March 2, 2009, p. 29.
13 Farrell, Maureen, "Weatherman," *Forbes*, March 16, 2009, p. 58.
14 DiColo, Jerry, "Chip Makers to Benefit From Utility Smart Meters," *The Wall Street Journal*, April 1, 2009, p. B6.
15 Clark, Don, "SanDisk Sees Leap for Data Storage Chips," *The Wall Street Journal*, February 10, 2009, p. B5.
16 Lawrence Livermore National Laboratory home page, *www.llnl.gov*, accessed on October 6, 2009.
17 IBM Web site, *www-03.ibm.com/systems/deepcomputing/bluegene*, accessed on October 16, 2009.
18 Clark, D. and Scheck, J., "High-Tech Companies Take Up Netbooks," *The Wall Street Journal*, January 6, 2009, p. B6.
19 Mossberg, Walter, "Some Favorite Apps," *The Wall Street Journal*, March 26, 2009, p. D1.

20 Staff, "Bringing Technology to the Bush," *The Australian Financial Review,* "August 31, 2009, p. 28.

21 Tham, Irene, "Changing the World, One Laptop at a Time," *The Straits Time,* July 16, 2009.

22 Wildstrom, Stephen, "Touch-Sensitive Desktops," *Business Week,* March 23, 2009, p. 97.

23 Urgo, Jacqueline, "Witness in Higbee Trial Describes 'Crazy Driving'," *The Philadelphia Inquirer,* May 12, 2009, p. B01.

24 Fab Lab, *http://fab.cba.mit.edu,* accessed August 25, 2007.

25 Hamm, Steve, "Tech Spending," *Business Week,* March 23, 2009, p. 72.

26 Burrows, Peter, "Will Windows 7 Reboot PC Sales?," *Business Week,* September 14, 2009, p. 20.

27 Scheck, Justin, "PC Makers Try Google, Challenging Microsoft," *The Wall Street Journal,* April 1, 2009, p. B1.

28 "Adobe Creative Suite 4," *www.adobe.com/products/creativesuite,* accessed December 10, 2009.

29 Wildstrom, Stephen, "Coming at You: 3D on Your PC," *Business Week,* January 19, 2009, p. 65.

30 Dearne, Karen, "Cheap Solution for Security," *The Australian,* May 26, 2009, p. 25.

31 Weier, Mary Hayes, "New Apps Help iPhones Get Down to Business," *Information Week,* March 23, 2009, p. 17.

32 Lai, Eric, "NYSE Turns to Appliances In BI Consolidation Effort," *Computerworld,* January 1, 2009, p. 3.

33 Delahunty, Steve, "Smarter, Not More, Storage," *Information Week,* April 13, 2009, p. 38.

34 Woolley, Scott, "Extraterrestrial Dreams," *Forbes,* April 13, 2009, p. 36.

35 Fulghum, David, "Unmanned Aircraft," *Aviation Week & Space Technology,* August 17, 2009, p. 20.

36 Kolbasuk McGee, Marianne, "Hospitals Getting Docs on Hosted E Records," *Information Week,* June 22, 2009, p. 17.

37 Hamm, Steve, "Cloud Computing's Big Bang for Business," *Business Week,* June 15, 2009, p. 42.

38 Foley, John, "10 Cloud Computing Predictions," *Information Week,* February 2, 2009, p. 20.

39 Mossberg, Walter, "The Latest Kindle," *The Wall Street Journal,* June 11, 2009, p. D1.

40 Worthen, Ben, "Mobile Banking Finds New Users," *The Wall Street Journal,* February 3, 2009, p. D1.

41 Angwin, Julia, "My New Twitter Flock," *The Wall Street Journal,* March 14, 2009, p. W2.

42 Boudreau, John, "Applications Change How We Use Mobile Devices," *Tampa Tribune,* April 6, 2009, p. 13.

43 Weier, Mary Hayes, "Collaboration Is Key to Increased Efficiency," *Information Week,* September 14, 2009, p. 90.

44 Grant, Tavia, "Workplace Democracy," *The Globe and Mail,* May 30, 2009, p. B14.

45 Staff, "Penske Launches Improved Extranet," *Bulk Transporter,* June 1, 2009, p. 50.

46 Staff, "CIO Profiles: Ken Silva," *Information Week,* September 7, 2009, p. 8.

47 Huang, M., "Marketing and Electronic Commerce," *Electronic Commerce Research and Applications,* October 16, 2009, p. 4.

48 Goldstein, Jacob, "Stimulus Funds for E-Records," *The Wall Street Journal,* March 24, 2009, p. B1.

49 Boudreau, John, "Applications Change How We Use Mobile Devices," *Tampa Tribune,* April 6, 2009, p. 13.

50 Silver, S. and Smith, E., "Two Services to Sell Tickets on Cellphones," *The Wall Street Journal,* April 1, 2009, p. B7.

51 Su, Y., "Why Are Enterprise Resource Planning Systems Indispensible?" *Journal of European Operations Research,* May 10, 2010, p. 81.

52 Mawson, Nicola, "Pick n Pay Goes for Growth in a Big Way," *Business Day,* April 24, 2009.

53 www.dell.com ,accessed on 10-12-09.

54 Fagerholt, Kjetil, et al, "An Ocean Of Opportunities," *OR/MS Today,* April 2009, p. 26.

55 Weier, Mary Hayes, "Business Gone Mobile," *Information Week,* March 30, 2009, p. 23.

56 Staff, "Healthland to Deliver Business Intelligence for Smaller Hospitals," *TechWeb,* September 8, 2009.

57 Conley, C., et al, "Factors Critical to Knowledge Management Success," *Advances in Developing Human Resources,* August 2009, p. 334.

58 Larger, Marshall, "Investing in Knowledge Management," *Customer Relationship Management,* "June 2009, p. 46.

59 Staff, "Science Advances Will Make Us All Cyborgs," *The Sun,* September 22, 2009, p. 8.

60 Kroeker, K. "Medical Nanobot," *Association for Computer Machinery,* September 2009, p. 18.

61 Rowley, Ian, "Drive, He Thought," *Business Week,* April 20, 2009, p. 10.

62 Steiner, Christopher, "A Bot in Time Saves Nine," *Forbes,* March 16, 2009, p. 40.

63 Staff, "Artificial Neural Networks," *Biotech Business Week,* October 5, 2009, p. 404.

64 Feng, W., Duan, Y., Fu, Z., & Mathews, "Understanding Expert Systems Applications from a Knowledge Transfer Perspective," *Knowledge Management Research & Practice,* June, 2009, p. 131.

65 Walton, T., "Virtual Reality is Reality," *Design Management Review,* Winter 2009, p. 6.

66 Boeing: Commercial Airplanes – 747 Home page, *www.boeing.com/commercial/787/family,* accessed October 7, 2009.

67 Copeland, Michael, "3-D Gets Down to Business," *Fortune,* March 30, 2009, p. 32.

68 Foust, Dean, "Top Performing Companies," *Business Week,* April 6, 2009, p. 40.

69 Kane, Yukare Iwatani, "Apple Woos Developers With New iPhone," *The Wall Street Journal,* March 18, 2009, p. B6.

70 Wildstrom, Stephen, "What to Entrust to The Cloud," *Business Week,* April 6, 2009, p. 89.

71 McGregor, Jena, "The Chore Goes Offshore," *Business Week,* March 23, 2009, p. 50.

72 Randall, David, "Be Green and Make A Buck," *Forbes,* March 2, 2009, p. 40.

73 Staff, "CIO Profiles: Phil Fasano," *InformationWeek,* May 25, 2009, p. 14.

74 Dong, S., et al, "Information Technology in Supply Chains," *Information Systems Research,* March 2009, p. 18.

75 Sanders, Peter, "Boeing Tightens Its Grip on Dreamliner Production," *The Wall Street Journal,* July 2, 2009, p. B1.

76 Yang, Jia Lynn, "Veggie Tales," *Fortune,* June 8, 2009, p. 25.

77 Huifen, Chen, "Courting the Small Enterprise," *The Business Times Singapore,* September 22, 2009.

78 Staff, "Duke Energy Signs Agreement with Convergys," *Telecomworld,* September 16, 2009.

79 Weier, Mary Hayes, "CRM as A Service," *InformationWeek,* February 2, 2009, p. 16.

80 Singer, S., et al, "Identifying Organizational Cultures That Promote Patient Safety," *Health Care Management Review,* "October-December 2009, p. 300.

81 Staff, "CIO Profiles: Ken Silva," *InformationWeek,* September 7, 2009, p. 8.

82 Christensen, Clayton, *The Innovator's Dilemma,* Harvard Business School Press, 1997, p. 225 and *The Inventor's Solution,* Harvard Business School Press, 2003.

83 Gibson, Ellen, "The School of Future Knocks," *BusinessWeek,* March 23, 2009, p. 44.

84 Conry-Murray, Andrew, "A Measure of Satisfaction," *InformationWeek,* January 26, 2009, p. 19.

85 Wixom, Barbara and Todd, Peter, "A Theoretical Integration of User Satisfaction and Technology Acceptance," *Information Systems Research,* March 2005, p. 85.

86   Bailey, J. and Pearson, W., "Development of a Tool for Measuring and Analyzing Computer User Satisfaction," *Management Science,* 29(5), 1983, p. 530.

87   Chaparro, Barbara, et al, "Using the End-User Computing Satisfaction Instrument to Measure Satisfaction with a Web Site," *Decision Sciences,* May 2005, p. 341.

88   Schwarz, A. and Chin, W., "Toward an Understanding of the Nature and Definition of IT Acceptance," *Journal of the Association for Information Systems,* April 2007, p. 230.

89   Davis, F., "Perceived Usefulness, Perceived Ease of Use, and User Acceptance of Information Technology," *MIS Quarterly,* 13(3) 1989, p. 319. Kwon, et al, "A Test of the Technology Acceptance Model," *Proceedings of the Hawaii International Conference on System Sciences,* January 4–7, 2000.

90   Ilie, V., et al, "Paper Versus Electronic Medical Records," *Decision Sciences,* May 2009, p. 213.

91   Barki, H., et al, "Information System Use-Related Activity," *Information Systems Research,* June 2007, p. 173.

92   Loch, Christoph and Huberman, Bernardo, "A Punctuated-Equilibrium Model of Technology Diffusion," *Management Science,* February 1999, p. 160.

93   Tornatzky, L. and Fleischer, M., "The Process of Technological Innovation," *Lexington Books,* Lexington, MA, 1990; Zhu, K. and Kraemer, K., "Post-Adoption Variations in Usage and Value of E-Business by Organizations," *Information Systems Research,* March 2005, p. 61.

94   Armstrong, Curtis and Sambamurthy, V., "Information Technology Assimilation in Firms," *Information Systems Research,* April 1999, p. 304.

95   Sykes, T. and Venkatesh, V., "Model of Acceptance with Peer Support," *MIS Quarterly,* June 2009, p. 371.

96   Agarwal, Ritu and Prasad, Jayesh, "Are Individual Differences Germane to the Acceptance of New Information Technology?" *Decision Sciences,* Spring 1999, p. 361.

97   Fuller, R. and Denis, A., "Does Fit Matter?" *Information Systems Research,* March 2009, p. 2.

98   Gupta, Aseem, "CIO Profiles: Aseem Gupta," *InformationWeek,* April 20, 2009, p. 14.

99   D'Amours, M., et al, "Optimization Helps Shermag Gain Competitive Advantage," *Interfaces,* July-August, 2009, p. 329.

100 Collins, Jim, *Good to Great,* Harper Collins Books, 2001, p. 300.

101 Murphy, Chris, "In for the Long Haul," *InformationWeek,* January 19, 2009, p. 38.

102 Porter, M. E., *Competitive Advantage: Creating and Sustaining Superior Performance,* New York: Free Press, 1985; *Competitive Strategy: Techniques for Analyzing Industries and Competitors,* The Free Press, 1980; and *Competitive Advantage of Nations,* The Free Press, 1990.

103 Porter, M. E. and Millar, V., "How Information Systems Give You Competitive Advantage," *Journal of Business Strategy,* Winter 1985. *See also* Porter, M. E., *Competitive Advantage* (New York: Free Press, 1985).

104 Staff, "American Diversified Holdings Enters into Broad Strategic Alliance with Leading Biotech Company," *Biotech Business Week,* September 21, 2009.

105 Goodhue, D., et al, "Addressing Business Agility Challenges with Enterprise Systems," *MIS Quarterly Executive,* June 2009, p. 73.

106 Brynjolfsson, Erik, et al, "The New, Faster Face of Innovation," *The Wall Street Journal,* August 17, 2009, p. R3.

107 Capell, Kerry, "Vodafone: Embracing Open Source with Open Arms," *BusinessWeek,* April 20, 2009, p. 52.

108 Reena, J., "Dusting Off a Big Idea in Hard Times," *BusinessWeek,* June 22, 2009, p. 44.

109 Mandel, Michael, "Innovation Interrupted," *BusinessWeek,* June 15, 2009, p. 35.

110 Murphy, Chris, "TransUnioin Finds Cost Savings, Seeks More," *InformationWeek,* March 23, 2009, p. 24.

111 Foley, John, "Cost Control," *InformationWeek,* March 2, 2009, p. 18.

112 Rai, S., et al, "LDP—O.R. Enhanced Productivity Improvements for the Printing Industry," *Interfaces,* January 2009, p. 69.

113 Staff, "CIO Profiles: Marty Colburn," *InformationWeek,* March 16, 2009, p. 16.

114 Tiwana, A., "Governance-Knowledge Fit in Systems Development Projects," *Information Systems Research,* June 2009, p. 180.

115 Staff, "Kodak Insite Campaign Manager," *Print Week,* July 10, 2009, p. 28.

116 Patterson S. and Ng, S., "NYSE's Fast-Trade Hub," *The Wall Street Journal,* July 30, 2009, p. C1.

117 Staff, "Hitachi's Quintet of Projectors," *AV Magazine,* September 1, 2009, p. 28.

118 Hoffman, Thomas, "9 Hottest Skills for 09," *Computerworld,* January 1, 2009, p. 26.

119 Staff, "100 Best Places to Work in IT in 2009," *Computerworld,* June 16, 2009.

120 Hoover, Nicholas, "$1 B Plan Includes New Data Center," *InformationWeek,* March 2, 2009, p. 26.

121 Farrell, Maureen, "Weatherman," *Forbes,* March 16, 2009, p. 58.

122 Top Coder Collegiate Challenge, *www.topcoder.com,* accessed September 2, 2007.

123 Preston, D., et al, "Examining the Antecedents and Consequences of CIO Strategic Decision-Making Authority," *Decision Sciences,* November 2008, p. 605.

124 May, Thornton, "CIOs Are Entering a Career Ice Age," *Computerworld,* January 12, 2009, p. 17.

125 Staff, "CIO Profiles: Peter Whatnell," *InformationWeek,* February 23, 2009, p. 12.

126 Staff, "CIO Profiles: Marty Colburn," *InformationWeek,* March 16, 2009, p. 16.

127 Pratt, Mary, "Steering Clear of Scandal," *Computerworld,* August 17, 2009, p. 22.

128 Staff, "CIO Profiles: Laxman Kumar Badiga," *InformationWeek,* March 2, 2009, p. 16.

129 Copeland, Michael, "Who's on the CTO Short List," *Fortune,* March 30, 2009, p. 18.

130 Hoover, Nick, "Fed CIO Scrutinizes Spending, Eyes Cloud," *InformationWeek,* March 16, 2009, p. 19.

131 *www.systest.com,* accessed November 9, 2007.

132 *www.ibm.com/services, www.eds.com,* and *www.accenture.com,* accessed November 15, 2009.

133 Garone, Elizabeth, "Growing a List of Opportunities," *The Wall Street Journal,* February 24, 2009, p. D5.

134 Anderson, Howard, "Project Triage: Skimpy Must Die," *InformationWeek,* March 16, 2009, p. 14.

135 King, Julia, "Your New-Age Workforce," *Computerworld,* January 1, 2009, p. 24.

136 Boyle, Matthew, "Enough to Make Monster Tremble," *BusinessWeek,* July 6, 2009, p. 43.

137 Needleman, Sarah, "A New Job Just a Tweet Away," *The Wall Street Journal,* September 8, 2009, p. B7.

138 Brandel, Mary, "Laid Off? Here's Your Net," *Computerworld,* August 17, 2009, p. 17.

139 Friedman, Thomas, *The World Is Flat,* Farrar, Straus and Giroux, 2005, p. 488.

140 European Commission Web site, *http://ec.europa.eu/trade/index_en.htm,* accessed December 10, 2009.

141 Foreign Trade Information Web site, www.sice.oas.org/agreements_e.asp, accessed December 10, 2009.

Technology

CHAPTER
· 2 ·

Hardware and Software

PRINCIPLES	LEARNING OBJECTIVES
• Computer hardware must be carefully selected to meet the evolving needs of the organization and its supporting information systems.	• Identify and discuss the role of the essential hardware components of a computer system. • Identify the characteristics of and discuss the usage of various classes of single-user and multiuser computer systems.
• The computer hardware industry and users are implementing green computing designs and products.	• Define the term green computing and identify the primary goals of this program.
• Systems and application software are critical in helping individuals and organizations achieve their goals.	• Identify and briefly describe the functions of the two basic kinds of software. • Outline the role of the operating system and identify the features of several popular operating systems.
• Organizations should not develop proprietary application software unless doing so will meet a compelling business need that can provide a competitive advantage.	• Discuss how application software can support personal, workgroup, and enterprise business objectives. • Identify three basic approaches to developing application software and discuss the pros and cons of each.
• Organizations should choose a programming language whose functional characteristics are appropriate for the task at hand, considering the skills and experience of the programming staff.	• Outline the overall evolution and importance of programming languages and clearly differentiate among the generations of programming languages.
• The software industry continues to undergo constant change; users need to be aware of recent trends and issues to be effective in their business and personal life.	• Identify several key software issues and trends that have an impact on organizations and individuals.

Information Systems in the Global Economy
Turboinštitut d.d., Slovenia

Exchanging Processing Power for Hydropower

Businesses are turning to various forms of renewable energy as they seek to reduce their carbon footprint and their impact on climate change. Among the most popular of renewable energy sources are wind and hydroelectric power. The process of turning wind and water into energy efficiently and affordably takes a considerable amount of science and engineering knowledge. Turboinštitut is a global company with a mission to promote water as an indispensable, sustainable, environmentally friendly source of energy and to continuously improve hydraulic machine performance. Turboinštitut employs 135 scientists, engineers, and staff at its research facility located in Ljubljana, the capital of Slovenia.

Turboinštitut was established in 1948 by the Yugoslav government as a strategic center for scientific and industrial research. Today it functions as an independent institute that has contributed its expertise in turbine research and development in the construction of hydro-fluid facilities (such as dams) in dozens of countries. The core of Turboinštitut's business is modeling, designing, and producing custom turbine hydraulics systems that meet the unique requirements of each installation and provide the maximum amount of energy from a water source. Turboinštitut's extensive research is carried out using computer hardware and software that enable simulations—more specifically, computational fluid dynamics (CFD)—to design highly successful hydro-energy installations.

Simulating the flow of water using CFD requires enormous amounts of computing capacity. As Turboinštitut has grown, its client base has expanded, and its turbine hydraulics systems have become more refined. More refined systems work smarter and can return more power from less water. However, they also require more advanced software and processing power during their design. Turboinštitut's success, combined with its outdated computer software and hardware, has created a backlog of CFD tasks requiring computation. The company decided that it needed new hardware. Not just any hardware, but its own energy efficient supercomputer that could deliver high performance with little power.

Whether the need is for calculating complex scientific models, for analyzing financial markets, or for supporting employees scattered around the globe, many businesses have needs similar to Turboinštitut's: to replace old, inefficient energy consuming computing centers with new, powerful, compact, energy-efficient computing centers. Turboinštitut opted to collaborate with IBM to implement a supercomputing system it named Ljubljana Supercomputer Center (LSC) ADRIA. ADRIA uses blade servers (a type of server technology) that pack many processing circuit boards (blades) together in a single rack. Turboinštitut's ADRIA uses 256 clustered blade servers for a total of 2,048 processing cores—supercomputing power to match Turboinštitut's computational needs.

Processors work closely with various forms of storage to deliver high-performance services. ADRIA uses 4,096 GB (billions of bytes) of RAM and 10 TB (trillions of bytes) of an IBM storage system that has additional servers designed to rapidly deliver data as needed. The resulting system can perform calculations 50 times faster than Turboinštitut's previous system in half the amount of physical space. ADRIA enables Turboinštitut to make more than 10,000 complicated CFD analyses per year. ADRIA's highly efficient processors and advanced cooling system require much less power than the previous system. Aleš Petan, chief information officer, is pleased that the company's environmentally friendly mission is fulfilled not only by Turboinštitut's hydraulic research, designs, and products, but also by the company's information system infrastructure.

Today Turboinštitut carries out its complex water turbine design simulations on highly energy-efficient servers in a fraction of the time required with its previous system. This enables the company to save time and money and get products to clients more quickly, giving Turboinštitut a huge competitive advantage. Like Turboinštitut, businesses in many industries depend on the latest hardware and software to provide higher levels of efficiencies with less overhead—both financial and environmental. Although companies most often invest in servers to provide information and computational services and storage, all forms of hardware can yield savings. When selecting hardware and software, businesses need technologies that provide the most power and speed and the highest level of efficiency and effectiveness.

As you read this chapter, consider the following:

- How does the type of hardware and software a company purchases affect the way the company operates?
- Businesses are in constant flux, growing, diversifying, acquiring, and working to reduce costs. How do these conditions and requirements affect the purchase of the hardware and software used to support the organization?
- In the quest for the most efficient and effective systems, businesses must invest in state-of-the-art hardware and software to win a competitive advantage. How might a business decide when it is best to upgrade its hardware and software?

Why Learn About Hardware and Software?

Organizations invest in computer hardware and software to improve worker productivity, increase revenue, reduce costs, and provide better customer service. Those that don't may be stuck with outdated hardware and software that is unreliable and cannot take advantage of the latest advances. As a result, obsolete hardware and software can place an organization at a competitive disadvantage. Managers, no matter what their career field and educational background, are expected to know enough about their business needs to be able to ask tough questions of those recommending the hardware and software to meet those needs. This is especially true in small organizations that might not have information systems specialists. Cooperation and sharing of information between business managers and IT managers is needed to make wise IT investments that yield real business results. Managers in marketing, sales, and human resources often help IS specialists assess opportunities to apply hardware and software and evaluate the various options and features. Managers in finance and accounting especially must also keep an eye on the bottom line, guarding against overspending, yet be willing to invest in computer hardware and software when and where business conditions warrant it.

Today's use of technology is practical—it's intended to yield real business benefits, as demonstrated by Turboinstitut. Employing information technology and providing additional processing capabilities can increase employee productivity, expand business opportunities, and allow for more flexibility. This chapter discusses the hardware and software components of a computer-based information system (CBIS), beginning with a definition of hardware.

Hardware refers to the physical components of a computer that perform the input, processing, storage, and output activities of the computer. When making hardware decisions, the overriding consideration of a business should be how hardware can support the objectives of the information system and the goals of the organization.

COMPUTER SYSTEMS: INTEGRATING THE POWER OF TECHNOLOGY

To assemble an effective and efficient system, you should select and organize components while understanding the trade-offs between overall system performance and cost, control, and complexity. For instance, in building a car, manufacturers try to match the intended use of the vehicle to its components. Racing cars, for example, require special types of engines, transmissions, and tires. Selecting a transmission for a racing car requires balancing how much engine power can be delivered to the wheels (efficiency and effectiveness) with how expensive the transmission is (cost), how reliable it is (control), and how many gears it has (complexity). Similarly, organizations assemble computer systems so that they are effective, efficient, and well suited to the tasks that need to be performed.

Because the business needs and their importance vary at different companies, the IS solutions they choose can be quite different.

People involved in selecting their organization's computer hardware must clearly understand current and future business requirements so they can make informed acquisition decisions. Consider the following examples of applying business knowledge to reach critical hardware decisions.

- The National Oceanic and Atmospheric Administration (NOAA) is a scientific agency whose mission is to understand and predict changes in the Earth's environment and conserve and manage costal and marine resources. As part of meeting its mission, the NOAA produces daily weather forecasts and severe storm warnings. The NOAA recognized the need for more computing and backup capacity. As a result, it recently upgraded from a single computer to two more powerful computer systems to improve weather forecasts by processing more detailed information and increasing the resolution of weather conditions. The two new computers, dubbed Stratus and Cirrus, have four times the processing power of the previous system. In addition, the Stratus computer is backed up by the Cirrus, which is housed at a separate location. If the Stratus fails for any reason, the Cirrus can take over in just a few minutes.[1]

- The North Carolina State Employees' Credit Union, with 1.5 million members and 223 branch offices, is the country's fourth largest credit union. The credit union continually looks for ways to reduce its ongoing costs and improve customer services. As a result, it made a considerable computer hardware investment to replace more than 3,000 personal computers with newer, lower-cost, more flexible systems.[2]

- The Seaport Hotel is a four-diamond rated, 426-room hotel located on Boston's scenic waterfront. Based on the results of focus groups and guest interviews, the hotel implemented touch-screen computers in each room, allowing guests to access current hotel and local area attraction information, get driving directions and weather reports, gain unlimited Internet and e-mail access, and place complimentary local and long-distance calls over the Internet.[3]

As these examples demonstrate, choosing the right computer hardware requires understanding its relationship to the information system and the needs of the organization. Furthermore, hardware objectives are subordinate to, but supportive of, the information system and the current and future needs of the organization.

HARDWARE COMPONENTS

Computer system hardware components include devices that perform the functions of input, processing, data storage, and output. See Figure 2.1.

Figure 2.1

Hardware Components

These components include the input devices, output devices, communications devices, primary and secondary storage devices, and the central processing unit (CPU). The control unit, the arithmetic/logic unit (ALU), and the register storage areas constitute the CPU.

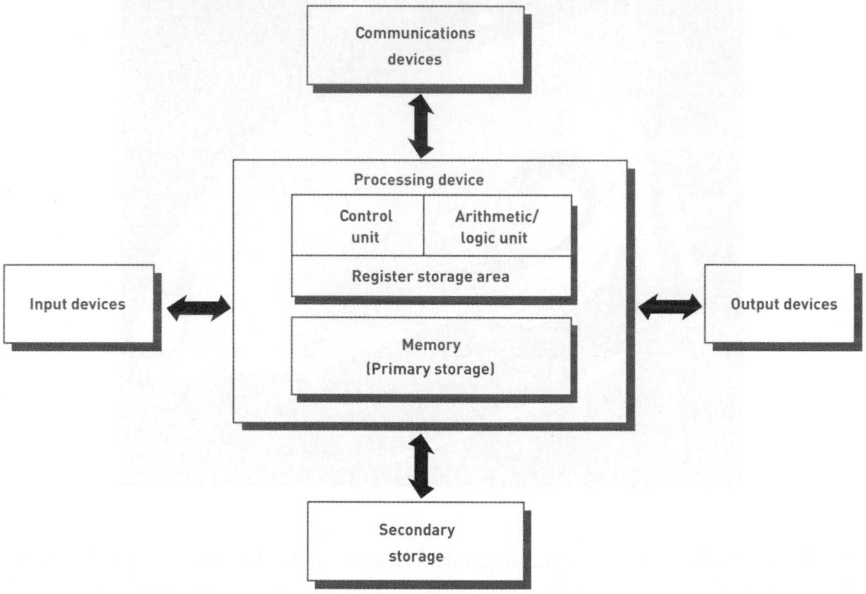

central processing unit (CPU)
The part of the computer that consists of three associated elements: the arithmetic/logic unit, the control unit, and the register areas.

arithmetic/logic unit (ALU)
The part of the CPU that performs mathematical calculations and makes logical comparisons.

control unit
The part of the CPU that sequentially accesses program instructions, decodes them, and coordinates the flow of data in and out of the ALU, the registers, the primary storage, and even secondary storage and various output devices.

The ability to process (organize and manipulate) data is a critical aspect of a computer system, in which processing is accomplished by an interplay between one or more of the central processing units and primary storage. Each **central processing unit** (CPU) consists of two primary elements: the arithmetic/logic unit and the control unit. The **arithmetic/logic unit (ALU)** performs mathematical calculations and makes logical comparisons. The **control unit** sequentially accesses program instructions, decodes them, and coordinates the flow of data in and out of the ALU, primary storage, and even secondary storage and various output devices. Primary memory, which holds program instructions and data, is closely associated with the CPU.

Now that you have learned about the basic hardware components and the way they function, you are ready to examine processing power, speed, and capacity. These three attributes determine the capabilities of a hardware device.

PROCESSING AND MEMORY DEVICES: POWER, SPEED. AND CAPACITY

The components responsible for processing—the CPU and memory—are housed together in the same box or cabinet, called the *system unit*. All other computer system devices, such as the monitor and keyboard, are linked either directly or indirectly into the system unit housing. As discussed previously, achieving IS objectives and organizational goals should be the primary consideration in selecting processing and memory devices. In this section, we investigate the characteristics of these important devices.

Processing Characteristics and Functions

Because efficient processing and timely output are important, organizations use a variety of measures to gauge processing speed. These measures include the time it takes to complete a machine cycle, clock speed, and others.

Clock Speed

Each CPU produces a series of electronic pulses at a predetermined rate, called the **clock speed**, which affects machine cycle time. The control unit executes an instruction in accordance with the electronic cycle, or pulses of the CPU "clock." Each instruction takes at least the same amount of time as the interval between pulses. The shorter the interval between pulses, the faster each instruction can be executed. The clock speed for personal computers is in the multiple gigahertz (GHz), or billions of cycles per second, range.

clock speed
A series of electronic pulses produced at a predetermined rate that affects machine cycle time.

Physical Characteristics of the CPU

CPU speed is also limited by physical constraints. Most CPUs are collections of digital circuits imprinted on silicon wafers, or chips, each no bigger than the tip of a pencil eraser. To turn a digital circuit within the CPU on or off, electrical current must flow through a medium (usually silicon) from point A to point B. The speed at which it travels between points can be increased by either reducing the distance between the points or reducing the resistance of the medium to the electrical current.

Memory Characteristics and Functions

Located physically close to the CPU (to decrease access time), memory provides the CPU with a working storage area for program instructions and data. The chief feature of memory is that it rapidly provides the data and instructions to the CPU.

Storage Capacity

Like the CPU, memory devices contain thousands of circuits imprinted on a silicon chip. Each circuit is either conducting electrical current (on) or not (off). Data is stored in memory as a combination of on or off circuit states. Usually 8 bits are used to represent a character, such as the letter A. Eight bits together form a **byte** (**B**). In most cases, storage capacity is measured in bytes, with 1 byte equivalent to one character of data. The contents of the Library of Congress, with more than 126 million items and 530 miles of bookshelves, would require about 20 petabytes of digital storage. Table 2.1 lists units for measuring computer storage capacity.

byte (B)
Eight bits that together represent a single character of data.

Types of Memory

Several forms of memory are available. Instructions or data can be temporarily stored in **random access memory** (**RAM**). RAM is temporary and volatile—RAM chips lose their contents if the current is turned off or disrupted (as in a power surge, brownout, or electrical noise generated by lightning or nearby machines). RAM chips are mounted directly on the computer's main circuit board or in chips mounted on peripheral cards that plug into the computer's main circuit board. These RAM chips consist of millions of switches that are sensitive to changes in electric current.

random access memory (RAM)
A form of memory in which instructions or data can be temporarily stored.

Table 2.1

Computer Storage Units

Name	Abbreviation	Number of Bytes
Byte	B	1
Kilobyte	KB	2^{10} or approximately 1,024 bytes
Megabyte	MB	2^{20} or 1,024 kilobytes (about 1 million)
Gigabyte	GB	2^{30} or 1,024 megabytes (about 1 billion)
Terabyte	TB	2^{40} or 1,024 gigabytes (about 1 trillion)
Petabyte	PB	2^{50} or 1,024 terabytes (about 1 quadrillion)
Exabyte	EB	2^{60} or 1,024 petabytes (about 1 quintillion)

read-only memory (ROM)
A nonvolatile form of memory.

Read-only memory (ROM), another type of memory, is usually nonvolatile. In ROM, the combination of circuit states is fixed, and therefore its contents are not lost if the power is removed. ROM provides permanent storage for data and instructions that do not change, such as programs and data from the computer manufacturer, including the instructions that tell the computer how to start up when power is turned on.

Multiprocessing

multiprocessing
The simultaneous execution of two or more instructions at the same time.

There are a number of forms of **multiprocessing**, which involves the simultaneous execution of two or more instructions.

Multicore Microprocessor

multicore microprocessor
A microprocessor that combines two or more independent processors into a single computer so they can share the workload and improve processing capacity.

A **multicore microprocessor** combines two or more independent processors into a single computer so that they can share the workload and boost processing capacity. A dual-core processor is like a four-lane highway—it can handle up to twice as many cars as its two-lane predecessor without making each car drive twice as fast. In addition, a dual-core processor enables people to perform multiple tasks simultaneously such as playing a game and burning a CD. Intel, AMD, and IBM are battling for leadership in the multicore processor marketplace.

Both Intel and AMD have improved on dual processors by introducing new quad-core chips. The Intel Core i7 integrates four processors onto a single chip and lets them share a common L3 cache.[4] AMD recently launched its quad-core Phenom II X4 Black Edition CPU, which operates at 3.4 GHz. Although the AMD clock speed is slightly faster than the Intel Core i7, comparing clock speed of two CPUs with different architectures means little. The entire computer system must be designed so that the CPU works well with main memory and the other components of the computer. The Intel Core i7 processor surpasses the AMD Phenom processor based on the results of various computing benchmarks.[5] IBM introduced its Power7 chip with eight processing cores. Because each core can process four tasks or threads, the chip provides a 32-core processor. With this powerful processor "electric utilities can move from processing less than one million meters per day, in a typical [power] grid, to more than 85 million reads per day in a smart grid."[6]

Parallel Computing

parallel computing
The simultaneous execution of the same task on multiple processors to obtain results faster.

Another form of multiprocessing, called **parallel processing**, speeds processing by linking several processors to operate at the same time, or in parallel. The most frequent uses for parallel computing include modeling, simulation, and analyzing large amounts of data. Parallel computing is used in medicine to develop new imaging systems to complete ultrasound scans in less time with greater accuracy, for example, enabling doctors to provide better diagnosis to patients. Instead of building physical models of new products, engineers can create a virtual model of them and use parallel computing to test how the products work and then change design elements and materials as needed. Clothing designers can simulate the look and movement of new designs on virtual models, reducing the development time for a seasonal clothing collection to just over one month from the traditional six-month period.[7]

Grid Computing

Grid computing is the use of a collection of computers, often owned by many people or organizations, to work in a coordinated manner to solve a common problem. Grid computing is one low-cost approach to parallel processing. The grid can include dozens, hundreds, or even thousands of computers that run collectively to solve extremely large parallel processing problems. Key to the success of grid computing is a central server that acts as the grid leader and traffic monitor. This controlling server divides the computing task into subtasks and assigns the work to computers on the grid that have (at least temporarily) surplus processing power. The central server also monitors the processing, and if a member of the grid fails to complete a subtask, it will restart or reassign the task. When all the subtasks are completed, the controlling server combines the results and advances to the next task until the whole job is completed.

IBM launched the World Community Grid project in 2004 to harness the unused computing power of personal and business computers into a large-scale public computing grid. Researchers at the University of Texas Medical Branch (UTMB) used the computer power of more than 1 million devices attached to the grid to test drug candidates for new and drug-resistant flu strains, such as H1N1. Stan Watowich, the lead researcher at UTMB claims that "we expect to identify new influenza drug candidates in less than a month. We can move from computer calculations into laboratory testing more quickly and with a sharper focus."[8]

grid computing
The use of a collection of computers, often owned by multiple individuals or organizations, to work in a coordinated manner to solve a common problem.

SECONDARY STORAGE AND INPUT AND OUTPUT DEVICES

As you have seen, memory is an important factor in determining overall computer system power. However, memory provides only a small amount of storage area for the data and instructions the CPU requires for processing. Computer systems also need to store larger amounts of data, instructions, and information more permanently than main memory allows. Secondary storage, also called *permanent storage,* serves this purpose.

Compared with memory, secondary storage offers the advantages of nonvolatility, greater capacity, and greater economy. Most forms of secondary storage are considerably less expensive than memory (see Table 2.2). Because of the electromechanical processes involved in using secondary storage, however, it is considerably slower than memory. The selection of secondary storage media and devices requires understanding their primary characteristics—access method, capacity, and portability.

Description	Cost	Storage Capacity (GB)	Cost Per GB
72 GB DAT 72 data cartridge	$14.95	72	$0.21
50 4.7 GB DVD+R disks	$21.99	235	$0.09
20 GB 4 mm backup data tape	$6.95	20	$0.35
500 GB portable hard drive	$113.95	500	$0.23
25 GB rewritable Blu-ray disk	$10.99	25	$0.44
9.1 GB write-once, read-many optical disk	$90.95	9.1	$9.99
4 GB flash drive	$9.99	4	$2.50
1 TB desktop external hard drive	119.99	1000	$0.12
2 GB DDR2 SDRAM memory upgrade	$49.99	2	$25.00

Table 2.2

Cost Comparison for Various Forms of Data Storage

(Source: Office Depot Web site, *www.officedepot.com*, December 9, 2009.)

Access Methods

Data and information access can be either sequential or direct. **Sequential access** means that data must be accessed in the order in which it is stored. For example, inventory data stored sequentially may be stored by part number, such as 100, 101, 102, and so on. If you want to retrieve information on part number 125, you need to read and discard all the data relating to parts 001 through 124.

Direct access means that data can be retrieved directly, without having to pass by other data in sequence. With direct access, it is possible to go directly to and access the needed data—such as part number 125—without reading through parts 001 through 124. For this reason, direct access is usually faster than sequential access. The devices used to sequentially access secondary storage data are simply called **sequential access storage devices (SASDs)**; those used for direct access are called **direct access storage devices (DASDs)**.

Secondary Storage Devices

The most common forms of secondary storage include magnetic tapes, magnetic disks, and optical discs. Some of these media (magnetic tape) allow only sequential access, while others (magnetic and optical discs) provide direct and sequential access.

Magnetic Tapes

One common secondary storage medium is **magnetic tape.** Similar to the kind of tape found in audio and video cassettes, magnetic tape is a Mylar film coated with iron oxide. Portions of the tape are magnetized to represent bits. Magnetic tape is a sequential access storage medium. Although access is slower, magnetic tape is usually less expensive than disk storage. Magnetic tape is often used to back up disk drives and to store data off-site for recovery in case of disaster. Technology is improving to provide tape storage devices with greater capacities and faster transfer speeds. Large, bulky tape drives have been replaced with much smaller tape cartridge devices measuring a few millimeters in diameter that take up much less floor space and allow hundreds of tape cartridges to be stored in a small area.

Magnetic Disks

Magnetic disks are also coated with iron oxide; they can be thin metallic platters (hard disks, see Figure 2.2) or Mylar film (diskettes). As with magnetic tape, magnetic disks represent bits using small magnetized areas. Magnetic disks are direct access storage devices that enable fast data retrieval and are used by companies that need to respond quickly to customer requests. For example, if a manager needs information on the credit history of a customer or the seat availability on a particular flight, the information can be obtained in seconds if the data is stored on a direct access storage device.

RAID

Companies' data storage needs are expanding rapidly. Today's storage configurations routinely entail many hundreds of gigabytes. However, putting the company's data online involves a serious business risk—the loss of critical business data can put a corporation out of operation. The concern is that the most critical mechanical components inside a disk storage device—the disk drives, the fans, and other input/output devices—can break.

Figure 2.2

Hard Disk

Hard disks provide direct access to stored data. The read/write head can move directly to the location of a desired piece of data, dramatically reducing access times as compared to magnetic tape.

(Source: Courtesy of Seagate Technology.)

Organizations now require their data-storage devices to be fault tolerant—they can continue with little or no loss of performance in the event of a failure of one or more key components. **Redundant array of independent/inexpensive disks (RAID)** is a method of storing data so that if a hard drive fails, the lost data on that drive can be rebuilt. With this approach, data is stored redundantly on different physical disk drives using a technique called *stripping* to evenly distribute the data. Point360 is a post-production company based in Burbank, California that edits, masters, reformats, and archives video files for its TV and film production clients. The firm implemented a RAID storage solution to provide a greater level of security and to manage the increasing amount of storage required by high-definition video and sophisticated visual effect rendering.[9]

Virtual Tape

Virtual tape is a storage technology for less frequently needed data so that it appears to be stored entirely on tape cartridges, although some parts might actually be located on faster hard disks. The software associated with a virtual tape system is sometimes called a *virtual tape server*. Virtual tape can be used with a sophisticated storage-management system that moves data to slower but less costly forms of storage media as people use the data less often. Virtual tape technology can decrease data access time, lower the total cost of ownership, and reduce the amount of floor space consumed by tape operations. The IS organization at Boston Medical Center is responsible for maintaining more than 400 TB of data associated with the operation of this 581-bed academic medical center. The organization adopted a virtual tape management system to cope with a 50 percent annual data growth rate while keeping data storage costs under control and meeting strict regulatory requirements.[10]

SAN

A **storage area network (SAN)** uses computer servers, distributed storage devices, and networks to tie everything together, as shown in Figure 2.3. To increase the speed of storing and retrieving data, high-speed communications channels are often used. Although SAN technology is relatively new, a number of companies are using SAN to successfully and efficiently store critical data. Austar is a major subscription TV provider in Australia that provides digital satellite services to some 713,000 subscribers.[11] The company employs a 60 terabyte SAN to keep a record of every transaction and interaction it has with customers. "It is all valuable information, says CIO Dean Walters. "Through those touch points we generate a lot of intelligence around the mood, openness, and attitudes to service of our customers. Importantly, we learn about a customer's propensity to leave the service."[12]

redundant array of independent/inexpensive disks (RAID)
A method of storing data that generates extra bits of data from existing data, allowing the system to create a "reconstruction map" so that if a hard drive fails, the system can rebuild lost data.

virtual tape
A storage device for less frequently needed data so that it appears to be stored entirely on tape cartridges, although some parts of it might actually be located on faster hard disks.

storage area network (SAN)
A special-purpose, high-speed network that provides high-speed connections among data-storage devices and computers over a network.

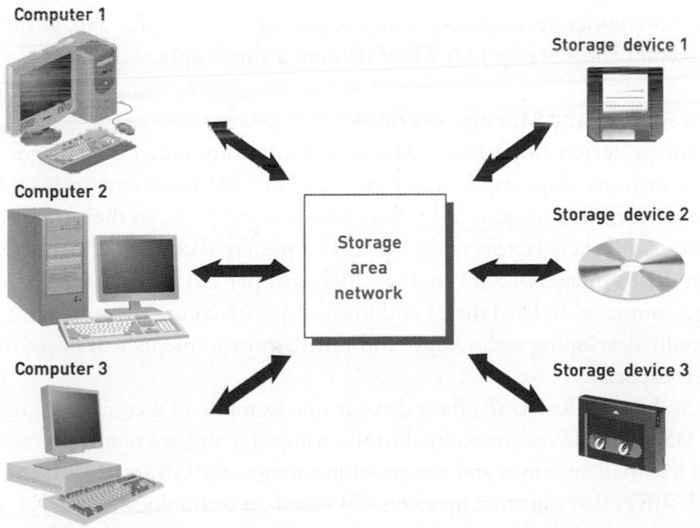

Figure 2.3

Storage Area Network

A SAN provides high-speed connections between data-storage devices and computers over a network.

Figure 2.4

Digital Video Disc and Player

DVDs look like CDs but have a greater storage capacity and can transfer data at a faster rate.

(Source: Courtesy of LaCie USA.)

Optical Discs

A common optical disc is the **compact disc read-only memory** (CD-ROM) with a storage capacity of 740 MB of data. After data is recorded on a CD-ROM, it cannot be modified—the disc is "read-only." A CD burner, the informal name for a CD recorder, is a device that can record data to a compact disc. *CD-recordable (CD-R)* and *CD-rewritable (CD-RW)* are the two most common types of drives that can write CDs, either once (in the case of CD-R) or repeatedly (in the case of CD-RW). CD-rewritable (CD-RW) technology allows PC users to back up data on CDs.

Digital Video Disc

A **digital video disc** (DVD) is a five-inch diameter CD-ROM look-alike with the ability to store about 135 minutes of digital video or several gigabytes of data (see Figure 2.4). Software programs, video games, and movies are common uses for this storage medium.

DVDs have replaced recordable and rewritable CD discs (CD-R and CD-RW) as the preferred format for sharing movies and photos. Whereas a CD can hold about 740 MB of data, a single-sided DVD can hold 4.7 GB, with double-sided DVDs having a capacity of 9.4 GB. Recordings can be made on record-once discs (DVD-R and DVD+R) or on rewritable discs (DVD-RW, DVD+RW, and DVD-RAM). Not all types of rewritable DVDs are compatible with other types.

The Blu-ray high-definition video disc format based on blue-laser technology stores at least three times as much data as a DVD now holds. The primary use for this new format is in home entertainment equipment to store high-definition video, though this format can also store computer data.

The Holographic Versatile Disc (HVD) is an advanced optical disc technology still in the development stage that would store more data than even the Blu-ray optical disc system. HVD devices are under development with the potential to transfer data at the rate of 1 to 20 GB per second and store up to 6 TB of data on a single optical disk.

Solid State Secondary Storage Devices

Solid state storage devices (SSDs) store data in memory chips rather than magnetic or optical media. These memory chips require less power and provide faster data access than magnetic data-storage devices. In addition, SSDs have few moving parts, so they are less fragile than hard disk drives. All these factors make the SSD a preferred choice for portable computers. Two current disadvantages of SSD are their high cost per GB of data storage (roughly a 5:1 disadvantage compared to hard disks) and lower capacity compared to current hard drives. SSD is a rapidly developing technology, and future improvements will lower their cost and increase their capacity.

A Universal Serial Bus (USB) flash drive is one example of a commonly used SSD. See Figure 2.5. USB flash drives are external to the computer and are removable and rewritable. Most weigh less than an ounce and can provide storage of 1 GB to 64 GB. SanDisk manufactures flash drives that can store up to 64 GB based on technology it calls X4, which stores 4 bits of data in each of the millions of tiny storage elements on a chip called cells.[13]

Figure 2.5

Flash Drive

Flash drives are solid state storage devices.

(Source: Image © 2010, Lipsky. Used under license from Shutterstock.com.)

Marketing Architects is a company of about 100 employees that provides radio, TV, and telephone marketing campaigns for its clients. The firm employs SSD technology to store data to provide faster data response time at lower storage costs compared to available alternatives.[14]

The overall trend in secondary storage is toward use of direct access methods, higher capacity, and increased portability. The business needs and needs of individual users should be considered when selecting a specific type of storage. In general, the ability to store large amounts of data and information and access it quickly can increase organizational effectiveness and efficiency.

Input Devices

Your first experience with computers is usually through input and output devices. These devices are the gateways to the computer system—you use them to provide data and instructions to the computer and receive results from it. Input and output devices are part of a computer's user interface, which includes other hardware devices and software that allow you to interact with a computer system.

As with other computer system components, an organization should keep their business goals in mind when selecting input and output devices. For example, many restaurant chains use handheld input devices or computerized terminals that let waiters enter orders efficiently and accurately. These systems have also cut costs by helping to track inventory and market to customers.

Literally hundreds of devices can be used for data input, ranging from special-purpose devices used to capture specific types of data to more general-purpose input devices. We will now discuss several.

Personal Computer Input Devices

A keyboard and a computer mouse are the most common devices used for entry of data such as characters, text, and basic commands. Some companies are developing newer keyboards that are more comfortable, adjustable, and faster to use. These keyboards, such as the split keyboard by Microsoft and others, are designed to avoid wrist and hand injuries caused by hours of keyboarding. Using the same keyboard, you can enter sketches on the touchpad and text using the keys.

A keyboard and mouse are two of the most common devices for computer input. Wireless mice and keyboards are now readily available.

(Source: Courtesy of Hewlett-Packard Company.)

You use a computer mouse to point to and click symbols, icons, menus, and commands on the screen. The computer takes a number of actions in response, such as placing data into the computer system.

Speech-Recognition Technology

speech-recognition technology
Input devices that recognize human speech.

Speech-recognition technology enables a computer equipped with a source of speech input such as a microphone to interpret human speech as an alternative means of providing data or instructions to the computer. The most basic systems require you to train the system to recognize your speech patterns or are limited to a small vocabulary of words. More advanced systems can recognize continuous speech without requiring you to break up your speech into discrete words. Very advanced systems used by the government and military can interpret a voice they have never heard and understand a rich vocabulary.

Companies that must constantly interact with customers are eager to reduce their customer support costs while improving the quality of their service. For example, SBI Funds Management is a fund management organization in India with more than 5.8 million investors. The organization deployed a speech-recognition system to replace all the services that live agents used to provide. Besides the cost savings generated by reducing staff, the system will eliminate the need for customers to wait for live agents to become available. Customers interact with the system using their natural voice and do not have to touch keys on their phone keypad.[15]

Digital Cameras

digital camera
An input device used with a PC to record and store images and video in digital form.

Digital cameras record and store images or video in digital form. When you take pictures, the images are electronically stored in the camera. You can download the images to a computer either directly or by using a flash memory card. After you store the images on the computer's hard disk, you can edit them, send them to another location, paste them into another application, or print them. For example, you can download a photo of your project team captured by a digital camera and then post it on a Web site or paste it into a project status report. Digital cameras have eclipsed film cameras used by professional photographers for photo quality and features such as zoom, flash, exposure controls, special effects, and even video-capture capabilities. With the right software, you can add sound and handwriting to the photo.

Canon, Casio, Nikon, Olympus, Panasonic, Pentax, Sony, and other camera manufacturers offer full-featured, high-resolution digital camera models for less than $250. Some manufacturers offer pocket-sized camcorders for less than $150.

Touch-Sensitive Screens

Advances in screen technology allow display screens to function as input as well as output devices. By touching certain parts of a sensitive screen, you can execute a program or cause the computer to take an action. Touch-sensitive screens are frequently used at gas stations for customers to select grades of gas and request a receipt, at fast-food restaurants for order clerks to enter customer choices, at information centers in hotels to allow guests to request facts about local eating and drinking establishments, and at amusement parks to provide directions to patrons. They also are used in kiosks at airports and department stores.

Optical Data Readers

You can use a special scanning device called an *optical data reader* to scan documents. The two categories of optical data readers are optical mark recognition (OMR) and optical character recognition (OCR). You use OMR readers for test scoring and other purposes when test takers use pencils to fill in boxes on OMR paper, which is also called a "mark sense form." OMR systems are used in standardized tests, including the SAT and GMAT tests, and are being considered as a means to capture voters' choices on Election Day. In comparison, most OCR readers use reflected light to recognize and scan various characters. With special software, OCR readers can convert handwritten or typed documents into digital data. After being entered, this data can be shared, modified, and distributed over computer networks to hundreds or thousands of people.

Missouri law requires that police officers capture data about each traffic stop and send this data to the State Attorney General. The St. Peters police department decided to comply with this by using OMR technology. The department designed a traffic stop data card to capture the required data, which the officer fills in during each traffic stop. At the end of each shift, the cards are turned in and scanned. The results are summarized in two forms: an overview report to provide the required data for the state attorney general and a more detailed analysis including charts and graphs to ensure the department is meeting the needs of the citizens.[16]

Magnetic Ink Character Recognition (MICR) Devices

In the 1950s, the banking industry became swamped with paper checks, loan applications, bank statements, and so on. To remedy this overload and process documents more quickly, the industry developed *magnetic ink character recognition (MICR),* a system for reading this data quickly. With MICR, data to help clear and route checks is placed on the bottom of a check or other form using a special magnetic ink. Data printed with this ink using a special character set can be read by both people and computers.

Pen Input Devices

By touching a touch screen with a pen input device, you can activate a command or cause the computer to perform a task, enter handwritten notes, and draw objects and figures. Pen input requires special software and hardware. Handwriting recognition software can convert handwriting on the screen into text. The Tablet PC from Microsoft and its hardware partners can transform handwriting into typed text and store the "digital ink" just the way a person writes it. Users can use a pen to write and send e-mail, add comments to Word documents, mark up PowerPoint presentations, and even hand-draw charts in a document. That data can then be moved, highlighted, searched, and converted into text. If perfected, this interface is likely to become widely used. Pen input is especially attractive if you are uncomfortable using a keyboard. The success of pen input depends on how accurately handwriting can be read and translated into digital form and at what cost.

Radio Frequency Identification

The purpose of a **Radio Frequency Identification (RFID)** system is to transmit data by a mobile device, called a tag, which is read by an RFID reader and processed according to the needs of an information system program (Figure 2.6). One popular application of RFID is to place a microchip on retail items and install in-store readers that track the inventory on the shelves to determine when shelves should be restocked. Recall that the RFID tag chip includes a special form of erasable programmable read-only memory (EPROM) that holds data about the item to which the tag is attached. A radio-frequency signal can update this memory as the status of the item changes. The data transmitted by the tag might provide identification, location information, or details about the product tagged, such as date manufactured, retail price, color, or date of purchase.

Radio Frequency Identification (RFID)
A technology that employs a microchip with an antenna to broadcast its unique identifier and location to receivers.

The Newmount Leeville Gold Mine in Nevada uses RFID technology to track miners, equipment, and vehicles. RFID tags are placed in the miners' cap lamps and mounted to vehicles and equipment to transmit real-time location data so the company can track where miners are working and quickly locate them in the event of an accident.[17]

Output Devices

Computer systems provide output to decision makers at all levels of an organization so they can solve a business problem or capitalize on a competitive opportunity. In addition, output from one computer system can provide input into another computer system. The desired form of this output might be visual, audio, or even digital. Whatever the output's content or form, output devices are designed to provide the right information to the right person in the right format at the right time.

Figure 2.6

RFID Tag

An RFID tag is small compared to current bar-coded labels used to identify items.

(Source: Courtesy of Intermec Technologies Corporation.)

RFID tag

pixel
A dot of color on a photo image or a point of light on a display screen.

plasma display
A type of display using thousands of smart cells (pixels) consisting of electrodes and neon and xenon gases that are electrically turned into plasma (electrically charged atoms and negatively charged particles) to emit light.

LCD display
Flat display that uses liquid crystals—organic, oil-like material placed between two polarizers— to form characters and graphic images on a backlit screen.

organic light-emitting diode (OLED) display
Flat display that uses a layer of organic material sandwiched between two conductors, which, in turn, are sandwiched between a glass top plate and a glass bottom plate so that when electric current is applied to the two conductors, a bright, electro-luminescent light is produced directly from the organic material.

Display Monitors

The display monitor is a device used to display the output from the computer. Because early monitors used a cathode-ray tube to display images, they were sometimes called *CRTs*. The cathode-ray tubes generate one or more electron beams. As the beams strike a phosphorescent compound (phosphor) coated on the inside of the screen, a dot on the screen called a pixel lights up. A **pixel** is a dot of color on a photo image or a point of light on a display screen. It appears in one of two modes: on or off. The electron beam sweeps across the screen so that as the phosphor starts to fade, it is struck and lights up again.

A **plasma display** uses thousands of smart cells (pixels) consisting of electrodes and neon and xenon gases that are electrically turned into plasma (electrically charged atoms and negatively charged particles) to emit light. The plasma display lights up the pixels to form an image based on the information in the video signal. Each pixel is made up of three types of light—red, green, and blue. The plasma display varies the intensities of the lights to produce a full range of colors. Plasma displays can produce high resolution and accurate representation of colors to create a high-quality image.

LCD displays are flat displays that use liquid crystals—organic, oil-like material placed between two polarizers—to form characters and graphic images on a backlit screen. These displays are easier on your eyes than CRTs because they are flicker-free, brighter, and they do not emit the type of radiation that concerns some CRT users. In addition, LCD monitors take up less space and use less than half of the electricity required to operate a comparably sized CRT monitor.

Organic light-emitting diode (OLED) uses a layer of organic material sandwiched between two conductors, which, in turn, are sandwiched between a glass top plate and a glass bottom plate. When electric current is applied to the two conductors, a bright, electro-luminescent light is produced directly from the organic material. OLEDs can provide sharper and brighter colors than LCDs and CRTs, and because they do not require a backlight, the displays can be half as thick as LCDs, and they are flexible. Another big advantage is that OLEDs do not break when dropped. OLED technology can also create 3D video displays by taking a traditional LCD monitor and then adding layers of transparent OLED films to create the perception of depth without the need for 3D glasses or laser optics. The iZ3D monitor is capable of displaying in both 2D and 3D mode. The manufacturer offered a 22-inch version of the monitor at a price of $300 to coincide with the debut of *Avatar*, a film directed by James Cameron.

Because most users leave their computers on for hours at a time, power usage is an important factor when deciding which type of monitor to purchase. Although the power usage varies from model to model, LCD monitors generally consume between 35 and 50 percent less power than plasma screens.[18] OLED monitors use even less power than LCD monitors.

Printers and Plotters

Hard copy is paper output from a device called a printer. Printers with different speeds, features, and capabilities are available. Some can be set up to accommodate different paper forms such as blank check forms, invoice forms, and so forth. Newer printers allow businesses to create customized printed output for each customer from standard paper and data input using full color.

The speed of the printer is typically measured by the number of pages printed per minute (ppm). Like a display screen, the quality, or resolution, of a printer's output depends on the number of dots printed per inch (dpi). A 600-dpi printer prints more clearly than a 300-dpi printer. A recurring cost of using a printer is the inkjet or laser cartridge that is used as pages are printed. Figure 2.7 shows an inkjet printer.

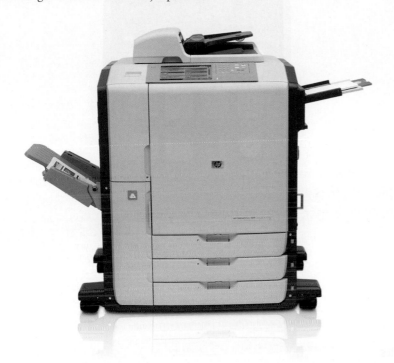

Figure 2.7

The Hewlett-Packard CM8060 Inkjet Printer

(Source: Courtesy of Hewlett-Packard Company.)

Laser printers are generally faster than inkjet printers and can handle more volume than inkjet printers. Laser printers print 25 to 60 ppm for black and white and 6 to 25 ppm for color. Inkjet printers that can print 12 to 40 ppm for black and white and 5 to 20 ppm for color are available for less than $200.

Plotters are a type of hard-copy output device used for general design work. Businesses typically use these devices to generate paper or acetate blueprints, schematics, and drawings of buildings or new products onto paper or transparencies. Standard plot widths are 24 inches and 36 inches, and the length can be whatever meets the need—from a few inches to several feet.

Digital Audio Player

A **digital audio player** is a device that can store, organize, and play digital music files. MP3 (MPEG-1 Audio Layer-3) is a popular format for compressing a sound sequence into a very small file while preserving the original level of sound quality when it is played. By compressing the sound file, it requires less time to download the file and less storage space on a hard drive.

You can use many different music devices about the size of a cigarette pack to download music from the Internet and other sources. These devices have no moving parts and can store hours of music. Apple expanded into the digital music market with an MP3 player (the iPod) and the iTunes Music Store, which allows you to find music online, preview it, and download it in a way that is safe, legal, and affordable. The Apple iPod has a 2.5-inch screen and can play video, including selected TV shows you can download from the iTunes Music Store. Other MP3 manufacturers include Dell, Sony, Samsung, Iomega, and Motorola, whose Rokr product is the first iTunes-compatible phone.

digital audio player
A device that can store, organize, and play digital music files.

The Apple iPod Touch, with a 3.5-inch wide screen, is a music player that also plays movies and TV shows, displays photos, and connects to the Internet. You can therefore use it to view YouTube videos, buy music online, check e-mail, and more. The display automatically adjusts the view when it is rotated from portrait to landscape. An ambient light sensor adjusts brightness to match the current lighting conditions.

The Apple's iPod Touch
(Source: Courtesy of Apple.)

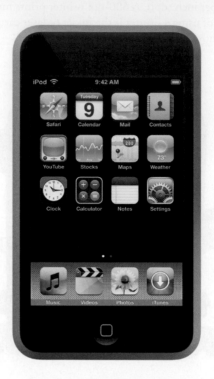

E-Books

The digital media equivalent of a conventional printed book is called an *e-book* (short for electronic book). The Project Gutenberg Online Book Catalog lists more than 30,000 free e-books and a total of more than 100,000 e-books available. E-books can be downloaded from Project Gutenberg (*www.gutenberg.org*) or many other sources onto personal computers or dedicated hardware devices known as e-book readers. The devices themselves cost around $250 to $350, and downloads of the bestselling books and new releases cost less than $10.00. The most current Amazon.com Kindle, the Barnes and Nobel Nook, the Samsung Papyrus, and the Sony Reader all use e-paper displays that look like printed pages, store contents without consuming power, and can be viewed using reflected light rather than the backlight required for LCD screens.[19] E-books weigh less than three-quarters of a pound, are around one-half inch thick, and come with a display screen ranging from 5 to 8 inches. See Figure 2.8. E-books are more compact than most paperbacks so they can be easily held in one hand. On many e-readers, the size of the text can be magnified for readers with poor vision.

Figure 2.8

E-book

(Source: Image © 2010, Photosani.
Used under license from
Shutterstock.com)

COMPUTER SYSTEM TYPES

Computer systems can range from desktop (or smaller) portable computers to massive supercomputers that require housing in large rooms. Let's examine the types of computer systems in greater detail. Table 2.3 shows general ranges of capabilities for various types of computer systems.

Table 2.3

Types of Computer Systems

Single-user computer systems can be divided into two groups—portable computers and nonportable computers.

Factor	Single-User Systems				
	Portable Computers				
	Handheld	**Laptop**	**Notebook**	**Netbook**	**Tablet**
Cost	$150–$400	$500–$1500	$200–$800	$200–$800	$750–$2,500
Weight (pounds)	<.30	< 7	< 5	< 2.5	< 5
Screen size (inches)	2.4–3.6	13.0–15.0	12.0–14.0	7.0–11.0	11.0–14.0
Typical use	Organize personal data	Improve worker productivity	Improve productivity of highly mobile worker	Access the Internet and e-mail	Capture data via pen input, improve worker productivity
	Nonportable Computers				
	Thin Client	**Desktop**	**Nettop**	**Workstation**	
Cost	$200–$800	$500–$2,500	<$300	$750–$5,000	
Weight (pounds)	< 1	< 30	< 5	< 35	
Screen size (inches)	10.0–15.0	13.0–27.0	Comes w/o monitor	13.0–27.0	
Typical use	Enter data and access the Internet	Improve worker productivity	Replace desktop with small, low-cost, low-energy computer	Perform engineering, CAD, and software development	

Factor	Multiple-User Computers		
	Server	**Mainframe**	**Supercomputer**
Cost	$500–$50,000	> $100,000	> $250,000
Weight (pounds)	> 25	> 100	> 100
Screen size (inches)	n/a	n/a	n/a
Typical use	Perform network and Internet applications	Perform computing tasks for large organizations and provide massive data storage	Run scientific applications; perform intensive number crunching

Table 2.3

(*Continued*)

Multiple-user computer systems include servers, mainframes, and supercomputers.

portable computer
A computer small enough to carry easily.

handheld computer
A single-user computer that provides ease of portability because of its small size.

smartphone
A phone that combines the functionality of a mobile phone, personal digital assistant, camera, Web browser, e-mail tool, and other devices into a single handheld device.

laptop computer
A personal computer designed for use by mobile users; it is small and light enough to sit comfortably on a user's lap.

notebook computer
Smaller than a laptop computer, an extremely lightweight computer that weighs less than 6 pounds and can easily fit in a briefcase.

netbook computer
The smallest, lightest, least expensive member of the laptop computer family.

tablet computer
A portable, lightweight computer with no keyboard that allows you to roam the office, home, or factory floor carrying the device like a clipboard.

Portable Computers

Many computer manufacturers offer a variety of **portable computers**, those that are small enough to carry easily. Portable computers include handheld computers, laptop computers, notebook computers, netbook computers, and tablet computers.

Handheld computers are single-user computers that provide ease of portability because of their small size—some are as small as a credit card. These systems often include a variety of software and communications capabilities. Most can communicate with desktop computers over wireless networks. Some even add a built-in GPS receiver with software that can integrate location data into the application. For example, if you click an entry in an electronic address book, the device displays a map and directions from your current location. Such a computer can also be mounted in your car and serve as a navigation system. One of the shortcomings of handheld computers is that they require a lot of power relative to their size.

A **smartphone** combines the functionality of a mobile phone, camera, Web browser, e-mail tool, MP3 player, and other devices into a single handheld device.

A **laptop computer** is a personal computer designed for use by mobile users. It is small and light enough to sit comfortably on a user's lap. Laptop computers use a variety of flat panel technologies to produce a lightweight and thin display screen with good resolution. In terms of computing power, laptop computers can match most desktop computers and come with powerful CPUs as well as large-capacity primary memory and disk storage. This type of computer is highly popular among students and mobile workers who carry their laptops to meetings and classes. Many personal computer users now prefer a laptop computer over a desktop because of its portability, lower energy usage, and smaller space requirements. Starting in 2008, more portable computers were sold in the U.S. than desktop computers.

A **notebook computer** is an extremely lightweight computer that weighs less than 6 pounds and can easily fit in a briefcase. It is smaller and lighter than a laptop computer. When Eddie Bauer commissioned Dave Hahn, fifteen-time conqueror of Mount Everest, and a production crew to climb the world's highest peak as a way to promote its line of First Ascent professional climbing gear, the team packed MacBooks notebook computers so they could blog and post photos and a few minutes of video edited on their computers.[20]

Netbook computers are the smallest (screen size of 7–10 inches), lightest (under 2.5 pounds), and least expensive ($200–$800) members of the laptop computer family. They are great for tasks that do not require a lot of computing power, such as sending and receiving e-mail or accessing the Internet. Many mobile workers have purchased them due to their portability and low cost. Netbook computers are not good for users who want to run demanding applications, have many applications open at one time, or need lots of data storage capacity.

Tablet computers are portable, lightweight computers with no keyboard that allow you to roam the office, home, or factory floor carrying the device like a clipboard. You can enter text with a writing stylus directly on the screen thanks to built-in handwriting recognition software. Other input methods include an optional keyboard or speech recognition. Tablet PCs that support input only via a writing stylus are called *slate computers*. The *convertible tablet PC* comes with a swivel screen and can be used as a traditional notebook or as a pen-based tablet PC.[21] Tablet computers are especially popular with students and are frequently used in the healthcare, retail, insurance, and manufacturing industries because of their versatility.

The Apple iPad is a tablet computer capable of running the same software that runs on the older Apple iPhone and iPod Touch devices, giving it a library of more than 140,000 applications. It also runs software developed specifically for it. The device has a 9.7-inch screen and an on-screen keypad. It weighs 1.5 pounds and supports Internet access over wireless networks. The initial version cannot support voice communications, nor does it have a built-in camera like the iPhone has.

Nonportable Single-User Computers

Nonportable single-user computers include thin client computers, desktop computers, nettop computers, and workstations.

A **thin client** is a low-cost, centrally managed computer with no extra drives (such as CD or DVD drives) or expansion slots. These computers have limited capabilities and perform only essential applications, so they remain "thin" in terms of the client applications they include. As stripped-down computers, they do not have the storage capacity or computing power of typical desktop computers, nor do they need it for the role they play. With no hard disk, they never pick up viruses or suffer a hard disk crash. Unlike personal computers, thin clients download data and software from a network when needed, making support, distribution, and updating of software applications much easier and less expensive.

Tokio Marine and Nichido Fire Insurance Company began deploying some 30,000 thin client systems in 2009 and estimates that the total cost of ownership of the thin clients will be about 30 percent less than deploying full-fledged computers. In addition, because no data is stored on the thin client systems, the firm will be able to implement a highly secure computing environment that carefully safeguards customer data.[22]

Desktop computers are single-user computer systems that are highly versatile. Named for their size, desktop computers can provide sufficient computing power, memory, and storage for most business computing tasks. The Apple iMac is a family of Macintosh desktop computers first introduced in 1998, in which all the components (including the CPU, the disk drives, and so on) fit behind the display screen.

A **nettop computer** is an inexpensive (less than $300) desktop computer designed to be smaller and lighter and to consume one-tenth the power of a traditional desktop computer.[23] It is designed to perform basic processing tasks such as Internet surfing, document processing, and audio/video playback. Unlike netbook computers, nettop computers are not designed to be portable; they come without a monitor, but they may include an optical drive (CD/DVD). Businesses are considering using nettop computers because they are inexpensive to buy and run, so they can improve an organization's profitability.

Workstations are more powerful than personal computers but still small enough to fit on a desktop. They are used to support engineering and technical users who perform heavy mathematical computing, computer-aided design (CAD), and other applications requiring

thin client

A low-cost, centrally managed computer with essential but limited capabilities and no extra drives (such as CD or DVD drives) or expansion slots.

desktop computer

A relatively small, inexpensive, single-user computer that is highly versatile.

nettop computer

An inexpensive desktop computer designed to be smaller, lighter, and consume much less power than a traditional desktop computer.

workstation

A more powerful personal computer used for mathematical computing, computer-aided design, and other high-end processing, but still small enough to fit on a desktop.

a high-end processor. Such users need very powerful CPUs, large amounts of main memory, and extremely high-resolution graphic displays. Workstations are typically much more expensive than the average desktop computer. Blue Sky Studios used powerful workstations for the rendering of the animated feature *Ice Age: Dawn of the Dinosaurs*. The new workstations provided Blue Sky animators with more powerful design tools to create images for the film in a much shorter time period.[24]

Multiple-User Computer Systems

Multiple-user computers are designed to support workgroups from a small department of two or three workers to large organizations with tens of thousands of employees and millions of customers. Multiple-user systems include servers, mainframe computers, and supercomputers.

A **server** is a computer used by many users to perform a specific task, such as running network or Internet applications. Servers typically have large memory and storage capacities, along with fast and efficient communications abilities. A Web server handles Internet traffic and communications. An enterprise server stores and provides access to programs that meet the needs of an entire organization. A file server stores and coordinates program and data files. Server systems consist of multiuser computers, including supercomputers, mainframes, and other servers. Often an organization will house a large number of servers in the same room where access to the machines can be controlled and authorized support personnel can more easily manage and maintain them from this single location. Such a facility is called a *server farm*.

The amazing 3D images of a futuristic world and the blue creatures from the movie *Avatar* were created by Weta Digital, Ltd, a visual effects company near Wellington, New Zealand. The 3D image rendering was performed on some 4,000 Hewlett-Packard blade servers in Weta's 10,000 square foot server farm.[25]

A **blade server** houses many computer motherboards that include one or more processors, computer memory, computer storage, and computer network connections. These all share a common power supply and air-cooling source within a single chassis. By placing many blades into a single chassis, and then mounting multiple chassis in a single rack, the blade server is more powerful but less expensive than traditional systems based on mainframes or server farms of individual computers. In addition, the blade server approach requires much less physical space than traditional server farms. Winn-Dixie is a Florida-based grocery chain that decided to consolidate all the stores' data processing onto 39 IBM blade servers located in its Jacksonville headquarters. Each blade server supports 16 individual stores. This centralized blade server solution avoided more than $5 million in capital costs that would have been required to upgrade each of the servers. It also eliminated administrative work at each store and greatly reduced each store's power usage and costs.[26]

A **mainframe computer** is a large, powerful computer shared by dozens or even hundreds of concurrent users connected to the machine over a network. The mainframe computer must reside in a data center with special heating, ventilating, and air-conditioning (HVAC) equipment to control temperature, humidity, and dust levels. In addition, most mainframes are kept in a secure data center with limited access to the room. The construction and maintenance of a controlled-access room with HVAC can add hundreds of thousands of dollars to the cost of owning and operating a mainframe computer.

The mainframe can handle the millions of daily transactions associated with airline, automobile, and hotel/motel reservation systems. It can process the tens of thousands of daily queries necessary to provide data to decision support systems. Its massive storage and input/output capabilities enable it to play the role of a video computer, providing full-motion video to multiple, concurrent users.

Supercomputers are the most powerful computers with the fastest processing speed and highest performance. They are *special-purpose machines* designed for applications that require extensive and rapid computational capabilities. Originally, supercomputers were used primarily by government agencies to perform the high-speed number crunching needed in weather forecasting and military applications. With recent reductions in the cost of these machines, they are now used more broadly for commercial purposes.

server
A computer used by many users to perform a specific task, such as running network or Internet applications.

blade server
A server that houses many individual computer motherboards that include one or more processors, computer memory, computer storage, and computer network connections.

mainframe computer
A large, powerful computer often shared by hundreds of concurrent users connected to the machine via terminals.

supercomputers
The most powerful computer systems with the fastest processing speeds.

IBM's Sequoia will be the fastest supercomputer in the world when it becomes operational in 2012 and can perform calculations at the rate of 20 petaflops—equivalent to an astounding 3 million computations by every human on the planet each second!

(Source: Courtesy of IBM Corporation.)

GREEN COMPUTING

Green computing is concerned with the efficient and environmentally responsible design, manufacture, operation, and disposal of IS-related products, including all types of computers, printers, and printer materials, including cartridges and toner. Business organizations recognize that going green is in their best interests in terms of public relations, safety of employees, and the community at large. They also recognize that green computing presents an opportunity to substantially reduce total costs over the life cycle of their IS equipment. Green computing has three goals: reduce the use of hazardous material, enable companies to lower their power-related costs (including potential cap and trade fees), and enable the safe disposal or recycling of some 700,000 tons of computers each year.

Computer manufacturers such as Apple, Dell, and Hewlett-Packard have long competed on the basis of price and performance. As the difference among the manufacturers in these two arenas narrows, support for green computing is emerging as a new business strategy for these companies to distinguish themselves from the competition. Apple claims to have the "greenest lineup of notebooks" and is making progress at removing toxic chemicals. Dell's new mantra is to become "the greenest technology company on Earth." Hewlett-Packard highlights its long tradition of environmentalism and is improving its packaging to reduce use of materials. Hewlett-Packard is also urging computer users around the world to shut down their computers at the end of the day to save energy and reduce carbon emissions.

We now turn to the other critical component of effective computer systems—software. Like hardware, software has made great technological leaps in a relatively short time span.

green computing
A program concerned with the efficient and environmentally responsible design, manufacture, operation, and disposal of IS-related products.

Electronics Manufacturers Face the Global E-Waste Problem

The world is consuming and discarding increasing amounts of electronics products every year. The United States Environmental Protection Agency estimates that "over 2 billion computers, televisions, wireless devices, printers, gaming systems, and other devices have been sold since 1980." Have you considered where those devices end up when they become obsolete or worn out? It turns out that a large percentage of them are shipped to third-world countries where they are dismantled in salvaging operations that strip out valuable metals and burn the remaining parts. Ghana, Nigeria, Pakistan, India, and China have become the world's primary dumpsters for electronics salvaging.

The real problem with electronics waste, or e-waste, lies in the use of toxic heavy materials such as mercury, cadmium, beryllium, and lead, along with PVC plastic and hazardous chemicals like brominated flame retardants (BFR) that are included in electronics components. When dumped in landfills, these toxic components leach into the land over time and are released into the atmosphere. Burning computer components also releases these toxic components into the atmosphere, with deadly results to the people doing the burning in the short term, and a gradual eroding of the global environment in the long term.

The third-world countries that salvage electronics rarely take environmental precautions. Children are often used as labor, becoming sick after prolonged exposure to the toxic waste.

Organizations such as Greenpeace are going to great lengths to call attention to this global problem. They are encouraging electronics manufacturers to eliminate toxic materials from the products they sell. They also encourage manufacturers to provide incentives for their customers to properly recycle electronics devices when they are through with them. In its annual "Guide to Greener Electronics," Greenpeace ranks technology companies on their level of "green-ness" based on the each company's manufacturing and recycling practices. For example, currently Nokia is ranked greenest due to its comprehensive voluntary take-back programs and recycling practices. Samsung is number two because it removed PVC from its LCD displays, BFR from some of its cell phones, and halogen from its chips and semiconductors. While Nokia and Samsung earned high points from Greenpeace, no company is currently ranked as being 100 percent green, and most are listed in the red with a lot of improvements still needed.

Some manufacturers are seeing green manufacturing practices as a method of gaining market share. The growing population of environmentally conscious consumers prefers to do business with green businesses. Apple has invested heavily in building its green reputation. It has worked to remove or dramatically reduce the amount of lead in its displays. It has eliminated or reduced dangerous chemicals, including arsenic and mercury, and compounds such as PVC and BFR. Apple operates recycling programs in 93 percent of the countries where Macs, iPhones, and iPods are sold. It even reduced packaging materials to a minimum to save trees and eliminate Styrofoam.

Apple is not unique in its efforts to reduce and safeguard e-waste but is considered the most progressive PC manufacturer by Greenpeace. Most other PC manufacturers are following suit. However, soon "going green" may not be a voluntary decision for electronics manufacturing. Legislation has been proposed in the U.S. to address the e-waste issue. One bill is named the "Electronic Device Recycling Research and Development Act." The bill addresses how to manage current e-waste, develop recycling programs, and eliminate the use of toxic materials in electronics. The U.S. has fallen behind other countries in responding to the problem. The European Union has already passed two directives to deal with e-waste: the Restriction on the Use of Hazardous Substances (RoHS) and the Waste Electrical and Electronic Equipment (WEEE).

With increased attention turning towards the e-waste problem, optimism is increasing. One study estimates that if recycling initiatives continue to expand, the e-waste problem should reach a peak global volume of 73 million metric tons by 2015, and then begin to decline.

Discussion Questions

1. What concerns have been raised over e-waste?
2. What actions can electronics manufacturers take to address the e-waste problem?

Critical Thinking Questions

1. What laws, if any, do you think are necessary to address this problem? Why?
2. How might the global community cooperate to help speed up recovery?

SOURCES: Johnston, Casey, "Legislation seeks to deal with growing piles of e-waste," *Ars Technica*, *http://arstechnica.com*, November 1, 2009; Lombardi, Candace, "Study: E-waste build-up will plateau by 2015," CNET news, *http://news.cnet.com*, May 6, 2009; Greenpeace E-Waste Web site, *www.greenpeace.org/international/campaigns/toxics/electronics*, accessed January 1, 2010; A Greener Apple Web page, *www.apple.com/hotnews/agreenerapple*, accessed January 1, 2010.

OVERVIEW OF SOFTWARE

As you learned in Chapter 1, software consists of computer programs that control the workings of computer hardware. **Computer programs** are sequences of instructions for the computer. Documentation describes the program functions to help the user operate the computer system. The program displays some documentation on screen, while other forms appear in external resources, such as printed manuals. People using commercially available software are usually asked to read and agree to End-User License Agreements (EULAs). After reading the EULA, you normally have to click an "I agree" button before you can use the software, which can be one of two basic types: systems software and application software.

computer programs
Sequences of instructions for the computer.

Application software has the greatest potential to affect processes that add value to a business because it is designed for specific organizational activities and functions.

(Source: © Jim West/Alamy.)

Systems software is the set of programs designed to coordinate the activities and functions of the hardware and various programs throughout the computer system. Each type of systems software is designed for a specific CPU design and class of hardware. Application software consists of programs that help users solve particular computing problems. In most cases, application software resides on the computer's hard disk before it is brought into the computer's main memory and run. Application software can also be stored on CDs, DVDs, and even flash or key chain storage devices that plug into a USB port. An increasing amount of application software is available on the Web. Sometimes referred to as a *rich Internet application (RIA)*, a Web-delivered software application combines hardware resources of the Web server and the PC to deliver valuable software services through a Web browser interface. Before a person, group, or enterprise decides on the best approach for acquiring application software, they should analyze their goals and needs carefully.

Supporting Individual, Group, and Organizational Goals

Every organization relies on the contributions of individuals, groups, and the entire enterprise to achieve business objectives. To help them achieve these objectives, the organization provides them with specific application software and information systems. One useful way of classifying the many potential uses of information systems is to identify the scope of the problems and opportunities addressed by a particular organization, called the sphere of influence. For most companies, the spheres of influence are personal, workgroup, and enterprise. Table 2.4 shows how various kinds of software support these three spheres.

Information systems that operate within the *personal sphere of influence* serve the needs of individual users. These information systems enable users to improve their personal effectiveness, increasing the amount of work that can be done and its quality. Such software is often referred to as *personal productivity software.* For example, MindManager software from Mindjet provides tools to help people diagram complex ideas and projects using an intuitive graphic interface.[27]

Software	Personal	Workgroup	Enterprise
Systems software	Personal computer and workstation operating systems	Network operating systems	Server and mainframe operating systems
Application software	Word processing, spreadsheet, database, graphics	Electronic mail, group scheduling, shared work, collaboration	General ledger, order entry, payroll, human resources

Table 2.4

Classifying Software by Type and Sphere of Influence

A *workgroup* is two or more people who work together to achieve a common goal. A workgroup may be a large, formal, permanent organizational entity such as a section or department or a temporary group formed to complete a specific project. An information system that operates in the *workgroup sphere of influence* supports a workgroup in the attainment of a common goal. Users of such applications must be able to communicate, interact, and collaborate to be successful. People can also use online calendar software such as Google Calendar to store personal appointments, but also to schedule meetings with others.[28]

Information systems that operate within the *enterprise sphere of influence* support the firm in its interaction with its environment. The surrounding environment includes customers, suppliers, shareholders, competitors, special-interest groups, the financial community, and government agencies. For example, many enterprises use IBM Cognos software as a centralized Web-based system where employees, partners, and stakeholders can report and analyze corporate financial data.[29]

Installing and Removing New Software

Before you can use any type of software, it must be installed on a computer. Installing new software usually involves only a few setup steps. Software for personal computers typically comes on CDs or is downloaded from the Web.

When possible, it is best to remove software using an add/remove software utility that comes with the operating system or that is part of some utility software such as Norton System Works and McAfee QuickClean. This will help ensure that all elements of unwanted software are removed.

SYSTEMS SOFTWARE

Controlling the operations of computer hardware is one of the most critical functions of systems software. Systems software also supports the application programs' problem-solving capabilities. Different types of systems software include operating systems and utility programs.

Operating Systems

An operating system (OS) is a set of computer programs that control the computer hardware and act as an interface with application programs. See Figure 2.9. Operating systems can control one computer or multiple computers, or they can allow multiple users to interact with one computer. The various combinations of OSs, computers, and users include the following:

- **Single computer with a single user.** This system is commonly used in a personal computer or a handheld computer that allows one user at a time.
- **Single computer with multiple users.** This system is typical of larger, mainframe computers that can accommodate hundreds or thousands of people, all using the computer at the same time.

- **Multiple computers with multiple users.** This system is typical of a network of computers, such as a home network with several computers attached or a large computer network with hundreds of computers attached around the world.
- **Special-purpose computers.** This type of system is typical of a number of computers with specialized functions, such as those that control sophisticated military aircraft, space shuttles, digital cameras, or home appliances.

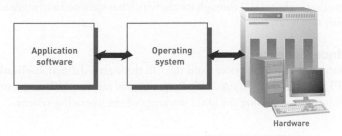

The OS, which plays a central role in the functioning of the complete computer system, is usually stored on disk. After a computer system is started, or "booted up," portions of the OS are transferred to memory as they are needed. You can also boot a computer from a CD, DVD, or even a thumb drive that plugs into a USB port. A storage device that contains some or all of the OS is often called a "rescue disk" because you can use it to start the computer if you have problems with the primary hard disk.

The collection of programs that make up the operating system performs a variety of activities, including the following:

- Performing common computer hardware functions
- Providing a user interface and input/output management
- Providing a degree of hardware independence
- Managing system memory
- Managing processing tasks
- Providing networking capability
- Controlling access to system resources
- Managing files

Common Hardware Functions

All applications must perform certain hardware-related tasks, such as the following:

- Get input from the keyboard or another input device
- Retrieve data from disks
- Store data on disks
- Display information on a monitor or printer

Each of these tasks requires a detailed set of instructions. The OS converts a basic request into the set of detailed instructions that the hardware requires. In effect, the OS acts as an intermediary between the application and the hardware. The typical OS performs hundreds of such tasks, translating each into one or more instructions for the hardware. The OS notifies the user if input or output devices need attention, if an error has occurred, and if anything abnormal happens in the system.

User Interface and Input/Output Management

One of the most important functions of any OS is providing a **user interface**, which allows people to access and interact with the computer system. The first user interfaces for mainframe and personal computer systems were command based.

A **command-based user interface** requires text commands to be given to the computer to perform basic activities. For example, the command ERASE 00TAXRTN would cause the computer to erase or delete a file called 00TAXRTN. RENAME and COPY are other examples of commands used to rename files and copy files from one location to another.

user interface
The element of the operating system that allows you to access and command the computer system.

command-based user interface
A user interface that requires you to give text commands to the computer to perform basic activities.

graphical user interface (GUI)
An interface that displays pictures
(icons) and menus that people use to
send commands to the computer
system.

A **graphical user interface (GUI)** displays pictures (called *icons*) and menus that people use to send commands to the computer system. Many people find that GUIs are easier to use because users intuitively grasp the functions. Today, the most widely used graphical user interface is Windows by Microsoft. As the name suggests, Windows is based on the use of a window, or a portion of the display screen dedicated to a specific application. The screen can display several windows at once. Building on the success of the iPhone, due in no small part to its unique and advanced multitouch user interface, Windows 7 also provides strong support for interacting with its GUI through touch, which has spawned a new generation of PCs being sold with touch displays.

application program interface (API)
An interface that allows applications
to make use of the operating
system.

Hardware Independence

To run, applications request services from the OS through a defined **application program interface (API)**, as shown in Figure 2.10. Programmers can use APIs to create application software without understanding the inner workings of the operating system.

Figure 2.10

Application Program Interface (API)

The API links application software to the operating system, providing hardware independence for software developers.

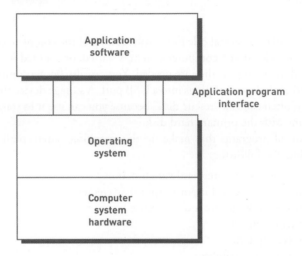

Memory Management

The OS also controls how memory is accessed and maximizes available memory and storage. Most newer OSs manage memory better than older OSs. The memory-management feature of many OSs allows the computer to execute program instructions effectively and to speed processing. One way to increase the performance of an old computer is to upgrade to a newer OS and increase the amount of memory.

Most OSs support virtual memory, which allocates space on the hard disk to supplement the immediate, functional memory capacity of RAM. Virtual memory works by swapping programs or parts of programs between memory and one or more disk devices—a concept called paging. This reduces CPU idle time and increases the number of jobs that can run in a given time span.

Processing Tasks

The task-management features of today's OSs manage all processing activities. Task management allocates computer resources to make the best use of each system's assets. Task-management software lets one user run several programs or tasks at the same time (multitasking) and allows several users to use the same computer at the same time (time sharing).

An OS with multitasking capabilities allows a user to run more than one application at the same time. Most computer users take advantage of multitasking OSs without realizing how innovative they are. Without having to exit a program, you can work in one application, easily pop into another, and then jump back to the first program, picking up where you left off. Better still, while you're working in the *foreground* in one program, one or more other applications can be churning away, unseen, in the *background,* sorting a database, printing a document, or performing other lengthy operations that otherwise would monopolize your computer and leave you staring at the screen unable to perform other work. Multitasking can save users a considerable amount of time and effort.

Time sharing allows more than one person to use a computer system at the same time. For example, 15 customer service representatives might be entering sales data into a computer system for a mail-order company at the same time. In another case, thousands of people might be simultaneously using an online computer service to get stock quotes and valuable business news.

The ability of the computer to handle an increasing number of concurrent users smoothly is called *scalability*. This feature is critical for systems expected to handle a large number of users such as a mainframe computer or a Web server. Because personal computer OSs are usually oriented toward single users, they do not need to manage multiple-user tasks often.

Networking Capability

Most operating systems include networking capabilities so that computers can join together in a network to send and receive data and share computing resources. PCs running Mac, Windows, or Linux operating systems allow users to easily set up home or business networks for sharing Internet connections, printers, storage, and data. Windows 7 includes a Home-Group feature that makes it easy to share photos, music, files, and printers with others on a home network. Operating systems for larger server computers are designed specifically for computer networking environments.

Access to System Resources and Security

Because computers often handle sensitive data that can be accessed over networks, the OS needs to provide a high level of security against unauthorized access to the users' data and programs. Typically, the OS establishes a logon procedure that requires users to enter an identification code, such as a user name, and a matching password. If the identification code is invalid or if the password does not match the identification code, the user cannot gain access to the computer. Some OSs require that user passwords change frequently—such as every 20 to 40 days. If the user is successful in logging on to the system, the OS restricts access to only portions of the system for which the user has been authorized. The OS records who is using the system and for how long and reports any attempted breaches of security.

File Management

The OS manages files to ensure that files in secondary storage are available when needed and that they are protected from access by unauthorized users. Many computers support multiple users who store files on centrally located disks or tape drives. The OS keeps track of where each file is stored and who can access them. The OS must determine what to do if more than one user requests access to the same file at the same time. Even on stand-alone personal computers with only one user, file management is needed to track where files are located, what size they are, when they were created, and who created them.

Current Operating Systems

Early OSs were very basic. Today, however, more advanced OSs have been developed, incorporating sophisticated features and impressive graphics effects. Table 2.5 classifies a number of current OSs by sphere of influence.

Microsoft PC Operating Systems

Since a small company called Microsoft developed PC-DOS and MS-DOS to support the IBM personal computer introduced in the 1980s, personal computer OSs have steadily evolved. *PC-DOS* and *MS-DOS* had command-driven interfaces that were difficult to learn and use. Each new version of OS has improved the ease of use, processing capability, reliability, and ability to support new computer hardware devices.

Windows XP (XP reportedly stands for the wonderful experience that you will have with your personal computer) was released in fall of 2001. Previous consumer versions of Windows were notably unstable and crashed frequently, requiring frustrating and time-consuming reboots. With XP, Microsoft sought to bring reliability to the consumer.

Table 2.5

Operating Systems Serving
Three Spheres of Influence

Personal	Workgroup	Enterprise
Microsoft Windows , Microsoft Windows Mobile	Microsoft Windows Server 2008	Microsoft Windows Server 2008
Mac OS X, Mac OS X iPhone	Mac OS X Server	
Linux	Linux	Linux
Google Android, Chrome OS		
Palm Web OS		
	UNIX	UNIX
	IBM i5/OS and z/OS	IBM i5/OS and z/OS
	HP-UX 11i	HP-UX 11i

Microsoft released *Windows Vista* in 2007 with the goal of providing a more secure and stable operating system. The new operating system includes a number of new features. The most advanced versions of Vista include a 3D graphics interface called Aero. However, the system requirements for Windows Vista with Aero require many users to purchase new, more powerful PCs. Another issue was that some software and hardware designed for Windows XP would not run on Vista.

The next version, *Windows 7*, was released in 2009 with improvements and new features. Most analysts classified Windows 7 as "Vista done right."[30] Besides addressing some of the flaws in Windows Vista, Windows 7 introduced new windows manipulation functionality that allows users to more easily find, access, and work with information in files. It also features improved home networking capabilities and improved applications. Windows 7 has strong support for touch displays and netbooks, ushering in a new era of mobile computing devices.

Apple Computer Operating Systems

Although IBM system platforms traditionally use one of the Windows OSs and Intel microprocessors (often called *Wintel* for this reason), Apple computers have used non-Intel microprocessors designed by Apple, IBM, and Motorola and a proprietary Apple OS—the *Mac OS*. Newer Apple computers, however, use Intel chips. Although Wintel computers hold the largest share of the business PC market, Apple computers are also quite popular, especially in the fields of publishing, education, graphic arts, music, movies, and media.

The Apple OSs have also evolved over a number of years and often provide features not available from Microsoft. Recently, however, Windows and Mac platforms have evolved to share many of the same features as they compete for users. In July 2001, Mac OS X was released as an entirely new OS for the Mac based on the UNIX operating system. It included a new user interface, which provided a new visual appearance for users—including luminous and semitransparent elements, such as buttons, scroll bars, windows, and fluid animation to enhance the user's experience.

Since its first release, Apple has upgraded OS X several times. Snow Leopard (OS X v10.6) is the most recent version of OS X, released in 2009 to compete with Windows 7. See Figure 2.11. Snow Leopard includes Time Machine, a powerful backup tool that allows users to view their system as it looked in the past and resurrect deleted files. Snow Leopard also includes multiple desktops, a video chat program that allows users to pose in front of imaginary landscapes, a powerful system search utility, and other updated software. Macs are also considered very secure, with no widespread virus or spyware infections to date.

Because Mac OS X runs on Intel processors, Mac users can set up their computer to run both Windows Vista and Mac OS X and select which platform they want to work with when they boot their computer. Such an arrangement is called *dual booting*. While Macs can dual boot into Windows, the opposite is not true. Apple does not allow OS X to be run on any machine other than an Apple. However, Windows PCs can dual boot with Linux and other OSs.

Figure 2.11

Mac OS X Snow Leopard

Linux

Linux is an OS developed by Linus Torvalds in 1991 as a student in Finland. The OS is distributed under the GNU General Public License, and its source code is freely available to everyone. It is, therefore, called an open-source operating system. This doesn't mean, however, that Linux and its assorted distributions are necessarily free—companies and developers can charge money for a distribution as long as the source code remains available. Linux is actually only the kernel of an OS, the part that controls hardware, manages files, separates processes, and so forth. Several combinations of Linux are available, with various sets of capabilities and applications to form a complete OS. Each of these combinations is called a *distribution* of Linux. Many distributions are available as free downloads.

Linux is available on the Internet and from other sources. Popular versions include Red Hat Linux and Caldera OpenLinux. Several large computer vendors, including IBM, Hewlett-Packard, and Intel, support the Linux operating system. For example, IBM has hundreds of programmers working with Linux. Linux is a popular OS for servers, distributed systems, and even supercomputers. Most computer science and engineering graduates are familiar with Linux, so there is no shortage of programmers and engineers for Linux-based systems. The flexibility of the open architecture also makes it easy to customize Linux for different needs in different environments.

Google has developed its own Linux-based operating system named Chrome OS. Chrome is designed for small mobile computers and netbooks with a focus on accessing Web-based information and services such as e-mail, Web browsing, social networks, and Google online applications.[31]

Workgroup Operating Systems

To keep pace with user demands, the technology of the future must support a world in which network usage, data-storage requirements, and data-processing speeds increase at a dramatic rate. This rapid increase in communications and data-processing capabilities pushes the boundaries of computer science and physics. Powerful and sophisticated OSs are needed to run the servers that meet these business needs for workgroups.

Windows Server

Microsoft designed *Windows Server* to perform a host of tasks that are vital for Web sites and corporate Web applications. For example, Microsoft Windows Server can be used to coordinate large data centers. The OS also works with other Microsoft products. It can be used to prevent unauthorized disclosure of information by blocking text and e-mails from being copied, printed, or forwarded to other people. Microsoft *Windows Server 2008* is the most recent version of Windows Server and delivers benefits such as a powerful Web server management system, virtualization tools that allow various operating systems to run on a single server, advanced security features, and robust administrative support.

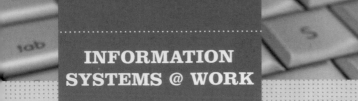

Blended Platforms at LinkedIn

Although Microsoft Windows dominates the business desktop OS market, Macs are beginning to make inroads. The popular business-focused social network, LinkedIn, finds that its employees generally have a strong preference for one operating system over another. Rather than forcing employees to use one standard operating system and compatible software, LinkedIn allows its employees to choose either the Windows or Mac platform and sometimes even Linux.

Big Web companies such as LinkedIn hire a variety of specialists ranging from Web developers and software engineers to graphic artists, designers, accountants, and executives. Often professionals from different disciplines prefer one platform over another because of specific software tools designed for that platform.

Macs are especially popular with so-called techies: Web developers, software engineers, and programmers. They like the Mac platform because of its power and because it is based on the UNIX kernel. UNIX is popular with programmers who like to work from the command line. LinkedIn provides all of its software engineers with state-of-the-art Mac Pros (desktop PCs) and MacBooks (notebook PCs). LinkedIn uses this equipment as an enticement to attract top-of-the-line engineers. According to LinkedIn, "the Mac factor" has a big impact on developers' decisions to join the company. Some developers even set up their Macs to run both Mac OS X and Linux so that they can develop LinkedIn software to run in browsers on the Linux platform.

Artists and graphic designers are typically divided between Windows and Macs depending on what software they prefer to use. Aperture is popular photo-editing software for the Mac. However, Photoshop and other popular graphics software from Adobe is available for both Windows and Mac. Generally speaking, Mac has a long history of appealing to digital media designers. It is especially popular with video and music producers.

For business applications, Microsoft Windows is typically king. It is rare to find a Mac in a business environment. LinkedIn product managers, accountants, human resources managers, executives, and other business staff have a choice of Microsoft Windows PCs or Macs. Surprisingly, 68 percent have chosen Macs. The general popularity of the Mac in Silicon Valley might be why LinkedIn has so many Mac users. Also, Microsoft Office and other business software are available for the Mac platform.

LinkedIn's IT department now provides services to all employees over the dual platform integrated network. Mac is completely compatible with Windows and has no problem sharing files and resources over a network. LinkedIn did not have to modify its network environment to accommodate both Macs and Windows PCs. As an increasing amount of computing takes place online rather than on the local PC, it is likely that the choice of PC platforms will become less important.

Discussion Questions

1. In what ways is LinkedIn unique in the options it provides its employees and in the choices it allows its employees to make?
2. If LinkedIn required all employees to use the same platform, how might that requirement detract from employee productivity?

Critical Thinking Questions

1. Do you think LinkedIn will serve as a trendsetter, with many businesses following suit?
2. What benefits does standardizing around one platform provide for businesses?

SOURCES: Apple staff, "LinkedIn. Not Just Your Ordinary Network," Apple Business Profiles, *www.apple.com/business/profiles/linkedin*, accessed October 12, 2009; LinkedIn Web site, *www.linkedin.com*, accessed February 11, 2010.

UNIX

UNIX is a powerful OS originally developed by AT&T for minicomputers—the predecessors of servers that are larger than PCs and smaller than mainframes. UNIX can be used on many computer system types and platforms, including workstations, servers, and mainframe computers. UNIX also makes it much easier to move programs and data among computers or to connect mainframes and workstations to share resources. There are many variants of UNIX—including HP/UX from Hewlett-Packard, AIX from IBM, UNIX SystemV from UNIX Systems Lab, Solaris from Sun Microsystems, and SCO from Santa Cruz Operations.

The online marketplace eBay uses Sun Microsystems servers, software, storage, and services to run its operations. Sun's Solaris operating system manages eBay's systems, including database servers, Web servers, tape libraries, and identity management systems. The online auction company found that when they switched to Sun and Solaris, system performance increased by 20 percent. [32] The Idaho National Laboratory also uses Solaris to conduct research in their work to design more efficient and safe nuclear reactors. [33]

Red Hat Linux

Red Hat Software offers a Linux network OS that taps into the talents of tens of thousands of volunteer programmers who generate a steady stream of improvements for the Linux OS. The *Red Hat Linux* network OS is very efficient at serving Web pages and can manage a cluster of up to eight servers. Linux environments typically have fewer virus and security problems than other OSs. Distributions such as SuSE and Red Hat have proven Linux to be a very stable and efficient OS.

Mac OS X Server

The *Mac OS X Server* is the first modern server OS from Apple Computer and is based on the UNIX OS. The most recent version is OS X Server 10.6 Snow Leopard. It includes support for 64-bit processing, along with several server functions and features that allow the easy management of network and Internet services such as e-mail, Web site hosting, calendar management and sharing, wikis, and podcasting.

Enterprise Operating Systems

Mainframe computers, often referred to as "Big Iron," provide the computing and storage capacity to meet massive data-processing requirements and offer many users high performance and excellent system availability, strong security, and scalability. In addition, a wide range of application software has been developed to run in the mainframe environment, making it possible to purchase software to address almost any business problem. As a result, mainframe computers remain a popular computing platform for mission-critical business applications for many companies. Examples of mainframe OSs include z/OS from IBM, HP-UX from Hewlett-Packard, and Linux.

z/OS

The *z/OS* is IBM's first 64-bit enterprise OS. It supports IBM's z900 and z800 lines of mainframes that can come with up to sixteen 64-bit processors. (The z stands for zero downtime.) The OS provides several new capabilities to make it easier and less expensive for users to run large mainframe computers. The OS has improved workload management and advanced e-commerce security. The IBM zSeries mainframe, like previous generations of IBM mainframes, lets users subdivide a single computer into multiple smaller servers, each of which can run a different application. In recognition of the widespread popularity of a competing OS, z/OS allows partitions to run a version of the Linux OS. This means that a company can upgrade to a mainframe that runs the Linux OS.

Germany's largest health insurance company, AOK, recently replaced its core systems with two IBM mainframe servers running z/OS. The company chose z/OS based on its reputation for high reliability and performance. AOK also uses IBM Tivoli software to assist in automating tasks in the mainframe infrastructure and storage management system. AOK is legally responsible for storing records for 30 years for its more than 25 million policy holders. [34] That level of responsibility requires the highest levels of system reliability.

HP-UX and Linux

HP-UX is a robust UNIX-based OS from Hewlett-Packard designed to handle a variety of business tasks, including online transaction processing and Web applications. It supports Internet database and a variety of business applications on server and mainframe enterprise system. It can work with Java programs and Linux applications. HP-UX supports Hewlett-Packard's computers and those designed to run Intel's Itanium processors. *Red Hat Enterprise Linux* for IBM mainframe computers is another example of an enterprise operating system.

Operating Systems for Small Computers, Embedded Computers, and Special-Purpose Devices

New OSs and other software are changing the way we interact with smartphones, cell phones, digital cameras, TVs, and other appliances. These OSs are also called *embedded operating systems* because they are typically embedded within a device, such as an automobile, TV recorder, or other device. Embedded software is a multibillion dollar industry. These OSs are also called *embedded operating systems*, or just *embedded systems*, because they are typically embedded within a device. Embedded systems are designed to perform specialized tasks. For example, an automotive embedded system might be responsible for controlling fuel injection. A digital camera's embedded system supports taking and viewing photos and may include a limited set of editing tools. An embedded system controlling an MRI machine controls a powerful magnetic field to acquire 3D images of the body. A GPS device uses an embedded system to help people find their way around town. See Figure 2.12. Some of the more popular OSs for devices are described in the following section.

Cell Phone Embedded Systems and Operating Systems

Cell phones have traditionally used embedded systems to provide communication and limited personal information management services to users. Symbian is the world's most widely used cell phone embedded OS and has traditionally provided voice and text communication, an address book, and a few other basic applications. When RIM introduced the BlackBerry smartphone in 2002, the mobile phone's capabilities were vastly expanded. Since then, cell phone embedded systems have been transformed into full-fledged personal computer OSs such as the iPhone OS, Google Android, and Microsoft Windows Mobile. Even traditional embedded systems such as Palm OS (now WebOS) and Symbian have evolved into PC operating systems, with APIs and software development kits that allow developers to design hundreds of applications providing a myriad of mobile services.

Windows Embedded

Windows Embedded is a family of Microsoft OSs included with or embedded into small computer devices. Windows Embedded includes several versions that provide computing power for TV set-top boxes, automated industrial machines, media players, medical devices, digital cameras, PDAs, GPS receivers, ATMs, gaming devices, and business devices such as cash registers. Microsoft Auto provides a computing platform for automotive software such as Ford Sync. The Ford Sync system uses an in-dashboard display and wireless networking technologies to link automotive systems with cell phones and portable media players. See Figure 2.13.

Figure 2.13

Microsoft Auto and Ford Sync

The Ford Sync system, developed on the Microsoft Auto operating system, is a communications and entertainment system that allows you to use voice commands with portable devices such as cell phones and media players.

(Source: Sam VarnHagen/Ford Motor Co.)

Proprietary Linux-Based Systems

Because embedded systems are typically designed for a specific purpose in a specific device, they are usually proprietary, or custom-created and owned by the manufacturer. Sony's Wii, for example, uses a custom-designed OS based on the Linux kernel. Linux is a popular choice for embedded systems because it is free and highly configurable. In October of 2009, Nokia released the N900 smartphone—the first Linux-based smartphone.[35] Linux has been used in many embedded systems, including e-book readers, ATM machines, cell phones, networking devices, and media players. At least nine distributions of Linux are designed for embedded systems. Linux is a major competitor to Symbian in the cell phone market and Microsoft Embedded in most other markets.

UTILITY PROGRAMS

Utility programs help to perform maintenance or correct problems with a computer system. For example, some utility programs merge and sort sets of data, keep track of computer jobs being run, compress files of data before they are stored or transmitted over a network (thus saving space and time), and perform other important tasks. Some utility programs can help computer systems run better and longer without problems.

Utility programs can also help to secure and safeguard data. For example, the publishing and motion picture industries use digital rights management (DRM) technologies to prevent copyright-protected books and movies from being unlawfully copied. The files storing the intellectual property are encoded so that software running on e-book readers and media players recognizes and plays only legally obtained copies. DRM has been criticized for infringing on the freedom and rights of customers. Record companies have already moved away from DRM technologies in an effort to win the appreciation of their customers.

Although many PC utility programs come installed on computers, you can also purchase utility programs separately. Table 2.6 provides examples of some common types of utilities.

utility program
Program that helps to perform maintenance or correct problems with a computer system

Table 2.6

Examples of Utility Programs

Personal	Workgroup	Enterprise
Software to compress data so that it takes less hard disk space	Software that maintains an archive of changes made to a shared document	Software to archive contents of a database by copying data from disk to tape
Software that assists in determining which files to delete to free up disk space	Software that monitors group activity to determine levels of participation	Software that monitors network traffic and server loads
Antivirus and antispyware software for PCs	Software that reports unsuccessful user logon attempts	Software that reports the status of a particular computer job

APPLICATION SOFTWARE

Application software applies the power of a computer to give individuals, workgroups, and the entire enterprise the ability to solve problems and perform specific tasks. Application programs interact with systems software, and the systems software directs the computer hardware to perform the necessary tasks.

Applications help you perform common tasks, such as creating and formatting text documents, performing calculations, or managing information, though some applications are more specialized. Application software is used throughout the medical profession to save and prolong lives. For example, Swedish Medical Center in Seattle, Washington uses content management software from Oracle to access patient records when and where they are needed.[36]

The functions performed by application software are diverse and range from personal productivity to business analysis. For example, application software can help sales managers track sales of a new item in a test market. Software from IntelliVid monitors video feeds from store security cameras and notifies security when a shopper is behaving suspiciously. Most of the computerized business jobs and activities discussed in this book involve application software. We begin by investigating the types and functions of application software.

Types and Functions of Application Software

The key to unlocking the potential of any computer system is application software. A company can either develop a one-of-a-kind program for a specific application (called **proprietary software**) or purchase and use an existing software program (sometimes called **off-the-shelf software**). It is also possible to modify some off-the-shelf programs, giving a blend of off-the-shelf and customized approaches. The relative advantages and disadvantages of proprietary software and off-the-shelf software are summarized in Table 2.7.

proprietary software
One-of-a-kind software designed for a specific application and owned by the company, organization, or person that uses it.

off-the-shelf software
Software mass-produced by software vendors to address needs that are common across businesses, organizations, or individuals.

Table 2.7

A Comparison of Proprietary and Off-the-Shelf Software

Proprietary Software		Off-the-Shelf Software	
Advantages	**Disadvantages**	**Advantages**	**Disadvantages**
You can get exactly what you need in terms of features, reports, and so on.	It can take a long time and significant resources to develop required features.	The initial cost is lower because the software firm can spread the development costs over many customers.	An organization might have to pay for features that are not required and never used.
Being involved in the development offers control over the results.	In-house system development staff may become hard pressed to provide the required level of ongoing support and maintenance because of pressure to move on to other new projects.	The software is likely to meet the basic business needs—you can analyze existing features and the performance of the package before purchasing.	The software might lack important features, thus requiring future modification or customization. This can be very expensive because users must adopt future releases of the software as well.
You can modify features that you might need to counteract an initiative by competitors or to meet new supplier or customer demands. A merger with or acquisition of another firm also requires software changes to meet new business needs.	The features and performance of software that has yet to be developed present more potential risk.	The package is likely to be of high quality because many customer firms have tested the software and helped identify its bugs.	The software might not match current work processes and data standards.

Many companies use off-the-shelf software to support business processes. In 2009, Forrester Research reported that 80 percent of enterprises use Microsoft Office.[37] Key questions for selecting off-the-shelf software include the following: (1) Will the software run on the OS and hardware you have selected? (2) Does the software meet the essential business requirements that have been defined? (3) Is the software manufacturer financially solvent and reliable? and (4) Does the total cost of purchasing, installing, and maintaining the software compare favorably to the expected business benefits?

Some off-the-shelf programs can be modified, in effect blending the off-the-shelf and customized approaches. For example, El Camino Hospital in Mountain View, California customized Microsoft's e-health management system, Amalga, to track patients with the H1N1 flu and those that may have been exposed to it.[38]

Another approach to obtaining a customized software package is to use an application service provider. An **application service provider** (ASP) is a company that can provide the software, support, and computer hardware on which to run the software from the user's facilities over a network. An ASP can also simplify a complex corporate software package so that it is easier for the users to set up and manage. ASPs provide contract customization of off-the-shelf software, and they speed deployment of new applications while helping IS managers avoid implementation headaches, reducing the need for many skilled IS staff members and decreasing project start up expenses. Such an approach allows companies to devote more time and resources to more important tasks. For example, Rapid Advance, a leading cash advance service for small to medium-sized businesses (SMBs), uses Business Objects and Crystal Reports, applications served by SAP, to manage its business intelligence (BI). The system provides real-time access to sales information, business partner information, and critical corporate reports.[39]

Using an ASP involves some risks—sensitive information could be compromised in a number of ways, including unauthorized access by employees or computer hackers; the ASP might not be able to keep its computers and network up and running as consistently as necessary; or a disaster could disable the ASP's data center, temporarily putting an organization out of business. These are legitimate concerns that an ASP must address.

The high overhead of an ASP designing, running, managing, and supporting many customized applications for many businesses has led to a new form of software distribution known as software as a service. **Software as a Service (SaaS)** allows businesses to subscribe to Web-delivered business application software by paying a monthly service charge or a per use fee. Like ASP, SaaS providers maintain software on their own servers and provide access to it over the Internet. SaaS usually uses a Web browser-based user interface. SaaS can reduce expenses by sharing its running applications among many businesses. For example, Sears, JC Penney, and Wal-Mart might use customer relationship management software provided by a common SaaS provider. Providing one high-quality SaaS application to thousands of businesses is much more cost-effective than custom designing software for each business.

SaaS and new Web development technologies have led to a new paradigm in computing called cloud computing. Cloud computing refers to the use of computing resources, including software and data storage, on the Internet (the cloud) rather than on local computers. The emergence of powerful Web programming languages and techniques, such as AJAX, lets developers create Web-based software that rivals traditional installed software. Rather than installing, storing, and running software on your own computer, with cloud computing, you use the Web browser to access software stored and delivered from a Web server. Typically, the data generated by the software is also stored on the Web server. For example, Tableau software allows users to import database or spreadsheet data to create powerful visualizations that provide useful information.[40] Cloud computing provides the benefit of being able to easily collaborate with others by sharing documents on the Internet.

Starbucks used cloud computing services from Salesforce.com when it designed its online community at *www.mystarbucksidea.com*. The site allows Starbucks to converse with its customers to find out how they feel about Starbucks and its products. The customer interactions are stored in a CRM system at Saleforce.com and accessed by Starbucks managers and executives using Salesforce.com's online reporting tools. The cloud computing solution has

application service provider (ASP)
A company that provides software, support, and the computer hardware on which to run the software from the user's facilities over a network.

Software as a Service (SaaS)
A service that allows businesses to subscribe to Web-delivered business application software by paying a monthly service charge or a per-use fee.

recorded 77,000 customer suggestions and hundreds of thousands of comments and votes, resulting in 25 new Starbucks products and services.[41]

Personal Application Software

Hundreds of computer applications can help individuals at school, home, and work. The features of personal application software are summarized in Table 2.8. In addition to these general-purpose programs, there are literally thousands of other personal computer applications to perform specialized tasks: to help you do your taxes, get in shape, lose weight, get medical advice, write wills and other legal documents, make repairs to your computer, fix your car, write music, and edit your pictures and videos. This type of software, often called *user software ox personal productivity software,* includes the general-purpose tools and programs that support individual needs.

Table 2.8

Examples of Personal Application Software

Type of Software	Explanation	Example
Word processing	Create, edit, and print text documents	Microsoft Word Corel WordPerfect Google Docs Apple Pages Sun Writer
Spreadsheet	Provide a wide range of built-in functions for statistical, financial, logical, database, graphics, and date and time calculations	Microsoft Excel IBM Lotus 1-2-3 Google Spreadsheet Apple Numbers Sun Calc
Database	Store, manipulate, and retrieve data	Microsoft Access IBM Lotus Approach Borland dBASE Sun Base
Graphics	Develop graphs, illustrations, and drawings	Adobe Illustrator Adobe FreeHand
Project management	Plan, schedule, allocate, and control people and resources (money, time, and technology) needed to complete a project according to schedule	Microsoft Project Symantec On Target Scitor Project Scheduler Symantec Time Line
Financial management	Provide income and expense tracking and reporting to monitor and plan budgets (some programs have investment portfolio management features)	Intuit Quicken
Desktop publishing (DTP)	Use with personal computers and high-resolution printers to create high-quality printed output, including text and graphics; various styles of pages can be laid out; art and text files from other programs can also be integrated into published pages	Quark XPress Microsoft Publisher Adobe PageMaker Corel Ventura Publisher Apple Pages

Word Processing

If you write reports, letters, or term papers, word-processing applications can be indispensable. The majority of personal computers in use today have word-processing applications installed. Such applications can be used to create, edit, and print documents. Most come with a vast array of features, including those for checking spelling, creating tables, inserting formulas, creating graphics, and much more. This book (and most like it) was entered into a word-processing application using a personal computer.

A team of people can use a word-processing program to collaborate on a project. The authors and editors who developed this book, for example, used the Track Changes and

Reviewing features of Microsoft Word to track and make changes to chapter files. With these features, you can add comments or make revisions to a document that a coworker can review and either accept or reject.

Spreadsheet Analysis

People use spreadsheets to prepare budgets, forecast profits, analyze insurance programs, summarize income tax data, and analyze investments. Whenever numbers and calculations are involved, spreadsheets should be considered. Features of spreadsheets include graphics, limited database capabilities, statistical analysis, built-in business functions, and much more. See Figure 2.14. The business functions include calculation of depreciation, present value, internal rate of return, and the monthly payment on a loan, to name a few. Optimization is another powerful feature of many spreadsheet programs. *Optimization* allows the spreadsheet to maximize or minimize a quantity subject to certain constraints. For example, a small furniture manufacturer that produces chairs and tables might want to maximize its profits. The constraints could be a limited supply of lumber, a limited number of workers that can assemble the chairs and tables, or a limited amount of various hardware fasteners that might be required. Using an optimization feature, such as Solver in Microsoft Excel, the spreadsheet can determine what number of chairs and tables to produce with labor and material constraints to maximize profits.

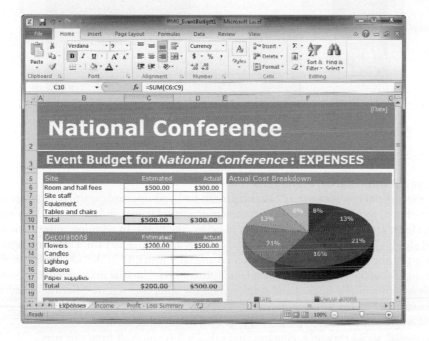

Figure 2.14

Spreadsheet Program

Spreadsheet programs such as Microsoft Excel should be considered when calculations are required.

Database Applications

Database applications are ideal for storing, manipulating, and retrieving data. These applications are particularly useful when you need to manipulate a large amount of data and produce reports and documents. Database manipulations include merging, editing, and sorting data. The uses of a database application are varied. You can keep track of a CD collection, the items in your apartment, tax records, and expenses. A student club can use a database to store names, addresses, phone numbers, and dues paid. In business, a database application can help process sales orders, control inventory, order new supplies, send letters to customers, and pay employees. Database management systems can be used to track orders, products, and customers; analyze weather data to make forecasts for the next several days; and summarize medical research results. A database can also be a front end to another application. For example, you can use a database application to enter and store income tax information, and then export the stored results to other applications, such as a spreadsheet or tax-preparation application.

Graphics Programs

With today's graphics programs, it is easy to develop attractive graphs, illustrations, and drawings. Graphics programs can be used to develop advertising brochures, announcements, and full-color presentations. If you are asked to make a presentation at school or work, you can use a graphics program to develop and display slides while you are making your talk. A graphics program can be used to help you make a presentation, a drawing, or an illustration. See Figure 2.15. Most presentation graphics programs come with many pieces of *clip art*, such as drawings and photos of people meeting, medical equipment, telecommunications equipment, entertainment, and much more.

Figure 2.15

Presentation Graphics Program

Presentation graphics programs such as Microsoft PowerPoint can help you make a presentation at school or work.

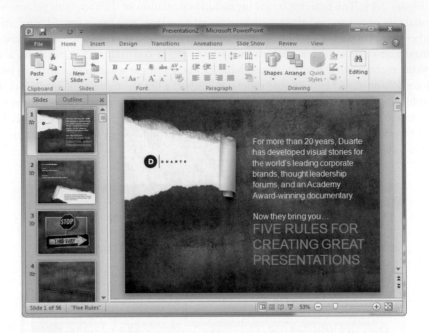

Personal Information Managers

Personal information managers (PIMs) help individuals, groups, and organizations store useful information, such as a list of tasks to complete or a list of names and addresses. They usually provide an appointment calendar and a place to take notes. In addition, information in a PIM can be linked. For example, you can link an appointment with a sales manager that appears in the calendar with information on the sales manager in the address book. When you click the appointment in the calendar, information on the sales manager from the address book is automatically opened and displayed on the computer screen. Microsoft Outlook is an example of a PIM software package.

iGoogle and other Web portals support PIM by allowing users to access calendars, to-do lists, e-mail, social networks, contacts, and other information all from one page.

Some PIMs allow you to schedule and coordinate group meetings. If a computer or handheld device is connected to a network, you can upload the PIM data and coordinate it with the calendar and schedule of others using the same PIM software on the network. You can also use some PIMs to coordinate e-mails to invite others to meetings. As users receive their invitations, they click a link or button to be automatically added to the guest list.

Software Suites and Integrated Software Packages

software suite
A collection of single programs packaged together in a bundle.

A **software suite** is a collection of single programs packaged together in a bundle. Software suites can include a word processor, spreadsheet, database management system, graphics program, communications tool, organizer, and more. Some suites support the development of Web pages, note taking, and speech recognition, whereby applications in the suite can accept voice commands and record dictation. Software suites offer many advantages. The software programs have been designed to work similarly, so after you learn the basics for one application, the other applications are easy to learn and use. Buying software in a bundled suite is cost-effective; the programs usually sell for a fraction of what they would cost individually.

Microsoft Office, Lotus Symphony, Corel WordPerfect Office, Lotus SmartSuite, Oracle StarOffice, Apple iWork, and Google Apps are examples of popular general-purpose software suites for personal computer users. Each of these software suites includes a spreadsheet program, word processor, database program, and graphics presentation software. All can exchange documents, data, and diagrams. See Table 2.9. In other words, you can create a spreadsheet and then cut and paste that spreadsheet into a document created using the word-processing application. Forrester Research reports that 80 percent of enterprise customers use some version of Microsoft Office. The latest version is Office 2010.

Personal Productivity Function	Microsoft Office	Lotus Symphony	Corel WordPerfect Office	Oracle StarOffice	Apple iWork	Google
Word Processing	Word	Documents	WordPerfect	Writer	Pages	Docs
Spreadsheet	Excel	Spreadsheets	Quattro Pro	Calc	Numbers	Spreadsheet
Presentation Graphics	PowerPoint	Presentations	Presentations	Impress	Keynote	Presentation
Database	Access		Paradox	Base		

Table 2.9

Major Components of Leading Software Suites

In addition to suites, some companies produce *integrated application packages* that contain several programs. For example, Microsoft Works is one program that contains basic word processing, spreadsheet, database, address book, calendar, and other applications. Although not as powerful as stand-alone software included in software suites, integrated software packages offer a range of capabilities for less money. Some integrated packages cost about $100.

Zoho, Google, and Thinkfree offer free Web-based productivity software suites that require no installation on the PC. Adobe has developed Acrobat.com, which features an impressive online suite including Buzzword for word processing, Tables for spreadsheet and database applications, and Presentations for presentation graphics. See Figure 2.16. Documents created with the software can be stored on the Web server. Currently these online applications are not as powerful and robust as installed software such as Microsoft Office. However, as the technology becomes more powerful and network connection speeds increase, users will probably need to install less software on their PCs and turn instead to using software online. After observing this trend, Microsoft responded with an online version of some of its popular Office applications. The online versions of Word, Excel, PowerPoint, and OneNote are tightly integrated with their desktop counterparts for easy sharing of documents among computers and collaborators.[42]

Figure 2.16

Web Suites

Adobe Acrobat.com provides a suite of online applications, including Buzzword, with cutting-edge interface designs.

Accounts payable	Invoicing
Accounts receivable	Manufacturing control
Airline industry operations	Order entry
Automatic teller systems	Payroll
Cash-flow analysis	Receiving
Check processing	Restaurant management
Credit and charge card administration	Retail operations
Distribution control	Sales ordering
Fixed asset accounting	Savings and time deposits
General ledger	Shipping
Human resource management	Stock and bond management
Inventory control	Tax planning and preparation

Organizations can no longer respond to market changes using nonintegrated information systems based on overnight processing of yesterday's business transactions, conflicting data models, and obsolete technology. Wal-Mart and many other companies have sophisticated information systems to speed processing and coordinate communications between stores and their main offices. Many corporations are turning to enterprise resource planning (ERP) software, a set of integrated programs that manage a company's vital business operations for an entire multisite, global organization. Thus, an ERP system must be able to support many legal entities, languages, and currencies. Although the scope can vary from vendor to vendor, most ERP systems provide integrated software to support manufacturing and finance. In addition to these core business processes, some ERP systems might support business functions such as human resources, sales, and distribution. The primary benefits of implementing ERP software include eliminating inefficient systems, easing adoption of improved work processes, improving access to data for operational decision making, standardizing technology vendors and equipment, and enabling supply chain management. In contrast, small businesses usually do not need complex enterprise application software. They rely on software such as Intuit QuickBooks and Microsoft Office Small Business Accounting for accounting and recording keeping.

Application Software for Information, Decision Support, and Specialized Purposes

Specialized application software for information, decision support, and other purposes is available in every industry. For example, many schools and colleges use Blackboard or other learning management software to organize class materials and grades.[44] Genetic researchers, as another example, are using software to visualize and analyze the human genome. Music executives use decision support software to help pick the next hit song. Sophisticated decision support software is also being used to increase the cure rate for cancer by analyzing about 100 scans of a cancerous tumor to create a 3D view of the tumor. Software can then consider thousands of angles and doses of radiation to determine the best program of radiation therapy. The software analysis takes only minutes, but the results can save years or decades of life for the patient. As you will see in future chapters, information, decision support, and specialized systems are used in businesses of all sizes and types to increase profits or reduce costs. But how are all these systems actually developed and built? The answer is through the use of programming languages, discussed next.

PROGRAMMING LANGUAGES

Both OSs and application software are written in coding schemes called *programming languages*. The primary function of a programming language is to provide instructions to the computer system so that it can perform a processing activity. IS professionals work with

programming languages, which are sets of keywords, symbols, and rules for constructing statements that people can use to communicate instructions to a computer. Programming involves translating what a user wants to accomplish into instructions that the computer can understand and execute. The desire to use the power of information processing efficiently in problem solving has pushed the development of literally thousands of programming languages, but only a few dozen are commonly used today. A brief summary of the various programming language generations is provided in Table 2.12.

programming languages
Sets of keywords, symbols, and rules for constructing statements that people can use to communicate instructions to a computer.

Table 2.12

The Evolution of Programming Languages

Generation	Language	Approximate Development Date	Sample Statement or Action
First	Machine language	1940s	00010101
Second	Assembly language	1950s	MVC
Third	High-level language	1960s	READ SALES
Fourth	Query and database languages	1970s	PRINT EMPLOYEE NUMBER IF GROSS PAY>1000
Beyond Fourth	Natural and intelligent languages	1980s	IF gross pay is greater than 40, THEN pay the employee overtime pay

Although many programming languages are used to write new business applications, more lines of code are written in COBOL in existing business applications than any other programming language. Today, programmers often use visual and object-oriented languages. In the future, they will likely be using artificial intelligence languages to a greater extent. In general, these languages are easier for nonprogrammers to use compared with older generation languages.

SOFTWARE ISSUES AND TRENDS

Because software is such an important part of today's computer systems, issues such as software bugs, licensing, and global software support have received increased attention.

Software Bugs

A software bug is a defect in a computer program that keeps it from performing as it is designed to perform. Some software bugs are obvious and cause the program to terminate unexpectedly. Other bugs are more subtle and allow errors to creep into your work. Computer and software vendors say that as long as people design and program hardware and software, bugs are inevitable. The following list summarizes tips for reducing the impact of software bugs.

- Register all software so that you receive bug alerts, fixes, and patches.
- Check the manual or read-me files for solutions to known problems.
- Access the support area of the manufacturer's Web site for patches.
- Install the latest software updates.
- Before reporting a bug, make sure that you can re-create the circumstances under which it occurs.
- After you re-create the bug, call the manufacturer's tech support line.
- Consider waiting before buying the latest release of software to give the vendor a chance to discover and remove bugs. Many schools and businesses don't purchase software until the first major revision with patches is released.

Copyrights and Licenses

Most software products are protected by law using copyright or licensing provisions. Those provisions can vary, however. In some cases, you are given unlimited use of software on one or two computers. This is typical with many applications developed for personal computers. In other cases, you pay for your usage—if you use the software more, you pay more. This approach is becoming popular with software placed on networks or larger computers. Most of these protections prevent you from copying software and giving it to others without restrictions. Some software now requires that you *register* or *activate* it before it can be fully used. Registration and activation sometimes put software on your hard disk that monitors activities and changes to your computer system.

Software Upgrades

Software companies revise their programs and sell new versions periodically. In some cases, the revised software offers new and valuable enhancements. In other cases, the software uses complex program code that offers little in terms of additional capabilities. In addition, revised software can contain bugs or errors. When software companies stop supporting older software versions or releases, some customers feel forced to upgrade to the newer software. Deciding whether to purchase the newest software can be a problem for corporations and people with a large investment in software. Should the newest version be purchased when it is released? Some users do not always get the most current software upgrades or versions, unless it includes significant improvements or capabilities. Instead, they might upgrade to newer software only when it offers vital new features. Software upgrades usually cost much less than the original purchase price.

Global Software Support

Large, global companies have little trouble persuading vendors to sell them software licenses for even the most far-flung outposts of their company. But can those same vendors provide adequate support for their software customers in all locations? Supporting local operations is one of the biggest challenges IS teams face when putting together standardized, company-wide systems. In slower technology growth markets, such as Eastern Europe and Latin America, there may be no official vendor presence at all. Instead, large vendors such as Sybase, IBM, and Hewlett-Packard typically contract out support for their software to local providers.

One approach that has been gaining acceptance in North America is to outsource global support to one or more third-party distributors. The software-user company may still negotiate its license with the software vendor directly, but it then hands over the global support contract to a third-party supplier. The supplier acts as a middleman between software vendor and user, often providing distribution, support, and invoicing. American Home Products Corporation handles global support for both Novell NetWare and Microsoft Office applications this way—throughout the 145 countries in which it operates. American Home Products negotiated the agreements directly with the vendors for both purchasing and maintenance, but fulfillment of the agreement is handled exclusively by Philadelphia-based Softsmart, an international supplier of software and services.

In today's computer systems, software is an increasingly critical component. Whatever approach individuals and organizations take to acquire software, it is important for everyone to be aware of the current trends in the industry. Informed users are wise consumers.

SUMMARY

Principle:

Computer hardware must be carefully selected to meet the evolving needs of the organization and its supporting information systems.

Hardware refers to the physical components of a computer that perform the input, processing, storage, and output activities of the computer. Processing is performed by an interplay between the central processing unit (CPU) and memory. Primary storage, or memory, provides working storage for program instructions and data to be processed and provides them to the CPU. Together, a CPU and memory process data and execute instructions.

Processing that uses several processing units is called multiprocessing. A multicore processor combines two or more independent processors into a single computer so that they can share the workload and boost processing capacity. Parallel processing involves linking several processors to work together to solve complex problems. Grid computing is the use of a collection of computers, often owned by multiple individuals or organizations, to work in a coordinated manner to solve a common problem.

Computer systems can store large amounts of data and instructions in secondary storage, which is less volatile and has greater capacity than memory. Storage media can be either sequential access or direct access. Common forms of secondary storage include magnetic tape, magnetic disk, video tape, optical disc storage, digital video disk, and solid state storage devices. Redundant array of independent/inexpensive disks (RAID) is a method of storing data that allows the system to more easily recover data in the event of a hardware failure. Storage area network (SAN) uses computer servers, distributed storage devices, and networks to provide fast and efficient storage.

Input and output devices allow users to provide data and instructions to the computer for processing and allow subsequent storage and output. These devices are part of a user interface through which humans interact with computer systems. Input and output devices vary widely, but they share common characteristics of speed and functionality.

A keyboard and computer mouse are the most common devices used for entry of data. Speech recognition technology enables a computer to interpret human speech as an alternative means of providing data and instructions. Digital cameras record and store images or video in digital form. Handwriting recognition software can convert handwriting on the screen into text. Radio-frequency identification (RFID) technology employs a microchip, called a tag, to transmit data that is read by an RFID reader. The data transmitted could include facts such as item identification number, location information, or other details about the item tagged.

Output devices provide information in different forms, from hard copy to sound to digital format. Display monitors are standard output devices; monitor quality is determined by size, number of colors that can be displayed, and resolution. Other output devices include printers, plotters, and e-books.

Portable single-user computers include handheld computers, laptops, notebook computers, netbook computers, and tablet computers. Nonportable single-user computers include thin clients, desktop computers, nettop computers, and workstations. Multiple-user computer systems include servers, mainframe computers, and supercomputers.

Grid computing is the use of a collection of computers, often owned by many people or organizations, to work in a coordinated manner to solve a common problem. Grid computing is one low-cost approach to parallel processing.

Principle:

The computer hardware industry and users are implementing green computing designs and products.

Green computing is concerned with the efficient and environmentally responsible design, manufacture, operation, and disposal of IS-related products.

Business organizations recognize that going green is in their best interests in terms of public relations, safety of employees, and the community at large. They also recognize that green computing presents an opportunity to substantially reduce total costs over the life cycle of their IS equipment.

Green computing has three goals: reduce the use of hazardous material, enable companies to lower their power-related costs, and enable the safe disposal or recycling of IS products.

Principle:

Systems and application software are critical in helping individuals and organizations achieve their goals.

Software consists of programs that control the workings of the computer hardware. The two main categories of software are systems software and application software. Systems software is a collection of programs that interacts between hardware and application software. Application software can be proprietary or off the shelf and enables people to solve problems and perform specific tasks.

An operating system (OS) is a set of computer programs that controls the computer hardware to support users' computing needs. An OS converts an instruction from an application into a set of instructions needed by the hardware.

This intermediary role allows hardware independence. An OS also manages memory, which involves controlling storage access and use by converting logical requests into physical locations and by placing data in the best storage space, perhaps virtual memory.

An OS manages tasks to allocate computer resources through multitasking and time-sharing. With multitasking, users can run more than one application at a time. Timesharing allows more than one person to use a computer system at the same time.

The ability of a computer to handle an increasing number of concurrent users smoothly is called scalability, a feature critical for systems expected to handle a large number of users.

An OS also provides a user interface, which allows users to access and command the computer. A command-based user interface requires text commands to send instructions; a graphical user interface (GUI), such as Windows, uses icons and menus.

Software applications use the OS by requesting services through a defined application program interface (API). Programmers can use APIs to create application software without having to understand the inner workings of the OS. APIs also provide a degree of hardware independence so that the underlying hardware can change without necessarily requiring a rewrite of the software applications.

Over the years, several popular OSs have been developed. These include several proprietary OSs used primarily on mainframes. Windows Vista, Windows XP, and Windows 7 are the most recent Microsoft Windows operating systems. Apple computers use proprietary OSs such as the Mac OS and Mac OS X. UNIX is a powerful OS that can be used on many computer system types and platforms, from personal computers to mainframe systems. UNIX makes it easy to move programs and data among computers or to connect mainframes and personal computers to share resources. Linux is the kernel of an OS whose source code is freely available to everyone. Several variations of Linux are available, with sets of capabilities and applications to form a complete OS, for example, Red Hat Linux, Caldera Open Linux, and Google Chrome. z/OS and HP-UX are OSs for mainframe computers. Some OSs have been developed to support consumer appliances such as Palm OS, Windows CE.Net, Windows XP Embedded, Pocket PC, and variations of Linux.

Symbian is the world's most widely used cell phone embedded OS and has traditionally provided voice and text communication, an address book, and a few other basic applications. When RIM introduced the BlackBerry smartphone in 2002, the mobile phone's capabilities were vastly expanded. Since then, cell phone embedded systems have transformed into full-fledged personal computer OSs such as the iPhone OS, Google Android, and Microsoft Windows Mobile.

Principle:

Organizations should not develop proprietary application software unless doing so will meet a compelling business need that can provide a competitive advantage.

Application software applies the power of the computer to solve problems and perform specific tasks. One useful way of classifying the many potential uses of information systems is to identify the scope of problems and opportunities addressed by a particular organization or its sphere of influence. For most companies, the spheres of influence are personal, workgroup, and enterprise.

User software, or personal productivity software, includes general-purpose programs that enable users to improve their personal effectiveness, increasing the quality and amount of work that can be done. Software that helps groups work together is often called workgroup application software and includes group scheduling software, electronic mail, and other software that enables people to share ideas. Enterprise software that benefits the entire organization can also be developed or purchased. Many organizations are turning to enterprise resource planning software, a set of integrated programs that manage a company's vital business operations for an entire multisite, global organization.

Three approaches to developing application software are to build proprietary application software, buy existing programs off the shelf, or use a combination of customized and off-the-shelf application software. Building proprietary software (in-house or on contract) has the following advantages: the organization will get software that more closely matches its needs; by being involved with the development, the organization has further control over the results; and the organization has more flexibility in making changes. The disadvantages include the following: it is likely to take longer and cost more to develop, the in-house staff will be hard pressed to provide ongoing support and maintenance, and there is a greater risk that the software features will not work as expected or that other performance problems will occur.

Purchasing off-the-shelf software has many advantages. The initial cost is lower, there is a lower risk that the software will fail to work as expected, and the software is likely to be of higher quality than proprietary software. Some disadvantages are that the organization might pay for features it does not need, the software might lack important features requiring expensive customization, and the system might require process reengineering.

Some organizations have taken a third approach—customizing software packages. This approach usually involves a mixture of the preceding advantages and disadvantages and must be carefully managed.

An application service provider (ASP) is a company that can provide the software, support, and computer hardware on which to run the software from the user's facilities over a network. ASPs provide contract customization of off-the-shelf software, and they speed deployment of new applications while helping IS managers avoid implementation headaches. Use of ASPs reduces the need for many skilled IS staff members and also lowers a project's start-up expenses.

Software as a service, or SaaS, allows business to subscribe to Web-delivered business application software by paying a monthly service charge or a per use fee.

Cloud computing refers to the use of computing resources, including software and data storage, on the Internet (the cloud) rather than on local computers.

Although hundreds of computer applications can help people at school, home, and work, the primary applications are word processing, spreadsheet analysis, database, graphics, and online services. A software suite, such as SmartSuite, WordPerfect, StarOffice, or Office, offers a collection of powerful programs.

Principle:

Organizations should choose a programming language whose functional characteristics are appropriate for the task at hand, considering the skills and experience of the programming staff.

All software programs are written in coding schemes called programming languages, which provide instructions to a computer to perform some processing activity. The several classes of programming languages include machine, assembly, high-level, query and database, and natural and intelligent languages.

Programming languages have changed since their initial development in the early 1950s. In the first generation, computers were programmed in machine language, and the second generation of languages used assembly languages. The third generation consists of many high-level programming languages that use English-like statements and commands. Fourth-generation languages include database and query languages such as SQL.

Users frequently use fourth generation and higher-level programming languages to develop their own simple programs.

Principle:

The software industry continues to undergo constant change; users need to be aware of recent trends and issues to be effective in their business and personal life.

Software bugs, software licensing and copyrighting, software upgrades, and global software support are all important software issues and trends.

A software bug is a defect in a computer program that keeps it from performing in the manner intended. Software bugs are common, even in key pieces of business software.

Software upgrades are an important source of increased revenue for software manufacturers and can provide useful new functionality and improved quality for software users.

Global software support is an important consideration for large, global companies putting together standardized, company-wide systems. A common solution is outsourcing global support to one or more third-party software distributors.

CHAPTER 2: SELF-ASSESSMENT TEST

Computer hardware must be carefully selected to meet the evolving needs of the organization and its supporting information systems.

1. A multicore processor combines two or more independent _____ into a single computer so that they can share the workload and boost processing capacity.

2. Executing an instruction by the CPU involves two phases: the instruction phase and the ____ phase.

3. Which of the following components performs mathematical calculations and makes logical comparisons?
   a. control unit
   b. register
   c. ALU
   d. main memory

The computer hardware industry and users are implementing green computing designs and products.

4. Green computing is concerned with the efficient and environmentally responsible design, manufacture, operation, and _____ of IS-related products.

Systems and application software are critical in helping individuals and organizations achieve their goals.

5. Systems software is a collection of programs that interacts between hardware and _____.

6. Application software can be _____ or off the shelf and enables people to solve problems and perform specific tasks.

Organizations should not develop proprietary application software unless doing so will meet a compelling business need that can provide a competitive advantage.

7. Software that enables users to improve their personal effectiveness, increasing the amount of work they can do and its quality, is called ___ .
   a. personal productivity software
   b. operating system software
   c. utility software
   d. graphics software

8. Optimization capabilities can be found in which type of application software?
   a. spreadsheets
   b. word-processing programs
   c. database programs
   d. graphics software

9. Programmers can use _____ to create application software without having to understand the inner workings of the OS.

Organizations should choose a programming language whose functional characteristics are appropriate for the task at hand, considering the skills and experience of the programming staff.

10. End users will never be able to master the complexities of programming and create their own applications. True or False?

The software industry continues to undergo constant change; users need to be aware of recent trends and issues to be effective in their business and personal life.

11. Outsourcing global support to one or more third-party software distributors is a common solution to global software support. True or False?

CHAPTER 2: SELF-ASSESSMENT TEST ANSWERS

(1) processors (2) execution (3) c (4) disposal (5) application software (6) proprietary (7) a (8) a (8) c (9) APIs (10) False (11) True

REVIEW QUESTIONS

1. When determining the appropriate hardware components of a new information system, what role must the user of the system play?
2. Identify two basic characteristics of RAM and ROM.
3. What is RFID technology? How does it work?
4. Identify the three components of the CPU and explain the role of each.
5. What is solid state storage technology? What advantages does it offer?
6. Identify and briefly describe the various classes of non-portable single-user computers.
7. Give three examples of recent operating systems.

8. What is Software as a Service (SaaS)? What advantages does it provide for meeting an organization's software needs?
9. What are the two basic types of software? Briefly describe the role of each.
10. What is cloud computing? What are the pros and cons of cloud computing?
11. What is an application service provider? What issues arise in considering the use of one?
12. What does the acronym API stand for? What is the role of an API?
13. Describe the term *enterprise resource planning (ERP)* system. What functions does such a system perform?

DISCUSSION QUESTIONS

1. Briefly discuss the advantages and disadvantages of frequent software upgrades from the perspective of the user of that software. How about from the perspective of the software manufacturer?
2. What would be the advantages for a university computer lab to install thin clients rather than standard desktop personal computers? Can you identify any disadvantages?
3. Which would you rather have—a handheld computer or a tablet computer? Why?

4. If cost were not an issue, describe the characteristics of your ideal computer. What would you use it for? Which operating system would you want it to run?
5. Identify the three spheres of influence and briefly discuss the software needs of each.
6. Identify the two fundamental sources for obtaining application software. Discuss the advantages and disadvantages of each source.
7. Define *Software as a Service*. Discuss some of the pros and cons of using software as a service.

8. In what ways is an operating system for a mainframe computer different from the operating system for a laptop computer? In what ways are they similar?

9. Discuss potential issues that can arise if an organization is not careful in selecting a reputable service organization to recycle or dispose of its IS equipment.

10. Briefly explain the difference between grid computing and cloud computing.

PROBLEM-SOLVING EXERCISES

1. Develop a spreadsheet that compares the features, initial purchase price, and ongoing operating costs for three laser printers. Now do the same for three inkjet printers. Write a brief memo on which printer you would choose and why. Cut and paste the spreadsheet into a document.

2. Use word-processing software to document what your needs are as a computer user and your justification for selecting either a desktop or laptop computer. Find a Web site that allows you to order and customize a computer and select those options that meet your needs in a cost-effective manner. Assume that you have a budget of $1,000. Enter the computer specifications into an Excel spreadsheet that you cut and paste into the document defining your needs. E-mail the document to your instructor.

3. Use a database program to enter five software products you are likely to use at work. List the name, vendor or manufacturer, cost, and features in the columns of a database table. Use a word processor to write a report on the software. Copy the database table into the word-processing program.

TEAM ACTIVITIES

1. With two or three of your classmates, visit a computer retail store and identify the most popular netbook computers. Interview members of the sales staff to find out why they think this particular laptop is popular.

2. With one or two of your classmates, visit a retail store that employs Radio Frequency Identification chips to track inventory. Interview an employee involved in inventory control, and document the advantages and disadvantages they see in this technology.

3. Divide your team into two groups. The first group should prepare a report using a word-processing program. Make sure to include a large number of spelling, grammatical, and similar errors in the document. The second group should use the word-processing program's features to locate and eliminate the errors. The entire team should write a report on the advantages and limitations of the spelling and grammar checking features of the word processing program you used. What additional features would you like to see in future word-processing programs?

WEB EXERCISES

1. Use the Web to research four productivity software suites from various vendors (see *http://en.wikipedia.org/wiki/Office_Suite*). Create a table in a word-processing document to show what applications are provided by the competing suites. Write a few paragraphs on which suite you think best matches your needs and why.

2. Do research on the Web to identify the current state of development and production of advanced technology secondary storage devices. What are some of the most promising devices? What issues are associated with mass producing these new devices? Write a brief report summarizing your findings.

3. Do research on the Web about application software that is used in an industry that is of interest to you. Write a brief report describing how the application software can be used to increase profits or reduce costs.

CAREER EXERCISES

1. Imagine that you are going to buy a single handheld device to improve your communication and organizational abilities. What tasks do you need it to perform? What features would you look for in this device? Visit a computer store or a consumer electronics store and see whether you can purchase such a device for less than $400.

2. Think of your ideal job. Describe five application software packages that could help you advance in your career. If the software package doesn't exist, describe the kinds of software packages that could help you in your career.

CASE STUDIES

Case One
Union Pacific Retires Its Big Iron

Union Pacific (UP) is the largest freight railroad franchise in North America. It has evolved due to multiple mergers among the six railway systems that laid the first tracks across the U.S. in the mid- to late-nineteenth century: the Western Pacific, the Southern Pacific, the Missouri Pacific, the Denver & Rio Grande Western, the Missouri-Kansas-Texas, and the Chicago & North Western Railroads. Although the UP has deep roots in American history, it was also quick to adopt new computing technologies when they became available in the 1960s.

In 1964, IBM introduced massive mainframe computers, referred to as "Big Iron," to corporate America, and UP was among the first to purchase one. The railroad used the mainframe to roll out one of the earliest transaction processing systems. UP was also the first railroad to develop a computerized car-scheduling system. Because it could efficiently schedule railroad car pick-ups and deliveries, the scheduling system gave UP a competitive advantage. UP customers with numerous cars to deliver to multiple locations could accurately inform their customers of delivery dates.

UP's state-of-the-art computing power spurred the company to grow to 8,400 locomotives traveling more than 32,000 miles of tracks to service 23 states across the western two-thirds of the country. The railroad links every major West Coast and Gulf Coast port and provides freight delivery to major gateway cities such as St. Louis and Chicago. The railway transports chemicals, coal, food, grain, metal, forestry products, automobiles, and auto parts. Companies such as APL Global Container Transport and General Motors use UP railways to transport products across the U.S., connecting with other railways in Mexico and Canada. UP also runs a passenger line in the Chicago area.

Over time, UP's mainframe system grew in value as it provided additional services. It was used as a centralized computer-aided dispatching system for the entire UP railway network. Locomotives were eventually fitted with GPS systems that reported their locations to the mainframe for more accurate scheduling. The programs that control UP's Big Iron information systems on the mainframe have grown to 11 million lines of assembly-language computer code. UP has certainly received its money's worth from its 1964 investment that still provides valuable services to the company today.

Recently, however, UP decided to move its information systems from the 40-year-old mainframe to a distributed system made up of thousands of blade servers. One reason for the change is the difficulty UP has experienced finding computer programmers that can work with the 11 million lines of antiquated assembler code that runs the mainframe. Moving to blade servers running the Linux operating system and reengineered code developed in today's popular programming languages will make it easy for the company to recruit new computer science graduates to work on the system.

Another reason for the change is that UP wants to offer its customers more detailed information about the movement of their shipments. Rather than just tracking cars, UP wants to track the palettes of crates within the cars. The new service would require a major reengineering of the code and significantly more computing power. The company decided the time is right to move to new technologies that can support new information services.

Such a massive change in fundamental information systems requires considerable planning so that service can remain uninterrupted. UP began its planning in 2004 and decided to develop the new blade system in parallel to the existing mainframe system, gradually moving operations from one to the other over the length of nearly a decade. The company could not find existing software that met its needs, so it is developing the systems that will run on the new servers. UP plans to complete the transition in 2014, when the final services will be transferred to the blade-based system.

UP is one of many corporations retiring Big Iron. IBM has recently suffered a 40 percent drop in mainframe revenues as businesses move to distributed systems. UP is replacing its

Big Iron with blade systems from Dell and HP at a total cost of more than $150 million. The company is designing its system to grow, or scale, with the business. The hope is that the system will serve the company for as many decades as the original system did.

Discussion Questions

1. When UP first acquired Big Iron, what services did the system provide that gave UP a competitive advantage?
2. Why does UP feel that now is the time to replace its Big Iron system?

Critical Thinking Questions

1. What benefits does a distributed system running on many blade computers offer over a system running on a single mainframe computer?
2. What should UP consider as it implements its new system to ensure that the new system can service the company for decades to come?

SOURCES: Wailgum, Thomas, "Now Departing: Union Pacific's 40-Year-Old Mainframe," *Computerworld*, April 12, 2009; Union Pacific Company Overview, *www.uprr.com/aboutup/corporate_info/uprrover.shtml*, accessed January 2, 2010; Hottman, Thomas, "Union Pacific Railroad replaces mainframe with SOA," itWorldCanada, *www.itworldcanada.com*, December 21, 2009; Burt, Jeffrey, "Union Pacific Railroad Moving from Mainframes to Blades," *eWeek.com*, August 17, 2009.

Case Two

Office Depot Gets SaaSy

Like most businesses, Office Depot has been searching for ways to boost its market share and increase revenue in times when money is tight. Glenn Trommer, director of e-commerce and implementation services at Office Depot, and his team work closely with the sales department looking for ways that technology can assist with sales.

One area identified as needing improvement was in Office Depot's ordering system for its business customers. Some of the transaction requests from business customers were not compatible with Office Depot's ordering software. In other words, customers could not specify their needs using the existing system. Office Depot's competitors, however, did not suffer from the same problem and were thus able to fill orders and satisfy customers that Office Depot could not.

Office Depot needed software that could better integrate the order-placing systems used by its customers with its own order-processing system. Glenn Trommer and his team had three options: they could design proprietary software themselves, they could purchase software off the shelf and customize it to their needs, or they could hire the services of a company that specializes in integration software. Because of its financial restraints, Glenn decided that investing in software as a service (SaaS) for its integration needs would allow the business to get the best return on its investment in the shortest amount of time. Glenn, his team, and Office Depot sales managers ultimately chose Hubspan, which provides a

cloud-based integration platform called Integration-as-a-service.

Hubspan's system was placed between Office Depot's and its business customers to interpret customer needs into a format that Office Depot could process. Glenn Trommer and his team worked with Hubspan to educate the service provider on Office Depot's selling approach, sales cycle, and data formats. In a short time, the new system was in place without any inconvenience to customers.

Since its adoption, Integration-as-a-service has provided Office Depot with a "significant" increase in incremental revenue—revenue generated by big corporate customers who had previously been driven to Office Depot competitors. The increase in revenue has led to an increase in market share for Office Depot. Glenn Trommer says, "In these tough economic times [it's] really helped us gain market share with very little investment."

Trommer appreciates that Hubspan's Integration-as-a-service is secure, reliable, and scalable. He is considering expanding its use to transactions with Office Depot's own vendors. The short implementation time, low technical requirements, flexibility, and low cost made SaaS a perfect solution for Office Depot's problem.

Discussion Questions

1. Why did Office Depot decide to use SaaS rather than developing their own software or purchasing off-the-shelf solutions?
2. What benefits do SaaS solutions provide to businesses like Office Depot?

Critical Thinking Questions

1. In general, which types of problems are best suited for SaaS solutions?
2. What risks, if any, is Office Depot taking by trusting its data and operations to Hubspan?

SOURCES: Violino, Bub, "Integration as a Service at Office Depot," *CIO Insight*, *www.cioinsight.com/c/a/Services/Integration-as-a-Service-at-Home-Depot-474099/1*, September 24, 2009; Hubspan Web site, *www.hubspan.com*, accessed January 24, 2010.

Questions for Web Case

See the Web site for this book to read about the Altitude Online case for this chapter. Following are questions concerning this Web case.

Altitude Online: Choosing Hardware

Discussion Questions

1. How might Altitude Online determine what new hardware devices it requires to support the service that its employees use?

2. How will Altitude Online determine the computing power and storage requirements of the new system?

Critical Thinking Questions

1. What should Altitude Online do with its old computer hardware as it is replaced with new hardware?
2. Why do you think Altitude Online decided to phase in new desktop computers but replace mobile devices all at once?

Altitude Online: Choosing Software

1. Why do you think Altitude Online uses two PC platforms— Windows and Mac—rather than standardizing on one? What are the benefits and drawbacks of their decision?

2. Why do you think a business is required to keep copies of all of its software licenses?

Critical Thinking Questions

1. How much freedom should a company like Altitude Online allow for its employees to choose their own personal application software? Why might a company prefer to standardize around specific software packages?
2. What benefits might be provided to an advertising media company like Altitude Online by upgrading to the latest media development and production software? How might upgrading provide the company with a competitive advantage?

NOTES

Sources for the opening vignette: IBM staff, "Turboinštitut powers green energy research with IBM BladeCenter," IBM Success Story, *www-01.ibm.com/software/success/cssdb.nsf/CS/ARBN-7T5LKQ? OpenDocument&Site=powersystems&cty=en_us*, June 26, 2009; Turboinštitut Web site, *www.turboinstitut.si*, accessed December 30, 2009; LSC ADRIA Web page, *www.turboinstitut.si/index.php? option=com_content&task=view&id=62*, accessed December 30, 2009.

1   Thibodeau, Patrick, "U.S. Buys Weather Supercomputer with Twin Backup," *Computerworld*, September 8, 2009.
2   IGEL Technology, "Computer Products Corporation," Press release, *www.igel.com/igel/,content_id,3413,navigation_id,1286,_psmand, 9.html*, accessed September 25, 2009.
3   IGEL Technology, "Seaport Hotel," Press release, *www.igel.de/ igel/,content_id,2153,navigation_id,245,_psmand,2.html*, accessed November 23, 2009.
4   Shah, Agam, "Intel's New Core i7 Chips Surface on Retail Sites," *Computerworld*, May 27, 2009.
5   Lai, Eric, "AMD's Latest Quad-Core Phenom CPU Ups Its Game," *Computerworld*, August 13, 2009.
6   Crothers, Brooke, "IBM Launches Power7 Chip, Systems," *Cnet*, February 7, 2010.
7   ACMA Computers, "Success Stories," Acma Computers Web site, *www.acma.com/acma/Casestudy.asp*, accessed January 16, 2010.
8   Shread, Paul, "Big Blue Takes on Swine Flu," *Grid Computing Planet*, May 12, 2009.
9   Staff, "Media Distributors Introduces Industry's Most Affordable RAID Storage Solution for Professional Video Production Customers," Green Technology Web site, *http://green.tmcnet.com/ news/2009/09/17/4375055.htm*, September 17, 2009.
10 Data Domain, "Boston Medical Center Presents on Reducing IT Costs and Simplifying Storage Management Using Data Domain," Press release, February 20, 2009.
11 Staff, "Company Profile," Austar Web site, *www.austarunited.com.au/about/profile*, accessed January 16, 2010.
12 Winterford, Brett, "Austar Breathes New Life into Old Data," *itnews for Australian Business*, September 16, 2009.
13 Clark, Don, "SanDisk Says New Chips Will Lower Production Costs," *Wall Street Journal*, October 13, 2009.

14 Staff, "Mid-Sized Enterprises Among the First to Adopt Solid State Storage Solutions," Reuters, July 7, 2009.
15 Nuance, "SBI Mutual Fund Deploys Nuance Powered Technology to Enhance Customer Experience," Press release, *www.nuance.com/ news/pressreleases/2009/20090721_sbiMutual.asp*, July 21, 2009.
16 Townsend, Captain Mark, "Remark-able Story: Remark Office OMR Helps Police Department Meet State Requirements," Remark Office Web site, *www.gravic.com/remark/officeomr/reviews/ remark3.html*.
17 Staff, "Newmont Gold - Leeville Mine In Nevada Goes Digital with MST," *Mining-Technology.com*, March 20, 2009.
18 Kondolojy, Amanda, "LCD Vs Plasma Monitors," eHow Web site, *www.ehow.com/about_4778386_lcd-vs- plasma-monitors.html*, accessed January 20, 2010.
19 Staff, "Solar Power—Yes, Life's Good," *The Daily Contributor*, October 12, 2009.
20 Ion, Florence, "Into Thin Air - Conquering Mt. Everest with a MacBook," *Mac/Life*, *www.maclife.com/print/5013*, October 2, 2009.
21 Stone, Brad and Vance, Ashlee, "Just a Touch Away, the Elusive Tablet PC," *The New York Times*, October 5, 2009.
22 NEC Corporation, "NEC Deploys One of Japan's Largest Thin Client Systems for Tokio Marine, Press release, *www.nec.co.jp/press/en/ 0909/2901.html*, September 29, 2009.
23 "Dell Studio Hybrid Intel Dual-Core 2.1 GHz Miniature Desktop (26GB/ 160GB) $299.99," Tech Bargains, October 20, 2009.
24 Sun Microsystems, "Blue Sky Studios Deploys Sun Solution for Ice Age: Dawn of the Dinosaurs Film," Press release, *www.sun.com/ aboutsun/pr/2009-07/sunflash.20090706.1.xml*, July 6, 2009.
25 Betts, Mitch, "Data Center Plays Supporting Role in Avatar," *Computerworld*, January 18, 2010.
26 IBM, "Winn-Dixie Strengthens Infrastructure with IBM Consolidation Solution, Press release, *www- 01.ibm.com/software/success/ cssdb.nsf/CS/DLAS-7VWRRS?OpenDocument&Site=bladecenter& cty=en_us*, September 15, 2009.
27 Boulton Clint, "Mindjet MindManager 8 for Mac Integrates with Apple Apps," *eweek*, January 27, 2010, *www.eweek.com*.
28 Dunn, Scott, "Master Your Schedule with Google Calendar," *PC World*, August 2008, p. 110.
29 IBM staff, "Leading energy management firm streamlines global reporting, increases synergies, thanks to powerful IBM Cognos

solution," IBM Case Study, *www-01.ibm.com/software/success/cssdb.nsf/cs/SANS-82GM69?OpenDocument&Site=cognos&cty=en_us*, accessed February 10, 2010.

30 Pogue, David, "Windows 7 Keeps the Good, Tries to Fix Flaws," *New York Times*, October 21, 2009.

31 Tweney, Dylan, "Google Chrome OS: Ditch Your Hard Drives, the Future Is the Web," *Wired*, *www.wired.com*, November 19, 2009.

32 Staff, "eBay Inc.," Sun Customer Snapshot Web site, *www.sun.com/customers/index.xml?c=ebay.xml&submit=Find*, accessed January 17, 2010.

33 Staff, "Idaho National Lab," Sun Customer Snapshot Web site, *www.sun.com/customers/index.xml?c=inl.xml&submit=Find*, accessed January 17, 2010.

34 IBM staff, "IBM eServer zSeries systems and GDPS help AOK Bavaria create a healthy environment for growth," IBM Success Stories, May 1, 2009, *www-01.ibm.com/software/success/cssdb.nsf/CS/JFTD-6VESNY?OpenDocument&Site=eserverzseries&cty=en_us*.

35 Kinnander, Ola, "Nokia to Roll Out Phone Based on Linux Software," *Wall Street Journal*, August 28, 2009, Technology Section, p. B4.

36 Oracle staff, "Swedish Medical Leverages Content Management Solution to Facilitate Patient-centric Care," Oracle Customer Snapshot, *www.oracle.com/customers/snapshots/swedish-medical-center-content-management- snapshot.pdf*, accessed January 17, 2010.

37 Montalbano, Elizabeth, "Forrester: Microsoft Office in No Danger From Competitors," *PC World*, *www.pcworld.com/businesscenter/article/166123/forrester_microsoft_office_in_no_danger_from_competitors.html ?tk=nl_dnx_h_crawl*, June 04, 2009.

38 Montalbano, Elizabeth, "Amalga Helps Hospital Keep Swine Flu in Check," *CIO*, *www.cio.com/article/494533/Amalga_Helps_Hospital_Keep_Swine_Flu_in_Check*, June 8, 2009.

39 Staff, "Customer: RapidAdvance," SAP Case Study *www.ondemand.com/customers/rapidadvance.asp*, accessed January 17, 2010.

40 Lai, Eric "Forget mashups: Tableau Software wants data junkies to do the 'viz'," *Computerworld*, *www.computerworld.com*, February 11, 2010.

41 Salesforce staff, "Powered By Salesforce CRM's Idea Community, My Starbucks Idea Brews Customer Feedback at Starbucks," Salesforce Success Story, *www.salesforce.com/customers/distribution-retail/starbucks.jsp*, accessed January 17, 2010.

42 Burrows, Peter, "Microsoft Defends Its Empire," *Businessweek*, July 6, 2009, pg 28.

43 Siegler, MG, "BarMax: The $1,000 iPhone App That Might Actually Be Worth It," *TechCrunch*, *http://techcrunch.com/2010/01/17/most-expensive-iphone-app-barmax*, January 17, 2010.

44 Schaffhauser, Dian, "Florida Virtual to Extend Use of LMS to Students," *The Journal*, *www.thejournal.com*, January 22, 2010.

CHAPTER
· 3 ·

Database Systems, Data Centers, and Business Intelligence

PRINCIPLES	LEARNING OBJECTIVES
▪ Data management and modeling are key aspects of organizing data and information.	▪ Define general data management concepts and terms, highlighting the advantages of the database approach to data management. ▪ Describe logical and physical database design considerations, the function of data centers, and the relational database model.
▪ A well-designed and well-managed database is an extremely valuable tool in supporting decision making.	▪ Identify the common functions performed by all database management systems, and identify popular database management systems.
▪ The number and types of database applications will continue to evolve and yield real business benefits.	▪ Identify and briefly discuss business intelligence, data mining, and other database applications.

Information Systems in the Global Economy
Aquent, United States

Leveraging Database Technology to Empower Marketing Professionals

Aquent is a global leader in marketing staffing. The company works with Fortune 500 marketing organizations to fill positions with highly qualified professionals drawn from a pool of thousands of marketing experts worldwide. Aquent works to place brand managers, copywriters, data analysts, Web designers, search-engine optimizers, and other specialists in full-time positions as well as short-term contract positions. According to Aquent, the industry is changing from one in which marketing experts join a company and work their way up the corporate ladder, to one in which marketing projects are hired out to specialists who move from company to company applying unique high-level skills to challenging projects. Aquent believes that it plays a key role in enabling this new era of marketing. It provides challenging projects for marketing professionals to hone their skills and advance in their field, while elevating the quality and effectiveness of marketing efforts within organizations.

Aquent has unique database and information system needs. Its clients are both large corporations and individuals. Although its primary business is staffing, it also provides ancillary services such as project management, translation and localization, and healthcare consulting. Its information systems must produce a wide range of reports to meet a variety of business needs. These needs include staffing levels and requirements, human resource usage, gross profit, pay rates, and many others. Because the company works with many organizations, it must manage diverse payroll schemes and schedules. Aquent also manages systems that allow it to provide insurance and retirement benefits to many of the marketing professional talents that it represents. The databases that support these wide-ranging and diverse systems are about as complicated as a business' databases can be.

To get a handle on all of its data, Aquent uses a database management system that collects operational data from around the world and stores it in a central data mart managed by the SAP Corporation. Each night, the system refreshes the data stored in the data mart with updates from data centers in Sydney, London, and Boston. A backup of the data is stored in Aquent's data center in Boston.

Aquent executives, managers, and personnel access the data through a Web-based system provided by SAP. SAP takes responsibility for storing and managing Aquent's database and providing a robust database management system (DBMS) accessed through a Web browser. This approach to database management, where a company outsources its DBMS to a service provider, is referred to as Database as a Service, or DaaS.

Aquent uses a business intelligence (BI) system to create ad-hoc and annual reports. Aquent regional managers run individual reports for Asia Pacific, Europe, and North America. They also run reports that cover all regions using common criteria to examine. Executives can get a high-level view of trends in corporate data and use BI tools to drill down into the data to discover specific areas of the business that require attention.

Aquent uses SAP data mining technology to examine data in the data mart and discover patterns and anomalies that cue decision makers to examine problems and opportunities. Predictive analysis tools help to provide managers with insight into the future based on an analysis of the past. Using these tools, Aquent can determine future demand for marketing professionals and ensure it can meet that demand. It may also determine a future lack of demand so that Aquent can advise some professionals to consider jobs in other related areas.

Using a central data mart and joining operations around the world, Aquent can more easily view itself as a multinational company. The distances among its global divisions are greatly reduced by its ability to combine corporate data and evaluate it both by region and in its totality.

As you read this chapter, consider the following:

- What role do databases play in the overall effectiveness of information systems?
- What techniques do businesses use to maximize the value of the information provided from databases?

Why Learn About Database Systems, Data Centers, and Business Intelligence?

A huge amount of data is entered into computer systems every day. Where does all this data go, and how is it used? How can it help you on the job? In this chapter, you will learn about database systems and business intelligence tools that can help you make the most effective use of information. If you become a marketing manager, you can access a vast store of data on existing and potential customers from surveys, their Web habits, and their past purchases. This information can help you sell products and services. If you become a corporate lawyer, you will have access to past cases and legal opinions from sophisticated legal databases. This information can help you win cases and protect your organization legally. If you become a human resource (HR) manager, you will be able to use databases and business intelligence tools to analyze the impact of raises, employee insurance benefits, and retirement contributions on long-term costs to your company. Regardless of your field of study in school, using database systems and business intelligence tools will likely be a critical part of your job. In this chapter, you will see how you can use data mining to extract valuable information to help you succeed. This chapter starts by introducing basic concepts of database management systems.

A database is an organized collection of data. Like other components of an information system, a database should help an organization achieve its goals. A database can contribute to organizational success by providing managers and decision makers with timely, accurate, and relevant information based on data. For example, Comic Relief, in London, England, raises money to assist the needy by hosting entertainment events featuring comedians. The organization uses a database to determine which clips in its televised fundraiser generate the highest emotional response from the public to determine whether the clip should be repeated.[1]

Databases also help companies generate information to reduce costs, increase profits, track past business activities, and open new market opportunities. In some cases, organizations collaborate in creating and using international databases. Six organizations, including the Organization of Petroleum Exporting Countries (OPEC), International Energy Agency (IEA), and the United Nations, use a database to monitor the global oil supply.

database management system (DBMS)
A group of programs that manipulate the database and provide an interface between the database and the user of the database and other application programs.

database administrator (DBA)
A skilled IS professional who directs all activities related to an organization's database.

A **database management system (DBMS)** consists of a group of programs that manipulate the database and provide an interface between the database and its users and other application programs. Usually purchased from a database company, a DBMS provides a single point of management and control over data resources, which can be critical to maintaining the integrity and security of the data. A database, a DBMS, and the application programs that use the data make up a database environment. A **database administrator (DBA)** is a skilled and trained IS professional who directs all activities related to an organization's database, including providing security from intruders. People hack into databases for various reasons. Consider the Latvian computer expert who hacked into a government database to make public the salary information of government officials. He intended to show the people that during the country's severe economic problems, government officials continued receiving high salaries.[2] In 2010, the names, birth dates, and Social Security numbers of 3.3 million students were stolen from a database owned by a student loan company.[3] Such data breaches have become commonplace for organizations because many databases are now accessible from the Internet. Data quality and accuracy also continue to be important issues for DBAs. For example, in Uckfield, England, government records for Pauline Grant and her farm became jumbled due to a land registry error. The mix-up resulted in a pig named Blossom on Grant's farm receiving mail encouraging her to vote in the upcoming election.[4]

Databases and database management systems are becoming even more important to businesses as they deal with increasing amounts of digital information. A report from IDC called "The Digital Universe Decade – Are you ready," estimates the size of the digital universe to be 1.2 zettabytes, or 1.2 trillion gigabytes.[5] If a tennis ball were one byte of information, a zettabyte-sized ball would be around the size of a million earths. Furthermore, between 2009 and 2020, the amount of information humanity creates will grow by a factor of 44, storage capacity will grow by a factor of 30, and the estimated investment in database infrastructure and administration will grow by only a factor of 1.4. IDC recommends that organizations move now to create policies, tools, and standards to accommodate the approaching tidal wave of digital data and information.

DATA MANAGEMENT

Without data and the ability to process it, an organization could not successfully complete most business activities. It could not pay employees, send out bills, order new inventory, or produce information to assist managers in decision making. As you recall, data consists of raw facts, such as employee numbers and sales figures. For data to be transformed into useful information, it must first be organized in a meaningful way.

The Hierarchy of Data

Data is generally organized in a hierarchy that begins with the smallest piece of data used by computers (a bit) and progresses through the hierarchy to a database. A bit (a binary digit) represents a circuit that is either on or off. Bits can be organized into units called *bytes*. A byte is typically eight bits. Each byte represents a **character**, which is the basic building block of most information. A character can be an uppercase letter (A, B, C... Z), lowercase letter (a, b, c ... z), numeric digit (0, 1, 2... 9), or special symbol (., !, +, -, /, ...).

Characters are put together to form a field. A **field** is typically a name, number, or combination of characters that describes an aspect of a business object (such as an employee, a location, or a truck) or activity (such as a sale). In addition to being entered into a database, fields can be computed from other fields. *Computed fields* include the total, average, maximum, and minimum value. A collection of data fields all related to one object, activity, or individual is called a **record**. By combining descriptions of the characteristics of an object, activity, or individual, a record can provide a complete description of it. For instance, an employee record is a collection of fields about one employee. One field includes the employee's name, another field contains the address, and still others the phone number, pay rate, earnings made to date, and so forth. A collection of related records is a **file**—for example, an employee file is a collection of all company employee records. Likewise, an inventory file is a collection of all inventory records for a particular company or organization. Some database software refers to files as tables.

At the highest level of this hierarchy is a *database*, a collection of integrated and related files. Together, bits, characters, fields, records, files, and databases form the **hierarchy of data**. See Figure 3.1. Characters are combined to make a field, fields are combined to make a record, records are combined to make a file, and files are combined to make a database. A database houses not only all these levels of data but also the relationships among them.

character
A basic building block of most information, consisting of uppercase letters, lowercase letters, numeric digits, or special symbols.

field
Typically a name, number, or combination of characters that describes an aspect of a business object or activity.

record
A collection of data fields all related to one object, activity, or individual.

file
A collection of related records.

hierarchy of data
Bits, characters, fields, records, files, and databases.

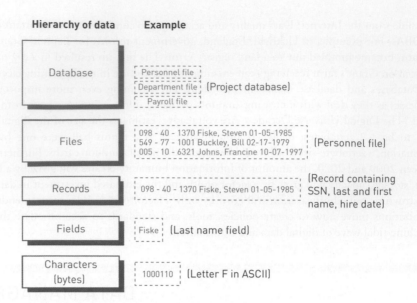

Data Entities, Attributes, and Keys

Entities, attributes, and keys are important database concepts. An **entity** is a general class of people, places, or things (objects) for which data is collected, stored, and maintained. Examples of entities include employees, inventory, and customers. Most organizations organize and store data as entities.

An **attribute** is a characteristic of an entity. For example, employee number, last name, first name, hire date, and department number are attributes for an employee. See Figure 3.2. The inventory number, description, number of units on hand, and location of the inventory item in the warehouse are attributes for items in inventory. Customer number, name, address, phone number, credit rating, and contact person are attributes for customers. Attributes are usually selected to reflect the relevant characteristics of entities such as employees or customers. The specific value of an attribute, called a **data item**, can be found in the fields of the record describing an entity.

Employee #	Last name	First name	Hire date	Dept. number
005-10-6321	Johns	Francine	10-07-1997	257
549-77-1001	Buckley	Bill	02-17-1979	632
098-40-1370	Fiske	Steven	01-05-1985	598

ENTITIES (records)

KEY FIELD

ATTRIBUTES (fields)

Most organizations use attributes and data items. Many governments use attributes and data items to help in criminal investigations. The United States Federal Bureau of Investigation is building a huge database of peoples' physical characteristics or biometrics.[6] At a cost of $1 billion, the database management system named Next Generation Identification will catalog digital images of faces, fingerprints, and palm prints of U.S. citizens and visitors. Each person in the database is an entity, each biometric category is an attribute, and each image is a data item. The information will be used as a forensics tool and to increase homeland security.

As discussed earlier, a collection of fields about a specific object is a record. A **key** is a field or set of fields in a record that identifies the record. A **primary key** is a field or set of fields that uniquely identifies the record. No other record can have the same primary key. For an employee record, such as the one shown in Figure 3.2, the employee number is an example of a primary key. The primary key is used to distinguish records so that they can be accessed, organized, and manipulated. Primary keys ensure that each record in a file is unique. For example, eBay assigns an "Item number" as its primary key for items to make sure that bids are associated with the correct item. See Figure 3.3.

key
A field or set of fields in a record that is used to identify the record.

primary key
A field or set of fields that uniquely identifies the record.

Figure 3.3

Primary Key

eBay assigns an "Item number" as a primary key to keep track of each item in its database.

Locating a particular record that meets a specific set of criteria might be easier and faster using a combination of secondary keys. For example, a customer might call a mail-order company to place an order for clothes. The order clerk can easily access the customer's mailing and billing information by entering the primary key—usually a customer number—but if the customer does not know the correct primary key, a secondary key such as last name can be used. In this case, the order clerk enters the last name, such as Adams. If several customers have a last name of Adams, the clerk can check other fields, such as address, first name, and so on, to find the correct customer record. After locating the correct customer record, the order can be completed and the clothing items shipped to the customer.

The Database Approach

At one time, information systems referenced specific files containing relevant data. For example, a payroll system would use a payroll file. Each distinct operational system used data files dedicated to that system. This approach to data management is called the **traditional approach to data management**.

Today, most organizations use the **database approach to data management**, whereby multiple information systems share a pool of related data. A database offers the ability to share data and information resources. Federal databases, for example, often include the results of DNA tests as an attribute for convicted criminals. The information can be shared with law enforcement officials around the country.

To use the database approach to data management, additional software—a database management system (DBMS)—is required. As previously discussed, a DBMS consists of a group of programs that can be used as an interface between a database and the user of the database. Typically, this software acts as a buffer between the application programs and the database itself. Figure 3.4 illustrates the database approach.

Table 3.1 lists some of the primary advantages of the database approach, and Table 3.2 lists some disadvantages.

traditional approach to data management
An approach to data management whereby each distinct operational system used data files dedicated to that system.

database approach to data management
An approach to data management whereby a pool of related data is shared by multiple information systems.

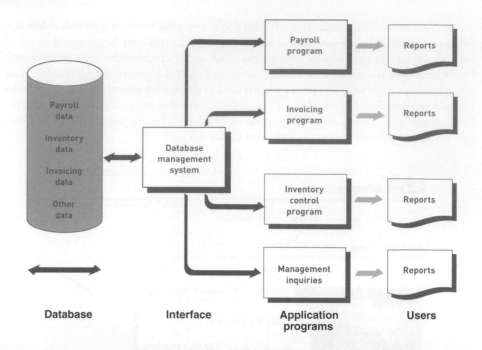

Advantages	Explanation
Improved strategic use of corporate data	Accurate, complete, up-to-date data can be made available to decision makers where, when, and in the form they need it. The database approach can also give greater visibility to the organization's data resources.
Reduced data redundancy	Data is organized by the DBMS and stored in only one location. This results in a more efficient use of system storage space.
Improved data integrity	With the traditional approach, some changes to data were not reflected in all copies of the data. The database approach prevents this problem because no separate files exist.
Easier modification and updating	The DBMS coordinates data modifications and updates. Programmers and users do not have to know where the data is physically stored. Data is stored and modified once. Modification and updating is also easier because the data is commonly stored in only one location.
Data and program independence	The DBMS organizes the data independently of the application program, so the application program is not affected by the location or type of data. Introduction of new data types not relevant to a particular application does not require rewriting that application to maintain compatibility with the data file.
Better access to data and information	Most DBMSs have software that makes it easy to access and retrieve data from a database. In most cases, users give simple commands to get important information. Relationships between records can be more easily investigated and exploited, and applications can be more easily combined.
Standardization of data access	A standardized, uniform approach to database access means that all application programs use the same overall procedures to retrieve data and information.
A framework for program development	Standardized database access procedures can mean more standardization of program development. Because programs go through the DBMS to gain access to data in the database, standardized database access can provide a consistent framework for program development. In addition, each application program need address only the DBMS, not the actual data files, reducing application development time.
Better protection of the data	Accessing and using centrally located data is easier to monitor and control. Security codes and passwords can ensure that only authorized people have access to particular data and information in the database, thus ensuring privacy.
Shared data and information resources	The cost of hardware, software, and personnel can be spread over many applications and users. This is a primary feature of a DBMS.

As you can see from Tables 3.1 and 3.2, the advantages of the database approach far outweigh the disadvantages. For that reason, nearly all businesses use databases of various types and sizes to collect important data that fuels information systems and decision making. Many modern databases serve entire enterprises, encompassing much of the data of the organization. Often, distinct yet related databases are linked to provide enterprise-wide databases. For example, many Wal-Mart stores include in-store medical clinics for customers. Wal-Mart uses a centralized electronic health records database that stores the information of all patients across all stores.[7] The database is interconnected with the main Wal-Mart database to provide information about customer's interactions with the clinics and stores.

Disadvantages	Explanation
More complexity	DBMSs can be difficult to set up and operate. Many decisions must be made correctly for the DBMS to work effectively. In addition, users have to learn new procedures to take full advantage of a DBMS.
More difficult to recover from a failure	With the traditional approach to file management, a failure of a file affects only a single program. With a DBMS, a failure can shut down the entire database.
More expensive	DBMSs can be more expensive to purchase and operate than traditional file management. The expense includes the cost of the database and specialized personnel, such as a database administrator, who is needed to design and operate the database. Additional hardware might also be required.

Table 3.2

Disadvantages of the Database Approach

DATA MODELING AND DATABASE CHARACTERISTICS

Because today's businesses have so many elements, they must keep data organized so that it can be used effectively. A database should be designed to store all data relevant to the business and provide quick access and easy modification. Moreover, it must reflect the business processes of the organization. When building a database, an organization must carefully consider these questions:

- *Content.* What data should be collected and at what cost?
- *Access.* What data should be provided to which users and when?
- *Logical structure.* How should data be arranged so that it makes sense to a given user?
- *Physical organization.* Where should data be physically located?

The U.S. federal government carefully considers what information it should make accessible and what information should remain private. The Bush administration kept a great deal of information private during the years following the bombing of the Twin Towers in New York City. When President Obama took office, he pledged to run a much more transparent government. He followed through with his promise by providing access to hundreds of government databases through the Web site *www.data.gov*. News agencies, research labs, and analysts can use this Web site to connect databases directly to government records to track government actions and information.[8] See Figure 3.5.

Figure 3.5

The U.S. federal government provides access to numerous data sets at *www.data.gov*.

(Source: *www.data.gov*.)

Data Center

Databases, and the systems that manipulate them, can be physically stored on computers as small as a PC or as large as mainframes and data centers. A **data center** is a climate-controlled building or set of buildings that house database servers and the systems that deliver mission-critical information and services. Data centers of large organizations are often distributed among several locations, but a recent trend has many organizations consolidating their data centers into a few large facilities. For example, the U.S. federal government is working to save billions of dollars by consolidating 1,100 data centers into a dozen facilities. The project is recognized as the largest data center consolidation in history.[9] The state of Texas is in the midst of a seven-year effort to consolidate its 31 data centers into two facilities in San Angelo and Austin.[10] Microsoft recently constructed a $550 million, 400,000-square-foot data center on 44 acres in San Antonio. Google invested $600 million for a mega data center in Lenoir, North Carolina, and $750 million for another in Goose Creek, South Carolina. Clearly, storing and managing data is a serious business.

Traditional data centers consist of warehouses filled with row upon row of server racks and powerful cooling systems to compensate for the heat generated by the processors. Microsoft,[11] Google,[12] and others have adopted a new modular data center approach, which uses large shipping containers like the ones that transport consumer goods around the world. The huge containers, such as the HP POD, are packed with racks of servers prewired and cooled to easily connect and set up. Microsoft recently constructed a 700,000-square-foot data center in Northlake, Illinois. It is considered to be one of the largest in the world, taking up 16 football fields of space. The mega facility is filled with 220 shipping containers packed with servers. Microsoft says that a new shipping container can be wheeled into place and connected to the Internet within hours.[13] See Figure 3.6.

Modular data centers are becoming popular around the world due to their convenience and efficiencies. Taiwan's Technology Research Institute is working to create standards for modular data centers in shipping containers that they say will reduce the costs of these units by half while increasing ease of use and reducing energy demands.[14]

While a company's data sits in large supercooled data centers, the people accessing that data are typically in offices spread across the country or around the world. In fact, the expectation of data center specialists such as Hewlett-Packard CEO Mark Hurd is that in the near future, the only personnel on duty at data centers will be security guards. Data centers are approaching the point of automation, whereby they can run and manage themselves while being monitored remotely. This is referred to as a "lights out" environment. The State of Vermont recently switched to a lights out approach for nights and weekends, reducing its staff by 40 percent and significantly reducing costs.[15] HP has moved to automated data centers, reducing its IT staffing needs by 3,000.[16]

As data centers continue to expand in terms of the quantity of data that they store and process, their energy demands are becoming an increasingly significant portion of the total energy demands of humanity. Businesses and technology vendors are working to develop green data centers that run more efficiently and require less energy for processing and cooling.

Data Modeling

When organizing a database, key considerations include determining what data to collect, who will have access to it, and how they might want to use it. After determining these details, an organization can create the database. Building a database requires two different types of designs: a logical design and a physical design. The *logical design* of a database is an abstract model of how the data should be structured and arranged to meet an organization's information needs. The logical design involves identifying relationships among the data items and grouping them in an orderly fashion. Because databases provide both input and output for information systems throughout a business, users from all functional areas should assist in creating the logical design to ensure that their needs are identified and addressed. The *physical design* starts from the logical database design and fine-tunes it for performance and cost considerations (such as improved response time, reduced storage space, and lower operating cost). The person who fine-tunes the physical design must have an in-depth knowledge of the DBMS. For example, the logical database design might need to be altered so that certain data entities are combined, summary totals are carried in the data records rather than calculated from elemental data, and some data attributes are repeated in more than one data entity. These are examples of **planned data redundancy**, which is done to improve the system performance so that user reports or queries can be created more quickly.

One of the tools database designers use to show the logical relationships among data is a data model. A **data model** is a diagram of entities and their relationships. Data modeling usually involves understanding a specific business problem and analyzing the data and information needed to deliver a solution. When done at the level of the entire organization, this is called enterprise data modeling. **Enterprise data modeling** is an approach that starts by investigating the general data and information needs of the organization at the strategic level, and then examines more specific data and information needs for the various functional areas and departments within the organization. Various models have been developed to help managers and database designers analyze data and information needs. An entity-relationship diagram is an example of such a data model.

Entity-relationship (ER) diagrams use basic graphical symbols to show the organization of and relationships between data. In most cases, boxes in ER diagrams indicate data items or entities contained in data tables, and diamonds show relationships between data items and entities. In other words, ER diagrams show data items in tables (entities) and the ways they are related.

ER diagrams help ensure that the relationships among the data entities in a database are correctly structured so that any application programs developed are consistent with business operations and user needs. In addition, ER diagrams can serve as reference documents after a database is in use. If changes are made to the database, ER diagrams help design them. Figure 3.7 shows an ER diagram for an order database. In this database design, one salesperson serves many customers. This is an example of a one-to-many relationship, as indicated by the one-to-many symbol (the "crow's-foot") shown in Figure 3.7. The ER diagram also shows that each customer can place one-to-many orders; each order includes one-to-many line items; and many line items can specify the same product (a many-to-one relationship). This database can also have one-to-one relationships. For example, one order generates one invoice.

planned data redundancy
A way of organizing data in which the logical database design is altered so that certain data entities are combined, summary totals are carried in the data records rather than calculated from elemental data, and some data attributes are repeated in more than one data entity to improve database performance.

data model
A diagram of data entities and their relationships.

enterprise data modeling
Data modeling done at the level of the entire enterprise.

entity-relationship (ER) diagrams
Data models that use basic graphical symbols to show the organization of and relationships between data.

Mega Data Centers and Their Environmental Impact

To keep up with the unprecedented amount of information being generated, businesses need to invest in larger and larger data centers. Many businesses find it more economical to outsource their data center needs. Dozens of mega data centers are being constructed around the world for a variety of uses.

Mega data centers typically cost hundreds of millions of dollars and consume acres of property. One of the world's largest was recently constructed for $301 million by Next Generation Data, outside of Newport in South Wales. The 750,000-square-foot (70,000-square-meter) facility has enough space to house 19,000 server racks that each hold a dozen servers. The facility hopes to serve hundreds of businesses, many located in nearby London. Its first two tenants, BT and Logica, signed contracts worth a combined $29 million.

Next Generation Data can provide its customers with certain guarantees of service and data protection. To guard against terrorist attacks, the data center has "triple-skinned walls, bomb-proof glass, prison-grade perimeter fencing, infrared detection, biometric recognition, and ex-special forces security guards." The data center's network is equally protected, and all systems have failback systems to guard against hardware or electrical failure.

The biggest environmental impact of mega data centers is their energy consumption for the processing, storage, and cooling required. The data center for Next Generation Data outside Newport has its own energy substation that provides 90 megavolt-amperes of electrical power. That's roughly equivalent to the requirements of a city of 400,000 people. Multiply this by the dozens of other mega data centers going online, including huge facilities such as Microsoft's new 700,000-square-foot center near Chicago, and the energy requirements increase around the world. Adding the energy needs of mega data centers to the increasing energy demands of developing countries with huge populations such as China and India results in unprecedented worldwide energy consumption.

When coal-burning power plants fulfill these energy demands, they add carbon to the atmosphere, which many scientists argue accelerates climate change. A number of efforts are underway to counteract the growing demand for data centers. Hardware manufacturers are producing servers that are more efficient, requiring half the energy as their predecessors to do twice the work. As new data centers go into operation, they are implementing new energy-efficient technologies. Gradually, as old wasteful systems break down, managers will migrate data to new green systems.

The Newport data center uses fresh air cooling and Energy Star rated equipment to help reduce its impact on the environment. The Environmental Protection Agency has recently released Energy Star standards for servers and is developing standards for enterprise storage as well. Such standards give hardware and software manufacturers targets to shoot for to keep systems running efficiently with less energy.

In light of environmental pressures and public sentiment, many companies are making pledges to reduce the energy requirements of information systems. Disney recently pledged to reduce its electricity consumption by 20 percent by 2013. By measuring Power Usage Effectiveness (PUE), companies can compare IS equipment power requirements to environmental power requirements. A PUE of 2—the industry average—indicates that processing and cooling are requiring equal amounts of energy. Disney and others hope to invest in technologies that have significantly lower energy requirements. Google discovered that adjusting thermostats in its data centers up from the frigid 60s to 80 degrees Fahrenheit helped to lower its PUE to 1.5.

Without a doubt, data centers will continue consuming increasing amounts of real estate. Through a combination of techniques and technologies that include consolidation, more efficient servers, more effective cooling techniques, and alternative energy sources, expanding data centers can reduce their impact on the environment.

Discussion Questions

1. Why is the increase in data center construction a concern for the environment?
2. What efforts can help to minimize the impact of data centers on the environment?

Critical Thinking Questions

1. Companies are finding it necessary to weigh the value of storing information against the value of affecting the environment. Write a few paragraphs outlining the importance of both and describing how companies might financially benefit from protecting both.
2. If you were a systems administrator for a data center, what steps would you take to create and manage a data center to store the maximum amount of valuable data with the minimum impact on the environment?

Sources: Niccolai, James, "750,000-sq.-ft. data center opens in Wales," *Computerworld*, March 15, 2010, *www.computerworld.com*; Lawson, Stephen, "EPA drafting Energy Star standards for enterprise storage," *Computerworld*, May 10, 2010, *www.computerworld.com*; Brodkin, Jon, "Disney, Verizon go green in the data center," *Computerworld*, October 6, 2009, *www.computerworld.com*; Niccolai, James, "Google: Crank up the heat in your data center," *Computerworld*, April 29, 2010, *www.computerworld.com*.

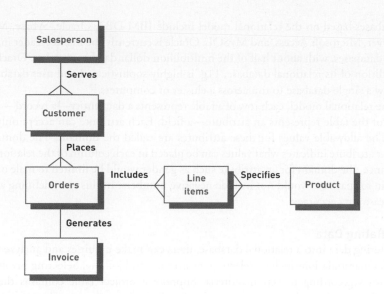

Figure 3.7

An Entity-Relationship (ER)
Diagram for a Customer Order
Database

Development of ER diagrams helps
ensure that the logical structure of
application programs is consistent
with the data relationships in the
database.

The Relational Database Model

Although there are a number of different database models, including flat files, hierarchical, and network models, the relational model has become the most popular, and use of this model will continue to increase. The **relational model** describes data using a standard tabular format; all data elements are placed in two-dimensional tables, called *relations*, which are the logical equivalent of files. The tables in relational databases organize data in rows and columns, simplifying data access and manipulation. It is normally easier for managers to understand the relational model than other database models. See Figure 3.8.

relational model
A database model that describes
data in which all data elements are
placed in two-dimensional tables,
called *relations*, which are the
logical equivalent of files.

Figure 3.8

A Relational Database Model

In the relational model, all data
elements are placed in two-
dimensional tables, or relations. As
long as they share at least one
common element, these relations
can be linked to output useful
information.

Data Table 1: Project Table

Project	Description	Dept. number
155	Payroll	257
498	Widgets	632
226	Sales manual	598

Data Table 2: Department Table

Dept.	Dept. name	Manager SSN
257	Accounting	005-10-6321
632	Manufacturing	549-77-1001
598	Marketing	098-40-1370

Data Table 3: Manager Table

SSN	Last name	First name	Hire date	Dept. number
005-10-6321	Johns	Francine	10-07-1997	257
549-77-1001	Buckley	Bill	02-17-1979	632
098-40-1370	Fiske	Steven	01-05-1985	598

Databases based on the relational model include IBM DB2, Oracle, Sybase, Microsoft SQL Server, Microsoft Access, and MySQL. Oracle is currently the market leader in general-purpose databases, with about half of the multibillion dollar database market. Oracle's most recent edition of its relational database, 11g, is highly sophisticated and uses database grids that allow a single database to run across a cluster of computers.[17]

In the relational model, each row of a table represents a data entity—a record—and each column of the table represents an attribute—a field. Each attribute can accept only certain values. The allowable values for these attributes are called the **domain**. The domain for a particular attribute indicates what values can be placed in each column of the relational table. For instance, the domain for an attribute such as gender would be limited to male or female. A domain for pay rate would not include negative numbers. In this way, defining a domain can increase data accuracy.

Manipulating Data

After entering data into a relational database, users can make inquiries and analyze the data. Basic data manipulations include selecting, projecting, and joining. **Selecting** involves eliminating rows according to certain criteria. Suppose a project table contains the project number, description, and department number for all projects a company is performing. The president of the company might want to find the department number for Project 226, a sales manual project. Using selection, the president can eliminate all rows but the one for Project 226 and see that the department number for the department completing the sales manual project is 598.

Projecting involves eliminating columns in a table. For example, a department table might contain the department number, department name, and Social Security number (SSN) of the manager in charge of the project. A sales manager might want to create a new table with only the department number and the Social Security number of the manager in charge of the sales manual project. The sales manager can use projection to eliminate the department name column and create a new table containing only the department number and SSN.

Joining involves combining two or more tables. For example, you can combine the project table and the department table to create a new table with the project number, project description, department number, department name, and Social Security number for the manager in charge of the project.

As long as the tables share at least one common data attribute, the tables in a relational database can be **linked** to provide useful information and reports. Being able to link tables to each other through common data attributes is one of the keys to the flexibility and power of relational databases. Suppose the president of a company wants to find out the name of the manager of the sales manual project and the length of time the manager has been with the company. Assume that the company has the manager, department, and project tables shown in Figure 3.8. A simplified ER diagram showing the relationship between these tables is shown in Figure 3.9. Note the crow's-foot by the project table. This indicates that a department can have many projects. The president would make the inquiry to the database, perhaps via a personal computer. The DBMS would start with the project description and search the project table to find out the project's department number. It would then use the department number to search the department table for the manager's Social Security number. The department number is also in the department table and is the common element that links the project table to the department table. The DBMS uses the manager's Social Security number to search the manager table for the manager's hire date. The manager's Social Security number is the common element between the department table and the manager table. The final result is that the manager's name and hire date are presented to the president as a response to the inquiry. See Figure 3.10.

domain
The allowable values for data attributes.

selecting
Manipulating data to eliminate rows according to certain criteria.

projecting
Manipulating data to eliminate columns in a table.

joining
Manipulating data to combine two or more tables.

linking
Data manipulation that combines two or more tables using common data attributes to form a new table with only the unique data attributes.

Figure 3.9

A Simplified ER Diagram Showing the Relationship Between the Manager, Department, and Project Tables

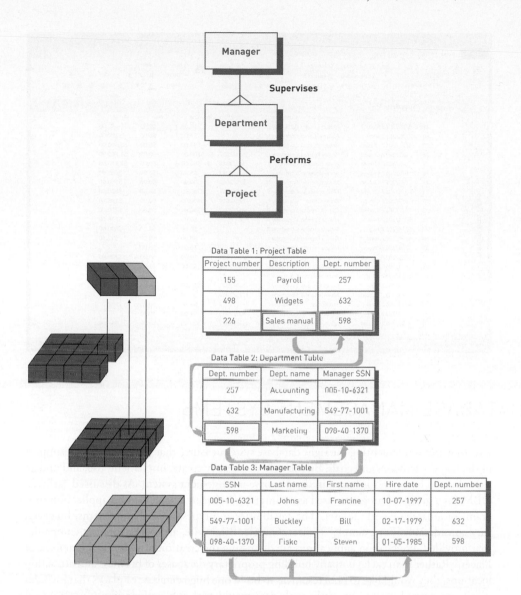

Figure 3.10

Linking Data Tables to Answer an Inquiry

In finding the name and hire date of the manager working on the sales manual project, the president needs three tables: project, department, and manager. The project description (Sales manual) leads to the department number (598) in the project table, which leads to the manager's SSN (098-40-1370) in the department table, which leads to the manager's name (Fiske) and hire date (01-05-1985) in the manager table.

One of the primary advantages of a relational database is that it allows tables to be linked, as shown in Figure 3.10. This linkage reduces data redundancy and allows data to be organized more logically. The ability to link to the manager's SSN stored once in the manager table eliminates the need to store it multiple times in the project table.

The relational database model is by far the most widely used. It is easier to control, more flexible, and more intuitive than other approaches because it organizes data in tables. As shown in Figure 3.11, a relational database management system, such as Access, provides tips and tools for building and using database tables. In this figure, the database displays information about data types and indicates that additional help is available. The ability to link relational tables also allows users to relate data in new ways without having to redefine complex relationships. Because of the advantages of the relational model, many companies use it for large corporate databases, such as those for marketing and accounting. The relational model can also be used with personal computers and mainframe systems. A travel reservation company, for example, can develop a fare-pricing system by using relational database technology that can handle millions of daily queries from online travel companies such as Expedia, Travelocity, and Orbitz.

Figure 3.11

Building and Modifying a
Relational Database

Relational databases provide many
tools, tips, and shortcuts to simplify
the process of creating and
modifying a database.

(Source: Courtesy of Microsoft
Corporation.)

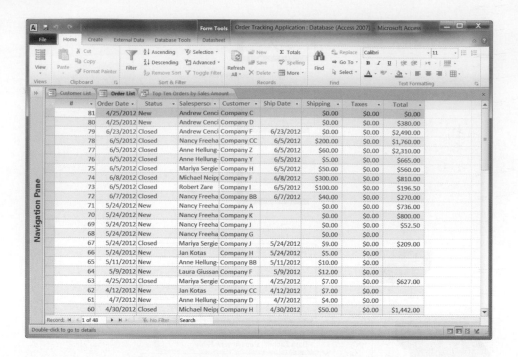

DATABASE MANAGEMENT SYSTEMS

Creating and implementing the right database system ensures that the database will support both business activities and goals. But how do we actually create, implement, use, and update a database? The answer is found in the database management system. As discussed earlier, a DBMS is a group of programs used as an interface between a database and application programs or a database and the user. The capabilities and types of database systems, however, vary considerably. For example, Twitter, Google, Brightkite, and other Internet companies that provide GPS location applications are discussing the creation of a "Unified Database of Places." Rather than each company building proprietary databases of business and attraction locations, they would like to pool resources to build one huge database of places that includes details on every location on earth; such data would fuel applications like Google Street View.[18] Indeed, DBMSs are used to manage all kinds of data for all kinds of purposes.

Overview of Database Types

Database management systems can range from small, inexpensive software packages to sophisticated systems costing hundreds of thousands of dollars. The following sections discuss a few popular alternatives. See Figure 3.12 for one example.

Flat File

A flat file is a simple database program whose records have no relationship to one another. Flat file databases are often used to store and manipulate a single table or file; they do not use any of the database models discussed previously, such as the relational model. Many spreadsheet and word-processing programs have flat file capabilities. These software packages can sort tables and make simple calculations and comparisons. Microsoft OneNote is designed to let people put ideas, thoughts, and notes into a flat file. In OneNote, each note can be placed anywhere on a page or in a box on a page, called a *container*. Pages are organized into sections and subsections that appear as colored tabs. After you enter a note, you can retrieve, copy, and paste it into other applications, such as word-processing and spreadsheet programs. ResMed, a medical firm that manufactures products to assist people with respiratory conditions, uses OneNote to collect new ideas for product improvements and track the status of those ideas through evaluation and implementation.[19] OneNote assists the company in its efforts to increase participation in reducing costs and becoming more efficient.

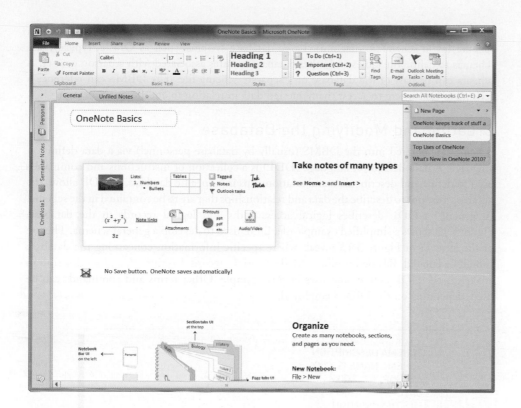

Figure 3.12

Microsoft OneNote

Microsoft OneNote lets you gather any type of information and then retrieve, copy, and paste the information into other applications, such as word-processing and spreadsheet programs.

(Source: Courtesy of Microsoft Corporation.)

Similar to OneNote, EverNote is a free online database service that can store notes and other pieces of information. Considering the amount of information today's high-capacity hard disks can store, the popularity of databases that can handle unstructured data will continue to grow.

Single User

A database installed on a personal computer is typically meant for a single user. Microsoft Office Access and FileMaker Pro are designed to support single-user implementations. Microsoft InfoPath is another example of a database program that supports a single user. This software is part of the Microsoft Office suite, and it helps people collect and organize information from a variety of sources. InfoPath has built-in forms that can be used to enter expense information, timesheet data, and a variety of other information.

Multiple Users

Small, midsize, and large businesses need multiuser DBMSs to share information throughout the organization over a network. These more powerful, expensive systems allow dozens or hundreds of people to access the same database system at the same time. Popular vendors for multiuser database systems include Oracle, Microsoft, Sybase, and IBM. Many single-user databases, such as Microsoft Access, can be implemented for multiuser support over a network, though they often are limited in the number of users they can support.

All DBMSs share some common functions, such as providing a user view, physically storing and retrieving data in a database, allowing for database modification, manipulating data, and generating reports. These DBMSs can handle the most complex data-processing tasks, and because they are accessed over a network, one database can serve many locations around the world. For example, the Linde Group is a global leader in industrial gases and hydrogen production. Its 50,000 employees, spread across 100 countries, all access a central database stored in a data center in Munich, Germany.[20]

Providing a User View

Because the DBMS is responsible for access to a database, one of the first steps in installing and using a large database involves "telling" the DBMS the logical and physical structure of

schema
A description of the entire database.

the data and the relationships among the data for each user. This description is called a **schema** (as in schematic diagram). Large database systems, such as Oracle, typically use schemas to define the tables and other database features associated with a person or user. A schema can be part of the database or a separate schema file. The DBMS can reference a schema to find where to access the requested data in relation to another piece of data.

Creating and Modifying the Database

data definition language (DDL)
A collection of instructions and commands used to define and describe data and relationships in a specific database.

Schemas are entered into the DBMS (usually by database personnel) via a data definition language. A **data definition language (DDL)** is a collection of instructions and commands used to define and describe data and relationships in a specific database. A DDL allows the database's creator to describe the data and relationships that are to be contained in the schema. In general, a DDL describes logical access paths and logical records in the database. Figure 3.13 shows a simplified example of a DDL used to develop a general schema. The use of the letter *X* in Figure 3.13 reveals where specific information concerning the database should be entered. File description, area description, record description, and set description are terms the DDL defines and uses in this example. Other terms and commands can be used, depending on the DBMS employed.

Figure 3.13

Using a Data Definition Language to Define a Schema

```
SCHEMA DESCRIPTION
SCHEMA NAME IS XXXX
AUTHOR      XXXX
DATE        XXXX
FILE DESCRIPTION
    FILE NAME IS XXXX
      ASSIGN XXXX
    FILE NAME IS XXXX
      ASSIGN XXXX
AREA DESCRIPTION
    AREA NAME IS XXXX
RECORD DESCRIPTION
    RECORD NAME IS XXXX
    RECORD ID IS XXXX
    LOCATION MODE IS XXXX
    WITHIN XXXX AREA FROM XXXX THRU XXXX
SET DESCRIPTION
    SET NAME IS XXXX
    ORDER IS XXXX
    MODE IS XXXX
    MEMBER IS XXXX
    .
    .
    .
```

data dictionary
A detailed description of all the data used in the database.

Another important step in creating a database is to establish a **data dictionary**, a detailed description of all data used in the database. The data dictionary contains the following information:

- Name of the data item
- Aliases or other names that may be used to describe the item
- Range of values that can be used
- Type of data (such as alphanumeric or numeric)
- Amount of storage needed for the item
- Notation of the person responsible for updating it and the various users who can access it
- List of reports that use the data item

A data dictionary can also include a description of data flows, the way records are organized, and the data-processing requirements. Figure 3.14 shows a typical data dictionary entry.

```
                    NORTHWESTERN MANUFACTURING

        PREPARED BY:          D. BORDWELL
        DATE:                 04 AUGUST 2010
        APPROVED BY:          J. EDWARDS
        DATE:                 13 OCTOBER 2010
        VERSION:              3.1
        PAGE:                 1 OF 1


        DATA ELEMENT NAME:    PARTNO
        DESCRIPTION:          INVENTORY PART NUMBER
        OTHER NAMES:          PTNO
        VALUE RANGE:          100 TO 5000
        DATA TYPE:            NUMERIC
        POSITIONS:            4 POSITIONS OR COLUMNS
```

Figure 3.14

A Typical Data Dictionary Entry

For example, the information in a data dictionary for the part number of an inventory item can include the following information:

- Name of the person who made the data dictionary entry (D. Bordwell)
- Date the entry was made (August 4, 2010)
- Name of the person who approved the entry (J. Edwards)
- Approval date (October 13, 2010)
- Version number (3.1)
- Number of pages used for the entry (1)
- Part name (PARTNO)
- Other part names that might be used (PTNO)
- Range of values (part numbers can range from 100 to 5,000)
- Type of data (numeric)
- Storage required (four positions are required for the part number)

A data dictionary is valuable in maintaining an efficient database that stores reliable information with no redundancy, and it makes it easy to modify the database when necessary. Data dictionaries also help computer and system programmers who require a detailed description of data elements stored in a database to create the code to access the data.

Storing and Retrieving Data

One function of a DBMS is to be an interface between an application program and the database. When an application program needs data, it requests the data through the DBMS. Suppose that to calculate the total price of a new car, a pricing program needs price data on the engine option—six cylinders instead of the standard four cylinders. The application program requests this data from the DBMS. In doing so, the application program follows a logical access path. Next, the DBMS, working with various system programs, accesses a storage device, such as disk drives, where the data is stored. When the DBMS goes to this storage device to retrieve the data, it follows a path to the physical location (physical access path) where the price of this option is stored. In the pricing example, the DBMS might go to a disk drive to retrieve the price data for six-cylinder engines. This relationship is shown in Figure 3.15.

This same process is used if a user wants to get information from the database. First, the user requests the data from the DBMS. For example, a user might give a command, such as LIST ALL OPTIONS FOR WHICH PRICE IS GREATER THAN 200 DOLLARS. This is the logical access path (LAP). Then, the DBMS might go to the options price section of a disk to get the information for the user. This is the physical access path (PAP).

Figure 3.15

Logical and Physical Access Paths

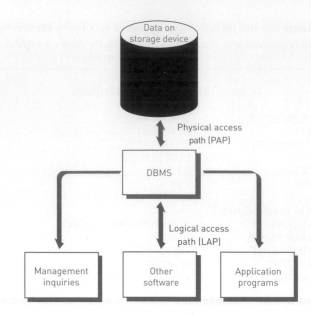

Two or more people or programs attempting to access the same record at the same time can cause a problem. For example, an inventory control program might attempt to reduce the inventory level for a product by ten units because ten units were just shipped to a customer. At the same time, a purchasing program might attempt to increase the inventory level for the same product by 200 units because inventory was just received. Without proper database control, one of the inventory updates might be incorrect, resulting in an inaccurate inventory level for the product. **Concurrency control** can be used to avoid this potential problem. One approach is to lock out all other application programs from access to a record if the record is being updated or used by another program.

Manipulating Data and Generating Reports

After a DBMS has been installed, employees, managers, and consumers can use it to review reports and obtain important information. For example, the Food Allergen and Consumer Protection Act, effective in 2006, requires that food manufacturing companies generate reports on the ingredients, formulas, and food preparation techniques for the public. Using a DBMS, a company can manage this requirement.

Some databases use *Query by Example (QBE)*, which is a visual approach to developing database queries or requests. Like Windows and other GUI operating systems, you can perform queries and other database tasks by opening windows and clicking the data or features you want. See Figure 3.16.

In other cases, database commands can be used in a programming language. For example, C++ commands can be used in simple programs that will access or manipulate certain pieces of data in the database. Here's another example of a DBMS query: SELECT * FROM EMPLOYEE WHERE JOB_CLASSIFICATION = "C2". The asterisk (*) tells the program to include all columns from the EMPLOYEE table. In general, the commands that are used to manipulate the database are part of the **data manipulation language (DML)**. This specific language, provided with the DBMS, allows managers and other database users to access and modify the data, make queries, and generate reports. Again, the application programs go through schemas and the DBMS before getting to the data stored on a device such as a disk.

In the 1970s, D. D. Chamberlain and others at the IBM Research Laboratory in San Jose, California, developed a standardized data manipulation language called *Structured Query Language (SQL)*, pronounced like the word *sequel* or spelled out as *SQL*. The EMPLOYEE query shown earlier is written in SQL. In 1986, the American National Standards Institute (ANSI) adopted SQL as the standard query language for relational databases. Since ANSI's acceptance of SQL, interest in making SQL an integral part of

concurrency control

A method of dealing with a situation in which two or more users or applications need to access the same record at the same time.

data manipulation language (DML)

A specific language, provided with a DBMS, which allows users to access and modify the data, to make queries, and to generate reports.

relational databases on both mainframe and personal computers has increased. SQL has many built-in functions, such as average (AVG), the largest value (MAX), the smallest value (MIN), and others. Table 3.3 contains examples of SQL commands.

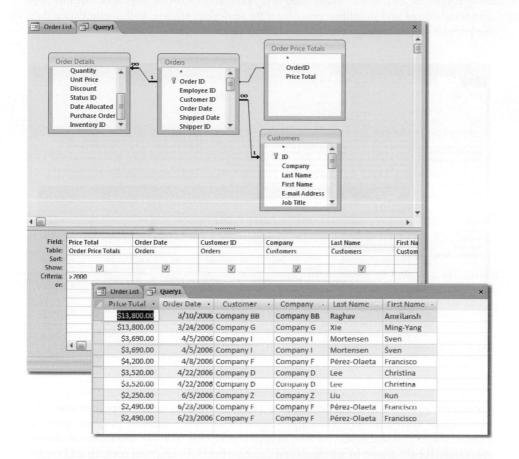

Figure 3.16

Query by Example

Some databases use Query by Example (QBE) to generate reports and information.

Table 3.3

Examples of SQL Commands

SQL Command	Description
SELECT ClientName, Debt FROM Client WHERE Debt > 1000	This query displays all clients (ClientName) and the amount they owe the company (Debt) from a database table called Client for clients who owe the company more than $1,000 (WHERE Debt > 1000).
SELECT ClientName, ClientNum, OrderNum FROM Client, Order WHERE Client.ClientNum=Order.ClientNum	This command is an example of a join command that combines data from two tables: the client table and the order table (FROM Client, Order). The command creates a new table with the client name, client number, and order number (SELECT ClientName, ClientNum, OrderNum). Both tables include the client number, which allows them to be joined. This is indicated in the WHERE clause, which states that the client number in the client table is the same as (equal to) the client number in the order table (WHERE Client.Client Num= Order.ClientNum).
GRANT INSERT ON Client to Guthrie	This command is an example of a security command. It allows Bob Guthrie to insert new values or rows into the Client table.

SQL lets programmers learn one powerful query language and use it on systems ranging from PCs to the largest mainframe computers. See Figure 3.17. Programmers and database users also find SQL valuable because SQL statements can be embedded into many programming languages, such as the widely used C++, Java, and COBOL languages. Because SQL uses standardized and simplified procedures for retrieving, storing, and manipulating data, the popular database query language can be easy to understand and use.

Figure 3.17

Structured Query Language

Structured Query Language (SQL) has become an integral part of most relational databases, as shown by this screen from Microsoft Access 2010.

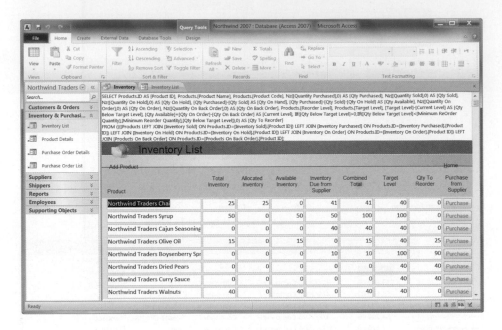

After a database has been set up and loaded with data, it can produce desired reports, documents, and other outputs. See Figure 3.18. These outputs usually appear in screen displays or hard-copy printouts. The output-control features of a database program allow you to select the records and fields you want to appear in reports. You can also make calculations specifically for the report by manipulating database fields. Formatting controls and organization options (such as report headings) help you to customize reports and create flexible, convenient, and powerful information-handling tools.

Figure 3.18

Database Output

A database application offers sophisticated formatting and organization options to produce the right information in the right format.

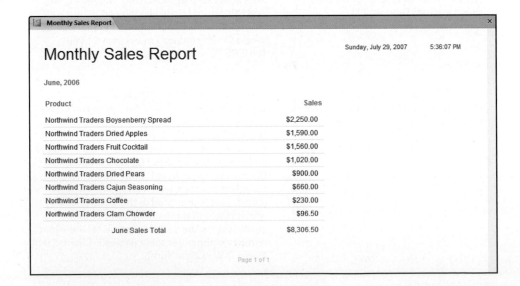

A DBMS can produce a wide variety of documents, reports, and other output that can help organizations achieve their goals. The most common reports select and organize data to present summary information about some aspect of company operations. For example, accounting reports often summarize financial data such as current and past-due accounts. Many companies base their routine operating decisions on regular status reports that show the progress of specific orders toward completion and delivery.

Database Administration

Database systems require a skilled database administrator (DBA), who is expected to have a clear understanding of the fundamental business of the organization, be proficient in the use of selected database management systems, and stay abreast of emerging technologies and new design approaches. The role of the DBA is to plan, design, create, operate, secure, monitor, and maintain databases. Typically, a DBA has a degree in computer science or management information systems and some on-the-job training with a particular database product or more extensive experience with a range of database products. See Figure 3.19.

Figure 3.19

Database Administrator

The role of the database administrator (DBA) is to plan, design, create, operate, secure, monitor, and maintain databases.

(Source: Image copyright 2010, Pinchuk Alexey. Used under license from Shutterstock.com.)

The DBA works with users to decide the content of the database—to determine exactly what entities are of interest and what attributes are to be recorded about those entities. Thus, personnel outside of IS must have some idea of what the DBA does and why this function is important. The DBA can play a crucial role in the development of effective information systems to benefit the organization, employees, and managers.

The DBA also works with programmers as they build applications to ensure that their programs comply with database management system standards and conventions. After the database is built and operating, the DBA monitors operations logs for security violations. Database performance is also monitored to ensure that the system's response time meets users' needs and that it operates efficiently. If there is a problem, the DBA attempts to correct it before it becomes serious.

A database failure can cause huge financial losses for a business. A failure due to mechanical problems, controller failures, viruses or attacks, or human failure can cause productivity in an organization to grind to a halt. Databases accessible from the Internet are at a higher level of risk from hackers and viruses than databases stored on private servers. For example, an SQL injection attack uses a Web form to issue SQL commands to a database over the Internet. The SQL command might prompt the database to reveal private data, or it might corrupt the data in the database. SQL injection attacks were used to steal 130 million credit and debit card numbers from databases owned by Heartland Payment Systems, TJX Companies, and other businesses in 2009.[21] In 2010, the Open Web Application Security Project (OWASP) listed injection attacks as the top security threat for Web applications. A large responsibility of a DBA is to protect the database from attack or other forms of failure. DBAs use security software, preventive measures, and redundant systems to keep data safe and accessible.

data administrator
A nontechnical position responsible for defining and implementing consistent principles for a variety of data issues.

Some organizations have also created a position called the **data administrator**, a nontechnical but important position responsible for defining and implementing consistent principles for a variety of data issues, including setting data standards and data definitions that apply across all the databases in an organization. For example, the data administrator would ensure that a term such as "customer" is defined and treated consistently in all corporate databases. This person also works with business managers to identify who should have read or update access to certain databases and to selected attributes within those databases. This information is then communicated to the database administrator for implementation. The data administrator can be a high-level position reporting to top-level managers.

Popular Database Management Systems

Some popular DBMSs for single users include Microsoft Access and FileMaker Pro. The complete DBMS market encompasses software used by professional programmers and that runs on midrange servers, mainframes, and supercomputers. The entire market generates billions of dollars per year in revenue by companies including IBM, Oracle, and Microsoft.

Like other software products, a number of open-source database systems are available, including PostgreSQL and MySQL. Open-source software was described in Chapter 2. In addition, many traditional database programs are now available on open-source operating systems. The popular DB2 relational database from IBM, for example, is available on the Linux operating system. The Sybase IQ database and other databases are also available on the Linux operating system.

A new form of database system is emerging that some refer to as *Database as a Service* (*DaaS*); others call it Database 2.0. DaaS is similar to Software as a Service (SaaS). Recall that a SaaS system is one in which the software is stored on a service provider's servers and is accessed by the client company over a network. In DaaS, the database is stored on a service provider's servers and accessed by the client over a network, typically the Internet. In DaaS, database administration is provided by the service provider. SaaS and DaaS are both part of the larger cloud computing trend. Cloud computing uses a giant cluster of computers that run high-performance applications. In cloud computing, all information systems and data are maintained and managed by service providers and delivered over the Internet. Businesses and individuals are freed from having to install, service, maintain, upgrade, and safeguard their systems.

More than a dozen companies are moving in the DaaS direction. They include Google, Microsoft, Oracle, Amazon, Intuit, MyOwnDB, and Trackvia. Oracle's DaaS combines cloud computing with grid computing and virtualization to provide cost-effective, reliable, and scalable database solutions.[22] Oracle provides both private clouds—accessible only to users on a private network—and public clouds—accessible to the public over the Internet. Razorfish, a digital advertising and marketing firm, uses Amazon's Elastic Cloud service to collect and analyze data (not personally identifiable) from browsing sessions using data mining techniques to develop effective marketing campaigns.[23] Procter and Gamble consolidated hundreds of projects into a single cloud database, which saved operational costs and reduced meeting time and data entry time for employees.[24]

Special-Purpose Database Systems

In addition to the popular database management systems just discussed, some specialized database packages are used for specific purposes or in specific industries. For example, Rex-Book from Urbanspoon is an iPad App designed for restaurants that utilizes a special-purpose online database to store and manage dining reservations.[25] Another unique special-purpose DBMS for biologists called Morphbank (*www.morphbank.net*) allows researchers from around the world to continually update and expand a library of more than 96,000 biological images to share with the scientific community and the public. Apple's iTunes software uses a special-purpose database system that includes fields for song name, rating, file size, time, artist, album, and genre. When iTunes users go to the iTunes store and search for an artist, they are actually querying the central iTunes database. See Figure 3.20.

Figure 3.20

iTunes Database

Apple's iTunes software uses a database to catalog and access music.

(Source: Courtesy of Apple.)

Selecting a Database Management System

The database administrator often selects the best database management system for an organization. The process begins by analyzing database needs and characteristics. The information needs of the organization affect the type of data that is collected and the type of database management system that is used. Important characteristics of databases include the following:

- *Database size*: The number of records or files in the database
- *Database cost*: The purchase or lease costs of the database
- *Concurrent users*: The number of people who need to use the database at the same time
- *Performance*: How fast the database can update records
- *Integration*: The ability to work seamlessly with other applications and databases
- *Vendor*: The reputation and financial stability of the database vendor

Using Databases with Other Software

Database management systems are often used with other software and with the Internet. A DBMS can act as a front-end application or a back-end application. A *front-end application* is one that people interact with directly. Marketing researchers often use a database as a front end to a statistical analysis program. The researchers enter the results of market questionnaires or surveys into a database. The data is then transferred to a statistical analysis program to determine the potential for a new product or the effectiveness of an advertising campaign. A *back-end application* interacts with other programs or applications; it only indirectly interacts with people or users. When people request information from a Web site, the Web site can interact with a database (the back end) that supplies the desired information. For example, you can connect to a university Web site to find out whether the university's library has a book you want to read. The Web site then interacts with a database that contains a catalog of library books and articles to determine whether the book you want is available. See Figure 3.21.

In some situations, front-end systems cannot connect directly to a back-end database due to compatibility issues. Middleware solutions, such as Oracle's Fusion software, are available to connect systems seamlessly, interpreting data from a variety of sources and translating it to a format compatible with the database.

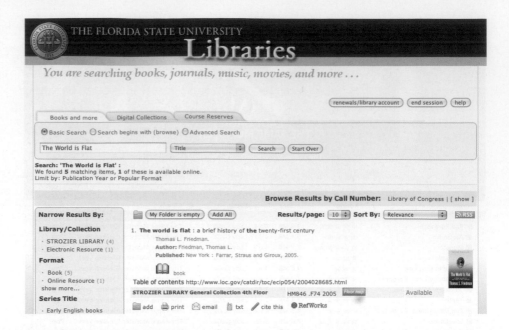

DATABASE APPLICATIONS

Today's database applications manipulate the content of a database to produce useful information. Common manipulations are searching, filtering, synthesizing, and assimilating the data, using a number of database applications. These applications allow users to link the company databases to the Internet, set up data warehouses and marts, use databases for strategic business intelligence, place data at different locations, use online processing and open connectivity standards for increased productivity, develop databases with the object-oriented approach, and search for and use unstructured data, such as graphics, audio, and video.

Linking the Company Database to the Internet

The ability to link databases to the Internet is one reason the Internet is so popular. A large percentage of corporate databases are accessed over the Internet through a standard Web browser. Being able to access bank account data, student transcripts, credit card bills, product catalogs, and a host of other data online is convenient for individual users and increases effectiveness and efficiency for businesses and organizations. Amazon.com, eHarmony.com, eBay, and many others have made billions of dollars by combining databases, the Internet, and smart business models.

Google, Microsoft, and others have developed Personal Health Record (PHR) systems designed to provide physicians and patients a single storage location for all medical records, accessed through a Web browser.[26] Google Health and Microsoft HealthVault provide "patient-centered" health records that empower patients to more easily participate in their own health care. President Obama is pushing to establish electronic health (e-health) records for all Americans prior to 2015 by making $17 billion available to e-health projects and programs. Database companies will be investing significant effort in developing health and medical databases systems that are accessible on the Internet.

Access to private medical information over the public Web has some privacy advocates concerned. However, the convenience that the system offers by dramatically reducing the number of paper forms to fill out and store, along with the reduction of clerical errors through streamlined data management procedures, has most in the field supporting the move to a centralized system. Encryption and authentication technologies will be used to make the systems as secure as possible.

Developing a seamless integration of databases with the Internet is sometimes called a *semantic Web*. A semantic Web provides metadata with all Web content using technology called the Resource Description Framework (RDF).[27] The result is a more organized Web that acts like one large database system. The World Wide Web Consortium (W3C) has established standards, including an RDF, for a semantic Web in hopes of bringing content providers onboard.

Data Warehouses, Data Marts, and Data Mining

The raw data necessary to make sound business decisions is stored in a variety of locations and formats. This data is initially captured, stored, and managed by transaction processing systems that are designed to support the day-to-day operations of the organization. For decades, organizations have collected operational, sales, and financial data with their online transaction processing (OLTP) systems. The data can be used to support decision making through data warehouses, data marts, and data mining.

Data Warehouses

A **data warehouse** is a database that holds business information from many sources in the enterprise, covering all aspects of the company's processes, products, and customers. The data warehouse provides business users with a multidimensional view of the data they need to analyze business conditions. Data warehouses allow managers to *drill down* to get more detail or *roll up* to take detailed data and generate aggregate or summary reports. A data warehouse is designed specifically to support management decision making, not to meet the needs of transaction processing systems. A data warehouse stores historical data that has been extracted from operational systems and external data sources. See Figure 3.22. This operational and external data is "cleaned up" to remove inconsistencies and integrated to create a new information database that is more suitable for business analysis.

data warehouse
A large database that collects business information from many sources in the enterprise, covering all aspects of the company's processes, products, and customers, in support of management decision making.

Figure 3.22

Elements of a Data Warehouse

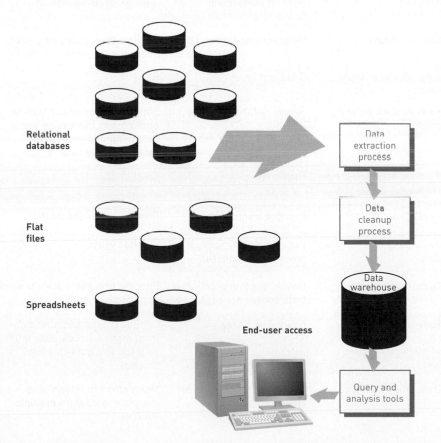

Data warehouses typically start out as very large databases, containing millions and even hundreds of millions of data records. As this data is collected from the various production systems, a historical database is built that business analysts can use to track changes in an organization over time and analyze current conditions. To keep it fresh and accurate, the data warehouse receives regular updates. Old data that is no longer needed is purged from the data warehouse. Updating the data warehouse must be fast, efficient, and automated, or the ultimate value of the data warehouse is sacrificed. It is common for a data warehouse to contain from three to ten years of current and historical data. Data-cleaning tools can merge data from many sources into one database, automate data collection and verification, delete unwanted data, and maintain data in a database management system.

Data warehouses can also acquire data from unique sources. Oracle's Warehouse Management software, for example, can accept information from Radio Frequency Identification (RFID) technology, which is being used to tag products as they are shipped or moved from one location to another. Honda Italia, the world leader in powered two-wheel vehicle manufacturing, uses RFID to feed its data warehouse with information about production. Each vehicle component is tagged with an RFID chip so it can be tracked through the entire production process. The RFID-based system provides highly detailed information to production managers who can tweak production to quickly identify problems and improve supply with little or no wasted effort or resources.[28]

The primary advantage of data warehousing is the ability to relate data in innovative ways. However, a data warehouse for a large organization can be extremely difficult to establish, with the typical cost exceeding $2 million. Table 3.4 compares online transaction processing (OLTP) and data warehousing.

Table 3.4

Comparison of OLTP and Data Warehousing

Characteristic	OLTP Database	Data Warehousing
Purpose	Support transaction processing	Support decision making
Source of data	Business transactions	Multiple files, databases—data internal and external to the firm
Data access allowed users	Read and write	Read only
Primary data access mode	Simple database update and query	Simple and complex database queries with increasing use of data mining to recognize patterns in the data
Primary database model employed	Relational	Relational
Level of detail	Detailed transactions	Often summarized data
Availability of historical data	Very limited—typically a few weeks or months	Multiple years
Update process	Online, ongoing process as transactions are captured	Periodic process, once per week or once per month
Ease of process	Routine and easy	Complex, must combine data from many sources; data must go through a data cleanup process
Data integrity issues	Each transaction must be closely edited	Major effort to "clean" and integrate data from multiple sources

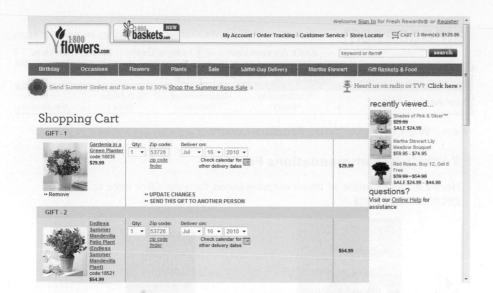

1-800-flowers.com uses a data warehouse to reference customer historical data to determine customer interests based on past interactions.[29]

(Source: 1800flowers.com.)

Data Marts

A **data mart** is a subset of a data warehouse. Data marts bring the data warehouse concept—online analysis of sales, inventory, and other vital business data that has been gathered from transaction processing systems—to small and medium-sized businesses and to departments within larger companies. Rather than store all enterprise data in one monolithic database, data marts contain a subset of the data for a single aspect of a company's business—for example, finance, inventory, or personnel. In fact, a specific area in the data mart might contain more detailed data than the data warehouse.

Data marts are most useful for smaller groups who want to access detailed data. A warehouse contains summary data that can be used by an entire company. Because data marts typically contain tens of gigabytes of data, as opposed to the hundreds of gigabytes in data warehouses, they can be deployed on less powerful hardware with smaller secondary storage devices, delivering significant savings to an organization. Although any database software can be used to set up a data mart, some vendors deliver specialized software designed and priced specifically for data marts. Companies such as Sybase, Software AG, Microsoft, and others have products and services that make it easier and cheaper to deploy these scaled-down data warehouses. The selling point: data marts put targeted business information into the hands of more decision makers. For example, the U.S. Department of Defense has created the Defense Health Services Systems' Clinical Data Mart (CDM) to deliver medical information to the more than 9 million military personnel worldwide. The system was developed in response to President Obama's call to "raise health care quality at lower costs."[30]

data mart
A subset of a data warehouse, used by small and medium-sized businesses and departments within large companies to support decision making.

Data Mining

Data mining is an information-analysis tool that involves the automated discovery of patterns and relationships in a data warehouse. Like gold mining, data mining sifts through mountains of data to find a few nuggets of valuable information. For example, Brooks Brothers, the oldest clothing retailer in the U.S., uses data mining to provide store managers with reports that help improve store performance and customer satisfaction.[31]

Data mining's objective is to extract patterns, trends, and rules from data warehouses to evaluate (i.e., predict or score) proposed business strategies, which will improve competitiveness, increase profits, and transform business processes. It is used extensively in marketing to improve customer retention; cross-selling opportunities; campaign management; market, channel, and pricing analysis; and customer segmentation analysis (especially one-to-one marketing). In short, data-mining tools help users find answers to questions they haven't thought to ask.

data mining
An information-analysis tool that involves the automated discovery of patterns and relationships in a data warehouse.

Amazon uses data mining to recommend products that will entice visitors.

(Source: *www.amazon.com*)

E-commerce presents another major opportunity for effective use of data mining. Attracting customers to Web sites is tough; keeping them there and ensuring they return is even tougher. For example, when retail Web sites launch deep-discount sales, they cannot easily determine how many first-time customers are likely to come back and buy again. Nor do they have a way of understanding which customers acquired during the sale are more likely to jump on future sales. As a result, companies are gathering data on user traffic through their Web sites and storing the data in databases. This data is then analyzed using data-mining techniques to personalize the Web site and develop sales promotions targeted at specific customers. Facebook has angered users on several occasions for sharing member data with commercial partners who then use the information in data mining to fuel targeted marketing campaigns. Some members feel that the practice is a violation of their privacy.[32]

predictive analysis

A form of data mining that combines historical data with assumptions about future conditions to predict outcomes of events, such as future product sales or the probability that a customer will default on a loan.

Predictive analysis is a form of data mining that combines historical data with assumptions about future conditions to predict outcomes of events, such as future product sales or the probability that a customer will default on a loan. Retailers use predictive analysis to upgrade occasional customers into frequent purchasers by predicting what products they will buy if offered an appropriate incentive. Genalytics, Magnify, NCR Teradata, SAS Institute, Sightward, SPSS, and Quadstone have developed predictive analysis tools. Predictive analysis software can be used to analyze a company's customer list and a year's worth of sales data to find new market segments that could be profitable.

American Airlines Consumer Research Department uses predictive analysis to guide its corporate decisions. Passengers on about 100 of the 3,300 flights that American flies each day are asked to fill out a brief survey. The data is loaded into a data warehouse where predictive analytics software processes the information to provide useful statistics. The statistics are analyzed to help determine flight timetables, flight staffing, in-flight services such as food and Internet, and other airline considerations.[33]

Traditional DBMS vendors are well aware of the great potential of data mining. Thus, companies such as Oracle, Sybase, Tandem, and Red Brick Systems are all incorporating data-mining functionality into their products. Table 3.5 summarizes a few of the most frequent applications for data mining.

Application	Description
Branding and positioning of products and services	Enable the strategist to visualize the different positions of competitors in a given market using performance (or other) data on dozens of key features of the product and then to condense all that data into a perceptual map of only two or three dimensions.
Customer churn	Predict current customers who are likely to switch to a competitor.
Direct marketing	Identify prospects most likely to respond to a direct marketing campaign (such as a direct mailing).
Fraud detection	Highlight transactions most likely to be deceptive or illegal.
Market basket analysis	Identify products and services that are most commonly purchased at the same time (e.g., nail polish and lipstick).
Market segmentation	Group customers based on who they are or on what they prefer.
Trend analysis	Analyze how key variables (e.g., sales, spending, promotions) vary over time.

Table 3.5

Common Data-Mining Applications

Business Intelligence

The use of databases for business-intelligence purposes is closely linked to the concept of data mining. **Business intelligence** (BI) involves gathering enough of the right information in a timely manner and usable form and analyzing it so that it can have a positive effect on business strategy, tactics, or operations. IMS Health, for example, provides a BI system designed to assist businesses in the pharmaceutical industry with custom marketing to physicians, pharmacists, nurses, consumers, government agencies, and nonprofit healthcare organizations.[34] BI turns data into useful information that is then distributed throughout an enterprise. It provides insight into the causes of problems and, when implemented, can improve business operations. For example, Puma North America, manufacturer of athletic footwear, uses SPSS software to provide business intelligence to its sales consultants. Puma's 70 independent sales consultants depend on SPSS for the information they need to make business decisions regarding orders, shipments, and product availability.[35]

Competitive intelligence is one aspect of business intelligence and is limited to information about competitors and the ways that knowledge affects strategy, tactics, and operations. Competitive intelligence is a critical part of a company's ability to see and respond quickly and appropriately to the changing marketplace. Competitive intelligence is not espionage—the use of illegal means to gather information. In fact, almost all the information a competitive-intelligence professional needs can be collected by examining published information sources, conducting interviews, and using other legal, ethical methods. Using a variety of analytical tools, a skilled competitive-intelligence professional can by deduction fill the gaps in information already gathered.

The term **counterintelligence** describes the steps an organization takes to protect information sought by "hostile" intelligence gatherers. One of the most effective counterintelligence measures is to define "trade secret" information relevant to the company and control its dissemination.

Data loss prevention (DLP) refers to systems designed to lock down data within an organization. DLP software from RSA, Symantec, Code Green, Safend, Trend Micro, Sophos, and others are designed to identify, monitor, and protect data wherever it may exist on a system. That includes data stored on disk, passing over a network, in databases, in files, in e-mail, and elsewhere. DLP is a powerful tool for counterintelligence and is a necessity in complying with government regulations that require companies to safeguard private customer data.[36]

business intelligence (BI)
The process of gathering enough of the right information in a timely manner and usable form and analyzing it to have a positive impact on business strategy, tactics, or operations.

competitive intelligence
One aspect of business intelligence limited to information about competitors and the ways that knowledge affects strategy, tactics, and operations.

counterintelligence
The steps an organization takes to protect information sought by "hostile" intelligence gatherers.

data loss prevention (DLP)
Systems designed to lock down—to identify, monitor, and protect—data within an organization.

The Database that Drives the Austrian Turnpike

ASFINAG Maut Service GmbH is the company responsible for planning, financing, building, maintaining, and operating the Austrian turnpike and highway system—all 2,100 kilometers of it. As with most European countries, Austria has relied on manually collected tolls to finance its highway system. Recently, the country turned to state-of-the-art database-driven systems to transport its highways into the twenty-first century.

Bernd Datler, head of system development for ASFINAG Maut Service GmbH, calls it "the world's first fully automated, free-flowing tolling system for commercial vehicles." ASFINAG hired Austrian IS service provider Raiffeisen Informatik GmbH to implement a system that would tag and track more than 700,000 commercial vehicles across Austrian roads, automatically billing each vehicle according to complicated specifications.

The database that supports this massive system would have to contend with a variety of data formats and high frequency of data input. It would feed numerous systems to serve a variety of needs.

The fully automated, free-flowing tolling system begins with driver registration. Drivers can register online, by phone, or at local sales centers. The registration process collects information about the driver, the vehicle, and the company that employs the driver. This information is fed into the database and is accessed by a customer relationship management (CRM) system as needed. Upon registration, drivers are provided with a radio transceiver box that is mounted to the dashboard of the commercial vehicle.

The 800 tollgates along Austrian highways were fitted with special microwave receivers that connect with the boxes on drivers' dashboards without the drivers needing to stop. As drivers pass a tollgate, data is continuously collected and entered into the database as transactions. Rather than billing a flat rate, the automated system allows for custom rates to be applied. Fees are calculated based on several criteria, including the size of the vehicle, whether it is full or empty, the time of day, and the vehicle's emission class. These last two criteria can be used to motivate drivers to travel at off-peak times and to use vehicles with low emissions.

The database also collects photographic data. Cameras mounted at tollgates photograph every vehicle to catch unregistered vehicles. The photos are used to see vehicle tags and registration to track down the vehicle's owner. The photo system is also used to collect tolls from noncommercial vehicles that use a registration sticker on the windshield.

Drivers can access their toll information online using a Web-based portal that delivers real-time reporting. Data is automatically transferred into a data warehouse, where ASFINAG managers have access to powerful business intelligence (BI) tools that allow them to generate reports on highway usage from multiple perspectives. The system manages 2.5 million transactions per week, without the need for any human intervention. Invoices are automatically generated in the customer's native language and are delivered electronically. The system was designed to be interoperable as well, reading not only Austrian-registered vehicles, but also vehicles registered in Switzerland, Germany, and Italy.

All in all, the project took more than 100 Raiffeisen Informatik IS professionals 18 months to complete. The team met its deadline and hit its goals in terms of quality, functionality, and costs.

Discussion Questions

1. What unique challenges did the ASFINAG project present for database installation, administration, and security?
2. How does the "world's first fully automated, free-flowing tolling system for commercial vehicles" benefit drivers and the highway system?

Critical Thinking Questions

1. What business functions are supported by the database at ASFINAG?
2. What database applications discussed in the chapter are used in conjunction with the ASFINAG database?

Sources: "Raiffeisen Informatik - SAP Software Powers Outsourced Toll-Collection System," SAP Customer Success Story, *www.sap.com/solutions/sap-businessobjects/customers*, accessed May 15, 2010; ASFiNAG Web site, *www.asfinag.at/en*, accessed May 15, 2010.

Distributed Databases

Distributed processing involves placing processing units at different locations and linking them via telecommunications equipment. A **distributed database**—a database in which the data can be spread across several smaller databases connected through telecommunications devices—works on much the same principle. A user in the Milwaukee branch of a clothing manufacturer, for example, might make a request for data that is physically located at corporate headquarters in Milan, Italy. The user does not have to know where the data is physically stored. See Figure 3.23.

Distributed databases give corporations and other organizations more flexibility in how databases are organized and used. Local offices can create, manage, and use their own databases, and people at other offices can access and share the data in the local databases. Giving local sites more direct access to frequently used data can improve organizational effectiveness and efficiency significantly. The New York City Police Department, for example, has thousands of officers searching for information located on servers in offices around the city.

Despite its advantages, distributed processing creates additional challenges in integrating different databases (information integration), maintaining data security, accuracy, timeliness, and conformance to standards. Distributed databases allow more users direct access at different sites; however, controlling who accesses and changes data is sometimes difficult. Also, because distributed databases rely on telecommunications lines to transport data, access to data can be slower.

To reduce telecommunications costs, some organizations build a replicated database. A **replicated database** holds a duplicate set of frequently used data. The company sends a copy of important data to each distributed processing location when needed or at predetermined times. Each site sends the changed data back to update the main database on an update cycle that meets the needs of the organization. This process, often called *data synchronization*, is used to make sure that replicated databases are accurate, up to date, and consistent with each other. A railroad, for example, can use a replicated database to increase punctuality, safety, and reliability. The primary database can hold data on fares, routings, and other essential information. The data can be continually replicated and downloaded on a read-only basis from the master database to hundreds of remote servers across the country. The remote locations can send back to the main database the latest figures on ticket sales and reservations.

distributed database
A database in which the data can be spread across several smaller databases connected via telecommunications devices.

replicated database
A database that holds a duplicate set of frequently used data.

For a clothing manufacturer,
computers might be located at
corporate headquarters, in the
research and development center,
in the warehouse, and in a company-
owned retail store.
Telecommunications systems link
the computers so that users at all
locations can access the same
distributed database, no matter
where the data is actually stored.

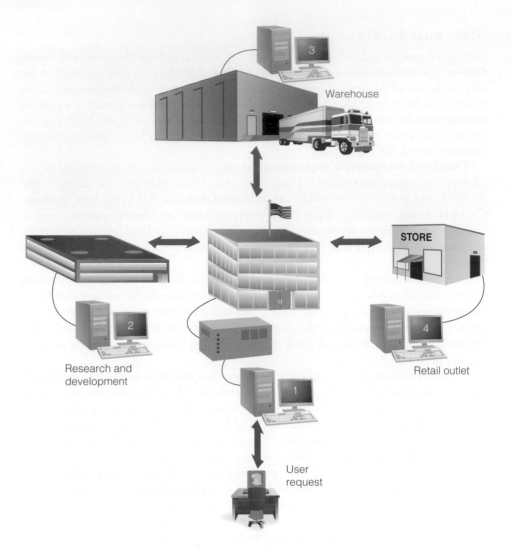

**online analytical processing
(OLAP)**
Software that allows users to
explore data from a number of
perspectives.

Online Analytical Processing (OLAP)

For nearly two decades, multidimensional databases and their analytical information display systems have provided flashy sales presentations and trade show demonstrations. All you have to do is ask where a certain product is selling well, for example, and a colorful table showing sales performance by region, product type, and time frame appears on the screen. Called **online analytical processing (OLAP)**, these programs are now being used to store and deliver data warehouse information efficiently. The leading OLAP software vendors include Microsoft, Cognos, SAP, Business Objects, MicroStrategy, Applix, Infor, and Oracle. Blue Mountain, Ontario's largest mountain resort, uses OLAP to allow its analysts, managers, and executives to quickly view and understand large sets of complex data. The resort includes 13 business lines, including restaurants, ski ticketing, call centers, and lodging. Decision makers use the OLAP system to view their data across multiple dimensions and drill down to access specifics.[37]

The value of data ultimately lies in the decisions it enables. Powerful information-analysis tools in areas such as OLAP and data mining, when incorporated into a data warehousing architecture, bring market conditions into sharper focus and help organizations deliver greater competitive value. OLAP provides top-down, query-driven data analysis; data mining provides bottom-up, discovery-driven analysis. OLAP requires repetitive testing of user-originated theories; data mining requires no assumptions and instead identifies facts and conclusions based on patterns discovered. OLAP, or multidimensional analysis, requires a great deal of human ingenuity and interaction with the database to find information in the database. A user of a data-mining tool does not need to figure out what questions to ask;

instead, the approach is, "Here's the data, tell me what interesting patterns emerge." For example, a data-mining tool in a credit card company's customer database can construct a profile of fraudulent activity from historical information. Then, this profile can be applied to all incoming transaction data to identify and stop fraudulent behavior, which might otherwise go undetected. Table 3.6 compares OLAP and data mining.

Characteristic	OLAP	Data Mining
Purpose	Supports data analysis and decision making	Supports data analysis and decision making
Type of analysis supported	Top-down, query-driven data analysis	Bottom-up, discovery-driven data analysis
Skills required of user	Must be very knowledgeable of the data and its business context	Must trust in data mining tools to uncover valid and worthwhile hypotheses

Table 3.6

Comparison of OLAP and Data Mining

Object-Relational Database Management Systems

An **object-oriented database** uses the same overall approach of objected-oriented programming that was discussed in Chapter 2. With this approach, both the data and the processing instructions are stored in the database. For example, an object oriented database could store monthly expenses and the instructions needed to compute a monthly budget from those expenses. A traditional DBMS might only store the monthly expenses. The popular Internet phone service Skype has been pleased with its object-oriented database from PostgreSQL. The object-oriented nature of the database allowed Skype to develop the database as the company grew and evolved.[38] Object-oriented databases are useful when a database contains complex data that needs to be processed quickly and efficiently.

In an object-oriented database, a *method* is a procedure or action. A sales tax method, for example, could be the procedure to compute the appropriate sales tax for an order or sale—for example, multiplying the total amount of an order by 7 percent, if that is the local sales tax. A *message* is a request to execute or run a method. For example, a sales clerk could issue a message to the object-oriented database to compute sales tax for a new order. Many object-oriented databases have their own query language, called *object query language* (*OQL*), which is similar to SQL, discussed previously.

An object-oriented database uses an **object-oriented database management system** (**OODBMS**) to provide a user interface and connections to other programs. Computer vendors who sell or lease OODBMSs include Versant and Objectivity. Many organizations are selecting object-oriented databases for their processing power. Versant's OODBMS, for example, is being used by companies in the telecommunications, defense, online gaming, and healthcare industries and by government agencies. The *Object Data Standard* is a design standard created by the *Object Database Management Group* (*www.odmg.org*) for developing object-oriented database systems.

An **object-relational database management system** (**ORDBMS**) provides a complete set of relational database capabilities plus the ability for third parties to add new data types and operations to the database. These new data types can be audio, images, unstructured text, spatial, or time series data that require new indexing, optimization, and retrieval features. Each of the vendors offering ORDBMS facilities provides a set of application programming interfaces to allow users to attach external data definitions and methods associated with those definitions to the database system. They are essentially offering a standard socket into which users can plug special instructions. DataBlades, Cartridges, and Extenders are the names applied by Oracle and IBM to describe the plug-ins to their respective products. Other plug-ins serve as interfaces to Web servers.

object-oriented database
A database that stores both data and its processing instructions.

object-oriented database management system (OODBMS)
A group of programs that manipulate an object-oriented database and provide a user interface and connections to other application programs.

object-relational database management system (ORDBMS)
A DBMS capable of manipulating audio, video, and graphical data.

Visual, Audio, and Other Database Systems

In addition to raw data, organizations are finding a need to store large amounts of visual and audio signals in an organized fashion. Credit card companies, for example, enter pictures of charge slips into an image database using a scanner. The images can be stored in the database and later sorted by customer name, printed, and sent to customers along with their monthly statements. Image databases are also used by physicians to store x-rays and transmit them to clinics away from the main hospital. Financial services, insurance companies, and government branches are using image databases to store vital records and replace paper documents. Drug companies often need to analyze many visual images from laboratories. Visual databases can be stored in some object-relational databases or special-purpose database systems. Many relational databases can also store images.

Combining and analyzing data from different databases is an increasingly important challenge. Global businesses, for example, sometimes need to analyze sales and accounting data stored around the world in different database systems. Companies such as IBM have developed *virtual database systems* to allow different databases to work together as a unified database system. The joining of separate databases into one is sometimes referred to as a *federated database system*. World-renowned insurer, Lloyd's of London, joined its database with that of International Underwriting Association (IUA), creating a virtual database that ensured improved customer service at a lower cost.[39]

In addition to visual, audio, and virtual databases, other special-purpose database systems meet particular business needs. *Spatial data technology* involves using a database to store and access data according to the locations it describes and to permit spatial queries and analysis. MapInfo software from Pitney Bowes allows businesses such as Home Depot, Sonic Restaurants, CVS Corporation, and Chico's to choose the optimal location for new stores and restaurants based on geospatial demographics. It also can be used to assist law enforcement agencies and emergency response teams to prepare for emergencies and provide community protection in an efficient manner.[40] The software provides information about local competition, populations, and traffic patterns to predict how a business will fare in a particular location. Builders and insurance companies use spatial data to make decisions related to natural hazards. Spatial data can even be used to improve financial risk management with information stored by investment type, currency type, interest rates, and time. Spatial data technology is a powerful tool that geographic information systems (GIS) use to plot information on a map.

Spatial data technology is used by law enforcement agencies to provide protection where it is most needed.

(Source: © David R. Frazier Photolibrary, Inc./Alamy.)

SUMMARY

Principle

Data management and modeling are key aspects of organizing data and information.

Data is one of the most valuable resources that a firm possesses. It is organized into a hierarchy that builds from the smallest element to the largest. The smallest element is the bit, a binary digit. A byte (a character such as a letter or numeric digit) is made up of eight bits. A group of characters, such as a name or number, is called a field (an object). A collection of related fields is a record; a collection of related records is called a file. The database, at the top of the hierarchy, is an integrated collection of records and files.

An entity is a generalized class of objects for which data is collected, stored, and maintained. An attribute is a characteristic of an entity. Specific values of attributes—called data items—can be found in the fields of the record describing an entity. A data key is a field within a record that is used to identify the record. A primary key uniquely identifies a record, while a secondary key is a field in a record that does not uniquely identify the record.

Traditional file-oriented applications are often characterized by program-data dependence, meaning that they have data organized in a manner that cannot be read by other programs. To address problems of traditional file-based data management, the database approach was developed. Benefits of this approach include reduced data redundancy, improved data consistency and integrity, easier modification and updating, data and program independence, standardization of data access, and more-efficient program development.

When building a database, an organization must consider content, access, logical structure, and physical organization of the database. Many enterprises build a data center to house the servers that physically store databases and the systems that deliver mission-critical information and services. One of the tools that database designers use to show the logical structure and relationships among data is a data model. A data model is a map or diagram of entities and their relationships. Enterprise data modeling involves analyzing the data and information needs of an entire organization. Entity-relationship (ER) diagrams can be used to show the relationships among entities in the organization.

The relational model places data in two-dimensional tables. Tables can be linked by common data elements, which are used to access data when the database is queried. Each row represents a record, and each column represents an attribute (or field). Allowable values for these attributes are called the domain. Basic data manipulations include selecting, projecting, and joining. The relational model is easier to control, more flexible, and more intuitive than the other models because it organizes data in tables.

Principle

A well-designed and well-managed database is an extremely valuable tool in supporting decision making.

A DBMS is a group of programs used as an interface between a database and its users and other application programs. When an application program requests data from the database, it follows a logical access path. The actual retrieval of the data follows a physical access path. Records can be considered in the same way: a logical record is what the record contains; a physical record is where the record is stored on storage devices. Schemas are used to describe the entire database, its record types, and their relationships to the DBMS.

A DBMS provides four basic functions: providing user views, creating and modifying the database, storing and retrieving data, and manipulating data and generating reports. Schemas are entered into the computer via a data definition language, which describes the data and relationships in a specific database. Another tool used in database management is the data dictionary, which contains detailed descriptions of all data in the database.

After a DBMS has been installed, the database can be accessed, modified, and queried via a data manipulation language. A more specialized data manipulation language is the query language, the most common being Structured Query Language (SQL). SQL is used in several popular database packages today and can be installed on PCs and mainframes.

Popular single-user DBMSs include Corel Paradox and Microsoft Access. IBM, Oracle, and Microsoft are the leading DBMS vendors. Database as a Service (DaaS), or Database 2.0, is a new form of database service in which clients lease use of a database on a service provider's site.

A database administrator (DBA) plans, designs, creates, operates, secures, monitors, and maintains databases. Attacks on databases such as SQL injection attacks are an all-too-common threat that DBAs must guard against. Selecting a DBMS begins by analyzing the information needs of the organization. Important characteristics of databases include the size of the database, the number of concurrent users, its performance, the ability of the DBMS to be integrated with other systems, the features of the DBMS, the vendor considerations, and the cost of the database management system.

Principle

The number and types of database applications will continue to evolve and yield real business benefits.

Traditional online transaction processing (OLTP) systems put data into databases very quickly, reliably, and efficiently, but they do not support the types of data analysis that today's businesses and organizations require. To address this need, organizations are building data warehouses, which are relational database management systems specifically designed to support management decision making. Data marts are subdivisions of data warehouses, which are commonly devoted to specific purposes or functional business areas.

Data mining, which is the automated discovery of patterns and relationships in a data warehouse, is a practical approach to generating hypotheses about the data that can be used to predict future behavior.

Predictive analysis is a form of data mining that combines historical data with assumptions about future conditions to forecast outcomes of events such as future product sales or the probability that a customer will default on a loan.

Business intelligence is the process of getting enough of the right information in a timely manner and usable form and analyzing it so that it can have a positive effect on business strategy, tactics, or operations. Competitive intelligence is one aspect of business intelligence limited to information about competitors and the ways that information affects strategy, tactics, and operations. Competitive intelligence is not espionage—the use of illegal means to gather information. Counterintelligence describes the steps an organization takes to protect information sought by "hostile" intelligence gatherers. Data loss prevention (DLP) refers to systems designed to lock down data within an organization.

With the increased use of telecommunications and networks, distributed databases, which allow multiple users and different sites access to data that may be stored in different physical locations, are gaining in popularity. To reduce telecommunications costs, some organizations build replicated databases, which hold a duplicate set of frequently used data.

Multidimensional databases and online analytical processing (OLAP) programs are being used to store data and allow users to explore the data from a number of different perspectives.

An object-oriented database uses the same overall approach of objected-oriented programming, first discussed in Chapter 2. With this approach, both the data and the processing instructions are stored in the database. An object-relational database management system (ORDBMS) provides a complete set of relational database capabilities, plus the ability for third parties to add new data types and operations to the database. These new data types can be audio, video, and graphical data that require new indexing, optimization, and retrieval features.

In addition to raw data, organizations are finding a need to store large amounts of visual and audio signals in an organized fashion. A number of special-purpose database systems are also being used.

CHAPTER 5: SELF-ASSESSMENT TEST

Data management and modeling are key aspects of organizing data and information.

1. A group of programs that manipulate the database and provide an interface between the database and the user of the database and other application programs is called a(n) _____.
   a. GUI
   b. operating system
   c. DBMS
   d. productivity software

2. A(n) _____ is a skilled and trained IS professional who directs all activities related to an organization's database.

3. A field is made up of multiple records. True or False?

4. A(n) _____ is a field or set of fields that uniquely identifies a database record.
   a. attribute
   b. data item
   c. key
   d. primary key

5. The _____ approach provides a pool of related data shared by multiple information systems.

6. Many businesses store their database and related systems in climate-controlled facilities called _____.

7. What database model places data in two-dimensional tables?
   a. relational
   b. network
   c. normalized
   d. hierarchical

A well-designed and well-managed database is an extremely valuable tool in supporting decision making.

8. _____ involves combining two or more database tables.

a. Projecting
b. Joining
c. Selecting
d. Data cleanup

9. Because the DBMS is responsible for providing access to a database, one of the first steps in installing and using a database involves telling the DBMS the logical and physical structure of the data and relationships among the data in the database. This description of an entire database is called a(n) _____.

10. The commands used to access and report information from the database are part of the _____.
    a. data definition language
    b. data manipulation language
    c. data normalization language
    d. schema

11. Access is a popular DBMS for _____ .
    a. personal computers
    b. graphics workstations
    c. mainframe computers
    d. supercomputers

12. A trend in database management, known as Database as a Service, places the responsibility of storing and managing a database on a service provider. True or False?

The number and types of database applications will continue to evolve and yield real business benefits.

13. A(n) _____ holds business information from many sources in the enterprise, covering all aspects of the company's processes, products, and customers.

14. An information-analysis tool that involves the automated discovery of patterns and relationships in a data warehouse is called _____.
    a. a data mart
    b. data mining
    c. predictive analysis
    d. business intelligence

15. _____ allows users to predict the future based on database information from the past and present.

16. The process of gathering information in a timely manner and in a usable form so that it positively affects business strategy, tactics, and operations is called _____.

CHAPTER 5: SELF-ASSESSMENT TEST ANSWERS

(1) c (2) database administrator (3) False (4) d (5) database (6) data centers (7) a (8) b (9) schema (10) b (11) a (12) True (13) data warehouse (14) b (15) Predictive analysis (16) business intelligence

REVIEW QUESTIONS

1. What is an attribute? How is it related to an entity?
2. Define the term *database*. How is it different from a database management system?
3. What is the hierarchy of data in a database?
4. What is a relation, and what is its importance to relational databases?
5. What is the purpose of a primary key? How is it useful in controlling data redundancy?
6. What is the purpose of data cleanup?
7. What are the advantages of the database approach over the traditional approach to database management?
8. What is data modeling? What is its purpose? Briefly describe three commonly used data models.
9. What is a data center, and why are they becoming increasingly important?
10. What is a database schema, and what is its purpose?
11. How can a data dictionary be useful to database administrators and DBMS software engineers?
12. Identify important characteristics in selecting a database management system.
13. What is the difference between a data definition language (DDL) and a data manipulation language (DML)?
14. What is the difference between projecting and joining?
15. What is a distributed database system?
16. What is a data warehouse, and how is it different from a traditional database used to support OLTP?
17. What is meant by the "front end" and the "back end" of a DBMS?
18. What is the relationship between the Internet and databases?
19. What is data mining? What is OLAP? How are they different?
20. What is an ORDBMS? What kind of data can it handle?
21. What is business intelligence? How is it used?
22. What is predictive analysis, and how does it assist businesses in gaining competitive advantage?
23. In what circumstances might a database administrator consider using an object-oriented database?

DISCUSSION QUESTIONS

1. You have been selected to represent the student body on a project to develop a new student database for your school. What is the first step in developing the database? What actions might you take to fulfill this responsibility to ensure that the project meets the needs of students and is successful?

2. Your company wants to increase revenues from its existing customers. How can data mining be used to accomplish this objective?

3. You are going to design a database for your school's outdoors club to track its activities. Identify the database characteristics most important to you in choosing a DBMS. Which of the database management systems described in this chapter would you choose? Why? Is it important for you to know what sort of computer the database will run on? Why or why not?

4. Make a list of the databases in which data about you exists. How is the data in each database captured? Who updates each database and how often? Is it possible for you to request a printout of the contents of your data record from each database? What data privacy concerns do you have?

5. If you were the database administrator for the iTunes store, how might you use predictive analysis to determine which artists and movies will sell most next year?

6. You are the vice president of information technology for a large, multinational consumer packaged goods company (such as Procter & Gamble or Unilever). You must make a presentation to persuade the board of directors to invest $5 million to establish a competitive-intelligence organization—including people, data-gathering services, and software tools. What key points do you need to make in favor of this investment? What arguments can you anticipate that the board might make?

7. Identity theft, whereby people steal personal information, continues to be a problem for consumers and businesses. Assume that you are the database administrator for a corporation with a large database that is accessible from the Web. What steps would you implement to prevent people from stealing personal information from the corporate database?

8. What roles do databases play in your most favorite online activities and Web sites?

PROBLEM-SOLVING EXERCISES

1. Develop a simple data model for the music you have on your digital music player or in your CD collection, in which each row is a song. For each row, what attributes should you capture? What will be the primary key for the records in your database? Describe how you might use the database to expand your music exposure and enjoyment.

2. A video movie rental store is using a relational database to store information on movie rentals to answer customer questions. Each entry in the database contains the following items: Movie Number (the primary key), Movie Title, Year Made, Movie Type, MPAA Rating, Number of Copies on Hand, and Quantity Owned. Movie Types are comedy, family, drama, horror, science fiction, and western. MPAA ratings are G, PG, PG-13, R, NC-17, and NR (not rated). Use a single-user database management system to build a data-entry screen to enter this data. Build a small database with at least ten entries.

3. To improve service to their customers, the salespeople at the video rental store have proposed a list of changes being considered for the database in the previous exercise. From this list, choose two database modifications and modify the data-entry screen to capture and store this new information. Proposed changes:

    a. To help store clerks locate the newest releases, add the date that the movie was first available.

    b. Add the director's name.

    c. Add the names of three primary actors in the movie.

    d. Add a rating of one, two, three, or four stars.

    e. Add the number of Academy Award nominations.

4. Your school maintains information about students in several interconnected database files. The student_contact file contains student contact information. The student_grades file contains student grade records, and the student_financial file contains financial records, including tuition and student loans. Draw a diagram of the fields these three files might contain, identify which field is a primary key in each file, and show which fields serve to relate one file to another. Use Figure 3.7 as a guide.

TEAM ACTIVITIES

1. In a group of three or four classmates, communicate with the person at your school who supervises information systems. Find out how many databases are used by your school and for what purpose. Also find out what policies and procedures are in place to protect the data stored from identity thieves and other threats.
2. As a team of three or four classmates, interview business managers from three different businesses that use databases. What data entities and data attributes are contained in each database? What database company did each company select to provide their database, and why? How do they access the database to perform analysis? Have they received training in any query or reporting tools? What do they like about their databases, and what could be improved? Do any of them use data-mining or OLAP techniques? Weighing the information obtained, select one of these databases as being most strategic for the firm and briefly present your selection and the rationale for the selection to the class.
3. Imagine that you and your classmates are a research team developing an improved process for evaluating loan applicants for automobile purchases. The goal of the research is to predict which applicants will become delinquent or for-feit their loan. Those who score well on the application will be accepted, and those who score exceptionally well will be considered for lower-rate loans. Prepare a brief report for your instructor addressing these questions:
   a. What data do you need for each loan applicant?
   b. What data might you need that is not typically requested on a loan application form?
   c. Where might you get this data?
   d. Take a first cut at designing a database for this application. Using the material in this chapter on designing a database, draw the logical structure of the relational tables for this proposed database. In your design, include the data attributes you believe are necessary for this database, and show the primary keys in your tables. Keep the size of the fields and tables as small as possible to minimize required disk drive storage space. Fill in the database tables with the sample data for demonstration purposes (ten records). After your design is complete, implement it using a relational DBMS.

WEB EXERCISES

1. Use a Web search engine to find information on specific products for one of the following topics: business intelligence, object-oriented databases, or Database as a Service. Write a brief report describing what you found, including a description of the database products and the companies that developed them.
2. More information is being produced than can currently be stored in data centers, and yet, existing data centers are consuming huge amounts of energy, putting a strain on the environment and on budgets. Go online to research "Green Data Center" to learn what can be done to store more data using fewer resources. Students with the most unique and useful suggestions may be awarded extra credit.

CAREER EXERCISES

1. What type of data is stored by businesses in a professional field that interests you? How many databases might be used to store that data? How would the data be organized within each database? How can techniques like data loss prevention (DLP) be used to protect critical databases?
2. How could you use business intelligence (BI) to do a better job at work? Give some specific examples of how BI can give you a competitive advantage.

CASE STUDIES

Case One

Managing International Trades with Powerful Database Systems

Internaxx is an international brokerage and banking service that services thousands of expatriates and international clients around the world. Based in Luxembourg, Germany, Internaxx provides international online brokerage services for private investors, companies, monetary and financial markets, and investment and pension funds. Internaxx provides real-time share trading at more than 15 stock exchanges around the world online or by phone.

Working in so many markets and with customers from 155 different countries, Internaxx must deal with large quantities of data. The majority of data collected by Internaxx originates from the many trading operations carried out by its customers. The ability to process that data efficiently is the key to Internaxx's success.

Internaxx uses its database to fuel business intelligence tools that allow it to process data both quantitatively and qualitatively. The company maintains a data warehouse on which it runs queries to provide insight into both its customers and market trends. While this insight helps to make the company more competitive, it is also required by regulations imposed by the European Union. The Market in Financial Instruments Directive (MiFid) requires financial institutions to acquire a detailed knowledge of customer behavior to better protect the customer's interests.

Internaxx's data warehouse provides data that feeds 50 annual reports updated daily, weekly, or monthly. These reports include analysis of trading operations, rate of customer conversion, total commissions received, and assorted financial, commercial, and marketing reports. Reports serve a variety of functions. One report calculates the distribution of customers around the world to enable effective marketing campaigns.

Additionally, the functional databases that feed into the data warehouse can provide real-time information on stock trades, which is essential for making wise trades in a fast-moving market. Customers around the world can watch the rise and fall of stock prices and market conditions as they occur, using powerful forecasting tools to ensure wise investments.

In summary, the Internaxx database fuels both wise investment decision-making for Internaxx customers and wise business decision-making for Internaxx executives. Rodolphe Marck, director information systems at Internaxx, says that due to the power of databases and the systems they fuel, Internaxx can "align our decisions and customer investment and acquisition strategies according to the requirements and trends of the international financial markets in real time."

Discussion Questions

1. What unique challenges do databases that deal with financial markets face?
2. How does Internaxx separate data in its database for annual reports from the data that fuels real-time analytics?

Critical Thinking Questions

1. In what ways does Internaxx use its database to provide the company with a competitive advantage?
2. What unique capabilities must the Internaxx database have in order to support trading in 15 stock exchanges around the world?

Sources: "Internaxx Profits from Customer Data Using Business Objects Enterprise," Business Objects Customers, *http://download.sap.com/download.epd?context=3DD10BD9CF3308AB173115F948E8B1EA10AAD28449AEF388654154F9D404D59B00468DCBB-B73B8A328A65D9F473D393CA243D6E6 731270B2*, accessed May 15, 2010; Internaxx Web site, *www.internaxx.lu*, accessed May 15, 2010.

Case Two

Using Databases to Map Human Migration

National Geographic was established in 1888 to advance human understanding of the world's cultural, historical, and natural resources. National Geographic is a nonprofit organization that has contributed greatly to scientific research. One of the most recent examples of its contribution to science is the Genographic Project.

According to the National Geographic Web site, the Genographic Project is a "landmark study of the human journey." Scientists believe that we all descended from a common group of ancestors who lived in Africa some 60,000 years ago. Over the millennia, that group reproduced and migrated to populate the entire globe. The Genographic Project intends to map that migration to allow individuals to trace their ancestry back through time and location using advanced DNA research.

Dozens of researchers around the world have been engaged in genetic research of indigenous cultures. DNA is being collected from select individuals and stored in a database where sequences can be automatically studied. Comparing commonalities between DNA strands allows researchers to draw conclusions about where an individual originated, and where his or her ancestors may have migrated to and from.

Stage 2 of the study brings in the general public. Volunteers from around the world can sign up online to become part of the study and trace their own heritages and family migrations. For $99, anyone can purchase a Genographic Project kit, which includes abundant information and a swab for gathering DNA from the inside of the mouth. The DNA sample is sent to the Genographic Project by mail. Participants can trace the progress of the research on their DNA online. Once analyzed, the participant is provided with historical information about their ancestors and migration paths dating back to that African community 60,000 years ago.

The same data-mining and business-intelligence tools used by businesses to find correlations between business data are applied to the DNA information stored by the Genographic Project. Advanced trend-mapping and analysis tools provide a deeper understanding of mutation rates and DNA-merging behavior. The participation of more than 300,000 volunteers has created one of the world largest repositories of genetic information, providing new insight into our migratory history and fueling collaboration on new projects.

Without the automation provided by database tools, this research would not be possible. Dr. Spencer Wells, National Geographic Explorer-in-Residence and Scientific Director of the Project, stated that "With hundreds of thousands of samples, researchers could easily become lost in our collected data. But, by working with IBM, we can distill this information into something useful—research breakthroughs and new findings."

Discussion Questions

1. What role do database and DBMS play in assisting with the Genographic Project?
2. What types of data are stored in the Genographic database, and how might it be organized into the data hierarchy discussed in this chapter?

Critical Thinking Questions

1. How is the manipulation of Genographic data similar to the manipulation of business data? What DBMS tools and techniques are shared by both?

2. How does National Geographics investment in this DBMS assist other researchers? How might this data be shared using the database concepts taught in this chapter?

Sources: "National Geographic, as part of its Genographic Project, tracks human migration across the millennia via DNA analysis," IBM Case Studies, July 28, 2009, *www-01.ibm.com/software/success/cssdb.nsf/CS/LMCM-7U7U29?OpenDocument&Site=wssoftware&cty=en_us*; Genographic Project Web site, *https://genographic.nationalgeographic.com*, accessed May 15, 2010

Questions for Web Case

See the Web site for this book to read about the Altitude Online case for this chapter. Following are questions concerning this Web case.

Altitude Online: Using Databases and Business Intelligence

Discussion Questions

1. What work is involved in merging multiple databases into one central database, as Altitude Online is doing?
2. Why do you think Altitude Online found it necessary to hire a database administrator? How will the ERP affect the responsibilities of IS personnel across the organization?

Critical Thinking Questions

1. In a major move such as this, what opportunities can Altitude Online take advantage of as it totally revamps its database system that it perhaps wouldn't consider before?
2. Why do you think Altitude Online is beginning work on its database prior to selecting an ERP vendor?

NOTES

Sources for the opening vignette: "Aquent: Staffing Firm Uses SAP BusinessObjects Software Tools to Deliver Talent," SAP Customer References, 2009, *www.sap.com/usa/solutions/sapbusinessobjects/customers*; "Aquent Enterprise Resource Technology Group,", *www.ertgroup.com/index.php/aquent*, accessed May 15, 2010; "Aquent – About Us,", *http://aquent.us/learn_more/about_us*, accessed May 15, 2010.

1   "Just-in-Time Intelligence Helps Comic Relief Boost Donations 25% While Reducing Failure Risk," Oracle Customer Snapshot, August 2009, *www.oracle.com/customers/snapshots/comic-relief-obiee-snapshot.pdf*.

2   Kirk, Jeremy, "Latvian police decline to hold database hacker," *Computerworld*, May 14, 2010, *www.computerworld.com/s/article/*

9176781/Latvian_police_decline_to_hold_database_hacker?source=rss_news.

3   Kirk, Jeremy, "Company says 3.3M student loan records stolen," *Computerworld*, March 20, 2010, *www.computerworld.com/s/article/9174312/Company_says_3.3M_student_loan_records_stolen?source=rss_news*.

4   UPI Staff, "Pig received voter registration reminder," UPI, May 12, 2010, *www.upi.com*.

5   "The Digital Universe Decade – Are You Ready?," *IDC*, May 2010, *www.emc.com/collateral/demos/microsites/idc-digital-universe/iview.htm*.

6   "Beyond fingerprints: The FBI's next generation database," *Home Security Newswire*, January 27, 2009, *http://*

homelandsecuritynewswire.com/beyond-fingerprints-fbis-next-generation-database.

7 Nucci, Cora, "Behind Walmart's E-Health Records Plans," *Information Week*, March 12, 2009, *www.informationweek.com/blog/main/archives/2009/03/behind_walmarts.html*.

8 Wadhwa, Vivek, "The Open Gov Initiative: Enabling Techiew to Solve Government Problems," TechCrunch, May 22, 2010, *www.techcrunch.com*.

9 Miller, Rich, "Feds Commence Huge Data Center Consolidation," *Data Center Knowledge*, March 1, 2010, *www.datacenterknowledge.com/archives/2010/03/01/feds-commence-huge-data-center-consolidation*.

10 Hoover, Nicholas, "Texas' Data Center Consolidation Project Under Scrutiny," *InformationWeek*, November 16, 2009, *www.informationweek.com/news/government/state-local/showArticle.jhtml?articleID=221800216*.

11 Miller, Rich, "Microsoft Goes All-in on Container Data Centers," *Data Center Knowledge*, December 2, 2008, *www.datacenterknowledge.com/archives/2008/12/02/microsoft-goes-all-in-on-container-data-centers*.

12 Miller, Rich, "Google Unveils Its Container Data Center," *Data Center Knowledge*, April 1, 2009, *www.datacenterknowledge.com/archives/2009/04/01/google-unveils-its-container-data-center*.

13 Miller, Rich, "Microsoft to Open Two Massive Data Centers," Data Center Knowledge, June 29, 2009, *www.datacenterknowledge.com*.

14 Nystedt, Dan, "Project aims to halve cost of a data center," InfoWorld, June 2, 2010, *www.infoworld.com*.

15 Lemos, Rob, "Vermont's Lights Out Data Center: No One's Home Nights or Weekends," *CIO*, October 9, 2009.

16 Thibodeau, Patrick, "HP job cuts point to shifting IT skills," Computerworld, June 1, 2010, www.computerworld.com.

17 Oracle Database Web site, *www.oracle.com/us/products/database*, accessed May 14, 2010.

18 Siegler, MG, "The Unified Database Of Places Is Coming Soon. Or Maybe Never," TechCrunch, May 8, 2010, *http://techcrunch.com/2010/05/08/place-database/?utm_source=feedburner&utm_medium=feed&utm_campaign=Feed:+Techcrunch+(TechCrunch)&utm_content=Google+Reader*.

19 Microsoft Staff, "Medical Firm Accelerates Manufacturing Improvements with Shared Notebooks," Microsoft Case Studies, May 12, 2010, *www.microsoft.com/casestudies/Case_Study_Detail.aspx?casestudyid=4000006922*.

20 Write, Janet, "The Linde Group – SAP Consulting Helps Gas and Engineering Giant Upgrade Smoothly," SAP Customer Reference, 2009, *http://download.sap.com/download.epd?context=39E32404E75EA4B1273E19E08A0B111B88A6FCFB35AF84CB1E6554F40DEF6AF732DFE610645F5982B448AB21CD663070DDDB0BAAAA55E33B*.

21 Vijayan, Jaikumar, "U.S. Says SQL Injection Caused Major Breaches," *Computerworld News Digest*, August 27/24, 2009, pg 4.

22 Oracle Cloud Computing Web site, *www.oracle.com/us/technologies/cloud*, accessed May 11, 2010.

23 "AWS Case Study: Razorfish," Amazon Web Services, 2010, *http://aws.amazon.com/solutions/case-studies/razorfish*.

24 McCann, Liz, "Case Study: Online Database Applications at Proctor & Gamble," Intuit Web site, January 28, 2010, *http://quickbase.intuit.com/blog/2010/01/28/case-study-online-database-applications-at-procter-gamble*.

25 Schonfeld, Erick, "Urbanspoon Wants To Challenege OpenTable With Its RezBook IPad App," *TechCrunch*, May 19, 2010, *www.techcrunch.com*.

26 Ackerman, Kate, "Personal Health Records May Not Be So Personal," *iHealthBeat*, May 14, 2010, *www.ihealthbeat.org/features/2010/personal-health-records-may-not-be-so-personal.aspx*.

27 Resource Description Framework (RDF) Web site, accessed May 11, 2010, *www.w3.org/RDF*.

28 IBM Staff, "Honda Italia Industriale teams with IBM to jump-start a significant business transformation project by adopting RFID technology for its production processes," IBM Case Study, April 26, 2010, *www-01.ibm.com/software/success/cssdb.nsf/cs/GMMY-84SLG2?OpenDocument&Site=gicss67snsr&cty=en_us*.

29 SAS Staff, "1-800-FLOWERS.COM customer connection blooms with SAS® Business Analytics," SAS Success Stories, *http://www.sas.com/success/1800flowersba.html*, accessed May 11, 2010.

30 "Case: Defense Health Services Systems," The Computerworld Honors Program, 2009, *www.cwhonors.org/CaseStudy/viewCaseStudy2009.asp?NominationID=324&Username=DHS324*.

31 "Brooks Brothers uses SAS® to drive global customer satisfaction and profitability," SAS Customer Success, accessed May 11, 2010, *www.sas.com/success/brooksbros.html*.

32 Staff, "Facebook loses friends over privacy settings," *The Economic Times*, May 15, 2010, *http://economictimes.indiatimes.com/infotech/internet/Facebook-loses-friends-over-privacy-settings/articleshow/5934675.cms*.

33 "SPSS Customers: American Airlines," SPSS Web site, *www.spss.com/success/template_view.cfm?Story_ID=43*, accessed May 11, 2010.

34 Guagenty, Matthew, "Mapping Patients To Drive Better Brand Decisions," Pharmaceutical Executive, March 2010, *www.imshealth.com/portal/site/imshealth/menuitem.0103f29c72c419cd88f611019418c22a/?vgnextoid=4549f7f4fc746210VgnVCM100000ed152ca2RCRD#*.

35 "PUMA North America," SPSS Customers, *www.spss.com/success/template_view.cfm?Story_ID=153*, accessed May 11, 2010.

36 George, Randy, "An Ounce of Loss Prevention," *Information Week*, July 6, 2009, pg 39.

37 IBM Staff, "Case Study: Blue Mountain," IBM Case Studies, May 10, 2010, *www-01.ibm.com/software/success/cssdb.nsf/CS/LWIS-7LPTPY?OpenDocument&Site=cognos&cty=en_us*.

38 "PostgreSQL at Skype," Skype Developer Zone Web site, *https://developer.skype.com/SkypeGarage/DbProjects/SkypePostgresqlWhitepaper*, accessed May 11, 2010.

39 "Mainframe Integration Works for Giants of the Insurance Industry," Attunity Customer Case Study, *www.attunity.com/Data/Uploads/Case%20Studies/Lloyds%20CS.pdf*, accessed May 11, 2010.

40 "Case Study – Cumberland County," Pitney Bowes Case Studies, *www.pbinsight.com/resources/case-studies/details/cumberland-county*, accessed May 11, 2010.

CHAPTER
• 4 •

Telecommunications, the Internet, Intranets, and Extranets

PRINCIPLES	LEARNING OBJECTIVES
■ A telecommunications system has many fundamental components that must be carefully selected and work together effectively to enable people to meet personal and organization objectives.	■ Identify and describe the fundamental components of a telecommunications system. ■ Identify several network types and describe the uses and limitations of each. ■ Name three basic processing alternatives for organizations that require two or more computer systems and discuss their fundamental features.
■ The Internet provides a critical infrastructure for delivering and accessing information and services.	■ Briefly describe how the Internet works, including alternatives for connecting to it and the role of Internet service providers.
■ Originally developed as a document-management system, the World Wide Web has grown to become a primary source of news and information, an indispensible conduit for commerce, and a popular hub for social interaction, entertainment, and communication.	■ Describe how the World Wide Web works and the use of Web browsers, search engines, and other Web tools.
■ The Internet and Web provide numerous resources for finding information, communicating and collaborating, socializing, conducting business and shopping, and being entertained.	■ Identify and briefly describe several applications associated with the Internet and the Web. ■ Outline a process and identify tools used to create Web content.
■ Popular Internet and Web technologies have been applied to business networks in the form of intranets and extranets.	■ Define the terms intranet and extranet and discuss how organizations are using them. ■ Identify several issues associated with the use of networks.

Information Systems in the Global Economy
Procter & Gamble, United States

Telepresence: The Next Best Thing to Being There

Procter & Gamble has invested heavily in a technology that allows its employees to attend a morning meeting in Kobe, Japan, have lunch with business partners in Barcelona, Spain, attend an afternoon round-table discussion in Istanbul, Turkey, and get home in time to see their children's soccer game in Cincinnati. No, P&G didn't purchase a fleet of X-15s (the fastest aircraft in the world), nor did it invent teleportation. What P&G did was to install telepresence equipment in more than 75 of its global locations.

When it comes to global enterprises, few are bigger than Procter and Gamble. P&G is an $80 billion company, with 135,000 employees spread across 80 countries, delivering well-known consumer brands such as Pampers, Puffs, Crest, Gillette, and two dozen others to more than 3 billion people in 180 countries.

P&G was founded on principles of collaboration. Its executives and managers are strong believers in the value of solutions developed through a collaborative effort. It's from these beliefs that its recent "Connect and Develop" strategy was born. Connect and Develop involves using new technologies to increase communication among all employees and especially those with expert knowledge.

It was in this spirit that P&G decided to invest in Cisco Telepresence technologies. Telepresence uses existing corporate high-speed networks to create video links between local or global endpoints. High-definition video cameras and microphones are installed in conference rooms to capture the participants sitting around a semicircular conference table. Large displays are mounted on a wall across from the people to display life-size images of remote participants. Stereo speakers match the participant's voice with the location of the participant's video image. This creates the illusion of several people seated around an oval table, when in reality, half of them are in some other location perhaps halfway around the world.

P&G calls its 75 telepresence conference rooms "Video Collaboration Studios." They range in capacity from a few people to dozens. While P&G won't divulge how much the total installation cost, a typical telepresence room can cost more than $250,000. P&G says that it is well worth the investment. It estimates that it saves $4 for every $1 invested due to savings in travel and productivity.

P&G employees use the Video Collaboration Studios as much as 80 percent of a 50-hour workweek. Participants can use Live Meeting software from Microsoft to enable any presenter to project his or her computer screen for others to see a spreadsheet, slide presentation, document, or photo. The system also makes it easier for P&G specialists and experts to share their knowledge with others across the company. P&G executives have even sealed partnership deals without needing to fly in their prospective partners.

The Cisco Telepresence system is easy to operate. Meetings are scheduled using calendaring software such as Microsoft Outlook. At the time of the meeting, video connections are established with a simple one-button click. P&G also arranged for Cisco to monitor its networks to make sure all runs smoothly. Cisco help is available via telepresence, also with a one-button click at any time.

Telepresence assists P&G in meeting its goals to become a green company. The reduction in travel helps reduce its carbon footprint, and sharing information electronically has reduced its need for paper. Telepresence has also assisted P&G in meeting its financial goals by increasing productivity, reducing the time employees spend in transit, and increasing communication and collaborative opportunities. Telepresence also helps P&G stay in close communication with its business partners and build new partnerships.

Like many companies, P&G leverages telecommunications technologies to streamline the flow of information between its employees, which improves productivity and increases market share and profits. Besides telepresence, P&G also uses instant messaging, wikis, phone chat, and other online communications to maximize the flow of information across the enterprise. Whether it's virtual travel via telepresence or a short text message, an appropriate telecommunications infrastructure is one of the big keys to success.

As you read this chapter, consider the following:

- What role do telecommunications play in connecting organizations and growing the global economy?
- In what ways are the Internet and Web used by individuals to improve our quality of life and by business to improve the bottom line?

Why Learn About Telecommunications and Networks?

Today's decision makers need to access data wherever it resides. They must be able to establish fast, reliable connections to exchange messages, upload and download data and software, route business transactions to processors, connect to databases and network services, and send output to printers. Regardless of your chosen major or future career field, you will need the communications capabilities provided by telecommunications and networks including the Internet, especially if your work involves the supply chain. Among all business functions, supply chain management might use telecommunications and networks the most because it requires cooperation and communications among workers in inbound logistics, warehouse and storage, production, finished product storage, outbound logistics, and most importantly, with customers, suppliers, and shippers. Many supply chain organizations make use of the Web to purchase raw materials, parts, and supplies at competitive prices. All members of the supply chain must work together effectively to increase the value perceived by the customer, so partners must communicate well. Other employees in human resources, finance, research and development, marketing, and sales positions must also use communications technology to communicate with people inside and outside the organization. To be a successful member of any organization, you must be able to take advantage of the capabilities that these technologies offer you. This chapter begins by discussing the importance of effective communications.

In today's high-speed global business world, organizations need always-on, always-connected computing for traveling employees and for network connections to their key business partners and customers. As we saw in the opening vignette, a forward-thinking company such as Procter & Gamble can improve communications among its many dealers, leading to cost reductions and increases in sales. Here are additional examples of organizations using telecommunications and networks to move ahead.

- In the face of a slowdown in the economy, many organizations are substituting video teleconferencing meetings for business travel in order to cut costs. British Telecom estimates that it saved $330 million per year on avoided travel costs and time saved. Microsoft estimates that it saves $90 million per year. Procter & Gamble and Deloitte have installed dozens of videoconferencing systems around the world and are saving millions each month in reduced travel expenses.[1]
- eBay, Google, Microsoft, and Yahoo!—all Web competitors—are promoting themselves by providing free Wi-Fi access to people at several airports, on various airlines, and at hotels. For example, people who connect to the Internet through a hot spot at an airport are shown a Web page that allows them to donate to a charity and have the donation matched by Google.[2]
- An estimated 54 percent of physicians own a smartphone and are increasingly using them as a valuable tool to provide patient services. Physicians can download applications to their phones that enable them to check drug references, perform common medical calculations, consult normal lab value charts, use decision-support tools, and view electronic medical records.[3]

- Thousands of companies use Webcasts to inform and educate potential customers about their products and services.
- The UNAIDS organization is using telecommunications to provide millions of people living in developing countries with access to healthcare information and services. Mobile phones are being used as low-cost tools for HIV data collection, epidemic tracking, and training of health workers.[4]

AN OVERVIEW OF TELECOMMUNICATIONS

Telecommunications refers to the electronic transmission of signals for communications, by such means as telephone, radio, and television. Telecommunications is creating profound changes in business because it lessens the barriers of time and distance. Telecommunications not only is changing the way businesses operate, but the nature of commerce itself. As networks are connected with one another and transmit information more freely, a competitive marketplace demands excellent quality and service from all organizations.

Figure 4.1 shows a general model of telecommunications. The model starts with a sending unit (1), such as a person, a computer system, a terminal, or another device, that originates the message. The sending unit transmits a signal (2) to a modem (3) that can perform many tasks, which can include converting the signal into a different form or from one type to another. The modem then sends the signal through a medium (4). A **telecommunications medium** is any material substance that carries an electronic signal to support communications between a sending and receiving device. Another modem (5) connected to the receiving device (6) receives the signal. The process can be reversed, and the receiving unit (6) can send another message to the original sending unit (1). An important characteristic of telecommunications is the speed at which information is transmitted, which is measured in bits per second (bps). Common speeds are in the range of thousands of bits per second (Kbps) to millions of bits per second (Mbps) and even billions of bits per second (Gbps).

telecommunications medium
Any material substance that carries an electronic signal to support communications between a sending and receiving device.

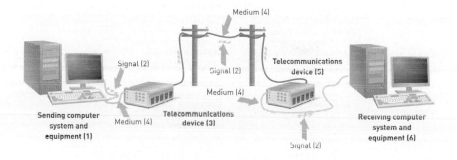

Medium (4)

Signal (2)   Signal (2)   Telecommunications device (5)

Medium (4)

Sending computer system and equipment (1)   Medium (4)   Telecommunications device (3)   Receiving computer system and equipment (6)

Signal (2)

Figure 4.1

Elements of a Telecommunications System

Telecommunications devices relay signals between computer systems and transmission media.

Advances in telecommunications technology allow us to communicate rapidly with clients and coworkers almost anywhere in the world. Telecommunications also reduces the amount of time needed to transmit information that can drive and conclude business actions.

Telecommunications technology enables businesspeople to communicate with coworkers and clients from remote locations.

(Source: © John Prescott.)

Channel Bandwidth

Telecommunications professionals consider the capacity of the communications path or channel when they recommend transmission media for a business. **Channel bandwidth** refers to the rate at which data is exchanged, usually measured in bits per second (bps)—the broader the bandwidth, the more information can be exchanged at one time. **Broadband communications** is a relative term but generally means a telecommunications system that can exchange data very quickly. For example, for wireless networks, broadband lets you send data at a rate greater than 1.5 Mbps. In general, today's organizations need more bandwidth for increased transmission speed to carry out their daily functions.

Communications Media

In designing a telecommunications system, the transmission media selected depends on the amount of information to be exchanged, the speed at which data must be exchanged, the level of concern for data privacy, whether or not the users are stationary or mobile, and many other business requirements. Transmission media can be divided into two broad categories: *guided transmission media,* in which communications signals are guided along a solid medium, and *wireless,* in which the communications signal is broadcast over airwaves as a form of electromagnetic radiation.

Guided Transmission Media Types

There are many different guided transmission media types. Table 4.1 summarizes the guided media types by physical media form. Common guided transmission media types are shown in Figure 4.2.

Table 4.1

Guided Transmission Media Types

Media Form	Description	Advantages	Disadvantages
Twisted-pair wire	Twisted pairs of copper wire, shielded or unshielded	Used for telephone service; widely available	Transmission speed and distance limitations
Coaxial cable	Inner conductor wire surrounded by insulation	Cleaner and faster data transmission than twisted-pair wire	More expensive than twisted-pair wire
Fiber-optic cable	Many extremely thin strands of glass bound together in a sheathing; uses light beams to transmit signals	Diameter of cable is much smaller than coaxial; less distortion of signal; capable of high transmission rates	Expensive to purchase and install
Broadband over power lines	Data is transmitted over standard high-voltage power lines	Can provide Internet service to rural areas where cable and phone service may be nonexistent	Can be expensive and may interfere with ham radios and police and fire communications

FiOS is a bundled set of communications services, including Internet, telephone, and TV, that operates over a total fiber-optic communications network. With this service, fiber-optic cable is run from the carrier's local exchange all the way to the customer's premises. (Cable networks often use fiber optic in their network backbone that connects their local exchanges, but typically do not run fiber optic to the customer's premises). FiOS is offered from Verizon in selected portions of the United States. At the top speed for FiOS of 50 Mbps, it would take just 16 minutes to download a two-hour movie (500 MB). This same task would take 53 minutes over a 15-Mbps cable network. [5] A shortcoming of this service is that if there is a power outage at the premises, there is no FiOS service. A battery backup unit is advisable to avoid this potential problem.

Many utilities, cities, and organizations are experimenting with *broadband over power lines (BPL)* to provide Internet access to homes and businesses over standard high-voltage power lines. IBM is working with communications provider International Broadband Electric Communications to implement BPL networks to provide broadband access to rural residents in Alabama, Indiana, Virginia, and Michigan. In addition to providing broadband service to electric customers, this new BPL connectivity will enable electric companies to better monitor, manage, and control the reliability of their electric grids.[6]

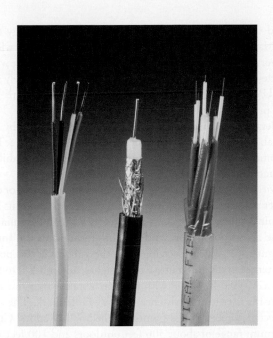

Figure 4.2

Types of Guided Transmission Media

Twisted-pair wire (left), coaxial cable (middle), fiber-optic cable (right)

(Source: © Greg Pease/Getty Images.)

Wireless Technologies

Wireless communications coupled with the Internet are revolutionizing how and where we gather and share information, collaborate in teams, listen to music or watch video, and stay in touch with our families and co-workers while on the road. With wireless capability, a coffee shop can become our living room, or the bleachers at a ball park can become our office. The many advantages and freedom provided by wireless communications are causing many organizations to consider moving to an all-wireless environment.

Wireless telecommunications involves the broadcast of communications in one of three frequency ranges: microwave, radio, and infrared, as shown in Table 4.2.

Technology	Description	Advantages	Disadvantages
Radio frequency range	Operates in the 3KHz–300 MHz range	Supports mobile users; costs are dropping	Signal highly susceptible to interception
Microwave—terrestrial and satellite frequency range	High-frequency radio signal (300 MHz–300 GHz) sent through atmosphere and space (often involves communications satellites)	Avoids cost and effort to lay cable or wires; capable of high-speed transmission	Must have unobstructed line of sight between sender and receiver; signal highly susceptible to interception
Infrared frequency range	Signals in the 300 GHz to 400 THz frequency range sent through air as lightwaves	Lets you move, remove, and install devices without expensive wiring	Must have unobstructed line of sight between sender and receiver; transmission effective only for short distances

Table 4.2

Frequency Ranges Used for Wireless Communications

Some of the more widely used wireless communications options are discussed next.

Near field communication (NFC) is a very short-range wireless connectivity technology designed for cell phones and credit cards. With NFC, consumers can swipe their credit cards or even cell phones within a few inches of NFC point-of-sale terminals to pay for purchases.

Bluetooth is a wireless communications specification that describes how cell phones, computers, printers, and other electronic devices can be interconnected over distances of 10–30 feet at a rate of about 2 Mbps. Bluetooth enables users of multifunctional devices to synchronize with information in a desktop computer, send or receive faxes, print, and, in general, coordinate all mobile and fixed computer devices. The Bluetooth technology is named after the tenth century Danish King Harald Blatand, or Harold Bluetooth in English. He had been instrumental in uniting warring factions in parts of what is now Norway, Sweden, and Denmark—just as the technology named after him is designed to allow collaboration between differing devices such as computers, phones, and other electronic devices.

near field communication (NFC)

A very short-range wireless connectivity technology designed for cell phones and credit cards.

Bluetooth

A wireless communications specification that describes how cell phones, computers, faxes, printers, and other electronic devices can be interconnected over distances of 10–30 feet at a rate of about 2 Mbps.

ultra wideband (UWB)
A form of short-range communications that employs extremely short electromagnetic pulses lasting just 50 to 1,000 picoseconds that are transmitted across a broad range of radio frequencies of several gigahertz.

Wi-Fi
A medium-range wireless telecommunications technology brand owned by the Wi-Fi Alliance.

Ultra wideband (UWB) is a wireless communications technology that transmits large amounts of digital data over short distances of up to 30 feet using a wide spectrum of frequency bands and very low power. Potential UWB applications include wirelessly connecting printers and other devices to desktop computers or enabling completely wireless home multimedia networks. UWB products are considered too expensive at this time to have broad market appeal, but prices are coming down.[7]

Wi-Fi is a wireless telecommunications technology brand owned by the Wi-Fi Alliance, which consists of about 300 technology companies, including AT&T, Dell, Microsoft, Nokia, and Qualcomm. The alliance exists to improve the interoperability of wireless local area network products.

With a Wi-Fi wireless network, the user's computer, smartphone, or other mobile device has a wireless adapter that translates data into a radio signal and transmits it using an antenna. A wireless access point, which consists of a transmitter with an antenna, receives the signal and decodes it. The access point then sends the information to the Internet over a wired connection. See Figure 4.3. When receiving data, the wireless access point takes the information from the Internet, translates it into a radio signal, and sends it to the device's wireless adapter. These devices typically come with built-in wireless transmitters and software to enable them to alert the user to the existence of a Wi-Fi network. The area covered by one or more interconnected wireless access points is called a "hot spot." Current Wi-Fi access points have a maximum range of about 300 feet outdoors and 100 feet within a dry-walled building. Wi-Fi has proven so popular that hot spots are popping up in places such as airports, coffee shops, college campuses, libraries, and restaurants. Many cities have implemented Wi-Fi based networks for their citizens and government workers.

The availability of free Wi-Fi within a hotel's premises has become very popular with the business traveler. The Aloft and Element hotels, part of the Starwood chain, cater to the extended stay traveler. Both hotels have recently added free Wi-Fi service in their rooms and lobbies. At the other end of the accommodations spectrum, The Four Seasons Hotels and Resorts also offer free Wi-Fi. Many other hotels offer or are considering adding free Wi-Fi.[8]

Figure 4.3

Wi-Fi Network

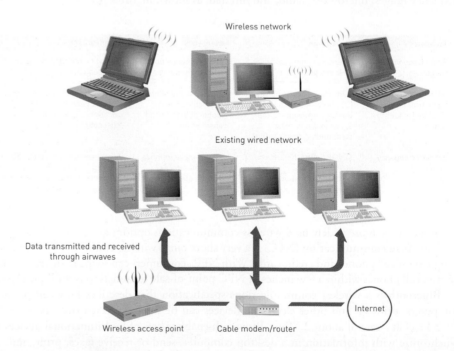

Microwave Transmission
Microwave is a high-frequency (300 MHz–300 GHz) signal sent through the air. Terrestrial (Earth-bound) microwaves are transmitted by line-of-sight devices, so the line of sight between the transmitter and receiver must be unobstructed. Typically, microwave stations are placed in a series—one station receives a signal, amplifies it, and retransmits it to the next

microwave transmission tower. Such stations can be located roughly 30 miles apart before the curvature of the Earth makes it impossible for the towers to "see" one another. Microwave signals can carry thousands of channels at the same time. Because they are line-of-sight transmission devices, microwave dishes are frequently placed in relatively high locations, such as mountains, towers, or tall buildings.

A communications satellite also operates in the microwave frequency range. See Figure 4.4. The satellite receives the signal from the Earth station, amplifies the relatively weak signal, and then rebroadcasts it at a different frequency. The advantage of satellite communications is that satellites can receive and broadcast over large geographic regions. Such problems as the curvature of the Earth, mountains, and other structures that block the line-of-sight microwave transmission make satellites an attractive alternative. Geostationary, low-Earth orbit, and small mobile satellite stations are the most common forms of satellite communications.

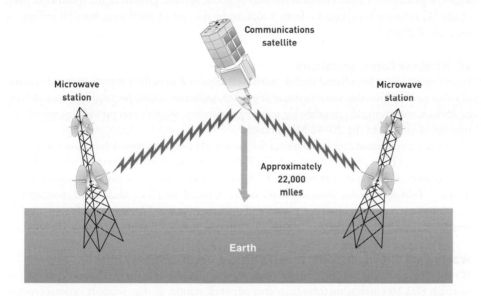

Figure 4.4

Satellite Transmission

Communications satellites are relay stations that receive signals from one Earth station and rebroadcast them to another.

A *geostationary satellite* orbits the Earth directly over the equator, approximately 22,300 miles above the Earth, so that it appears stationary. The U.S. National Weather Service relies on the Geostationary Operational Environmental Satellite program for weather imagery and quantitative data to support weather forecasting, severe storm tracking, and meteorological research.

A *low earth orbit (LEO) satellite* system employs many satellites, each in an orbit at an altitude of less than 1,000 miles. The satellites are spaced so that, from any point on the Earth at any time, at least one satellite is on a line of sight. Iridium Communications Inc. provides a global communications network that spans the entire Earth using 66 satellites in a near polar orbit at an altitude of 485 miles. Calls are routed among the satellites to create a reliable connection between call participants that cannot be disrupted by natural disasters such as earthquakes, tsunamis, or hurricanes that knock out ground-based wireless towers and wire or cable-based networks.[9] Iridium phones were the primary source of communication in New Orleans after Hurricane Katrina in 2005 and after the major earthquake in China in 2008.[10]

A *very small aperture terminal (VSAT)* is a satellite ground station with a dish antenna smaller than 3 meters in diameter. The U.S. Army is buying thousands of VSATs under a five-year $5 billion contract. Paul Brown, technical adviser for the Defense Communications and Army Transmission Systems, explains: "In a war zone, a soldier who needs to order parts would typically have to get into his vehicle and drive from one outlying camp to another. Providing them with a VSAT solution allows them to stay within the wire of their base and communicate directly to order those parts. VSATs save lives; that's a critical piece."[11]

3G Wireless Communications

3G wireless communications supports wireless voice and broadband speed data communications in a mobile environment. It is called 3G for third generation of solutions for wireless voice and data communications. Additional capabilities include mobile video, mobile e-commerce, location-based services, mobile gaming, and the downloading and playing of songs.

The ITU established a single standard for cellular networks in 1999. The goal was to standardize future digital wireless communications and allow global roaming with a single handset. Called IMT-2000, now referred to as 3G, this standard provides for faster transmission speeds in the range of 2–4 Mbps. Originally, 3G was supposed to be a single, unified, worldwide standard, but the 3G standards effort split into several different standards.

The four primary 3G communications carriers in the U.S. are AT&T, Sprint Nextel, T-Mobile, and Verizon. China Mobil operates the world's largest 3G network that covered some 70 percent of China's cities at the end of 2009. A rapid growth in the number of users of the 3G network is expected—from 3 million by the end of 2009 to at least 50 million by the end of 2010.[12]

4G Wireless Communications

Fourth-generation broadband mobile wireless is expected to deliver more advanced versions of enhanced multimedia, smooth streaming video, universal access, portability across all types of devices, and eventually, worldwide roaming capability. 4G will also provide increased data transmission rates in the 20–40 Mbps range.

AT&T announced that it will have a fully ready 4G cellular network based on Long Term Evolution (LTE) in 2011. This puts AT&T about one year behind Verizon, which plans to start its own 4G LTE network in 2010. LTE has the potential to download data at up to 100 Mbps.[13] T-Mobile is also planning a 4G network based on LTE. Meanwhile, telecommunications operator Teliasonera implemented a 4G network in Norway in 2009. In a public demonstration, the network downloaded a 7 megabyte MP3 file in just one second.[14]

Worldwide Interoperability for Microwave Access (WiMAX)

Worldwide Interoperability for Microwave Access (WiMAX)
The common name for a set of IEEE 802.16 wireless metropolitan area network standards that support various types of communications access.

Worldwide Interoperability for Microwave Access (WiMAX) is the common name for a set of IEEE 802.16 wireless metropolitan area network standards that support various types of communications access. In many respects, WiMAX operates like Wi-Fi, only over greater distances and at faster transmission speeds. A WiMAX tower connects directly to the Internet via a high-bandwidth, wired connection. A WiMAX tower can also communicate with another WiMAX tower using a line-of-sight, microwave link. The distance between the WiMAX tower and an antenna can be as great as 30 miles. WiMAX can support data communications at a rate of 70 Mbps. Fewer WiMAX base stations are required to cover the same geographical area than when Wi-Fi technology is used.

Clearwire Corporation is a wireless Internet service provider founded by cellular phone pioneer Craig McCaw. Sprint and Clearwire pooled their WiMAX assets and obtained more than $3.2 billion in capital from Intel, Comcast, and Google to create a new, independent company (Clearwire Communications) whose goal is to build a national mobile broadband network across the U.S. Clearwire operates the 4G mobile WiMAX network, and Sprint and Comcast sell airtime on the network. This enables the companies to compete with the wireless data offerings of AT&T and Verizon.[15]

Clear residential modems provide up to 6 Mbps download speeds, while mobile Internet customers can expect to receive up to 4 Mbps download speeds. Sprint is working with Intel, Motorola, Nokia, and Samsung to provide WiMAX-capable PC cards, gaming devices, laptops, cameras, and even phones.

WiMAX is a key component of Intel's broadband wireless strategy to deliver innovative mobile platforms for "anytime, anywhere." Centrino is a marketing initiative from Intel. Intel touts that its Centrino laptops deliver improved performance, have a longer battery life, and can access a variety of wireless networks, including WiMAX. Dell, Fujitsu, Lenovo, Panasonic, Samsung, and Toshiba are among the manufacturers that produce laptop computers with the Centrino label.[16]

Most telecommunications experts agree that WiMAX is an attractive option for developing countries with little or no wireless telephone infrastructure. However, it is not clear whether WiMAX will be as successful in developed countries such as the United States, where regular broadband is plentiful and cheap, and where 3G wireless networks already cover most major metropolitan areas.

Telecommunications Hardware

Networks require various telecommunications hardware devices to operate, including smartphones, modems, multiplexers, front-end processors, private branch exchanges, switches, bridges, routers, and gateways.

Smartphones

A smartphone combines the functionality of a mobile phone, camera, Web browser, e-mail tool, MP3 player, and other devices into a single handheld device. For example, the Apple iPhone is a combination mobile phone, widescreen iPod, and Internet access device capable of supporting e-mail and Web browsing. An iPhone user can connect to the Internet either via Wi-Fi or AT&T's Edge data network.

Apple iPhone

The iPhone is a combination mobile phone, widescreen iPod, and Internet access device.

(Source: Courtesy of Apple.)

Smartphones have their own software operating systems and are capable of running applications that have been created for their particular operating system. As a result, the capabilities of smartphones will continue to evolve as new applications become available. The Apple iPhone (iPhone OS), BlackBerry (RIM OS), and Palm (Palm OS) smartphones come with their own proprietary operating systems. The Android, Microsoft Windows Mobile, and Symbian operating systems are used on various manufacturers' smartphones.

Smartphone applications are developed by the manufacturers of the handheld device, by the operators of the communications network on which they operate, and by third-party software developers. More than 125,000 Apple iPhone applications can be downloaded from the Apple Apps for iPhone online store. BlackBerry, Palm, and Android-based phones sell their apps through their own online stores—BlackBerry App World, Palm Pre Applications, and Android Market.

The Droid smartphone is manufactured by Motorola and uses the Android operating system from Google. The Nexus One is another smartphone that uses Google's Android mobile operating system. The device is manufactured according to Google's specifications by Taiwan's HTC Corporation. The phone is similar in size and shape to the iPhone but boasts a higher resolution, enhanced 3D graphics, and speech-to-text features that enable the user to dictate Facebook and Twitter updates. The Nexus One can be bought directly from Google over the Web and comes "unlocked," meaning it is not restricted to use on a single network provider. As of this writing, Google offers it for use on the T-Mobile network, with later versions expected for use on Verizon, Vodafone, and AT&T.[17]

Each of the smartphones listed in Table 4.3 comes with phone capabilities, a digital camera, video recorder, GPS tracking capability, digital player, Internet browser, and e-mail tool.

Feature	Palm Pre	BlackBerry Curve 8900	Apple iPhone 3 GS
Approximate cost (Fall 2009)	$150	$400	$300
Screen size (inches)	3.2	2.4	3.5
Height (inches)	3.9	4.3	4.5
Depth (inches)	0.6	0.5	0.4
Width (inches)	2.3	2.4	2.4
Weight (ounces)	4.7	3.9	4.8

Table 4.4 identifies several common telecommunications hardware devices and describes the function they perform.

Table 4.4

Common Telecommunications Devices

Device	Function
Modem	Translates data from a digital form (as it is stored in the computer) into an analog signal that can be transmitted over ordinary telephone lines.
Fax modem	Facsimile devices, commonly called fax devices, allow businesses to transmit text, graphs, photographs, and other digital files via standard telephone lines. A fax modem is a very popular device that combines a fax with a modem, giving users a powerful communications tool.
Multiplexer	Allows several telecommunications signals to be transmitted over a single communications medium at the same time, thus saving expensive long-distance communications costs.
PBX	A communications system that manages both voice and data transfer within a building and to outside lines. In a PBX system, switching PBXs can be used to connect hundreds of internal phone lines to a few phone company lines.
Front-end processor	Special-purpose computer that manages communications to and from a computer system serving many people.
Switch	Uses the physical device address in each incoming message on the network to determine which output port it should forward the message to in order to reach another device on the same network
Bridge	Connects one LAN to another LAN that uses the same telecommunications protocol.
Router	Forwards data packets across two or more distinct networks toward their destinations through a process known as routing. Often an Internet service provider (ISP) installs a router in a subscriber's home that connects the ISP's network to the network within the home
Bridge	Connects one LAN to another LAN that uses the same telecommunications protocol.

Merrill Lynch, the global financial management and advisory company, implemented an advanced VoIP voice trading system. More than 4,000 IQ/MAX next-generation trading desktops were installed on its two largest trading floors in New York and London plus other trading floors around the globe.[11]

NETWORKS AND DISTRIBUTED PROCESSING

A **computer network** consists of communications media, devices, and software needed to connect two or more computer systems or devices. The computers and devices on the networks are also called *network nodes*. After they are connected, the nodes can share data, information, work processes and allow employees to collaborate on projects. If a company uses networks effectively, it can grow into an agile, powerful, and creative organization, giving it a long-term competitive advantage. Organizations can use networks to share hardware, programs, and databases. Networks can transmit and receive information to improve organizational effectiveness and efficiency. They enable geographically separated workgroups to share documents and opinions, which fosters teamwork, innovative ideas, and new business strategies.

computer network
The communications media, devices, and software needed to connect two or more computer systems or devices.

Network Types

Depending on the physical distance between nodes on a network and the communications and services it provides, networks can be classified as personal area, local area, metropolitan area and wide area network.

Personal Area Networks

A **personal area network (PAN)** is a wireless network that connects information technology devices close to one person. With a PAN, you can connect a laptop, digital camera, and portable printer without cables. You can download digital image data from the camera to the laptop and then print it on a high-quality printer—all wirelessly. Bluetooth is the industry standard for PAN communications.

personal area network (PAN)
A network that supports the interconnection of information technology within a range of 33 feet or so.

Local Area Networks

A network that connects computer systems and devices within a small area, such as an office, home, or several floors in a building is a **local area network (LAN)**. Typically, LANs are wired into office buildings and factories, as shown in Figure 4.5. Although LANs often use unshielded twisted-pair wire, other media—including fiber-optic cable—is also popular. Increasingly, LANs are using some form of wireless communications. You can build LANs to connect personal computers, laptop computers, or powerful mainframe computers.

local area network (LAN)
A network that connects computer systems and devices within a small area, such as an office, home, or several floors in a building.

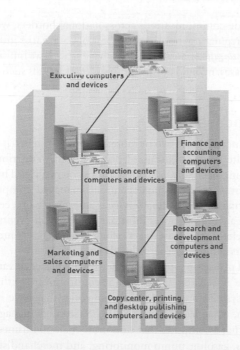

Executive computers and devices

Production center computers and devices

Finance and accounting computers and devices

Research and development computers and devices

Marketing and sales computers and devices

Copy center, printing, and desktop publishing computers and devices

Figure 4.5

Typical LAN

All network users within an office building can connect to each other's devices for rapid communication. For instance, a user in research and development could send a document from her computer to be printed at a printer located in the desktop publishing center.

Docklands Light Railway is one of the first light rail systems in Britain and carries over 67 million passengers per year. Serco Docklands, the operator of the railway, is upgrading its existing LAN to support the railway's rollout of new services and applications, including a novel public address system that services 32 railway stations and is controlled by personal computer workstations at a single control center. The system will also provide passengers with audio/visual updates to keep track of the comings and goings of the various rail cars.[18]

With more people working at home, connecting home computing devices and equipment into a unified network is on the rise. Small businesses are also connecting their systems and equipment. A home or small business can connect network, computers, printers, scanners, and other devices. A person working on one computer, for example, can use data and programs stored on another computer's hard disk. In addition, several computers on the network can share a single printer. To make home and small business networking a reality, many companies are offering standards, devices, and procedures.

Metropolitan Area Networks

metropolitan area network (MAN)
A telecommunications network that connects users and their computers in a geographical area that spans a campus or city.

A **metropolitan area network** (**MAN**) is a telecommunications network that connects users and their computers in a geographical area that spans a campus or city. A MAN might redefine the many networks within a city into a single larger network or connect several LANs into a single campus LAN. Grupo AMSA S.A. is a privately owned textile manufacturing company located in Lima, Peru. It operates three production plants that are separated by roughly 10 km from one another. The company implemented a MAN to support wireless communications among the three plants. The network provides high-speed data, voice, and Internet traffic among the locations, enabling effective planning, tracking, and controlling of operations at all three facilities.[19]

Wide Area Networks

wide area network (WAN)
A telecommunications network that connects large geographic regions.

A **wide area network** (**WAN**) is a telecommunications network that connects large geographic regions. A WAN might be privately owned or rented and includes public (shared users) networks. When you make a long-distance phone call or access the Internet, you are using a WAN. WANs usually consist of computer equipment owned by the user, together with data communications equipment and telecommunications links provided by various carriers and service providers.

China Datang Corporation is a state-owned corporation with 88 subsidiaries, including several large power generation plants. Its operations span 21 provinces of China and required the implementation of a WAN covering virtually the entire country. The network ensures high-quality voice, video, and data communications among the sites connected to the network.[20]

WANs often provide communications across national borders, which involves national and international laws regulating the electronic flow of data across international boundaries, often called *transborder dataflow*. Some countries have strict laws limiting the use of telecommunications and databases, making normal business transactions such as payroll costly, slow, or even impossible.

Distributed Processing

When an organization needs to use two or more computer systems, it can implement one of three basic processing alternatives: centralized, decentralized, or distributed. With **centralized processing**, all processing occurs in a single location or facility. This approach offers the highest degree of control because a single centrally managed computer performs all data processing. The Ticketmaster reservation service is an example of a centralized system. One central computer with a database stores information about all events and records the purchases of seats. Ticket clerks at various ticket selling locations can enter order data and print the results, or customers can place orders directly over the Internet.

centralized processing
An approach to processing wherein all processing occurs in a single location or facility.

With **decentralized processing**, processing devices are placed at various remote locations. Each processing device is isolated and does not communicate with any other processing device. Decentralized systems are suitable for companies that have independent operating units, such as 7-Eleven, where each of its 5,800 U.S. stores is managed to meet local retail conditions. Each store has a computer that runs more than 50 business applications, such as cash register operations, gasoline pump monitoring, and merchandising.

decentralized processing
An approach to processing wherein processing devices are placed at various remote locations.

With **distributed processing**, processing devices are placed at remote locations but are connected to each other via a network. One benefit of distributed processing is that managers can allocate data to the locations that can process it most efficiently. Kroger operates more than 2,400 supermarkets, each with its own computer to support store operations such as customer checkout and inventory management. These computers are connected to a network so that sales data gathered by each store's computer can be sent to a huge data repository on a mainframe computer for efficient analysis by marketing analysts and product supply chain managers.

Ongoing terrorist attacks around the world and the heightened sensitivity to natural disasters (such as the 2010 earthquakes in Chile and Haiti and the series of blizzards in the mid-Atlantic and northeast sections of the U.S. in early 2010) have motivated many companies to distribute their workers, operations, and systems much more widely, a reversal of the previous trend toward centralization. The goal is to minimize the consequences of a catastrophic event at one location while ensuring uninterrupted systems availability.

> **distributed processing**
> An approach to processing wherein processing devices are placed at remote locations but are connected to each other via a network.

Client/Server Systems

Users can share data through file server computing, which allows authorized users to download entire files from certain computers designated as file servers. After downloading data to a local computer, a user can analyze, manipulate, format, and display data from the file.

In **client/server architecture**, multiple computer platforms are dedicated to special functions, such as database management, printing, communications, and program execution. These platforms are called *servers*. Each server is accessible by all computers on the network. Servers can be computers of all sizes; they store both application programs and data files and are equipped with operating system software to manage the activities of the network. The server distributes programs and data to the other computers (clients) on the network as they request them. An application server holds the programs and data files for a particular application, such as an inventory database. The client or the server can do the processing.

LaserPro is a client/server application used by lenders in more than 2,800 of the approximately 7,500 banks in the U.S. to generate the paperwork needed to offer consumer, real estate, and commercial loans. Based on loan officer responses to system prompts, this document assembly system creates the loan documents by selecting and customizing text from more than 13,000 standard paragraphs. The system includes a legal and compliance knowledge base of state and federal loan regulations to ensure that the documents it generates conform to all laws.[21] The system increases loan officer productivity and ensures that all documents meet the rapidly evolving set of regulations for loans.

> **client/server architecture**
> An approach to computing wherein multiple computer platforms are dedicated to special functions, such as database management, printing, communications, and program execution.

Telecommunications Software

A **network operating system (NOS)** is systems software that controls the computer systems and devices on a network and allows them to communicate with each other. The NOS performs similar functions for the network as operating system software does for a computer, such as memory and task management and coordination of hardware. When network equipment (such as printers, plotters, and disk drives) is required, the NOS makes sure that these resources are used correctly. Novell NetWare, Windows 2000, Windows 2003, and Windows 2008 are common network operating systems. MySpace, the popular social networking Web site that offers an interactive, user-submitted network of friends, personal profiles, blogs, music, and videos internationally, was one of the first very busy Web sites to adopt the use of Windows Server 2008.

Software tools and utilities are available for managing networks. With **network-management software**, a manager on a networked personal computer can monitor the use of individual computers and shared hardware (such as printers); scan for viruses; and ensure compliance with software licenses. Network-management software also simplifies the process of updating files and programs on computers on the network—a manager can make changes through a communications server instead of having to visit each individual computer. In addition, network-management software protects software from being copied, modified, or downloaded illegally. It can also locate telecommunications errors and potential network problems. Some of the many benefits of network-management software include fewer hours spent on routine tasks (such as installing new software), faster response to problems, and greater overall network control.

> **network operating system (NOS)**
> Systems software that controls the computer systems and devices on a network and allows them to communicate with each other.

> **network-management software**
> Software that enables a manager on a networked desktop to monitor the use of individual computers and shared hardware (such as printers); scan for viruses; and ensure compliance with software licenses.

Data Synch Systems provides remote network monitoring for more than 50 customers. The organization minimized customer interruptions and saved more than $80,000 in site visit expenses by using network-management software.[22]

Now that we have covered many of the basics of telecommunications, let's discuss the use of the Internet.

USE AND FUNCTIONING OF THE INTERNET

The Internet is the world's largest computer network; it is actually a collection of inter-connected networks, all freely exchanging information using a common set of protocols. (A communications protocol is a set of rules that govern the exchange of information over a communications channel.) The Internet began as an experiment linking together research institutes and universities. Over time, more organizations were connected to the Internet, followed by homes and businesses, until today, when it is difficult to find a computer that is not connected to the Internet. More than 750 million computers, or hosts,[23] make up today's Internet, supporting nearly 2 billion users. Those numbers are expected to continue growing.[24] Figure 4.6 shows the staggering growth of the Internet, as measured by the number of Internet host sites, or domain names. Domain names are discussed later in the chapter.

Figure 4.6

Internet Growth: Number of Internet Domain Names

(Source: Data from "The Internet Domain Survey," *www.isc.org*.)

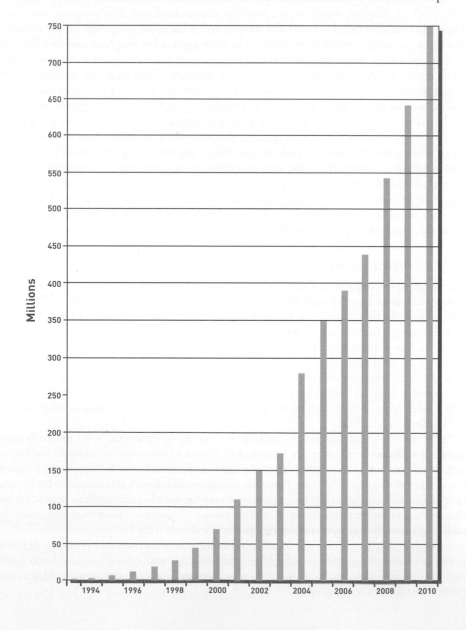

The Internet is truly international in scope, with users on every continent—including Antarctica. More than 220 million people in the United States (76.3 percent of the population) use the Internet. Although the United States has high Internet penetration among its population, it does not constitute the majority of people online. Of all the people using the Internet, citizens of Asian countries make up 42.4 percent, Europeans 23.6 percent, and North Americans only 14.4 percent.[25] China has by far the most Internet users at 384 million, which is more people than the total U.S. population, but only 29 percent of China's total population. Use of the Internet is growing around the globe, though at differing rates for each country. For example, most Internet use in South Korea is through high-speed broadband connections, and more than 71 percent of the population is online. In contrast, north of the border in North Korea, the government prohibits Internet use and other civil liberties. Being connected to the Internet provides global economic opportunity to individuals, businesses, and countries.

The freedom of expression provided by the Internet has been a source of controversy in many countries. China, for example, blocks Internet content that it feels is subversive or threatening to "national unity."[26] Ireland, France, Australia, and several other countries have blocked Internet access to individuals caught illegally downloading copyrighted material.[27] Other countries have blocked content that is considered blasphemous to their national religions.[28] Even the U.S. has experimented with blocking "indecent" Web content.[29] However, the Internet includes so many connections that completely controlling its flow of information is next to impossible.

China has more than 384 million Internet users online, which is only 29 percent of its population.

(Source: Reuters/Landov.)

The ancestor of the Internet was the **ARPANET**, a project started by the U.S. Department of Defense (DoD) in 1969. The ARPANET was both an experiment in reliable networking and a means to link DoD and military research contractors, including many universities doing military-funded research. *(ARPA* stands for the Advanced Research Projects Agency, the branch of the DoD in charge of awarding grant money. The agency is now known as DARPA—the added *D* is for *Defense)* The ARPANET was highly successful, and every university in the country wanted to use it. This wildfire growth made it difficult to manage the ARPANET, particularly its large and rapidly growing number of university sites. So, the ARPANET was broken into two networks: MILNET, which included all military sites, and a new, smaller ARPANET, which included all the nonmilitary sites. The two networks remained connected, however, through use of the **Internet Protocol** (**IP**), which enables traffic to be routed from one network to another as needed. All the networks connected to the Internet use IP, so they all can exchange messages.

Today, people, universities, and companies are attempting to make the Internet faster and easier to use. To speed Internet access, a group of corporations and universities called the University Corporation for Advanced Internet Development (UCAID) is working on a faster, alternative Internet. Called Internet2 (I2), Next Generation Internet (NGI), or

ARPANET
A project started by the U.S. Department of Defense (DoD) in 1969 as both an experiment in reliable networking and a means to link the DoD and military research contractors, including many universities doing military-funded research.

Internet Protocol (IP)
A communication standard that enables computers to route communications traffic from one network to another as needed.

Abilene, depending on the universities or corporations involved, the new Internet offers the potential of faster Internet speeds, up to 2 Gbps per second or more.[30] The *National Lamb-daRail* (*NLR*) is a cross-country, high-speed (10 Gbps) fiber-optic network dedicated to research in high-speed networking applications.[31] The NLR provides a "unique national networking infrastructure" to advance networking research and next-generation network-based applications in science, engineering, and medicine. This new high-speed fiber-optic network will support the ever-increasing need of scientists to gather, transfer, and analyze massive amounts of scientific data.

How the Internet Works

In the early days of the Internet, the major telecommunications (telecom) companies around the world agreed to connect their networks so that users on all the networks could share information over the Internet. The cables, routers, switching stations, communication towers, and satellites that make up these networks are the hardware over which Internet traffic flows. These large telecom companies are called *network service providers* (*NSPs*). Examples include Verizon, Sprint, British Telecom, and AT&T. The combined backbones of these and other NSPs—the fiber-optic cables that span the globe over land and under sea, make up the Internet backbone.

The Internet transmits data from one computer (called a *host*) to another. See Figure 4.7. If the receiving computer is on a network to which the first computer is directly connected, it can send the message directly. If the receiving and sending computers are not directly connected to the same network, the sending computer relays the message to another computer that can forward it. The message is typically sent through one or more routers to reach its destination. It is not unusual for a message to pass through a dozen or more routers on its way from one part of the Internet to another.

Figure 4.7

Routing Messages over the Internet

The Internet routes data packets over the network backbone from router to router to reach their destinations.

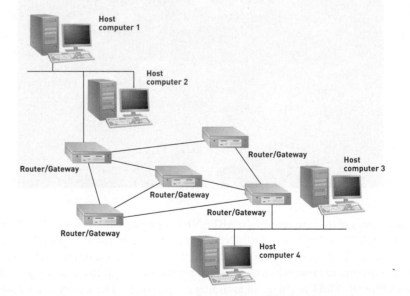

The various telecommunications networks that are linked to form the Internet work much the same way—they pass data around in chunks called packets, each of which carries the addresses of its sender and its receiver along with other technical information. The set of rules used to pass packets from one host to another is the IP protocol. Many other communications protocols are used in connection with IP. The best known is the Transmission Control Protocol (TCP). Many people use TCP/IP as an abbreviation for the combination of TCP and IP used by most Internet applications. After a network following these standards links to the Internet's backbone, it becomes part of the worldwide Internet community.

An **IP address** is a 64-bit number (e.g. 208.77.201.209) assigned to each computer to identify it on the Internet. The Internet will soon be upgraded to IPv6, which uses 128-bit addresses to provide for many more devices.[32] Because people prefer to work with words

IP address

A 64-bit number that identifies a computer on the Internet.

rather than numbers, a system called the Domain Name System (DNS) was created. Domain names such as www.cengage.com are mapped to IP addresses such as 69.32.133.79 using the DNS. If you type either www.cengage.com or 69.32.133.79 into your Web browser, you will access the same Web site.

A **Uniform Resource Locator (URL)** is a Web address that specifies the exact location of a Web page using letters and words that map to an IP address and a location on the host. The URL gives those who provide information over the Internet a standard way to designate where Internet resources such as servers and documents are located. Consider the URL for Course Technology, http://www.cengage.com/coursetechnology.

The "http" specifies the access method and tells your software to access a file using the Hypertext Transport Protocol. This is the primary method for interacting with the Internet. In many cases, you don't need to include http:// in a URL because it is the default protocol. The "www" part of the address sometimes, but not always, signifies that the address is associated with the World Wide Web service. The URL *www.cengage.com* is the domain name that identifies the Internet host site. The part of the address following the domain name—/coursetechnology—specifies an exact location on the host site.

Domain names must adhere to strict rules. They always have at least two parts, with each part separated by a dot (period). For some Internet addresses, the far right part of the domain name is the country code, such as au for Australia, ca for Canada, dk for Denmark, fr for France, de (Deutschland) for Germany, and jp for Japan. Many Internet addresses have a code denoting affiliation categories, such as com for business sites and edu for education sites. (Table 4.5 contains a few popular categories.) The far left part of the domain name identifies the host network or host provider, which might be the name of a university or business. Other countries outside the United States use different top-level domain affiliations from the ones described in the table.

Uniform Resource Locator (URL)
A Web address that specifies the exact location of a Web page using letters and words that map to an IP address and a location on the host.

Table 4.5

U.S. Top-Level Domain Affiliations

Affiliation ID	Affiliation
com	Business sites
edu	Educational sites
gov	Government sites
net	Networking sites
org	Nonprofit organization sites
mobi	Mobile-compatible sites for smartphones

Note that some other countries outside the United States use different top-level domain affiliations from the ones described in Table 4.5.

The Internet Corporation for Assigned Names and Numbers (ICANN) is responsible for managing IP addresses and Internet domain names. One of its primary concerns is to make sure that each domain name represents only one individual or entity——the one that legally registers it. For example, if your teacher wanted to use *www.course.com* for a course Web site, he or she would discover that domain name has already been registered by Course Technology and is not available. ICANN uses companies called *accredited domain name registrars* to handle the business of registering domain names. For example, you can visit *www.namecheap.com*, an accredited registrar, to find out if a particular name has already been registered; If not, you can register the name for around $9 per year. Once you do so, ICANN will not allow anyone else to use that domain name as long as you pay the yearly fee.

Accessing the Internet

You can connect to the Internet in numerous ways. See Figure 4.8. Which access method you choose is determined by the size and capability of your organization or system.

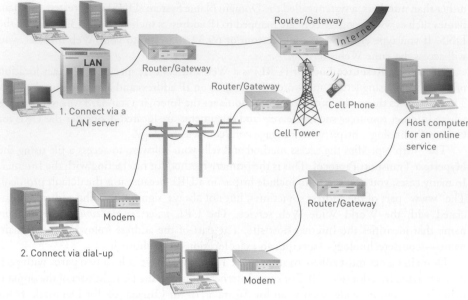

Connect via LAN Server

This approach is used by businesses and organizations that manage a local area network (LAN). By connecting a server on the LAN to the Internet using a router, all users on the LAN are provided access to the Internet. Business LAN servers are typically connected to the Internet at very fast data rates, sometimes in the hundreds of Mbps. In addition, you can share the higher cost of this service among several dozen LAN users to allow a reasonable cost per user.

Connecting via Internet Service Providers

Internet service provider (ISP)
Any organization that provides
Internet access to people.

Companies and residences unable to connect directly to the Internet through a LAN server must access the Internet through an Internet service provider. An **Internet service provider (ISP)** is any organization that provides Internet access to people. Thousands of organizations serve as ISPs, ranging from universities that make the Internet available to students and faculty, to small Internet businesses, to major telecommunications giants such as AT&T and Comcast. To connect to the Internet through an ISP, you must have an account with the service provider (for which you usually pay) along with software (such as a browser) and devices (such as a computer or smartphone) that support a connection via TCP/IP.

Perhaps the least expensive but slowest connection provided by ISPs is a dial-up connection. A *dial-up Internet connection* uses a modem and standard phone line to "dial-up" and connect to the ISP server. Dial-up is considered the slowest of connections because it is restricted by the 56 Kbps limitation of traditional phone line service. A dial-up connection also ties up the phone line so that it is unavailable for voice calls. While dial-up was originally the only way to connect to the Internet from home, it is rapidly becoming replaced by high-speed services.

Several "high-speed" Internet services are available for home and business. They include cable modem connections from cable television companies, DSL connections from phone companies (a telecommunications service that delivers high-speed Internet access to homes and small businesses over the existing phone lines of the local telephone network), and satellite connections from satellite television companies.

In addition to connecting to the Internet through wired systems such as phone lines and television cables, wireless Internet over cellular and Wi-Fi networks has become common. Thousands of public Wi-Fi services are available in coffee shops, airports, hotels, and elsewhere, where Internet access is provided free, for an hourly rate, or for a monthly subscription fee. In 2010, McDonalds became the biggest provider of free wireless in the U.S. when it began offering free Wi-Fi at 11,000 of its 13,000 restaurants.[33] Many businesses have followed suit using free Wi-Fi access as a tool to attract customers. Wi-Fi is even making its way into aircraft, allowing business travelers to be productive during air travel by accessing e-mail and corporate networks.

Cell phone carriers also provide Internet access for handsets, notebooks, and tablets. New 4G mobile phone services rival wired high-speed connections enjoyed at home and work. Sprint, Verizon, AT&T, and other popular carriers are working to bring 4G service to subscribers, beginning in large metropolitan areas. Sprint's WiMAX 4G network provides speeds up to 4 Mbps, while T-Mobile's HSPA+ 4G network provides up to 5.4 Mbps[34]. AT&T and Verizon are rolling out LTE networks that promise to be even faster.[35] By purchasing data plans, users can connect to these networks with smartphones and computers. Table 4.6 compares data transfer speeds and popular Internet services based on 2010 quotes from Comcast, Sprint, AT&T and HughesNet.

Service	Download Speed Per Monthly Cost: Standard Plan	Download Speed Per Monthly Cost: Premium Plan
Cable	15 Mbps/$42.95	50 Mpbs/$99.95
DSL	1 Mbps/$19.95	6 Mbps/$35
Satellite	1 Mbps/$59.99	5 Mbps/$349.99
Cellular	1 Mbps/$59.99	6 Mbps/$69.99
Dial-up	56 Kbps/$12.50	N/A

Table 4.6

Internet Service Options Compared

When Apple introduced the iPhone, one of its slogans was the "Internet in your pocket." The iPhone serves to prove the popularity of, and potential for, Internet services over a handset. Many other smartphones followed hot on the heels of the iPhone offering similar services on all of the cellular networks. More recently, the iPhone 4 brought video calling into vogue, while the iPad and other tablets provided anywhere, anytime access to all types of Internet services on a larger display.

Connecting Wirelessly

The iPad connects to the Internet over cellular or Wi-Fi networks.

(Source: © Stefan Sollfors/Alamy.)

Cloud Computing

Cloud computing refers to a computing environment where software and storage are provided as an Internet service and accessed with a Web browser. As Internet connection speeds increase and wireless Internet access becomes pervasive, computing activities are increasingly being delivered over the Internet rather than from installed software on PCs. Google and Yahoo! store the e-mail of many users, along with calendars, contacts, and to-do lists. Facebook provides social interaction and can store personal photos, as can Flickr and a dozen other photo sites. Pandora delivers music, and Hulu and YouTube deliver movies. Google Docs, Microsoft Web Apps, Zoho, 37signals, Flypaper, Adobe Buzzword, and others provide Web-delivered productivity and information management software. Soon, it seems, most computing will take place on Internet servers through the Web browser. All your friends, photos, documents, music, and media will be available to you from any Internet-connected device. This is the world of cloud computing.

The term cloud computing comes from the use of a cloud in network diagrams to represent the Internet in an abstract sense. See Figure 4.9. Cloud computing service providers manage their services much like a utility company manages its resources. The processing and storage requirements of all of its clients can be spread over numerous servers. As business grows, more servers are added. If one server fails, others pick up the slack. Cloud computing is extremely scalable and often takes advantage of virtualization technologies.

Cloud computing offers tremendous advantages to businesses.[36] By outsourcing business information systems to the cloud, a business saves on system design, installation, and maintenance. Employees can also access corporate systems from any Internet-connected computer using a standard Web browser. For example, RezBook is a cloud computing application (app for short) designed by Urbanspoon for the iPad to manage reservations for restaurants. Hosts and wait staff in a restaurant use their iPads to access the system stored on an Internet server to check for open tables, track each table's progress through a meal, and store and access customer reservations.[37] Some companies, including Microsoft and Google, provide free online storage (with capacity limitations) for access from any Internet-connected computer. Dropbox, SugarSync, and others include file backup and synchronization services to their cloud storage offerings.

ETHICAL AND SOCIETAL ISSUES

Danger in the Cloud

The general public has embraced cloud computing more readily than many businesses. Millions of people trust cloud computing technologies from Google to store their e-mail, appointment calendars, and address books. They trust Facebook to store their photos and personal information. Businesses have been more hesitant to trust Internet firms with their valuable corporate information—and with good cause.

Cloud computing services do fail, leaving users unable to access programs or data. It is not uncommon for Google, Twitter, Microsoft, Facebook, and other online companies to experience server outages. In other cases, cloud computing services have lost customer data. Perhaps the most notable was the catastrophe with T-Mobile's Sidekick smartphone service. In October 2009, T-Mobile informed its thousands of Sidekick users that it had lost their data and might not be able to recover it. It advised the users not to turn off their cell phones, as the data stored on them would be irrecoverable. The Sidekick uses a cloud computing data service from a Microsoft subsidiary ironically named Danger, Inc. to back up user data from smartphones. The data stored includes user contacts, calendars, notes, photos, text messages, and other data typically stored on mobile phones. The cloud storage system for this data failed and had no backup system in place. As angry Sidekick users posted comments online, the failure gained the attention of businesses and consumers.

In the end, much of the data was recovered, and those who lost data were compensated with a $100 credit. Still, the incident is considered a black eye for T-Mobile and cloud computing. Similar incidents such as Gmail outages cause businesses to be leery about trusting cloud computing with important data.

The City of Los Angeles recently decided to trust Google and its online applications rather than using traditional software such as Microsoft Office. The decision was not made lightly. The $75 million contract came with several stipulations. Google has agreed to pay a considerable penalty if a security breach occurs. Google is legally responsible for any release of data in violation of a nondisclosure agreement. The city's data must also be encrypted, stored on a dedicated server, and kept in the U.S. with limited access.

Such assurances are essential if cloud computing is going to live up to its potential for businesses. Microsoft is lobbying Congress to create new laws designed to offer protection for data stored in the cloud and to enact stiffer penalties for hackers who attempt to illegally access it. Microsoft hopes that government support for cloud computing will spread globally so that data can be safe wherever it is stored within the global business infrastructure. Companies around the world that provide outsourced cloud storage will be held accountable by the same laws that govern U.S. firms. "We need a free trade agreement for data," says Brad Smith, senior vice president and general counsel at Microsoft.

Without government assistance, open standards, and international cooperation, some fear that cloud computing will be controlled by two or three big companies, leaving smaller companies unable to compete. Users may be locked into one service provider's proprietary system. Open standards, on the other hand, would allow customers to easily transfer their cloud computing services from one company to another. Smith calls for cloud computing vendors to band together to establish open standards for data storage that provide transparency for security and privacy. "Simply put, it should not be enough for service providers simply to say that their services are private and secure," Smith said. "There needs to be some transparency about why this is the case."

Jonathan Rochelle, a group product manager at Google, suggests that cloud computing isn't any more dangerous than storing data on your own PC or server. "While it feels more comfortable, the same way the money under your mattress feels more comfortable, it may not be the best way to manage your information," suggests Rochelle. The point is that the public is willing to trust banks and companies with their financial well being, so why not trust the companies that provide cloud computing services with your information?

Discussion Questions

1. Would you be comfortable storing all of your data, including personal data, media, and professional data, in the cloud rather than on your own PC? Why or why not?
2. What assurances and practices do you feel are necessary from cloud computing firms to earn the trust of businesses and the public?

Critical Thinking Questions

1. What role can government(s) play in helping cloud computing realize its potential?
2. Why do you think Microsoft, a company that has been historically opposed to open standards, is now lobbying for them?

SOURCES: Weinschenk, Carl, "T-Mobile Resumes Sidekick Sales Despite Costly Risk," NewsFactor, *www.newsfactor.com/story.xhtml?story_id=70139&full_skip=1*, November 17, 2009; Resende, Patricia, "L.A. Cloud Contract Goes to Google Over Microsoft," NewsFactor, *www.newsfactor.com/story.xhtml?story_id=69765*, October 28, 2009; Thibodeau, Patrick, "Microsoft Seeks Legal Protections for Data Stored in Cloud," *CIO, www.cio.com/article/520724/Microsoft_Seeks_Legal_Protections_for_Data_Stored_in_Cloud?source=rss_all*, January 21, 2010; Gross, Grant, "Microsoft calls for cloud computing transparency," *ITWorld, www.itworld.com/government/93452/microsoft-calls-cloud-computing-transparency?utm_source=feedburner&utm_medium=feed&utm_campaign=Feed%3A+ItworldToday+%28ITworld+Today%29*, January 20, 2010; Dunn, John E., "Internet Heading for 'Perfect Storm'," *CIO, www.cio.com/article/519770/Internet_Heading_for_Perfect_Storm_?source=rss_all*, January 20, 2010.

In business, cloud computing is often referred to as Software as a Service (SaaS) and the vendors that provide the software are application service providers (ASPs). Salesforce.com is an example of SaaS as applied to customer relationship management (CRM). Salesforce.com services more than 50,000 businesses, providing hundreds of CRM applications for a wide variety of business types. Employees of those businesses can access customer data from desktops, notebooks, and cell phones using a Web browser. Businesses that use Salesforce.com don't have to worry about supporting complicated CRM software on their servers, installing updates and security patches, and troubleshooting problems. The SaaS provider manages it all.

The popularity of netbooks, small notebook computers designed primarily for accessing Web applications, and nettops, their desktop equivalent, are an indication of the direction of computing. As cloud computing grows, the power and storage capacity of users' computers can diminish. High-speed wireless access and a decent keyboard and display are all you really need for computing in the cloud.

THE WORLD WIDE WEB

The World Wide Web was developed by Tim Berners-Lee at CERN, the European Organization for Nuclear Research in Geneva. He originally conceived of it as an internal document-management system. From this modest beginning, the Web has grown to become a primary source of news and information, an indispensible conduit for commerce, and a popular hub for social interaction, entertainment, and communication.

How the Web Works

Web
Server and client software, the hypertext transfer protocol (http), standards, and mark-up languages that combine to deliver information and services over the Internet.

hyperlink
Highlighted text or graphics in a Web document that, when clicked, opens a new Web page containing related content.

Web browser
Web client software such as Internet Explorer, Firefox, Chrome, Safari, and Opera used to view Web pages.

While the terms Internet and Web are often used interchangeably, technically, the two are different technologies. The Internet is the infrastructure on which the Web exists. The Internet is made up of computers, network hardware such as routers and fiber-optic cables, software, and the TCP/IP protocols. The **Web**, on the other hand, consists of server and client software, the hypertext transfer protocol (http), standards, and mark-up languages that combine to deliver information and services over the Internet.

The Web was designed to make information easy to find and organize. It connects billions of documents, which are now called Web pages, stored on millions of servers around the world. These are connected to each other using **hyperlinks**, specially denoted text or graphics on a Web page, that, when clicked, open a new Web page containing related content. Using hyperlinks, users can jump between Web pages stored on various Web servers—creating the illusion of interacting with one big computer. Because of the vast amount of information available on the Web and the wide variety of media, the Web has become the most popular means of information access in the world today.

In short, the Web is a hyperlink-based system that uses the client/server model. It organizes Internet resources throughout the world into a series of linked files, called pages, accessed and viewed using Web client software called a **Web browser**. Internet Explorer, Firefox, Chrome, Safari, and Opera are five popular Web browsers. See Figure 4.10. A collection of pages on one particular topic, accessed under one Web domain, is called a Web site. The Web was originally designed to support formatted text and pictures on a page. It has evolved to support many more types of information and communication including user interactivity, animation, and video. Web *plug-ins* help provide additional features to standard Web sites. Adobe Flash and Real Player are examples of Web plug-ins.

Figure 4.10

Mozilla Firefox

Web browsers such as Firefox let you access Internet resources such as this customizable Web portal from Google.

Hypertext Markup Language (HTML) is the standard page description language for Web pages. HTML is defined by the World Wide Web Consortium (referred to as "W3C") and has developed through numerous revisions. It is currently in its fifth revision—HTML5. HTML tells the browser how to display font characteristics, paragraph formatting, page layout, image placement, hyperlinks, and the content of a Web page. HTML uses **tags**, which are codes that tell the browser how to format the text or graphics: as a heading, list, or body text, for example. Web site creators "mark up" a page by placing HTML tags before and after one or more words. For example, to have the browser display a sentence as a heading, you place the <h1> tag at the start of the sentence and an </h1> tag at the end of the sentence. When you view this page in your browser, the sentence is displayed as a heading. HTML also provides tags to import objects stored in files, such as photos, pictures, audio, and movies, into a Web page. In short, a Web page is made up of three components: text, tags, and references to files. The text is your Web page content, the tags are codes that mark the way words will be displayed, and the references to files insert photos and media into the Web page at specific locations. All HTML tags are enclosed in a set of angle brackets (< and >), such as <h2>. The closing tag has a forward slash in it, such as </b> for closing bold. Consider the following text and tags:

```
<html>

<head>

<title>Table of Contents</title>

<link href="style.css" rel="stylesheet" type="text/css" />

</head>

<body style="background-color:#333333">

<div id="container">

<p><img src="header.png" width="602" height="78" /></p>

<h1 align=center>Principles of Information Systems</h1>

<ol>

<li>An Overview</li>

<li>Information Technology Concepts</li>
```

Hypertext Markup Language (HTML)
The standard page description language for Web pages.

HTML tags
Codes that tell the Web browser how to format text—as a heading, as a list, or as body text—and whether images, sound, and other elements should be inserted.

```
<li>Business Information Systems</li>

<li>Systems Development</li>

<li>Information Systems in Business and Society</li>

</ol>

</div>

</body>

</html>
```

The <html> tag identifies this as an HTML document. HTML documents are divided into two parts: the <head> and the <body>. The <body> contains everything that is viewable in the Web browser window, and the <head> contains related information such as a <title> to place on the browser's title bar. The background color of the page is specified in the <body> tag using a hexadecimal code. The heading "Principles of Information Systems" is identified as the largest level 1 heading with the <h1> tag, typically a 16–18 point font, centered on the page. The <ol> tag indicates an ordered list, and the <li> tags indicate list items. The resulting Web page is shown in Figure 4.11.

Figure 4.11

HTML Code Interpreted by a Browser

The example HTML code as interpreted by the Firefox Web browser on a Mac.

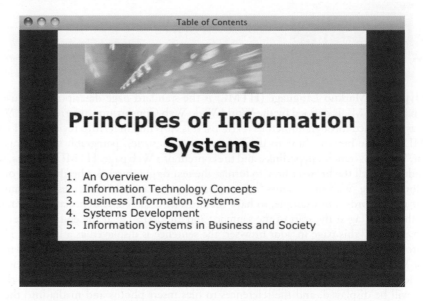

Although HTML is the standard page description language for Web pages, some other Web standards have become nearly equal to HTML in importance, including Extensible Markup Language (XML), Cascading Style Sheets (CSS), and Wireless Markup Language (WML).

Extensible Markup Language (XML) is a markup language for Web documents containing structured information, including words and pictures. XML does not have a predefined tag set. With HTML, for example, the <hl> tag always means a first-level heading. The content and formatting are contained in the same HTML document. XML Web documents contain the content of a Web page. The formatting of the content is contained in a style sheet. A few typical instructions in XML follow:

Extensible Markup Language (XML)

The markup language designed to transport and store data on the Web.

```
<book>

<chapter>Hardware</chapter>

<topic>Input Devices</topic>

<topic>Processing and Storage Devices</topic>

<topic>Output Devices</topic>

</book>
```

A **Cascading Style Sheet** (CSS) is a file or portion of an HTML file that defines the visual appearance of content in a Web page. Using CSS is convenient because you only need to define the technical details of the page's appearance once, rather than in each HTML tag. For example, the visual appearance of the preceding XML content may be contained in the following style sheet. This style sheet specifies that the chapter title "Hardware" is displayed on the Web page in a large Arial font (18 points). "Hardware" will also appear in bold blue text. The "Input Devices" title will appear in a smaller Arial font (12 points) and italic red text.

Cascading Style Sheet (CSS)
A markup language for defining the visual design of a Web page or group of pages.

```
chapter: (font-size: 18pt; color: blue; font-weight: bold;
display: block; font-family: Arial;

margin-top: 10pt; margin-left: 5pt)

topic: (font-size: 12pt; color: red; font-style: italic; display:
block; font-family: Arial;

margin-left: 12pt)
```

XML is extremely useful for organizing Web content and making data easy to find. Many Web sites use CSS to define the design and layout of Web pages, XML to define the content, and HTML to join the content (XML) with the design (CSS). See Figure 4.12. This modular approach to Web design allows you to change the visual design without affecting the content, or to change the content without affecting the visual design.

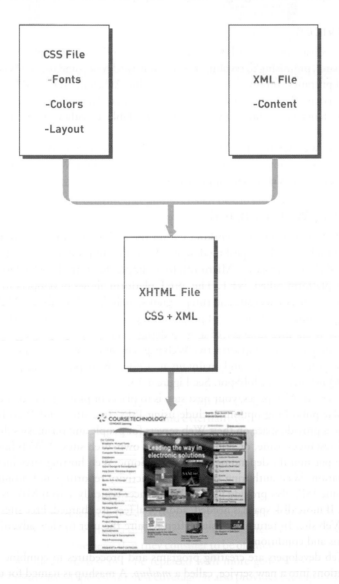

Figure 4.12

XML, CSS, and XHTML

Today's Web sites are created using XML to define content, CSS to define the visual style, and XHTML to put it all together.

deeper analysis of the subject matter. For example, during military conflicts overseas, online news services provide news articles in text, audio, and video coverage. Clicking links allows you to drill down and find out more about geographic regions by viewing maps, for example; you could also link to historical coverage of international relations and learn about the battle equipment being deployed.

Most newspaper, radio, and television news services have expanded to provide online news coverage. Text and photos are supported by the HTML standard. Video, sometimes called a Webcast, and audio are provided in the browser through plug-in technology and in podcasts. See Figure 4.14. Bringing the news to the Web is eliminating the lines of distinction between traditional newspaper, radio, and television news sources.

Figure 4.14

News Webcast

Online news is available in text, audio, and video formats, providing the ability to drill down into stories.

(Source: *http://abcnews.go.com*.)

While traditional news sources migrate to the Web, new sources are emerging from online companies. News Web sites from Google, Yahoo!, Digg, and Newsvine provide the most popular or interesting news stories from a variety of news sources.

In a trend some refer to as social journalism or citizen journalism, ordinary citizens are more involved in reporting the news than ever before. The online community is taking journalism into its hands and reporting the news from each person's perspective using an abundance of online tools. Although social journalism provides important news not available elsewhere, its sources may not be as reliable as mainstream media sources. It is sometimes difficult to discern news from opinion. News from nonprofessional journalists, reporting without the strict guidelines of formal news agencies, may be biased, misrepresented, mistaken, or perhaps even deliberately misleading. Many citizen journalists would be quick to point out that mainstream media may also be biased in its reporting of the news.

Education and Training

As a tool for sharing information and a primary repository of information on all subjects, the Web is ideally suited for education and training. Advances in interactive Web technologies further support important educational relationships between teacher and student and among students. See Figure 4.15. The Web can play a major role in education from pre-K through adult continuing education. In today's highly competitive and rapidly changing professional environments, more professionals are turning to the Web to learn skills that will enhance their professional value.

Even before children enter school, they are engaged with educational content on the Web at sites such as JumpStart.com. Primary schools use the Web to inform parents of school schedules and activities. Teachers give elementary school students research exercises in the classroom and at home that use Web resources. By high school, students have integrated the Web into daily study habits. Teachers manage class Web pages that contain information and links for students to use in homework exercises.

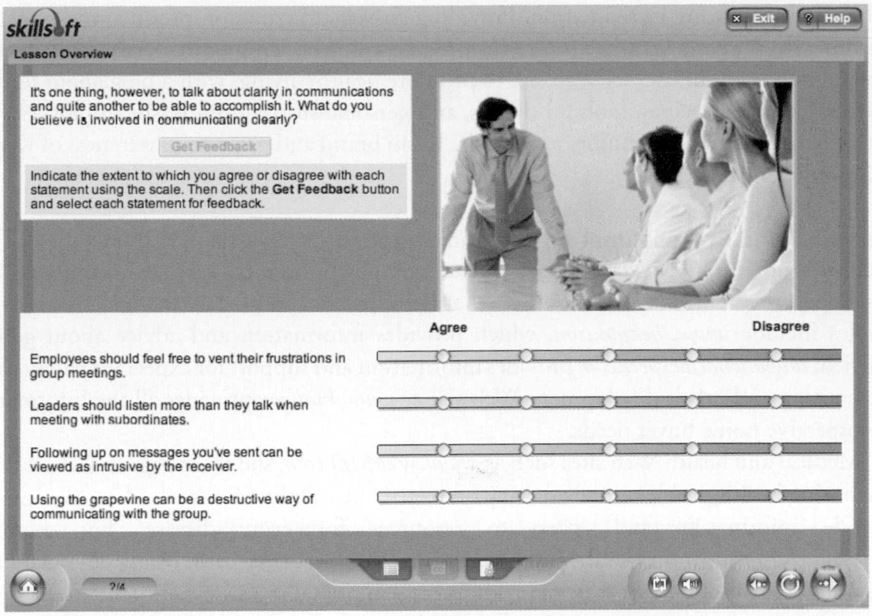

Figure 4.15

Skillsoft Online Professional Training and Certification

The Internet supports education from pre-K to lifelong learning.

(Source: *http://skillsoft.com.*)

Most college-level courses rely on the Web to enhance learning. Educational support products, such as Blackboard, provide an integrated Web environment that includes virtual chat for class members; a discussion group for posting questions and comments; access to the class syllabus and agenda, student grades, and class announcements; and links to class-related material. Some course Web sites even deliver filmed lectures using Webcasting technology. Such environments can complement the traditional classroom experience or be the sole method of course delivery.

Conducting classes over the Web with no physical class meetings is called *distance education.* Many colleges and universities offer distance education classes, which provide a convenient method for nontraditional students to attend college. Nontraditional students include older students who have job or family obligations that might otherwise prohibit them from attending college. Distance education offers them a way of working through class material on a flexible schedule. Some schools offer entire degree programs through distance education.

In a program it calls Open Courseware, the Massachusetts Institute of Technology (MIT) offers all of its courses free online. Students who take courses via Open Courseware do not earn credit toward a degree or have access to teachers, but they can benefit from the knowledge gained. Since MIT's move online, many other schools have followed suit. Organizations such as the Open Courseware Consortium and the Center for Open Sustainable Learning have been established to support open education around the world.

Beyond traditional education, corporations such as Skillsoft offer professional job-skills training over the Web. Job seekers often use these services to acquire specialized business or technical training. Some of the training leads to certification. Certification verifies a person's skill and understanding in a particular area. It has become very important, especially for some technical skill sets, to assure an employer that a job applicant truly has the skills claimed. Some corporations and organizations contract with Skillsoft to provide on-the-job training for current employees to expand their skills.

Business Information

Businesses often use Internet and Web-based systems for knowledge management within the enterprise. Information systems within the organization may be accessed through Web portals and dashboards that provide a single source for all business-related news and information. Such portals may extend beyond the walls of the office to notebooks and smartphones through Internet connections.

Providing news and information about a business and its products through the company's Web site and online social media can assist in increasing a company's exposure to the general public and improving its reputation. Providing answers to common product questions and customer support online can help keep customers coming back for more. For example, natural food company Kashi uses its Web site to promote healthy living, with a blog about leading a natural lifestyle, recipes, tools for dieters, and personal stories from Kashi employees. The Web site helps build community around the Kashi brand and promotes awareness of Kashi's philosophy and products.[38]

Personal and Professional Advice and Support

Web sites now support every subject and activity of importance. As people confront life's challenges, they can find Web resources that can educate and prepare them to succeed. Examples include *www.theknot.com*, which provides information and advice about getting married; *www.whattoexpect.com* provides information and support for expectant parents; and the housing and urban development Web site at *www.hud.gov* provides all the information a prospective home buyer needs.

Medical and health Web sites such as *www.WebMD.com*, shown in Figure 4.16, assist in diagnosing health problems and advising on treatments. Online forums and support groups provide information and access to resources for every disease. For example, *www.LiveStrong.org* provides free support resources for cancer victims. Many physicians and hospitals provide abundant educational information online to assist their patients.

Figure 4.16

WebMD

Web sites such as WebMD help to diagnose health problems and advise on treatments.

(Source: *www.webmd.com*.)

The Web is an excellent source of job-related information. People looking for their first jobs or seeking information about new job opportunities can find a wealth of information on the Web. Search engines, such as Google or Bing (discussed next) can be a good starting point for searching for specific companies or industries. You can use a directory on Yahoo's home page, for example, to explore industries and careers. Most medium and large companies have Web sites that list open positions, salaries, benefits, and people to contact for further information. The IBM Web site, *www.ibm.com*, has a link to "Jobs." When you click this link, you can find information on jobs with IBM around the world. Some sites can help you develop a résumé and assist you during your job search. They can also help you develop an effective cover letter for a résumé, prepare for a job interview, negotiate an employment contract, and more. In addition, several Internet sites specialize in helping you find job information and even apply for jobs online, including *www.monster.com*, *www.hotjobs.com*, and *www.careerbuilder.com*.

Search Engines and Web Research

The fundamental purpose of the Web is to make it easier to find related documents from diverse Internet sources by following hyperlinks. However, the Web has become so large that many complain of information overload, or the inability to find the information they need due to the overabundance of information. To relieve the strain of information overload, Web developers have provided Web search engines to help organize and index Web content.

A **search engine** is a valuable tool that enables you to find information on the Web by specifying words or phrases known as keywords, which are related to a topic of interest. You can also use operators such as OR and NOT for more precise search results.

Search engines have become the biggest application on the Web. Search giant Google has become one of the world's most profitable companies, with more than $16.5 billion in annual revenue. Web search has become such a profitable business because it's the one application that everyone uses. Search engine companies make money through advertisements. Because everyone uses search, advertisers are keen to pay to have their ads posted on search pages.

Most search engines use an automated approach that scours the Web with "bots" (automated programs) called spiders that follow all Web links in an attempt to catalog every Web page by topic. Each Web page is analyzed and ranked using unique algorithms and the resulting information is stored in a database. Google maintains more than 4 billion indexed Web pages on 30 clusters of up to 2,000 computers each, totaling over 30 petabytes of data.

A keyword search at Yahoo!, Bing, or Google isn't a search of the Web but rather a search of a database that stores information about Web pages. The database is continuously checked and refreshed so that it is an accurate reflection of the current status of the Web.

The Bing and Cuil search engines have attempted to innovate with their design. Bing refers to itself as a decision engine, providing more than just a long list of links in its search results. Bing also includes media—music, videos, and games—in its search results.[39] See Figure 4.17. Cuil takes it one step further with easy-to-navigate search results and tabs that show related topics.

Figure 4.17

Microsoft Bing Decision Engine

Microsoft calls its search engine a decision engine to distinguish it from other search software.

(Source: *www.bing.com*.)

Search engines have become important to businesses as tools to draw visitors to the business' Web site. Many businesses invest in search engine optimization (SEO)—a process for driving traffic to a Web site by using techniques that improve the site's ranking in search results. SEO is based on the understanding that Web page links listed on the first page of search results, as high on the list as possible, have a far greater chance of being clicked. SEO professionals study the algorithms employed by search engines, altering the Web page contents and other variables to improve the page's chance of being ranked number one. SEO professionals use *Web analytics software* to study detailed statistics about visitors to their sites.

In addition to search engines, you can use other Internet sites to research information. Wikipedia, an encyclopedia with more than 3.3 million English-language entries created and edited by millions of users, is another example of a Web site that can be used to research information. See Figure 4.18. In Hawaiian, *wiki* means quick, so a wikipedia provides quick access to information. The Web site is both open source and open editing, which means that people can add or edit entries in the encyclopedia at any time. Besides being self-regulating, Wikipedia articles are vetted by around 1,700 administrators. However, even with so many administrators, it is possible that some entries are inaccurate and biased.

Figure 4.18

Wikipedia

Wikipedia captures the knowledge of tens of thousands of experts.

(Source: *en.wikipedia.org*.)

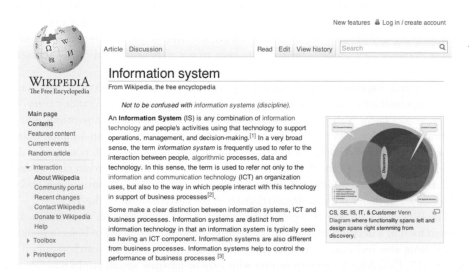

While Wikipedia is the best-known, general-purpose wiki, many other wikis are designed for special purposes. Wikimedia, the nonprofit organization behind Wikipedia, has wikis for books, news, media, and open learning. Zoho, Wikispaces, Wetpaint, and others provide tools to create wikis for any use. Wiki.com is a search engine designed for searching thousands of wikis. A wiki can be used in an enterprise to enable the sharing of information between employees, training for new employees, or customer support for products.[40]

Besides online card catalogs, libraries typically provide links to public and sometimes private research databases on the Web. Online research databases allow visitors to search for information in thousands of journal, magazine, and newspaper articles. Information database services are valuable because they offer the best in quality and convenience. They conveniently provide full-text articles from reputable sources over the Web. College and public libraries typically subscribe to many databases to support research. One of the most popular private databases is LexisNexis Academic Universe. LexisNexis provides access to full-text documents from over 5,900 news, business, legal, medical, and reference publications. You can access the information through a standard keyword search engine. See Figure 4.19.

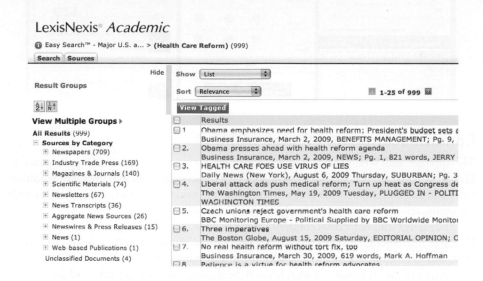

Figure 4.19

LexisNexis

A search at LexisNexis on "Health Care Reform" yields hundreds of full-text articles.

(Source: *www.lexisnexis.com.*)

Web Portals

A **Web portal** is a Web page that combines useful information and links and acts as an entry point to the Web—they typically include a search engine, a subject directory, daily headlines, and other items of interest. Many people choose a Web portal as their browser's home page (the first page you open when you begin browsing the Web), so the two terms are used interchangeably. Portals provide a convenient starting point for Web exploration in a general or a specific context. They allow users to have convenient access to their most frequently used Web resources.

Many Web pages have been designed to serve as Web portals. iGoogle, Yahoo!, AOL, and MSN are examples of horizontal portals; "horizontal" refers to the fact that these portals cover a wide range of topics. My Yahoo! and iGoogle allow users to custom design their page, selecting from hundreds of widgets—small applications that deliver information and services. Yahoo also integrates with Facebook so that Facebook users can access their friends and news streams from the My Yahoo portal.[41] See Figure 4.20.

Vertical portals are pages that provide information and links for special-interest groups. For example, the portal at *www.iVillage.com* focuses on items of interest to women, and *www.AskMen.com* is a vertical portal for men.

Many businesses set up corporate portals for their employees. Corporate portals (sometimes called dashboards) provide access to work-related resources such as corporate news and information, along with access to business tools, databases, and communication tools to support collaboration. Some businesses use a corporate portal to provide employees with work-related online content and to limit access to other Web content.

Web portal

A Web page that combines useful information and links and acts as an entry point to the Web— they typically include a search engine, a subject directory, daily headlines, and other items of interest. Many people choose a Web portal as their browser's home page (the first page you open when you begin browsing the Web).

Figure 4.20

MyYahoo! Personalized Portal

Personalized portals contain custom designs and widgets.

(Source: *www.yahoo.com.*)

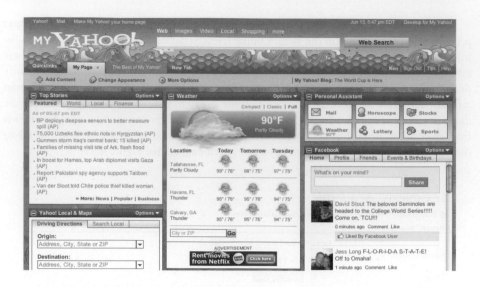

Communication and Collaboration

The Internet and Web provide many applications for communication and collaboration. From text-based applications such as e-mail, to telepresence systems that use high-definition video and audio that allow individuals from around the world to meet around a table, Internet communications supports many levels of communication.

The many forms of Internet communication include instant messaging, chat, virtual chat, blogging, microblogging, status updates, Internet phone, video chat and conferencing, virtual chat, and Web conferencing. When selecting a communication method, first consider the importance of the exchange and what needs to be shared. The most effective and meaningful communications are typically the least convenient. While the value of an in-person, face-to-face meeting should not be underestimated, many communications benefit from the convenience provided by the Internet.

E-Mail

E-mail is a useful form of Internet communication that supports text communication, HTML content, and sharing documents as e-mail attachments. E-mail is accessed through Web-based systems or through dedicated e-mail applications such as Outlook and Thunderbird. E-mail can also be distributed through enterprise systems to desktop computers, notebook computers, and smartphones.

Many people use online e-mail services such as Hotmail, MSN, and Gmail to access e-mail. See Figure 4.21. Online e-mail services store messages on the server, so users need to be connected to the Internet to view, send, and manage e-mail. Other people prefer to use software such as Microsoft Outlook, Apple Mail, or Mozilla Thunderbird, which retrieve e-mail from the server and deliver it to the user's PC. Post Office Protocol (POP) is used to transfer messages from e-mail servers to your PC. E-mail software typically includes more information management features than online e-mail services, and let you save your e-mail on your own PC, making it easier to manage and organize messages and to keep the messages private and secure. Another protocol called Internet Message Access Protocol (IMAP) allows you to view e-mail using Outlook or other e-mail software without downloading and storing the messages locally. Some users prefer this method because it allows them to view messages from any Internet-connected PC.

Business users that access e-mail from smartphones such as the BlackBerry take advantage of a technology called push e-mail. Push e-mail uses corporate server software that transfers, or pushes, e-mail to the handset as soon as it arrives at the corporate e-mail server. To the BlackBerry user, it appears as though e-mail is delivered directly to the handset. Push e-mail allows the user to view e-mail from any mobile or desktop device connected to the corporate server. This arrangement allows users flexibility in where, when, and how they access and manage e-mail.

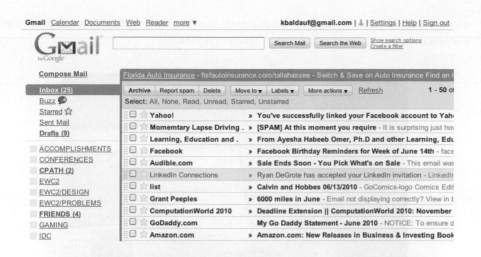

Figure 4.21

Gmail

Gmail is one of several free online e-mail services.

BlackBerry users have instant access to e-mail sent to their business accounts.

(Source: Courtesy of Marvin Woodyatt/Photoshot/Landov.)

Instant Messaging

Instant messaging is online, real-time communication between two or more people who are connected to the Internet. See Figure 4.22. With instant messaging, participants build buddy lists, or contact lists, that let them see which contacts are currently logged on to the Internet and available to chat. You can send messages to one of your online buddies, which opens a small dialog box on your buddy's computer and allows the two of you to chat via the keyboard. Although chat typically involves exchanging text messages with one other person, more advanced forms of chat are emerging. Today's instant messaging software supports not only text messages, but also sharing images, sounds, files, and voice communications. Popular instant messaging services include America Online Instant Messenger (AIM), MSN Messenger, Google Talk, and Yahoo!.

instant messaging
A method that allows two or more people to communicate online in real time using the Internet.

Microblogging, Status Updates, and News Feeds

Twitter is a Web application that allows members to report on what they are doing throughout the day. Referred to as a microblogging service, Twitter allows users to send short text updates (up to 140 characters) from a cell phone or Web account to their Twitter followers. While Twitter has been hugely successful for personal use, businesses are finding value in the service as well. Businesspeople use Twitter to stay in close touch with associates by sharing their location and activities throughout the day. Businesses also find Twitter to be a rich source of consumer sentiment that can be tapped to improve marketing, customer relations, and product development. Many businesses have a presence on Twitter, dedicating personnel to communicate with customers by posting announcements and reaching out to individual users. Village Books, an independent bookstore, uses Twitter to build relationships with its customers and to make them feel part of their community.[42]

The popularity of Twitter has caused social networks such as Facebook, LinkedIn, and MySpace to include Twitter-like news feeds. Previously referred to as Status Updates, Facebook users share their thoughts and activities with their friends by posting messages to Facebook's News Feed.

Conferencing

Some Internet technologies support real-time online conferencing. Teleconferences have been a popular form of remote conferencing for many years. Participants dial into a common phone number to share a multi-party phone conversation. The Internet has made it possible for those involved in teleconferences to share computer desktops. Using services such as WebEx or GoToMeeting, conference participants log on to common software that allows them to broadcast their computer display to the group. This is quite useful for presenting with PowerPoint, demonstrating software, training, or collaborating on documents. Participants verbally communicate by phone or PC microphone. Some conferencing software uses Web cams to broadcast video of the presenter and group participants. For example, Papa Johns uses GoToMeeting to conduct training sessions for managers around the world. Five online training sessions a year saves the company $50,000 in travel expenses.[43]

Telepresence takes video conferencing to the ultimate level. Telepresence systems such as those from Cisco and HP use high-resolution video and audio with high-definition displays to make it appear that conference participants are actually sitting around a table. See Figure 4.23. Participants enter a telepresence studio where they sit at a table facing display screens that show other participants from a variety of geographic locations. Cameras and microphones collect high-quality video and audio at all locations, and transmit them over high-speed network connections to provide an environment that replicates actual physical presence. Document cameras and computer software are used to share views of computer screens and documents with all participants. Baxter Healthcare Corporation finds that using telepresence systems enables "faster decision making and problem solving...streamlining the way projects move through the system."[44]

You don't need to be a big business to enjoy the benefits of video conversations. Free software is available to make video chat easy to use for anyone with a computer, a Web cam, and a high-speed Internet connection. Online applications such as Google Chat and Microsoft Messenger support video connections between Web users. For spontaneous, random video chat with strangers, you can use *www.Chatroulette.com* and Internet Conga Line.[45] Software such as Apple iChat and Skype provide computer-to-computer video chat so users can speak to each other face to face. In addition to offering text, audio, and video chat on computers, Skype offers its video phone service over Internet-connected TVs. Recent Internet-connected sets from Panasonic and Samsung ship with the Skype software preloaded. You attach a Web cam to your TV to video-chat from your sofa.

Figure 4.23

Halo Collaboration Meeting Room

HP's Halo telepresence system allows people at various locations to meet as though they were gathered around a table.

(Source: Courtesy of Hewlett-Packard.)

Web 2.0 and the Social Web

Over the past several years, the Web has evolved from a one-directional resource where users only obtain information to a two-directional resource where users obtain and contribute information. Consider Web sites such as YouTube, Wikipedia, and Facebook as examples. The Web has also grown in power to support full-blown software applications such as Google Docs and is becoming a computing platform itself. These two major trends in how the Web is used and perceived have created dramatic changes in how people, businesses, and organizations use the Web, creating a paradigm shift to **Web 2.0.**

The original Web, now referred to as Web 1.0, provided a platform for technology-savvy developers and the businesses and organizations that hired them to publish information for the general public to view. The introduction of user-generated content supported by Wikipedia, blogging, and podcasting made it clear that those using the Web were also interested in contributing to its content. This led to the development of Web sites with the sole purpose of supporting user-generated content and user feedback.

Web sites such as YouTube and Flickr allow users to share video and photos with other people, groups, and the world. Microblogging sites such as Twitter allow people to post thoughts and ideas throughout the day for friends to read. Social bookmarking sites such as *www.digg.com* and *www.delicious.com* allow users to pool their votes to determine what online news stories and Web pages are most interesting at any given time of the day. Similarly, Epinions and many retail Web sites allow consumers to voice their opinions about products. All of these popular Web sites serve as examples of how the Web has transformed to become the town square where people share information, ideas, and opinions; meet with friends; and make new acquaintances.

Businesses can observe social network users to determine their tastes and interests. Such data can be mined to discover consumer trends to guide product design and offerings.[46] Consumer information can also be used for targeted advertising.[47]

Some businesses are including social networking features in their products. The use of social media in business is called Enterprise 2.0. Enterprise 2.0 applications such as Salesforce's Chatter bring Facebook-like interaction to the workplace. Employees post profiles, making it easy to find colleagues with knowledge that is useful to the work environment. News feeds provide a constant patter of interaction and discussion around work-related topics. While many see Enterprise 2.0 applications as revitalizing forces in the workplace, others worry that they are distracting.[48]

Web 2.0
The Web as a computing platform that supports software applications and the sharing of information among users.

Rich Internet Applications

The introduction of powerful Web-delivered applications such as Google Docs, Adobe Photoshop Express, Xcerion Web-based OS, and Microsoft Office Web Apps have elevated the Web from an online library to a platform for computing. Many of the computer activities traditionally provided through software installed on a PC can now be carried out using rich Internet applications (RIAs) in a Web browser without installing any software. A **rich Internet application** is software that has the functionality and complexity of traditional application software, but runs in a Web browser and does not require local installation. See Figure 4.24. RIAs are the result of continuously improving programming languages and platforms designed for the Web.

rich Internet application (RIA)
Software that has the functionality and complexity of traditional application software, but does not require local installation and runs in a Web browser.

Figure 4.24

Rich Internet Application

SlideRocket is a rich Internet application for creating vibrant online presentations.

(Source: *www.sliderocket.com*.)

Blogging and Podcasting

A **Web log**, typically called a **blog**, is a Web site that people can create and use to write about their observations, experiences, and opinions on a wide range of topics. The community of blogs and bloggers is often called the *blogosphere*. A *blogger* is a person who creates a blog, while *blogging* refers to the process of placing entries on a blog site. A blog is like a journal. When people post information to a blog, it is placed at the top of the blog page. Blogs can include links to external information and an area for comments submitted by visitors. Video content can also be placed on the Internet using the same approach as a blog. This is often called a *video log* or *vlog*.

Internet users may subscribe to blogs using a technology called Really Simple Syndication (RSS). RSS is a collection of Web technologies that allow users to subscribe to Web content that is frequently updated. With RSS, you can receive a blog update without actually visiting the blog Web site. You can also use RSS to receive other updates on the Internet, for example, from news Web sites such as *nyt.com*, which provides the daily news from the New York Times. Software used to subscribe to RSS feeds is called *aggregator software*. Google Reader is a popular aggregator for subscribing to blogs. Blog search engines include Technorati, Feedster, and Blogdigger. You can also use Google to locate a blog.

A corporate blog can be useful for communicating with customers, partners, and employees. However, companies and their employees need to be cautious about the legal risks of blogging. Blogging can expose a corporation and its employees to charges of defamation, copyright and trademark infringement, invasion of privacy, and revealing corporate secrets.

A *podcast* is an audio broadcast over the Internet. The name podcast originated from Apple's *iPod* combined with the word *broadcast*. A podcast is like an audio blog. Using PCs,

Web log (blog)
A Web site that people can create and use to write about their observations, experiences, and opinions on a wide range of topics.

recording software, and microphones, you can record podcast programs and place them on the Internet. Apple's iTunes provides free access to tens of thousands of podcasts, sorted by topic and searchable by keyword. See Figure 4.25. After you find a podcast, you can download it to your PC (Windows or Mac), to an MP3 player such as the iPod, or any smartphone. You can also subscribe to podcasts using RSS software included in iTunes and other digital audio software.

Figure 4.25

iTunes Podcasts

iTunes provides free access to tens of thousands of podcasts.

(Source: *www.apple.com/itunes/ podcasts*.)

People and corporations can use podcasts to listen to audio material, increase revenues, or advertise products and services. You can listen to podcasts of radio programs, including some programs from National Public Radio (NPR), while you are driving, walking, making a meal, or doing most other activities. Colleges and universities often use blogs and podcasts to deliver course material to students.

Online Media and Entertainment

Music, movies, television program episodes, user-generated videos, e-book, audio books are all available online to download and purchase or stream.

Content streaming is a method of transferring large media files over the Internet so that the data stream of voice and pictures plays more or less continuously as the file is being downloaded. For example, rather than wait for an entire 5 MB video clip to download before they can play it, users can begin viewing a streamed video as it is being received. Content streaming works best when the transmission of a file can keep up with the playback of the file.

content streaming
A method for transferring large media files over the Internet so that the data stream of voice and pictures plays more or less continuously as the file is being downloaded.

Music

The Internet and the Web have made music more accessible than ever, with artists distributing their songs through online radio, subscription services, and download services. The Web has had a dramatic impact on the music industry, causing unprecedented changes in marketing and distribution. Digital distribution has allowed artists to distribute music directly to fans without the need for record companies. It has also opened the door to music piracy and illegal distribution of music that is legally protected by copyright.

Internet radio is similar to local AM and FM radio except that it is digitally delivered to your computer over the Internet, providing many more choices of stations. For example, Live365 provides access to thousands of radio stations in more than 200 musical genre categories. Some stations charge a subscription fee, but most do not.

Compressed music formats such as MP3 have made music swapping over the Internet a convenient and popular activity. File-sharing software such as BitTorrent provide a means by which some music fans copy and distribute music, often without consideration of copyright law. The result is a popular music distribution system that is largely illegal and difficult to control, and cuts deeply into the recording industry's profits. In addition, it is not always safe to swap files with strangers. Downloaded music files may actually be viruses renamed to look like MP3 files. Music industry giants have pulled together to win back customers by offering legal and safe electronic music distribution at a reasonable price that provides services and perks not offered by file-sharing networks.

Several legal music download services are available. Apple's iTunes was one of the first online music services to find success. Microsoft, Amazon.com, Wal-Mart, and other retailers also sell music online. Downloaded music may include digital rights management (DRM) technology that prevents or limits the user's ability to make copies and play the music on multiple players.

Movies, Video, and Television

With increasing amounts of Internet bandwidth going to more homes, streaming video and television are becoming commonplace. Once content with small, low-quality YouTube videos, the public now craves professionally produced video. Some are connecting their computers to high-definition television sets to access television and even motion picture quality programming. The Web and TV are rapidly merging into a single integrated system available from home entertainment systems, PCs, and cell phones.

Television is expanding to the Web in leaps and bounds. Web sites such as Hulu and Internet-based television platforms like Joost provide television programming from hundreds of providers, including most mainstream television networks. See Figure 4.26. Joost attracts an estimated 67 million viewers per month.[49] Hulu provides a premium service called Hulu Plus that provides an extended menu of programming to iPhones, iPads, TVs, and more for $10 per month. Hulu CEO Jason Kilar says the service compliments rather than replaces cable TV. But most analysts agree that it's only a matter of time until cable is challenged by Internet-based television programming.[50] Many TV networks offer online full-length streamed episodes of popular programs, including season premieres that are released before airing on television. Clicker.com serves as a television guide to most online programming.

Figure 4.26

Hulu

Hulu provides online access to thousands of prime-time television hits and movies.

(Source: *www.hulu.com*.)

Motion pictures are also making their way to Internet distribution. Subscription services such as Netflix allow members to rent DVDs by mail or stream movies over the Web to their computers. Apple iTunes users can rent or purchase movies from the iTunes store and download them to a computer.

No discussion of Internet video would be complete without mentioning YouTube. YouTube supports the online sharing of user-created videos. Every day, people upload hundreds of thousands of videos to YouTube and view hundreds of millions of these videos. YouTube videos are relatively short and cover a wide range of categories from the nonsensical to college lectures. See Figure 4.27. Other video streaming sites include Google Video, Yahoo! Video, Metacafe, and AOL Video.

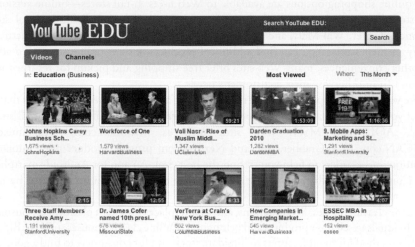

Figure 4.27

YouTube EDU

YouTube EDU provides thousands of educational videos from hundreds of universities.

(Source: *www.youtube.com/edu*.)

E-Books and Audio Books

An e-book is a book stored digitally, rather than on paper, and read on a display using e-book reader software. E-books have been available for quite a while, nearly as long as computers. However, it wasn't until the introduction of Amazon's eBook reading device, the Kindle, in 2007 that they gained more widespread acceptance. Several features of the Kindle appeal to the general public. First, the Kindle features ePaper, a display that does not include back-lighting like traditional displays. Some feel that ePaper is less harsh on your eyes than using a backlit display. Second, the Kindle is light and compact, similar in size and weight to a paperback book, but thinner than most books. Finally, Amazon created a vast library of eBooks that could be purchased and downloaded to the Kindle over whispernet—a wireless network provided free of charge by Sprint.

As an eBook reader, the Apple's iPad functions much like the Kindle; however, the iPad provides thousands of applications in addition to e-books. Apple offers users much the same selection of books as Kindle users. Amazon has even developed a Kindle application that runs on the iPad so users can access their Kindle libraries on the iPad.

Google has partnered with libraries to digitize more than 10 million books in its Google Books Library Project. A Google search includes searching for information in its Google Books Library. Google Books includes digital copies of in-copyright books, out-of-copyright books, and books from publishers with whom Google has partnered. Search results from Google Books may include snippets of information about the book and where to buy it, or, in the case of out-of-copyright books, the entire contents. Google has made an effort to open its own eBook store, but so far they've been blocked by U.S. courts because of worries that Google would obtain monopoly power over the industry.

Online Games

Video games have become a huge industry with more than $25.3 billion in total U.S. sales annually in recent years.[51] Many video games are available online. They include single-user, multiuser, and massively multiuser games. The Web offers a multitude of games for all ages. From Nickelodeon's Sponge Bob games to solitaire to massively multiplayer online role-playing games (MMORPG), a wide variety of offerings suits every taste. Of course, the Web provides a medium for downloading single-player games to your desktop, notebook, hand-held, or cell phone device, but the power of the Web is most apparent with multiplayer games.

Game consoles such as the Wii, Xbox, and PlayStation provide multiplayer options for online gaming over the Internet. Subscribers can play with or against other subscribers in 3D virtual environments. They can even talk to each other using a microphone headset. Microsoft's Xbox LIVE provides features that allow users to keep track of their buddies online and match up with other players who are of the same skill level.

Shopping Online

Many online shopping options are available to Web users. E-tail stores—online versions of retail stores—provide access to many products that may be unavailable in local stores. JCPenney, Target, Wal-Mart, and many others only carry a percentage of their actual inventory in their stores. You can find additional products at their online stores. To add to their other conveniences, many Web sites offer free shipping and pickup for returned items that don't fit or otherwise meet a customer's needs.

Like your local shopping mall, cybermalls provide access to a collection of stores that aim to meet your every need. Cybermalls are typically aligned with popular Web portals such as Yahoo!, AOL, and MSN. Web sites such as mySimon.com, DealTime.com, PriceSCAN.com, PriceGrabber.com, and NexTag.com provide product price quotations from numerous e-tailers to help you to find the best deal. An app for Android smartphones called Compare Everywhere allows users to compare the price of an item offered by many retailers.

Online clearinghouses, Web auctions, and marketplaces provide a platform for businesses and individuals to sell their products and belongings. Online clearinghouses such as uBid.com provide a method for manufacturers to liquidate stock and for consumers to find a good deal. Outdated or overstocked items are put on the virtual auction block, and users bid on the objects. The highest bidder(s) when the auction closes gets the merchandise—often for less than 50 percent of the advertised retail price.

The most popular online auction or marketplace is *www.eBay.com*. See Figure 4.28. eBay provides a public platform for global trading where anyone can buy, sell, or trade practically anything. eBay offers a wide variety of features and services that enable members to buy and sell on the site quickly and conveniently. Buyers have the option to purchase items at a fixed price or in auction-style format, where the highest bid wins the product. Information about auction items on eBay includes how much time is left in the auction, the current highest bid, as well as details about the item and seller. On any given day, millions of items are listed on eBay across thousands of categories.

Figure 4.28

eBay

eBay.com provides an online marketplace where anyone can buy, sell, or trade practically anything.

(Source: *www.ebay.com*.)

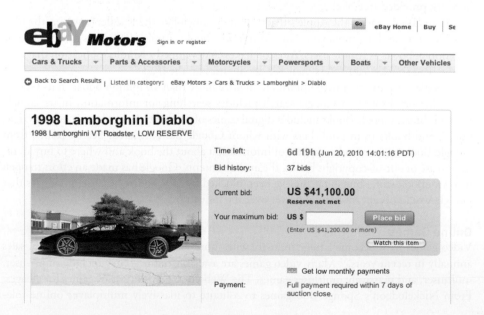

Craigslist is a network of online communities that provides free online classified advertisements. It is a popular online marketplace for purchasing items from local individuals. Many shoppers turn to Craigslist rather than going to the classifieds in the local paper.

Businesses benefit from shopping online as well. *Global supply management* (*GSM*) online services provide methods for businesses to find the best deals on the global market for raw materials and supplies needed to manufacture their products. *Electronic exchanges* provide an industry-specific Web resource created to deliver a convenient centralized platform for B2B e-commerce among manufacturers, suppliers, and customers. You can read more about this topic in Chapter 5.

Travel, Geolocation, and Navigation

Travel Web sites such as *www.travelocity.com*, *www.expedia.com*, and *www.priceline.com* help travelers find the best deals on flights, hotels, car rentals, vacation packages, and cruises. Priceline offers a slightly different approach from Travelocity and Expedia. It allows shoppers to name a price they're willing to pay for a ticket and then works to find an airline that can meet that price. After flights have been reserved, travelers can use these Web sites to book hotels and rental cars, often at discounted prices.

Travel agencies, resorts, airlines, cruise lines, and all businesses associated with travel have a strong online presence. Map Web sites such as Bing Maps and Google Maps are invaluable for finding your way to and around destinations; you can even view your destination from street view. See Figure 4.29. Web sites like *www.tripit.com* allow you to organize all of your travel plans, including flights, car rentals, hotel reservations, restaurants, and landmarks in one easy-to-access Web page. Today, most travel begins on the Web.

Figure 4.29

Bing Maps

Mapping software such as Bing Maps provide streetside views of Times Square.

(Source: *www.bing.com/maps*.)

Google Maps also provides extensive location-specific business information, satellite imagery, up-to-the-minute traffic reports, and Street View. Street View is the result of Google employees driving the streets of the world's cities in vehicles with high-tech camera gear, taking 360-degree images. These images are integrated into Google Maps to allow users to get a "street view" of an area that can be manipulated as if they were actually walking down the street looking around. Bing Maps takes it a step further with high-resolution aerial photos and street-level 3D photographs.

Selling Real Estate with Google Maps

The REA Group, headquartered in Australia, started its business in 1995 with $24,000 AUD (Australian dollars) and the belief that the Web would grow to become a dominant tool for the real estate industry. The company worked with Australian realtors to develop effective online real estate advertising. Before long, its Web site, *www.realestate.com.au*, became the most popular real estate portal in Australia. Today, the REA Group has 18 Web portals and operations in 12 countries, including Belgium, France, Germany, Hong Kong, Italy, Luxembourg, New Zealand, and the United Kingdom, with an annual revenue of more than AUD $156 million.

As online map software such as Google Maps arrived on the scene, the REA Group immediately saw its potential for the real estate market. CEO Simon Baker believes that shoppers want to see as much information about a property as possible before contacting an agent. The REA Group purchased simple mapping software to add to its real estate Web portals. Unfortunately, the software was slow to load, and key features such as zooming in and out worked inconsistently in different Web browsers. Even worse, the software failed to plot some coordinates accurately.

Fortunately, mapping technology improved over time, and the REA Group became impressed with the features provided by Google Maps. In particular, the company appreciated that Google Maps could be customized to meet the needs of the real estate industry and could easily be embedded into Web pages. The REA Group dropped their previous mapping technology and moved to Google Maps.

The API Premier version of Google Maps allows businesses to associate information with map coordinates. Google Maps embedded in *www.realestate.com.au* allows shoppers to see home locations on a map using a map view, a satellite view, hybrid view, terrain view, or street view. Shoppers can click the house icons on the map to view details and price. Google Maps also provides detailed property boundary lines. Those shopping for real estate can combine Google Maps and interior photos of the home to get a fairly thorough inspection before taking time for a physical visit. The REA Group believes that the embedded map is the most important feature of the online shopping experience.

Additionally, shoppers can perform structured and geographic searches for property. They can also use the embedded Google search bar to look for nearby schools, restaurants, attractions, and businesses. The REA Group plans to extend the capabilities of Google Maps with an assortment of overlays that provide information on a particular geographic region.

After the REA Group added Google Maps to its Australian site, it witnessed a steady boost in traffic and sales leads. Rather than competing with real estate agents, the software provides the agents with well-informed customers ready to buy, allowing the agents to make more sales with less effort. More recently, the company has rolled out the technology to its UK and Luxembourg sites and is currently continuing to extend the software to its remaining sites.

In 2009, the REA Group provided more than 1.5 billion maps through its real estate portals, making it one of the largest map providers in the world. CEO Simon Baker says that "With the mapping technology from Google, we're able to enhance the user experience, increase site stickiness, and improve the quality and accuracy of the data being displayed."

Discussion Questions

1. How has Google Maps improved the online shopping experience for visitors to *www.realestate.com.au*?
2. How does Google Maps help real estate agents in their work?

Critical Thinking Questions

1. What might the REA Group have done to avoid the problems it experienced with its first mapping software?
2. How might a company like the REA Group make money?

SOURCES: Staff, "Global real estate portal network adds Google Maps to enhance customers' house hunting experience," Google Maps Case Study, *www.google.com/enterprise/maps/reagroup.html*, accessed January 23, 2010; The REA Group Web site, *www.rea-group.com*, accessed January 24, 2010; *realestate.com.au*, accessed January 24, 2010.

INTRANETS AND EXTRANETS

An intranet is an internal corporate network built using Internet and World Wide Web standards and products. Employees of an organization use it to gain access to corporate information. After getting their feet wet with public Web sites that promote company products and services, corporations are seizing the Web as a swift way to streamline—even transform—their organizations. These private networks use the infrastructure and standards of the Internet and the World Wide Web. Using an intranet offers one considerable advantage: Many people are already familiar with Internet technology, so they need little training to make effective use of their corporate intranet.

An intranet is an inexpensive yet powerful alternative to other forms of internal communication, including conventional computer setups. One of an intranet's most obvious virtues is its ability to reduce the need for paper. Because Web browsers run on any type of computer, the same electronic information can be viewed by any employee. That means that all sorts of documents (such as internal phone books, procedure manuals, training manuals, and requisition forms) can be inexpensively converted to electronic form on the Web and be constantly updated. An intranet provides employees with an easy and intuitive approach to accessing information that was previously difficult to obtain. For example, it is an ideal solution to providing information to a mobile salesforce that needs access to rapidly changing information.

A growing number of companies offer limited network access to selected customers and suppliers. Such networks are referred to as extranets, which connect people who are external to the company. An **extranet** is a network that links selected resources of the intranet of a company with its customers, suppliers, or other business partners. Again, an extranet is built around Web technologies.

Security and performance concerns are different for an extranet than for a Web site or network-based intranet. User authentication and privacy are critical on an extranet so that information is protected. Obviously, the network must perform well to provide quick response to customers and suppliers. Table 4.7 summarizes the differences between users of the Internet, intranets, and extranets.

extranet
A network based on Web technologies that links selected resources of a company's intranet with its customers, suppliers, or other business partners.

Type	Users	Need User ID and Password?
Internet	Anyone	No
Intranet	Employees	Yes
Extranet	Business partners	Yes

Table 4.7

Summary of Internet, Intranet, and Extranet Users

Secure intranet and extranet access applications usually require the use of a *virtual private network* (*VPN*), a secure connection between two points on the Internet. VPNs transfer information by encapsulating traffic in IP packets and sending the packets over the Internet, a practice called **tunneling**. Most VPNs are built and run by ISPs. Companies that use a VPN from an ISP have essentially outsourced their networks to save money on wide area network equipment and personnel.

tunneling
The process by which VPNs transfer information by encapsulating traffic in IP packets over the Internet.

SUMMARY

Principle:

A telecommunications system has many fundamental components that must be carefully selected and work together effectively to enable people to meet personal and organization objectives.

Telecommunications refers to the electronic transmission of signals for communications, including telephone, radio, and television. Telecommunications is creating profound changes in business because it removes the barriers of time and distance.

The elements of a telecommunications system include the sending and receiving devices, modems, the transmission media, and the message. The sending unit transmits a signal to a modem, which performs a number of functions such as converting the signal into a different form or from one type to another. The modem then sends the signal through a medium, which is anything that carries an electronic signal and serves as an interface between a sending device and a receiving device. The signal is received by another modem that is connected to the receiving computer. The process can then be reversed, and another message can pass from the receiving unit to the original sending unit. A communications channel is the transmission medium that carries a message from the source to its receivers.

The telecommunications media that physically connect data communications devices can be divided into two broad categories: guided transmission media, in which a communications signal travels along a solid medium, and wireless media, in which the communications signal is sent over airwaves. Guided transmission media include twisted-pair wire cable, coaxial cable, fiber-optic cable, and broadband over power lines. Wireless telecommunications involves the broadcast of communications in one of three frequency ranges: microwave, radio, and infrared.

Wireless communications options include near field communications, Bluetooth, ultra wideband, Wi-Fi, 3G, 4G, and WiMAX.

Telecommunications uses various devices, including smartphones, modems, multiplexers, PBXs, front-end processors, switches, bridge, routers, and gateways.

The effective use of networks can turn a company into an agile, powerful, and creative organization, giving it a long-term competitive advantage. Networks let users share hardware, programs, and databases across the organization. They can transmit and receive information to improve organizational effectiveness and efficiency. They enable geographically separated workgroups to share documents and opinions, which fosters teamwork, innovative ideas, and new business strategies.

The physical distance between nodes on the network and the communications and services provided by the network determines whether it is called a personal area network (PAN), local area network (LAN), metropolitan area network (MAN), or a wide area network (WAN).

The electronic flow of data across international and global boundaries is often called transborder data flow.

When an organization needs to use two or more computer systems, it can follow one of three basic data processing strategies: centralized, decentralized, or distributed.

A client/server system is a network that connects a user's computer (a client) to one or more server computers (servers). A client is often a PC that requests services from the server, shares processing tasks with the server, and displays the results.

A communications protocol is a set of rules that govern the exchange of information over a communications channel.

A network operating system controls the computer systems and devices on a network, allowing them to communicate with one another. Network-management software enables a manager to monitor the use of individual computers and shared hardware, scan for viruses, and ensure compliance with software licenses.

Principle:

The Internet provides a critical infrastructure for delivering and accessing information and services.

The Internet is truly international in scope, with users on every continent. It is the world's largest computer network. Actually, it is a collection of interconnected networks, all freely exchanging information. The Internet transmits data from one computer (called a host) to another. The set of conventions used to pass packets from one host to another is known as the Internet Protocol (IP). Many other protocols are used with IP. The best known is the Transmission Control Protocol (TCP). TCP is so widely used that many people refer to the Internet protocol as TCP/IP, the combination of TCP and IP used by most Internet applications. Each computer on the Internet has an assigned IP address for easy identification. A Uniform Resource Locator (URL) is a Web address that specifies the exact location of a Web page using letters and words that map to an IP address, and a location on the host.

Cloud computing refers to a computing environment where software and storage are provided as an Internet service and accessed with a Web browser. Computing activities are increasingly being delivered over the Internet rather than from installed software on PCs. Cloud computing offers tremendous advantages to businesses. By outsourcing business information systems to the cloud, a business saves on system design, installation, and maintenance. Employees can also access corporate systems from any Internet-connected computer using a standard Web browser.

People can connect to the Internet backbone in several ways: via a LAN whose server is an Internet host, or via a dial-up connection, high-speed service, or wireless service. An Internet service provider is any company that provides access to the Internet. To use this type of connection, you must have an account with the service provider and software that allows a direct link via TCP/IP.

Principle:

Originally developed as a document-management system, the World Wide Web has grown to become a primary source of news and information, an indispensible conduit for commerce, and a popular hub for social interaction, entertainment, and communication.

The Web is a collection of tens of millions of servers that work together as one in an Internet service providing information via hyperlink technology to billions of users worldwide. Thanks to the high-speed Internet circuits connecting them and hyperlink technology, users can jump between Web pages and servers effortlessly—creating the illusion of using one big computer. Because of its ability to handle multimedia objects and hypertext links between distributed objects, the Web is emerging as the most popular means of information access on the Internet today.

As a hyperlink-based system that uses the client/server model, the Web organizes Internet resources throughout the world into a series of linked files, called pages, accessed and viewed using Web client software, called a Web browser. Internet Explorer, Firefox, Chrome, Opera, and Safari are popular Web browsers. A collection of pages on one particular topic, accessed under one Web domain, is called a Web site.

Hypertext Markup Language (HTML) is the standard page description language for Web pages. The HTML lays let the browser know how to format the text. HTML also indicates where images, sound, and other elements should be inserted. Some newer Web standards are gaining in popularity, including Extensible Markup Language (XML), Cascading Style Sheets (CSS), and Wireless Markup Language (WML).

Web 2.0 refers to the Web as a computing platform that supports software applications and the sharing of information between users. The Web is changing from a one-directional resource where users find information to a two-directional resource where users find and share information. The Web has also grown in power to support complete software applications and is becoming a computing platform in and of itself. A rich Internet application (RIA) is software that has the functionality and complexity of traditional application software, but runs in a Web browser and does not require local installation. Java, PHP, AJAX, MySQL, .NET, and Web application frameworks are all used to create interactive Web pages

Principle:

The Internet and Web provide numerous resources for finding information, communicating and collaborating, socializing, conducting business and shopping, and being entertained.

The most common and popular uses for the Internet and Web can be categorized as publishing information, assisting users in finding information, supporting communication and collaboration, building online community, providing software applications, providing a platform for expressing ideas and opinions, delivering media of all types, providing a platform for commerce, and supporting travel and navigation.

The Web has become the most popular medium for distributing and accessing information. It is a powerful tool for keeping informed about local, state, national, and global news. As a tool for sharing information and a primary repository of information on all subjects, the Web is ideally suited for education and training. Museums, libraries, private businesses, government agencies, and many other types of organizations and individuals offer educational materials online for free or a fee. Many businesses use the Web browser as an interface to corporate information systems. Web sites have sprung up to support every subject and activity of importance.

A search engine is a valuable tool that enables you to find information on the Web by specifying words that are key to a topic of interest—known as keywords. Some search companies have experimented with human-powered and human-assisted search. In addition to search engines, you can use other Internet sites to research information. Wikipedia, an encyclopedia with more than 3.3 million English-language entries created and edited by millions of users, is another example of a Web site that can be used to research information. There are other wikis are designed for special purposes. Online research is also greatly assisted by traditional resources that have migrated from libraries to Web sites such as online databases.

A Web portal is a Web page that combines useful information and links and acts as an entry point to the Web—the first page you open when you begin browsing the Web. A Web portal typically includes a search engine, a subject directory, daily headlines, and other items of interest. They can be general or specific in nature.

The Internet and Web provide many applications for communication and collaboration. E-mail is an incredibly useful form of Internet communication that not only supports text communication but also supports HTML content, and sharing documents as email attachments. Instant messaging is online, real-time communication between two or more people who are connected to the Internet. Referred to as a microblogging service, Twitter allows users to send short text updates (up to 140 characters long) from a cell phone or the Web to their Twitter followers. There are a number of Internet technologies that support real-time online conferencing. The Internet has made it possible for those involved in teleconferences to share computer desktops. Using services such as WebEx or GoToMeeting, conference participants log on to common software that allows them to broadcast their computer display to the group. Telepresence systems such as those from Cisco and HP use high-resolution video and audio with high-definition displays to make it appear that conference participants are actually sitting around a table.

Web sites such as YouTube and Flickr allow users to share video and photos with other people, groups, and the world. Microblogging sites such as Twitter allow people to post thoughts and ideas throughout the day for friends to read. Social bookmarking sites such as *www.digg.com* and *www.delicious.com* allow users to pool their votes to determine what online news stories and Web pages are most interesting at any given time of the day. Similarly, Epinions and many retail Web sites allow consumers to voice their opinions about products. Social networking Web sites provide Web-based tools for users to share information about themselves with people on the Web and to find, meet, and converse with other members.

Many of the computer activities traditionally provided through software installed on a PC can now be carried out using RIAs in a Web browser without installing any software. RIAs are the result of continuously improving programming languages and platforms designed for the Web.

A Web log, or blog, is a Web site that people can create and use to write about their observations, experiences, and opinions on a wide range of topics. Internet users may subscribe to blogs using a technology called Really Simple Syndication (RSS). RSS is a collection of Web technologies that allow users to subscribe to Web content that is frequently updated. With RSS, you can receive a blog update without actually visiting the blog Web site. A *podcast* is an audio broadcast over the Internet.

Like news and information, all forms of media and entertainment have followed their audiences online. The Internet and the Web have made music more accessible than ever, with artists distributing their songs through online radio, subscription services, and download services. With increasing amounts of Internet bandwidth going to more homes, streaming video and television are becoming commonplace. E-books have been available for quite a while, nearly as long as computers. eBook reading devices are

gaining in user acceptance. Online games include the many different types of single-user, multiuser, and massively multiuser games played on the Internet and the Web.

Many online shopping options are available to Web users. E-tail stores—online versions retail stores—provide access to many products that may be unavailable in local stores. Like your local shopping mall, cybermalls provide access to a collection of stores that aim to meet your every need. Online clearinghouses, Web auctions, and marketplaces provide a platform for businesses and individuals to sell their products and belongings.

The Web has had a profound effect on the travel industry and the way people plan and prepare for trips. From getting assistance with short trips across town to planning long holidays abroad, travelers are turning to the Web to save time and money and overcome much of the risk involved in visiting unknown places. Mapping and geolocation tools are among the most popular and successful Web applications. MapQuest, Google Maps, and Bing Maps are examples. Geo-tagging is technology that allows for tagging information with an associated location.

Principle:

Popular Internet and Web technologies have been applied to business networks in the form of intranets and extranets.

An intranet is an internal corporate network built using Internet and World Wide Web standards and products. Because Web browsers run on any type of computer, the same electronic information can be viewed by any employee. That means that all sorts of documents can be converted to electronic form on the Web and can constantly be updated.

An extranet is a network that links selected resources of the intranet of a company with its customers, suppliers, or other business partners. It is also built around Web technologies. Security and performance concerns are different for an extranet than for a Web site or network-based intranet. User authentication and privacy are critical on an extranet. Obviously, the network must perform well to provide quick response to customers and suppliers.

CHAPTER 4: SELF-ASSESSMENT TEST

A telecommunications system has many fundamental components that must be carefully selected and work together effectively to enable people to meet personal and organization objectives.

1. Any material that carries an electronic signal to support communications is called a(n) _____.

2. A set of rules that govern the exchange of information over a communications channel is called a _____.
   a. network channel
   b. network bandwidth
   c. communications protocol
   d. circuit switching

3. ___ involves the broadcast of communications in one of three frequency ranges: microwave, radio, or infrared.

4. A telecommunications service that delivers high-speed Internet access to homes and small businesses over existing phone lines is called ___.
   a. BPL
   b. DSL
   c. Wi-Fi
   e. Ethernet

The Internet provides a critical infrastructure for delivering and accessing information and services.

5. The _____ was the ancestor of the Internet, and was developed by the U.S. Department of Defense.

6. On the Internet, what enables traffic to flow from one network to another?
   a. Internet Protocol
   b. ARPANET
   c. Uniform Resource Locator
   d. LAN server

7. _____is a computing environment where software and storage are provided as an Internet service and accessed with a Web browser.
   a. Cloud computing
   b. Internet Society (ISOC)
   c. The Web
   d. America Online (AOL)

8. A(n) _____ is a company that provides people and organizations with access to the Internet.

Originally developed as a document-management system, the World Wide Web has grown to become a primary source of news and information, an indispensible conduit for commerce, and a popular hub for social interaction, entertainment, and communication.

9. CSS is a markup language designed to transport and store data on the Web. True or False?

10. Which technology was developed to assist in easily specifying the visual appearance of Web pages in a Web site?
    a. HTML
    b. XHTML
    c. XML
    d. CSS

11. What is the standard page description language for Web pages?
    a. Home Page Language
    b. Hypermedia Language
    c. Java
    d. Hypertext Markup Language (HTML)

The Internet and Web provide numerous resources for finding information, communicating and collaborating, socializing, conducting business and shopping, and being entertained.

12. Digg (*www.digg.com*) and del.icio.us (*www.delicious.com*) are examples of _____ Web sites.
    a. media sharing
    b. social network
    c. social bookmarking
    d. content streaming

13. A(n) _____ is a valuable tool that enables you to find information on the Web by specifying words or phrases related to a topic of interest, known as keywords.

14. _____ uses high-resolution video and audio with high-definition displays to make it appear that conference participants are actually sitting around a table.

Popular Internet and Web technologies have been applied to business networks in the form of intranets and extranets.

15. A(n) _____ is a network based on Web technology that links customers, suppliers, and others to the company.

16. An intranet is an internal corporate network built using Internet and World Wide Web standards and products. True or False?

CHAPTER 4: SELF-ASSESSMENT TEST ANSWERS

(1) communications media (2) c. (3) wireless communications (4) b. (5) ARPANET (6) a. (7) a. (8) Internet Service Provider (ISP) (9) False (10) d. (11) d. (12) c. (13) search engine (14) telepresence (15) extranet (16) True

REVIEW QUESTIONS

1. What is meant by broadband communications?
2. Describe the elements and steps involved in the telecommunications process.
3. What is a telecommunications protocol?
4. What are the names of the three primary frequency ranges used in wireless communications?
5. What is VPN? How do organizations use this technology?
6. What is the difference between near field communication and ultra wideband?

7. What is the difference between Wi-Fi and WiMAX communications?
8. What roles do the bridge, router, gateway, and switch play in a network?
9. Distinguish between a PAN, LAN, MAN, and WAN.
10. What is TCP/IP? How does it work?
11. Explain the naming conventions used to identify Internet host computers.
12. What is a Web browser? Provide four examples.
13. What is cloud computing?
14. Briefly describe three ways to connect to the Internet. What are the advantages and disadvantages of each approach?
15. What is an Internet service provider? What services do they provide?
16. How do Web application frameworks assist Web developers?
17. What is a podcast?
18. What is content streaming?
19. What is instant messaging?
20. What is the Web? Is it another network like the Internet or a service that runs on the Internet?
21. What is a URL and how is it used?
22. What is an intranet? Provide three examples of the use of an intranet.
23. What is an extranet? How is it different from an intranet?

DISCUSSION QUESTIONS

1. Why is an organization that employs centralized processing likely to have a different management decision-making philosophy than an organization that employs distributed processing?
2. Briefly discuss the evolution of wireless 3G and 4G communications.
3. Identify the fundamental differences between a file server system and a client/server system.
4. What are the pros and cons of distributed processing versus centralized processing for a large retail chain?
5. Social networks are being widely used today. Describe how this technology could be used in a business setting. Are there any drawbacks or limitations to using social networks in a business setting?
6. Why is it important to have an organization that manages IP addresses and domain names?
7. What are the benefits and risks involved in using cloud computing?
8. Describe how a company could use a blog and podcasting.
9. Briefly describe how the Internet phone service operates. Discuss the potential impact that this service could have on traditional telephone services and carriers.
10. Why is XML an important technology?
11. How do HTML, CSS, and XML work together to create a Web page?
12. What are the defining characteristics of a Web 2.0 site?
13. Name four forms of Internet communication describing the benefits and drawbacks of each.
14. What social concerns surround geolocation technologies?
15. One of the key issues associated with the development of a Web site is getting people to visit it. If you were developing a Web site, how would you inform others about it and make it interesting enough that they would return and tell others about it?
16. Downloading music, radio, and video programs from the Internet is easier than in the past, but some companies are still worried that people will illegally obtain copies of this programming without paying the artists and producers royalties. If you were an artist or producer, what would you do?
17. How could you use the Internet if you were a traveling salesperson?
18. Briefly summarize the differences in how the Internet, a company intranet, and an extranet are accessed and used.

PROBLEM-SOLVING EXERCISES

1. As the CIO of a hospital, you are convinced that installing a wireless network and providing portable computers to nurses and doctors is a necessary step to reduce costs and improve patient care. Use PowerPoint or similar software to make a convincing presentation to management for adopting such a program. Your presentation must identify benefits and potential issues that must be overcome to make such a program a success.
2. Think of a business that you might like to establish. Use a word processor to define the business in terms of what product(s) or service(s) it provides, where it is located, and its name. Go to *www.godaddy.com* and find an appropriate domain name for your business that is not yet taken. Shop

around online for the best deal on Web site hosting. Write a paragraph about your experience finding a name, and why you chose the name that you did.

3. You have been hired to research the use of a blog for a company. Develop a brief report on the advantages and disadvantages of using a blog to advertise corporate products and services. Using a graphics program, prepare a slide show to help you make a verbal presentation.

TEAM ACTIVITIES

1. Form a team to identify the public locations (such as an airport, public library, or café) in your area where wireless LAN connections are available. Visit at least two locations and write a brief paragraph discussing your experience at each location trying to connect to the Internet.

2. With your teammates, identify a company that is making effective use of Web 2.0 technologies on its Web site. Write a review of the site and why you believe it is effective?

3. Use Flickr.com to have a photo contest. Each group member should post four favorite photos that they personally took. Share account information between group members and use photo comment boxes to vote on your favorite photos. The photo with the most favorable comments wins.

WEB EXERCISES

1. Do research on the Web to identify the latest 4G communications developments. In your opinion, which carrier's 4G network (AT&T, Sprint Nextel, T-Mobil, or Verizon) is the most widely deployed? Write a short report on what you found.

2. Research some of the potential disadvantages of using the Internet, such as privacy, fraud, or unauthorized Web sites. Write a brief report on what you found.

3. Set up an account on *www.twitter.com* and invite a few friends to join. Use Twitter to send messages to your friends on their cell phones, keeping everyone posted on what you are doing throughout the day. Write a review of the service to submit to your instructor.

CAREER EXERCISES

1. Consider a future job position with which you are familiar through work experience, coursework, or a study of industry performance. How might you use some of the telecommunications and network applications described in this chapter in this job?

2. Do research to assess potential career opportunities in the telecommunications or networking industry. Consider resources such as the Bureau of Labor Statistics list of fastest growing positions, *Network World*, and *Computerworld*. Are there particular positions within these industries that offer good opportunities? What sort of background and education is required for candidates for these positions? You might be asked to summarize your findings for your class in a written or oral report.

CASE STUDIES

Case One

Adidas Turns to Cellular Network for Inventory Data

The Adidas Group is a global leader in the sporting goods industry, offering a large portfolio of products in virtually every country in the world. Adidas Group brands include Adidas, Reebok, and TaylorMade – Adidas Golf. The Adidas Group headquarters is in Herzogenaurach, Germany, with 170 subsidiaries located around the world.

Adidas America is the U.S. division of the Adidas Group, employing around 2,000 people at its offices and distribution centers across the country. Recently, Adidas America discovered an inexpensive way to use telecommunications technologies to help their salesforce become more productive.

Adidas sales representatives spend a lot of time traveling, paying visits to customers that range from small private sporting shops to large franchises. One major inconvenience for the salesforce and the customers it calls on lies in checking current inventory levels for order placement. For years, a salesperson was required to phone Adidas customer service representatives (CSRs) to find out which products were in stock and which were sold out. Because Adidas inventory fluctuates frequently, this important information is needed to keep customers from being frustrated by back-ordered products. Having to phone for the information was frustrating for sales representatives as well as customers who would have to wait while the salesperson talked on the phone. It also over-burdened the CSRs, whose main job wasn't inventory reporting but assisting customers.

The Adidas America information systems group thought they had a solution. They set up a VPN to enable customer representatives to securely connect to the corporate data from their laptops and access the inventory data directly. However, what looks good on paper doesn't always work well when implemented. Customer service employees were not comfortable visiting customers with their laptop computers, because they had to make the customer wait while they booted up and looked up information. Instead, they would typically take the customer's order, then check inventory later, getting back to the customer if there were product shortages. This certainly wasn't the intended workflow, and the solution ended up being as bad as the original problem.

Adidas Sales Force Automation Manager, Tim Oligmueller, was struck by a solution while attending a telecommunications trade show. He realized that the BlackBerry smartphones the salesforce used could do much more than deliver e-mail and make phone calls. He saw demonstrations of BlackBerry phones that ran sophisticated business information systems. Tim realized that the solution was in the salespeople's pockets all along. All that was required was for Adidas to invest in BlackBerry's Mobile Data System and Enterprise Server and for the information system group to develop their own Sales Force and CRM applications.

Within two weeks and with an investment of less than $10,000, Tim and his team had deployed a mobile information system that now allows salespeople to quickly and effortlessly access inventory data without inconveniencing customers or CSRs. Now salespeople can provide customers with the information they need to make quicker purchasing decisions on the spot. Salespeople and CSRs spend less time on the phone with each other, freeing both sides to be more productive. Customers are impressed with the improved service and technological innovativeness of the company. Salespeople can e-mail catalogues and order information to the customer as they converse. The new system has increased customers' confidence in doing business with Adidas.

Adidas is looking to many other wireless applications that can assist the salesforce and others in the organization. As with all of the examples provided in this chapter, telecommunications technologies provide Adidas with rapid access to information when and where it is needed, empowering the company to maximize its productivity and potential and gain a competitive advantage.

Discussion Questions

1. Describe the problem that plagued Adidas' salesforce that was addressed through a telecommunications solution.
2. Why do you think the handheld solution was better than the laptop computer solution?

Critical Thinking Questions

1. What other types of information, besides inventory, might salespeople like to have access to over their BlackBerry phones?
2. What security concerns might arise over mobile access to private corporate information?

SOURCES: BlackBerry Staff, "Adidas America Equips Sales Team with Powerful Wireless Sales Tool," BlackBerry Case Study, www.blackberry.com/products/pdfs/Adidas_LOB_CS_Final.pdf, accessed February 20, 2010; About Adidas Web page, www.adidas.com/us/shared/aboutadidas.asp, accessed February 20, 2010; Schultz, Beth, "Realizing Rapid ROI Through Mobility," Computerworld White Paper, www.slideshare.net/PingElizabeth/rapid-roire-alizing-rapid-roi-through-mobility, accessed February 20, 2010.

Case Two

Barriers to Enterprise 2.0

Web 2.0 and social media such as Facebook, Wikipedia, and YouTube have transformed life for many people, providing new ways to connect with friends and share information and media. While social Web 2.0 sites have become popular with the general public, businesses have had a difficult time deciding how these technologies can benefit their employees.

The use of Web 2.0 technologies and social media in the enterprise has been dubbed Enterprise 2.0. In most instances, Enterprise 2.0 is an extension of a corporate intranet; it is sealed off for access to employees only. VPN technology may be used to allow employee access from any Internet-connected device using a corporate login.

Some companies have been extremely successful with the implementation of Enterprise 2.0. Cisco, one of the world's largest technology companies, has implemented Enterprise 2.0 for its 65,000 employees. Cisco is the global leader in the design, manufacturing, and sales of networking and communications technologies and services. Cisco created a Facebook-like application for its employees to assist in finding subject matter experts within the organization. Like Facebook users, Cisco employees create profiles within the system that include their professional areas of expertise. When other employees need assistance with a problem, a quick search of the system will lead them to an expert within the organization.

Cisco also provides a video wiki used for training on different products and technologies and a Wikipedia-like application for sharing knowledge across the Enterprise. A number of mash-up applications have been developed to draw information from the main Enterprise 2.0 applications to address specific needs. For example, one mash-up can be used to quickly contact technical support staff.

Initially Cisco experienced some cultural pushback in the planning stage of its Enterprise 2.0 applications. A highly skeptical group of engineers thought that the company was wasting resources on needless technologies. Ultimately, the technologies have created real improvements for Cisco's business models.

Cisco is in a vast minority of companies that have successfully implemented Enterprise 2.0; however, it is likely that many companies will be following Cisco's lead. Traditional business culture often acts as a barrier to the adoption of Web 2.0 technologies. There is often a prevailing notion that posting to social networks is wasting time rather than being productive, and doesn't constitute "real work." In reality, people are most productive when interacting with networks of colleagues.

Another barrier to Enterprise 2.0 is a concern that social networks act as gateways to chaos, generating an unmanageable amount of mostly worthless data. The response to this argument is to provide tools that allow users to filter out the junk to get an optimal signal-to-noise ratio. Those that have been successful with Enterprise 2.0 find that getting users to generate as much noise and activity as possible creates the most valuable information to mine.

Yet another barrier to implementing Enterprise 2.0 is fear that the social network will be used as a "digital soapbox for disgruntled employees." This issue, as with the others, can be handled with proper management of the system and employees. If employees have grounds for complaining, management can more easily address those issues through the open forum of a social network.

In general, Enterprise 2.0 has been slow to take off due to an inability to easily show a return on investment. Social networks within an enterprise provide a soft return that is sometimes difficult to quantify. Successful Enterprise 2.0 implementations typically have two things in common: they are built to support key business processes, and they are not expected to show a ROI.

Discussion Questions

1. What Web 2.0 applications can provide benefits to employees in a business environment?
2. What barriers exist in some businesses that hamper the adoption of Enterprise 2.0?

Critical Thinking Questions

1. How might an information system administrator make a case for the implementation of Enterprise 2.0 when no ROI can be easily demonstrated?
2. In a large global enterprise, how might Enterprise 2.0 applications be organized so as to provide local benefits as well as global benefits?

SOURCES: Bennett, Elizabeth, "Web 2.0 in the Enterprise 2.0," CIO Insight, July 13, 2009; Gardner, W. David, "Enterprise 2.0: How Cloud Computing Is Challenging CIOs," Information Week, June 15, 2010, www.informationweek.com.

Questions for Web Case

See the Web site for this book to read about the Altitude Online case for this chapter. Following are questions concerning this Web case.

Altitude Online: Telecommunications and Networks

Discussion Questions

1. What telecommunications equipment is needed to fulfill Altitude Online's vision?
2. Why is it necessary to lease a line from a telecommunication company?

Critical Thinking Questions

1. What types of services will be provided over Altitude Online's network?
2. What considerations should Jon and his team take into account as they select telecommunications equipment?

Altitude Online: The Internet, Web, Intranets, and Extranets

Discussion Questions

1. What impact will the new ERP system have on Altitude Online's public-facing Web site? How will it affect its intranet?
2. What types of applications will be available from the employee dashboard?

Critical Thinking Questions

1. Altitude Online employees have various needs, depending on their position within the enterprise. How might the dashboard and intranet provide custom support for individual employee needs?
2. What Web 2.0 applications should Altitude Online consider for its dashboard? Remember that the applications must be available only on the secure intranet.

NOTES

Sources for the opening vignette: Cisco staff, "Procter & Gamble Revolutionizes Collaboration with Cisco Telepresence," Cisco Web site, *www.cisco.com/en/US/solutions/ns669/vid_pg.html*, accessed February 14, 2010; Hamblen, Matt, "PepsiCo to deploy telepresence from Cisco and BT globally," *Computerworld*, *www.computerworld.com*, February 2, 2010; Hamblen, Matt, "Firms use collaboration tools to tap the ultimate IP – worker ideas," *Computerworld*, *www.computerworld.com*, September 2, 2009; Computerworld staff, "Video Collaboration Studios," Computerworld Honors Program, 2008, *www.cwhonors.org/viewCaseStudy2008.asp?NominationID=697*.

1   Winston, Andrew, "Will Videoconferencing Kill Business Class Travel?" *Harvard Business Publishing*, August 3, 2009.

2   Svensson, Peter, "Google, Yahoo, eBay, and Microsoft Use Free Wi-Fi Hot Spots as Marketing Lure," *Associated Press*, November 10, 2009.

3   Dolan, Pamela Lewis, "Knowledge on Call: Finding New Uses for Smartphones," *American Medical News*, January 5, 2009.

4   Staff, "Telecom: Tools Connecting the World and Communicating About HIV," UNAIDS Web page, *www.unaids.org/en/KnowledgeCentre/Resources/FeatureStories/archive/2009/20091005_telecom.asp*, accessed October 5, 2009.

5   Verizon FiOS Internet Web page, *www22.verizon.com/Residential/FiOSInternet/FiOSvsCable/FiOSvsCable.htm*.

6   Reed, Brad, "IBM Uses BPL to Extend Broadband Coverage to Rural Areas," *Network World*, February 18, 2009.

7   Fleishman, Greg, "UWB Groups Shutters, Sends Tech to Bluetooth, USB Groups," *Ars Technica*, March 16, 2007.

8   Staff, "Hotel Chatter Annual Wi-Fi Report 2009," *www.hotelchatter.com/special/Best_WiFi_Hotels_2009*, accessed January 6, 2010.

9   Iridium Everywhere Web page, *www.iridium.com/about/globalnetwork.php*, accessed November 18, 2009.

10 Wells, Jane, "Iridium Satellite Is Back and Ready for Liftoff," *CNBC.com*, February 26, 2009.

11 Rosenberg, Barry, "VSAT Proves Crucial to Battlefield Communications," *Defense Systems*, June 5, 2009.

12 Nystedt, Dan, "China Mobile's 3G Coverage to Rapidly Expand," *Computerworld*, November 18, 2009.

13 Staff, "AT&T Promises Formal 4G Service in 2011," *electronista*, February 17, 2009.

14 Staff, "Success Stories—Teliasonera in Norway Pioneers the 4G Wireless Broadband Services in the World," Huawei Web site, *www.huawei.com/publications/view.do?id=6035&cid=11341&pid=2043*, accessed March 12, 2010.

15 Goldstein, Phil, "Intel: Clearwire Has Sufficient Capital," FierceWireless Web site, *www.fiercewireless.com*, February 12, 2009.

16 "Intel to Launch Centrino WiMAX 6250 Codenamed Kilmer Peak in Q1 2010," *Going Wimax.com*, *www.goingwimax.com/intel-to-launch-centrino-wimax-6250-codenamed kilmer-peak-in-q1-2010*, November 10, 2009.

17 Jaroslovsky, Rich, "Google Phone Threatens Droid More than iPhone: Rich Jaroslovsky," *BusinessWeek*, January 11, 2010.

18 Staff, "Docklands Light Railway on Track for Growth with New Cisco LAN and VoIP-Based Public Address System," *New Blaze*, February 10, 2009.

19 Staff, "Netkrom Customer Success Stories Metropolitan Backbone Network," *www.netkrom.com/success_stories_lima.html*, accessed December 29, 2009.

20 Staff, "NE20 Serves Datang Power WAN," *www.huawei.com/products/datacomm/catalog.do?id=361*, accessed December 29, 2009.

21 Staff, "LaserPro," Harland Financial Solutions Web site, *www.harlandfinancialsolutions.com/search.php?searchStr=laserpro*, accessed March 12, 2010.

22 Packet Trap Networks home page, *www.packettrap.com*, accessed January 20, 2010.

23 The ISC Domain Survey, *http://www.isc.org/solutions/survey*, accessed June 8, 2010.

24 Internet Usage Statistics, *http://www.internetworldstats.com/stats.htm*, accessed June 8, 2010.

25 Internet Usage World Stats Web site, *www.internetworldstats.com*, accessed June 8, 2010.

26 McDonald, Scott, "China Vows to Keep Blocking Online Content," NewsFactor, June 8, 2010, *www.newsfactor.com*.

27 Anderson, Nate, "Internet disconnections come to Ireland, starting today," *Ars Technica*, May 24, 2010, *www.arstechnica.com*.

28 Greene, Patrick Allen, "Pakistan blocks Facebook over 'Draw Mohammed Day'," CNN, May 19, 2010, *www.cnn.com*.

29 "Communications Decency Act," Wikipedia, *http://en.wikipedia.org*, accessed July 3, 2010.

30 Internet2 Web site, *www.internet2.edu*, accessed June 8, 2010.

31 National LambdaRail Web site, *www.nlr.net*, accessed June 8, 2010.

32 Miller, Rich, "Facebook Begins Deploying IPv6," Data Center Knowledge, June 10, 2010 *www.datacenterknowledge.com*.

33 Levine, Barry, "You Want Wi-Fi with That? McDonalds to Make Wi-Fi Free," NewsFactor, December 16, 2009, *www.newsfactor.com*.

34 Stimek, Blake, "Data Speed Showdown: Sprint 4G vs T-Mobile HSPA +," IntoMobile, *www.intomobile.com*, June 4, 2010.

35 Cox, John, "AT&T vs. Verizon," *PCWorld*, *www.pcworld.com*, June 7, 2010.

36 Gaskin, James, "Clouds Now Strong Enough to Support Your Business," *Computerworld, www.computerworld.com,* October 6, 2009.

37 Schonfeld, Erick, "Urbanspoon Wants to Challenge OpenTable With Its Rezbook iPad App," TechCrunch, *www.techcrunch.com,* May 19, 2010.

38 Kashi Web site, *www.kashi.com,* accessed July 6, 2010

39 Pegoraro, Rob, "Bing adds music, videos, games," *The Washington Post, http://voices.washingtonpost.com,* June 23, 2010.

40 Porter, Alan, "Getting over the barriers to wiki adoption," *Ars Technica, www.arstechnica.com.* , February 13, 2010

41 Noyes, Katherine, "Facebook Quickens Yahoo's Pulse," E-Commerce Times, *www.ecommercetimes.com,* June 7, 2010.

42 Cohen, Jason, "Socializing the Storefront, Part 1: Give Them Something to Talk About," E-Commerce Times, *www.ecommercetimes.com,* June 21, 2010.

43 "Case Study: Papa John's International," GoToMeeting, *www.gotomeeting.com/fec/images/pdf/caseStudies/ GoToMeeting_Papa_Johns_Case_Study.pdf,* accessed July 6, 2010.

44 "ROI with Cisco Telepresence," Cisco Telepresence Case Study, *www.cisco.com/en/US/prod/collateral/ps7060/ps8329/*

ps8330/9599/TelePresence_Research_Brief_Final_03_20_09 .pdf, March 2009.

45 Butcher, Mike, "Internet Conga Line – a new take on ChatRoulette which feels safer," TechCrunch, *www.techcrunch.com,* June 23, 2010.

46 Naone, Erica, "Getting a Grip on Online Buzz," Technology Review, *www.technologyreview.com.* , February 9, 2010

47 Wauters, Robin, "Measuring The Value Of Social Media Advertising," TechCrunch, *www.techcrunch.com.* , April 20, 2010

48 Nakano, Chelsi, "Enterprise 2.0 Roll-up: Social Computing Is Destroying the Enterprise, Yet Solutions Arrive in Waves," CRM Newswire, *www.crmnewswire.com,* June 24, 2010.

49 Wauters, Robin, "Joost Video Network Stuns with Big Reach: 67 Million Viewers Per Month," TechCrunch, *www.techcrunch.com,* April 14, 2010.

50 Hollister, Sean, "Hulu CEO: We're 'complimentary' to cable," Engadget, *www.engadget.com,* July 1, 2010.

51 Cressman, Jordan, "2009 Video Game Sales Eclipse $25 Billion in US," I4U News, *www.i4u.com,* May 11, 2010.

Business Information Systems

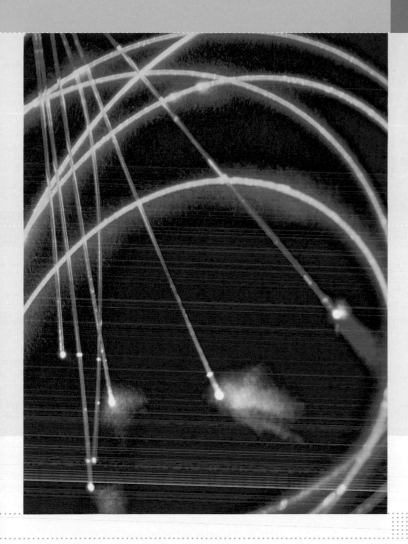

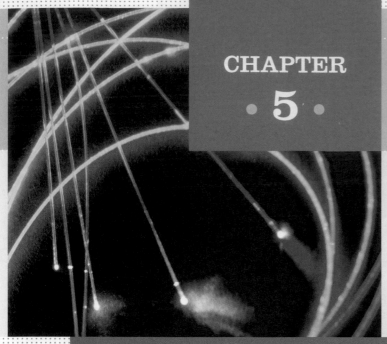

PRINCIPLES	LEARNING OBJECTIVES
▪ Electronic and mobile commerce are evolving, providing new ways of conducting business that present both potential benefits and problems.	▪ Describe the current status of various forms of e-commerce, including B2B, B2C, C2C, and m-commerce. ▪ Identify several e-commerce and m-commerce applications. ▪ Identify several advantages associated with the use of e-commerce and m-commerce.
▪ E-commerce and m-commerce require the careful planning and integration of a number of technology infrastructure components.	▪ Identify the key components of technology infrastructure that must be in place for e-commerce and m-commerce to work. ▪ Discuss the key features of the electronic payment systems needed to support e-commerce and m-commerce.
▪ An organization must have information systems that support the routine, day-to-day activities that occur in the normal course of business and help a company add value to its products and services.	▪ Identify the basic activities and business objectives common to all transaction processing systems
▪ A company that implements an enterprise resource planning system is creating a highly integrated set of systems, which can lead to many business benefits.	▪ Discuss the advantages and disadvantages associated with the implementation of an enterprise resource planning system.
▪ A company that implements a customer relationship management system is building a source of information about customers that can improve sales, marketing, and customer service.	▪ State the objective of a customer relationship management system and describe several of its basic functions.
▪ There are many potential international issues associated with the operation of enterprise systems.	▪ Identify the challenges that multinational corporations face in planning, building, and operating their enterprise systems.

Information Systems in the Global Economy
Tommy Hilfiger, United States

Hilfiger Leaves E-Commerce to the Experts

Tommy Hilfiger is one of the world's leading brands of premium lifestyle clothing. Hilfiger clothing reflects "Classic American Cool" styling that is popular around the world. Hilfiger clothing is distributed in more than 65 countries through high-class fashion shops, department stores, and more than 900 of its own retails stores.

In 2009, Hilfiger decided it was time to focus on improving its online sales. Its Web store sales had been respectable but had not implemented any of the latest e-commerce tools for driving sales.

Executives at Hilfiger recognize that the company's core competencies lie in fashion, marketing, and merchandising. Hilfiger has no interest in becoming an IT company, building an IT department, or investing in servers and computing equipment. It didn't take long for the company to decide that its e-commerce systems should be outsourced. Whatever system Hilfiger used would have to be powerful and yet easy for customers to interact with.

Hilfiger director of e-commerce, Tom Davis, reviewed several popular e-commerce hosting services. E-commerce hosts provide everything needed for an e-commerce site, including hardware, software, databases, networking, and support, for an annual fee. Of all the companies Davis evaluated, ATG struck him as being the best fit for Hilfiger's needs. ATG's Commerce OnDemand platform is a fully hosted e-commerce system that includes the latest and most advanced e-commerce tools and techniques. The ATG Business Control Center is an easy-to-use interface that allows companies to enter products into an online catalog. It is also used to set price points, arrange products within the site, build promotions, and set shipping fees.

Tom Davis worked with engineers at ATG to implement the new Hilfiger site within 120 days. During the first months of operation, Hilfiger merchandising staff experimented with the site, introducing more products each week and working to improve conversion rates (the percentage of visitors that purchase products). In order to boost cross-sales, the staff manually entered one-to-one product relationships so that when a customer views an item or checks out, another related item is recommended.

The process of hand-coding the related item recommendations required a significant time investment. After three months, Davis discovered that the recommendations had no significant impact on sales and so he gave up the effort. ATG came to the rescue with an e-commerce tool called ATG Recommendations. ATG Recommendations works dynamically to recommend products that a customer is likely to find appealing, using a sophisticated predictive algorithm that combines data from the merchandise catalog, the site structure, historical and current shopper behavior, and aggregate shopper behavior.

Implementing the new ATG Recommendations system took less than four weeks and required minimal changes to the site. Hilfiger found that the software was well worth the investment. After only one month, the Recommendations system increased online sales by 16 percent. You can view the Recommendations system at the bottom of every product page as a series of merchandise photos that "people who viewed this also viewed."

Hilfiger learned that its approach to manually adding recommendations failed because it was primarily recommending products that it was keen to sell. ATG Recommendations was successful because it recommends items that it calculates a customer would want, without regard to Hilfiger's priorities. The only business rule that governs the Recommendations system is that recommendations are not made that might reduce an order's value.

Over time, Hilfiger expanded its use of ATG Recommendations, tying it to e-mail campaigns, a Top Sellers page, and a Gift Guide feature; integrating it into higher level pages; and offering Last Chance Recommendations at checkout. Through ATG Recommendations, Hilfiger can expose customers to a wider breadth of products on each page as well as products that the customer is likely to find appealing.

Hilfiger's new e-commerce site has been incredibly successful. In its first holiday season, the new site doubled the company's online revenue. Davis says that "the likelihood of checkout is three times higher for those shoppers who interacted with Recommendations than those who did not."

Outsourcing the e-commerce business to an experienced e-commerce company has freed Hilfiger employees to concentrate on what they do best: planning promotions and strategies to optimize their products' exposure and further increase sales.

As you read this chapter, consider the following:

- What advantages do e-commerce and m-commerce offer sellers and vendors over traditional shopping venues?
- What types of information systems are critical to the success of a business and how are the systems related to one another?

Why Learn About Electronic and Mobile Commerce and Enterprise Systems?

Electronic and mobile commerce and enterprise systems have transformed many areas of our lives and careers. One fundamental change has been the manner in which companies interact with their suppliers, customers, government agencies, and other business partners. As a result, most organizations today have or are considering setting up business on the Internet and implementing integrated enterprise systems. To be successful, all members of the organization need to participate in that effort. As a sales or marketing manager, you will be expected to help define your firm's e-commerce business model. Customer service employees can expect to use enterprise systems to provide improved customer service. As a human resource or public relations manager, you will likely be asked to provide content for a Web site directed to potential employees and investors. Analysts in finance need to know how to use enterprise systems to capture and report the data needed to manage and control the firm's operations. Clearly, as an employee in today's organization, you must understand the potential role of e-commerce and enterprise systems, how to capitalize on their many opportunities, and how to avoid their pitfalls. The emergence of m-commerce adds an exciting new dimension to these opportunities and challenges. This chapter begins by providing a brief overview of the dynamic world of e-commerce and defines its various components.

AN INTRODUCTION TO ELECTRONIC COMMERCE

electronic commerce
Conducting business activities (e.g., distribution, buying, selling, marketing, and servicing of products or services) electronically over computer networks.

Electronic commerce is the conducting of business activities (e.g., distribution, buying, selling, marketing, and servicing of products or services) electronically over computer networks. This includes any business transaction executed electronically between companies (business-to-business), companies and consumers (business-to-consumer), consumers and other consumers (consumer-to-consumer), public sector and business (government-to-business), and the public sector to citizens (government-to-citizen).

Business activities that are strong candidates for conversion to e-commerce are ones that are paper based, time consuming, and inconvenient for customers. During the past decade, it is estimated that the Internet created 1.2 million jobs, including half a million jobs in e-commerce companies and companies that deliver the physical goods, and another 200,000 jobs at Internet service providers. E-commerce contributes 2.1 percent, or $300 billion, to the total U.S. gross domestic product.[1]

Business-to-Business (B2B) E-Commerce

Business-to-business (B2B) e-commerce is a subset of e-commerce in which all the participants are organizations. B2B e-commerce is a useful tool for connecting business partners in a virtual supply chain to cut resupply times and reduce costs. Although the business-to-consumer market grabs more of the news headlines, the B2B market is considerably larger and is growing more rapidly.

B2B Computer Products is a nationwide B2B reseller of computer hardware and software from hundreds of manufacturers. It also sells complete phone solutions for voice, data, and Internet service from leading providers. Its engineers configure, install, and manage the products it sells.[2]

Business-to-Consumer (B2C) E-Commerce

Business-to-consumer (B2C) e-commerce organizations sell their products directly to consumers. More than just a tool for placing orders, the Internet is an extremely useful way to compare prices, features, value, and other customers' opinions. Internet shoppers can, for example, unleash shopping bots or access sites such as eBay Shopping.com, Google Froogle, Shopzilla, PriceGrabber, Yahoo! Shopping, or Excite to browse the Internet and obtain lists of items, prices, and merchants. Many B2C merchants have added what is called "social commerce" to their Web sites by creating a section where shoppers can go to see only those products that have been reviewed and listed by other shoppers.

By using B2C e-commerce to sell directly to consumers, producers or providers of consumer products can eliminate the middlemen, or intermediaries, between them and the consumer. In many cases, this squeezes costs and inefficiencies out of the supply chain and can lead to higher profits and lower prices. The elimination of intermediate organizations between the producer and the consumer is called *disintermediation*.

Dell is an example of a manufacturer that has successfully embraced this model to achieve a strong competitive advantage. People can specify a unique computer online, and Dell assembles the components and ships the computer directly to the consumer within five days.

> **business-to-business (B2B) e-commerce**
> A subset of e-commerce in which all the participants are organizations.

> **business-to-consumer (B2C) e-commerce**
> A form of e-commerce in which customers deal directly with an organization and avoid intermediaries.

Dell sells its products through the Dell.com Web site.

(Source: *www.dell.com*.)

Following a successful pilot project involving more than 5,000 consumers, Procter and Gamble (P&G), the consumer goods marketer with annual sales of more than $80 billion, launched a business-to-consumer Web site. This initiative, called the estore, represents a major change for the company as its products are normally sold indirectly to consumers through major retailers such as Wal-Mart, Kroger, and Target. While most sales will continue to be through retailers, P&G will use the estore to gain a better understanding of its consumers as well as build stronger consumer relationships. On the other hand, the Web site has generated some ill will between P&G and the major retailers, as it reduces their in-store sales of P&G products. The site is owned and operated by PFSWeb and will sell exclusively P&G products to U.S. consumers.[3]

Consumer-to-Consumer (C2C) E-Commerce

Consumer-to-consumer (C2C) e-commerce is a subset of e-commerce that involves electronic transactions between consumers using a third party to facilitate the process. eBay is an example of a C2C e-commerce site; customers buy and sell items to each other through the site. Other examples of C2C sites include Bidzcom, Craigslist, eBid, ePier, Ibidfree, Kijiji, Ubid, and Tradus. Etsy is a C2C Web site that specializes in the buying and selling of handmade and vintage items, including art, bath and beauty products, craft supplies, clothing, jewelry, quilts, and toys. The site allows sellers to set up personal storefronts where they display the items they have for sale. Etsy facilitates sales worth around $10 to $13 million each month. It earns revenue by charging a listing fee of $.20 and receiving a 3.5 percent commission on each sale.[4]

Table 5.1 summarizes the key factors that differentiate among B2B, B2C, and C2C e-commerce.

Table 5.1

Differences Between B2B, B2C, and C2C

Factors	B2B	B2C	C2C
Value of sale	Thousands or millions of dollars	Tens or hundreds of dollars	Tens of dollars
Length of sales process	Days to months	Days to weeks	Hours to days
Number of decision makers involved	Several people to a dozen or more	One or two	One or two
Uniformity of offer	Typically a uniform product offering	More customized product offering	Single product offering, one of a kind
Complexity of buying process	Extremely complex, much room for negotiation on price, payment and delivery options, quantity, quality, options and features	Relatively simple, limited discussion over price and payment and delivery options	Relatively simple, limited discussion over payment and delivery options; negotiation over price
Motivation for sale	Driven by a business decision or need	Driven by an individual consumer's need or emotion	Driven by an individual consumer's need or emotion

e-Government

e-Government is the use of information and communications technology to simplify the sharing of information, speed formerly paper-based processes, and improve the relationship between citizens and government. Government-to-citizen (G2C), government-to-business (G2B), and government-to-government (G2G) are all forms of e-Government, each with different applications.

Citizens can use G2C applications to submit their state and federal tax returns online, renew auto licenses, apply for student loans, and make campaign contributions. At the Recovery.gov Web site, citizens can view where federal stimulus money is being allocated by state, county, zip code, and congressional district.[5] New York City created a health and human services multilingual Web portal to enable its 8 million residents to determine their eligibility for 35 city, state, and federal human service benefit programs, print application forms, search for office locations, and create an account to access their information.[6]

G2B applications support the purchase of materials and services from private industry by government procurement offices, enable firms to bid on government contracts, and help businesses receive current government regulations related to their operations. Business.gov allows businesses to access information about laws and regulations and relevant forms needed to comply with federal requirements for their business.

G2G applications are designed to improve communications among the various levels of government. For example, the E-Vital initiative establishes common electronic processes for federal and state agencies to collect, process, analyze, verify, and share death record information.[7] Geospatial One-Stop's Web portal, GeoData.gov, makes it easier, faster, and less expensive to find, create, share, and access geographic data and maps among all levels of government.[8]

MOBILE COMMERCE

As discussed briefly in Chapter 1, mobile commerce (m-commerce) relies on the use of mobile, wireless devices, such as cell phones and smartphones, to place orders and conduct business. Handset manufacturers such as Ericsson, Motorola, Nokia, and Qualcomm are working with communications carriers such as AT&T, Cingular, Sprint/Nextel, and Verizon to develop such wireless devices, related technology, and services. The Internet Corporation for Assigned Names and Numbers (ICANN) created a .mobi domain to help attract mobile users to the Web. mTLD Top Level Domain Ltd of Dublin, Ireland, administers this domain and helps to ensure that the .mobi destinations work quickly, efficiently, and effectively with user handsets.

Mobile Commerce in Perspective

The market for m-commerce in North America is maturing much later than in Western Europe and Japan for several reasons. In North America, responsibility for network infrastructure is fragmented among many providers, consumer payments are usually made by credit card, and many Americans are unfamiliar with mobile data services. In most Western European countries, communicating via wireless devices is common, and consumers are much more willing to use m-commerce. Japanese consumers are generally enthusiastic about new technology and are much more likely to use mobile technologies for making purchases.

Nearly 450 million users worldwide accessed the Internet via mobile devices in 2009. It is estimated that the number of mobile devices accessing the Internet will exceed 1 billion by 2013.[9]

According to ABI Research, m-commerce spending in the United States grew from $369 million in sales in 2008 to $1.2 billion in 2009. The firm projects m-commerce sales for 2010 of $2.4 billion.[10] The number of mobile Web sites is expected to grow rapidly because of advances in wireless broadband technologies, the development of new and useful applications, and the availability of less costly but more powerful handsets. Indeed, the Web site Mobil Mammoth highlights a new mobile Web site every day.[11] Experts point out that the relative clumsiness of mobile browsers and security concerns must be overcome to ensure rapid m-commerce growth.

As with any new technology, m-commerce will succeed only if it provides users with real benefits. Companies involved in m-commerce must think through their strategies carefully and ensure that they provide services that truly meet customers' needs.

ELECTRONIC AND MOBILE COMMERCE APPLICATIONS

E-commerce and m-commerce are being used in innovative and exciting ways. This section examines a few of the many B2B, B2C, C2C, and m-commerce applications in retail and wholesale, manufacturing, marketing, advertising, price comparison, couponing, investment and finance, banking, and e-boutiques.

Retail and Wholesale

electronic retailing (e-tailing)
The direct sale of products or services by businesses to consumers through electronic storefronts, typically designed around an electronic catalog and shopping cart model.

E-commerce is being used extensively in retailing and wholesaling. **Electronic retailing**, sometimes called *e-tailing*, is the direct sale of products or services by businesses to consumers through electronic storefronts, which are typically designed around the familiar electronic catalog and shopping cart model. Companies such as Office Depot, Wal-Mart, and many others have used the same model to sell wholesale goods to employees of corporations. Tens of thousands of electronic retail Web sites sell everything from soup to nuts. Table 5.2 lists the top-rated B2C Web sites according to the Top 100 Online Retail Satisfaction Index from ForeSee Results and FGI Research. The measurement results are based on the American Consumer Satisfaction Index methodology developed by the University of Michigan.[12]

Table 5.2

Top-Rated B2C Web Sites (Spring 2009)

Source: Burns, Enid, "Study: Consumer Satisfaction in E-Commerce Slips," ClickZ, May 11, 2009, *www.clickz.com/3633686*, accessed March 14, 2010.

Web Site	ACSI Index	Products Sold
Netflix.com	85	DVDs by mail
Amazon.com	84	Books, music, DVDs, and more
Avon.com	81	Beauty, health, and fitness products
DrsFosterSmith.com	81	Pet supplies
Newegg.com	81	Computers and computer-related products
QVC.com	81	Fashion, beauty, jewelry, and home products
TigerDirect.com	79	Computers and computer-related products
HPShopping.com	78	Computers and computer-related products
LLBean.com	78	Men's and women's clothing
Shutterfly.com	78	Photo sharing
VictoriaSecret.com	78	Lingerie and women's clothing

cybermall
A single Web site that offers many products and services at one Internet location.

Cybermalls are another means to support retail shopping. A **cybermall** is a single Web site that offers many products and services at one Internet location—similar to a regular shopping mall. An Internet cybermall pulls multiple buyers and sellers into one virtual place, easily reachable through a Web browser. For example, 1StopTireShop allows Internet shoppers to compare and select tires from some 18 different tire manufacturers. Etailers Mall allows shoppers to shop at dozens of bath, body, candle, cosmetics, and jewelry e-tailers on the Internet.

A key sector of wholesale e-commerce is spending on manufacturing, repair, and operations (MRO) goods and services—from simple office supplies to mission-critical equipment, such as the motors, pumps, compressors, and instruments that keep manufacturing facilities running smoothly. MRO purchases often approach 40 percent of a manufacturing company's total revenues, but the purchasing system can be haphazard, without automated controls. In addition to these external purchase costs, companies face significant internal costs resulting from outdated and cumbersome MRO management processes. For example, studies show that a high percentage of manufacturing downtime is often caused by not having the right part at the right time in the right place. The result is lost productivity and capacity. E-commerce software for plant operations provides powerful comparative searching capabilities to enable managers to identify functionally equivalent items, helping them spot opportunities to combine purchases for cost savings. Comparing various suppliers, coupled with consolidating more spending with fewer suppliers, leads to decreased costs. In addition, automated workflows are typically based on industry best practices, which can streamline processes.

McMaster-Carr is a supplier of products used to maintain industrial and commercial facilities specializing in the next day delivery of MRO materials and supplies. Its Web site offers more than 480,000 items for sale. Customers can use its search feature to quickly find the items they need.

Manufacturing

One approach taken by many manufacturers to raise profitability and improve customer service is to move their supply chain operations onto the Internet. Here, they can form an **electronic exchange** to join with competitors and suppliers alike to buy and sell goods, trade market information, and run back-office operations, such as inventory control, as shown in Figure 5.1. This approach has greatly speeded up the movement of raw materials and finished products among all members of the business community and has reduced the amount of inventory that must be maintained. It has also led to a much more competitive marketplace and lower prices. Private exchanges are owned and operated by a single company. The owner uses the exchange to trade exclusively with established business partners. Public exchanges are owned and operated by industry groups. They provide services and a common technology platform to their members and are open, usually for a fee, to any company that wants to use them.

electronic exchange
An electronic forum where manufacturers, suppliers, and competitors buy and sell goods, trade market information, and run back-office operations.

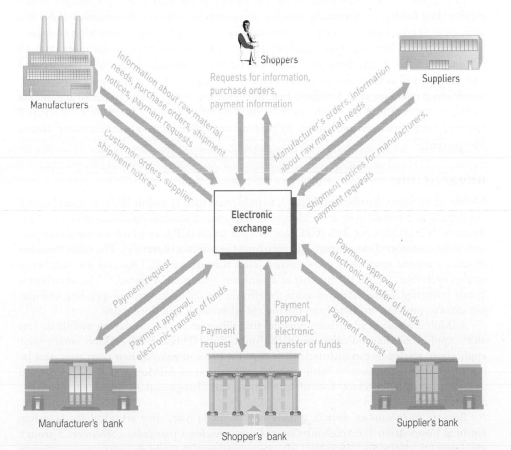

Figure 5.1

Model of an Electronic Exchange

Avendra is a private exchange that provides access to a $3 billion supply chain of goods and services from more than 900 suppliers to about 4,500 customers in the hospitality industry. The exchange was formed by ClubCorp, Fairmont Hotels, Hyatt, Intercontinental Hotels Group, and Marriott International in 2001.[13] Not only do customers benefit from cost savings generated by Avendra's volume purchasing, they can draw on Avendra's extensive hospitality expertise to select those products that best meet their needs and budgets.

Several strategic and competitive issues are associated with the use of exchanges. Many companies distrust their corporate rivals and fear they might lose trade secrets through participation in such exchanges. Suppliers worry that online marketplaces will drive down the prices of goods and favor buyers. Suppliers also can spend a great deal of money in the setup to participate in multiple exchanges. For example, more than a dozen new exchanges have appeared in the oil industry, and the printing industry is up to more than 20 online marketplaces. Until a clear winner emerges in particular industries, suppliers are more or less forced to sign on to several or all of them. Yet another issue is potential government scrutiny of exchange participants—when competitors get together to share information, it raises questions of collusion or antitrust behavior.

Marketing

The nature of the Web enables firms to gather more information about customer behavior and preferences as customers and potential customers gather information and make their purchase decisions. Analysis of this data is complicated because of the Web's interactivity and because each visitor voluntarily provides or refuses to provide personal data such as name, address, e-mail address, telephone number, and demographic data. Internet advertisers use the data to identify specific portions of their markets and target them with tailored advertising messages. This practice, called **market segmentation**, divides the pool of potential customers into subgroups usually defined in terms of demographic characteristics, such as age, gender, marital status, income level, and geographic location.

In the past, market segmentation has been difficult for B2B marketers because firmographic data (addresses, financials, number of employees, industry classification code) was difficult to obtain. Now, however, Nielsen, the marketing and media information company, has developed its Business-Facts database that provides this information for more than 13 million businesses. Using this data, analysts can estimate potential sales for each business and rank it against all other prospects and customers. Windstream Communications, a telecommunications company with 3 million customers, worked with Nielsen to perform a market segmentation and used the results to drive its marketing strategy. As a result, direct mail response rates have risen more than 50 percent, and telemarketing sales increased almost 500 percent.[14]

Advertising

Mobile ad networks distribute mobile ads to publishers such as mobile Web sites, application developers, and mobile operators. Mobile ad impressions are generally bought at a cost per thousand (CPM), cost per click (CPC), or cost per action (CPA, in which the advertiser pays only if the customer clicks through and then buys the product or service). The main measures of success are the number of users reached, click through rate (CTR), and the number of actions users take, such as the number of downloads prompted by the ad.[15] The advertiser is keenly interested in this data to measure the effectiveness of its advertising spending and may pay extra to purchase the data from the mobile ad network or a third party.

AdMob is a mobile advertising provider that serves up ads for display on mobile devices and in applications like those that run on the Android and iPhone. With AdMob, smartphone application developers can distribute their apps for free and recover their costs over time by payments from advertisers.[16] With Google's acquisition of AdMob and Apple's new iAd mobile advertising product, lots of innovation and change can be expected in mobile advertising.

Because m-commerce devices usually have a single user, they are ideal for accessing personal information and receiving targeted messages for a particular consumer. Through m-commerce, companies can reach individual consumers to establish one-to-one marketing

market segmentation
The identification of specific markets to target them with advertising messages.

relationships and communicate whenever it is convenient—in short, anytime and anywhere. According to one recent study, 51 percent of the consumers in 11 countries during the 2009 holiday season used their mobile phones to perform in-store activities, including shopping, seeking peer feedback on products, obtaining product information, and capturing coupons.[17]

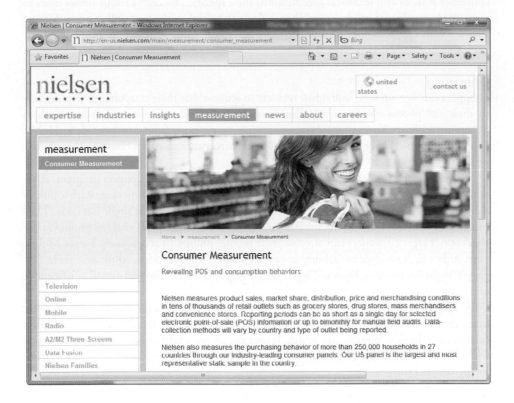

Nielsen is a major marketing company that measures and analyzes how consumers acquire information, consume media, and buy goods and services.

(Source: http://en-us.nielsen.com.)

Consumers are increasingly using mobile phones to purchase goods and perform other transactions online.

(Source. © ICP / Alamy.)

Price Comparison

A growing number of companies provide a mobile phone service that enables shoppers to compare prices and products on the Web. Google Product Search works for iPhone and Android handsets. The shopper enters the name of the product into the Google search field and clicks "See Shopping Results" to display a list of suppliers and prices. You can also request consumer reviews and technical specifications for a specific choice.[18] Frucall allows users to enter the bar code of a product, and then it finds and displays the best online prices for any product with that bar code. You can also read reviews or purchase the item immediately by clicking a button.[19]

Couponing

Shoppers can sign up with individual retailers to request that their coupons are sent directly to their cell phone. Shoppers can also subscribe to mobile coupon aggregators such as 8coupons, Cellfire, Yowza, and Zavers to receive promotions from many retailers. Mobile coupons are more likely to be redeemed (15–20 percent) than paper coupons (less than 1 percent).[20]

Target was the first national retailer to offer a scannable mobile coupon program. After opting into this program (via m.target.com), shoppers receive a text message with a link to a mobile Web page that contains multiple offers, all accessible under a single bar code. They redeem the coupons by scanning the bar code displayed on their Web-enabled phones at the checkout. Target customers can also use their mobile phones to access their Target Mobile GiftCards, check product availability at various Target store locations, administer their Target gift registries and lists, browse Target weekly ads, and receive text and e-mail notifications of deals.[21]

Valpak has launched a free mobile coupon application for smartphone users that delivers more than 17,000 offers. The application allows users to search for coupons by categories such as auto, beauty, dining, and health. It uses the phone's GPS to identify stores near you with offers, sorts the stores by distance from your current location, and provides directions to any selected store.[22]

Investment and Finance

The Internet has revolutionized the world of investment and finance. Perhaps the changes have been so significant because this industry had so many built-in inefficiencies and so much opportunity for improvement.

The brokerage business adapted to the Internet faster than any other arm of finance. The allure of online trading that enables investors to do quick, thorough research and then buy shares in any company in a few seconds and at a fraction of the cost of a full-commission firm has brought many investors to the Web.

Banking

Online banking customers can check balances of their savings, checking, and loan accounts; transfer money among accounts; and pay their bills. These customers enjoy the convenience of not writing checks by hand, tracking their current balances, and reducing expenditures on envelopes and stamps. In addition, paying bills online is good for the environment because it reduces the amount of paper used, thus saving trees and reducing greenhouse gases.

All of the major banks and many of the smaller banks in the U.S. enable their customers to pay bills online; many support bill payment via cell phone or other wireless device. Banks are eager to gain more customers who pay bills online because such customers tend to stay with the bank longer, have higher cash balances, and use more of the bank's products and services. To encourage the use of this service, many banks have eliminated all fees associated with online bill payment.

Consumers who have enrolled in mobile banking and downloaded the mobile application to their cell phones can check their credit card balances before making major purchases and can avoid credit rejections. They can also transfer funds from savings to checking accounts to avoid an overdraft.

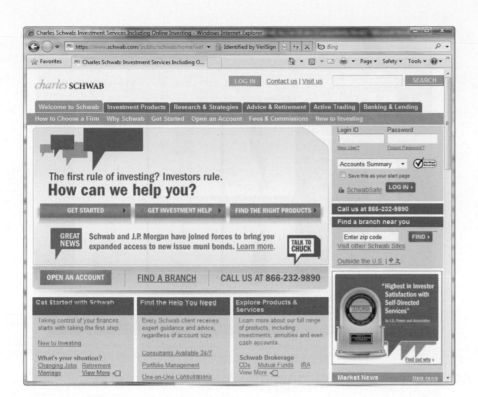

The Charles Schwab & Co. Web site provides online information and tools for investors.

(Source: *www.schwab.com*.)

Bank of America offers mobile banking via an application download. The application is custom designed for the unique features and capabilities of the iPhone, Blackberry, and Android smartphones. The user can check balances, transfer funds, pay bills, and use the geolocator feature to find the nearest ATM or banking center. The application uses advanced encryption technology to protect against unauthorized access. [23]

E-Boutiques

An increasing number of Web sites offer personalized shopping consultations for shoppers interested in upscale, contemporary clothing—dresses, sportswear, denim, handbags, jewelry, shoes, and gifts. Key to the success of Web sites such as Charm Boutique and ShopLaTiDa is a philosophy of high customer service and strong, personal client relationships. Online boutique shoppers complete a personal shopping profile by answering questions about body measurements, profession, interests, preferred designers, and areas of shopping where they would welcome assistance. Shoppers are then given suggestions on what styles and designers might work best and where they can be found—online or in brick-and-mortar shops.

Gilt is a private (invited members only), limited-time sales Web site where shoppers compete against each other for exclusive designer apparel and accessories. Items on the Web site typically sell out within 36 hours. The exclusive membership helps Gilt know who its customers are and what they like and don't like. The Web site has 1.6 million members. Those who use an iPad can see all the key information on one screen and then quickly tap through the sales process and drag items from the screen to their cart.[24] Although the Web site can be accessed by multiple mobile devices, an iPad with its larger screen size, along with its ability to zoom and flip through images with the swipe of a finger, makes Gilt look more like a magazine.[25]

Gilt is a private (invited members only), limited-time sales Web site where shoppers compete against each other for exclusive designer apparel and accessories.

(Source: Courtesy of Gilt Groupe, Inc.)

Advantages of Electronic and Mobile Commerce

Conversion to an e-commerce or m-commerce system enables organizations to reduce the cost of doing business, speed the flow of goods and information, increase the accuracy of order processing and order fulfillment, and improve the level of customer service. These advantages are summarized in Table 5.3.

Table 5.3

Advantages of Electronic and Mobile Commerce

Advantages	Explanation
Provides global reach	Allows manufacturers to buy at a low cost worldwide and offers enterprises the chance to sell to a global market right from the very start-up of their business.
Reduces costs	Eliminates time-consuming and labor-intensive steps throughout the order and delivery process so that more sales can be completed in the same period and with increased accuracy.
Speeds flow of goods and information	The flow of information is accelerated because of the established electronic connections and communications processes.
Increased accuracy	Enables buyers to enter their own product specifications and order information directly so that human data-entry error is eliminated.
Improves customer service	Increased and more detailed information about delivery dates and current status increases customer loyalty.

Now that we've examined several e-commerce and m-commerce applications, let's look at the key components of technology infrastructure that must be in place to make this all work.

Virtual Models Sell Clothes at Sears.com

Sears & Roebuck was one of first businesses to develop a mail-order catalog business. For decades, the Sears catalog has been a popular method of ordering products for those unable or unwilling to travel to a department store. There isn't much difference between ordering from a catalog and ordering from many e-commerce sites. Customers peruse a catalog filled with photos and descriptions of products, then phone in or enter their order and await delivery. More recently, however, digital technologies have been developed that provide e-commerce sites with powerful tools to drive sales—especially for products that have traditionally been difficult to sell online.

While books, music, and airline tickets are relatively easy to sell online, physical merchandise—especially clothing—is much more challenging. Shoppers like to try on clothing and see how outfit components look together. Trying on clothes and experimenting with clothing combinations has been impossible to do online, so most shoppers prefer to shop in stores for their clothes.

Recently, 3D technologies have provided online shoppers with the next best thing to being there. IBM and a company named My Virtual Model have partnered to provide powerful e-commerce tools for online clothing stores such as Sears.

My Virtual Model provides a custom designed 3D virtual model for shoppers to use to see how clothes look when tried on. Each customer uses online tools to create a custom virtual model that closely resembles his or her own physical characteristics. Body size and shape, skin color, hair color and style, and facial features can be customized to look like the customer. You can even upload a photo of your face to create an exact virtual twin. Once created, the virtual model can be used to try on clothing from the online catalog. Mix and match shirts, pants, shoes, skirts, hats, and all kinds of apparel to find a combination that suits your tastes and appearance. The model can be rotated to view the clothes from all angles.

Sears combined My Virtual Model with IBM's WebSphere Commerce software to provide a visual catalog of apparel that can be dragged onto a virtual model. The system provides additional social networking features that allow shoppers to share images of themselves in various outfits with friends online to solicit their feedback and suggestions. Sears and other retailers have found that shoppers are more likely to feel confident about a purchase when friends offer encouragement. Retailers that use My Virtual Model with social networking functionality have seen a 30 to 40 percent increase in sales conversion rates.

Some analysts believe that the use of social networking and 3D technologies will propel the next generation of e-commerce and assist retailers in invigorating sales after the economic downturn. One study found that 81 percent of consumers who use social networks seek shopping advice from friends and followers. 74 percent of social network users say that their social network influences their buying decisions. It is only logical for businesses to pursue social networks for ecommerce.

The impact of social networks on sales is sometimes referred to as the "Twitter effect." Bad publicity for a product on Twitter can have a devastating effect on sales. Increasing numbers of shoppers rely on social networks for advice while shopping online and in stores. It is not unusual to find shoppers in store aisles consulting their social network via an iPhone or other smartphone prior to tossing an item into their shopping cart.

Companies such as CrossView specialize in what they call cross-channel enablement: developing strategies for improving the shopping experience across all shopping environments and covering brand reputation across all online influences. Businesses are realizing that consumers are drawing information that shapes their shopping decisions from numerous sources, including social networks, television and newspaper ads, in-store promotions, and elsewhere. Businesses are responding to this by taking a holistic approach to advertising and marketing. Companies such as Sears are combining the latest technologies like 3D virtual models with social networking tools and traditional marketing techniques to create new and powerful ways to influence and win over consumers.

Discussion Questions

1. What recent technologies are being harnessed to invigorate e-commerce sales?
2. Why is it important for retailers to turn to social networking as a tool for building positive brand recognition?

Critical Thinking Questions

1. How might a company respond to negative publicity on social networks like Twitter?
2. What other types of products might benefit from the integration of virtual 3D technologies within e-commerce sites?

SOURCES: Walsh, Lawrence, "Solution Providers Transform Retail, E-Commerce," Channel Insider, September 1, 2009, *www.channelinsider.com/c/a/ IBM/Solution-Providers-Transform-Retail-eCommerce-122473*; IBM Staff, "Sears Transforms the Online Shopping Experience with Help From IBM and My Virtual Model," IBM Press Release, September 17, 2008; My Virtual Model Web site, *http://corpo.mvm.com*, accessed April 1, 2010.

E-COMMERCE AND M-COMMERCE TECHNOLOGY INFRASTRUCTURE

Successful implementation of e-commerce requires significant changes to existing business processes and substantial investment in IS technology. These technology components must be chosen carefully and be integrated to support a large volume of transactions with customers, suppliers, and other business partners worldwide. Online consumers complain that poor Web site performance (e.g., slow response time, inadequate customer support, and lost orders) drives them to abandon some e-commerce sites in favor of those with better, more reliable performance. This section provides a brief overview of the key technology infrastructure components. See Figure 5.2.

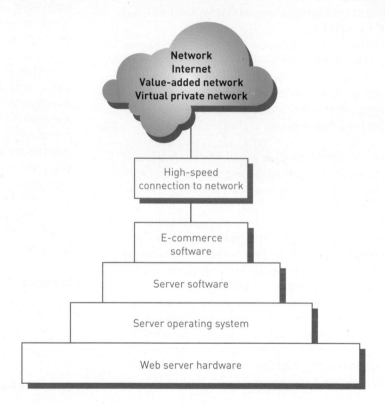

Hardware

A Web server hardware platform complete with the appropriate software is a key e-commerce infrastructure ingredient. The amount of storage capacity and computing power required of the Web server depends primarily on two things: the software that must run on the server and the volume of e-commerce transactions that must be processed. Although IS staff can sometimes define the software to be used, they can only estimate how much traffic the site will generate. As a result, the most successful e-commerce solutions are designed to be highly scalable so that they can be upgraded to meet unexpected growth in user traffic.

A key decision facing new e-commerce companies is whether to host their own Web site or to let someone else do it. Many companies decide that using a third-party Web service provider is the best way to meet initial e-commerce needs. The third-party company rents space on its computer system and provides a high-speed connection to the Internet, which minimizes the initial out-of-pocket costs for e-commerce start-up. The third party can also provide personnel trained to operate, troubleshoot, and manage the Web server. Of course, many companies decide to take full responsibility for acquiring, operating, and supporting the Web server hardware and software themselves, but this approach requires considerable up-front capital and a set of skilled and trained workers. No matter which approach a company takes, it must have adequate hardware backup to avoid a major business disruption in case of a failure of the primary Web server.

Web Server Software

In addition to the Web server operating system, each e-commerce Web site must have Web server software to perform fundamental services, including security and identification, retrieval and sending of Web pages, Web site tracking, Web site development, and Web page development. The two most widely used Web server software packages are Apache HTTP Server (51 percent market share) and Microsoft Internet Information Services (35 percent market share).[26]

E-Commerce Software

After you have located or built a host server, including the hardware, operating system, and Web server software, you can begin to investigate and install e-commerce software to support five core tasks: catalog management to create and update the product catalog, product configuration to help customers select the necessary components and options, shopping cart facilities to track the items selected for purchase, e-commerce transaction processing, and Web traffic data analysis to provide details to adjust the operations of the Web site. See Figure 5.3.

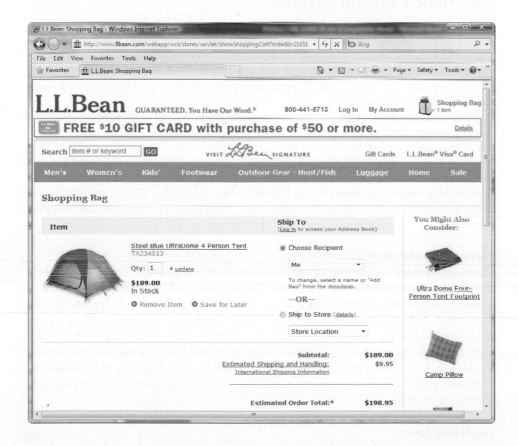

Figure 5.3

Electronic Shopping Cart

An electronic shopping cart (or bag) allows online shoppers to view their selections and add or remove items.

(Source: *www.llbean.com*.)

Mobile Commerce Hardware and Software

For m-commerce to work effectively, the interface between the wireless, handheld device and its user must improve to the point that it is nearly as easy to purchase an item on a wireless device as it is to purchase it on a PC. In addition, network speed must improve so that users do not become frustrated. Security is also a major concern, particularly in two areas: the security of the transmission itself and the trust that the transaction is being made with the intended party. Encryption can provide secure transmission. Digital certificates, discussed later in this chapter, can ensure that transactions are made between the intended parties.

The handheld devices used for m-commerce have several limitations that complicate their use. Their screens are small, perhaps no more than a few square inches, and might be able to display only a few lines of text. Their input capabilities are limited to a few buttons, so entering data can be tedious and error prone. They also have less processing power and less bandwidth than desktop computers, which are usually hardwired to a high-speed LAN. They also operate on limited-life batteries. For these reasons, it is currently impossible to directly access many Web sites with a handheld device. Web developers must rewrite Web applications so that users with handheld devices can access them.

To address the limitations of wireless devices, the industry has undertaken a standardization effort for their Internet communications. The Wireless Application Protocol (WAP) is a standard set of specifications for Internet applications that run on handheld, wireless devices. It effectively serves as a Web browser for such devices.

Electronic Payment Systems

Electronic payment systems are a key component of the e-commerce infrastructure. Current e-commerce technology relies on user identification and encryption to safeguard business transactions. Actual payments are made in a variety of ways, including electronic cash, electronic wallets, and smart, credit, charge, and debit cards. Web sites that accept multiple payment types convert more visitors to purchasing customers than merchants who offer only a single payment method.

Authentication technologies are used by many organizations to confirm the identity of a user requesting access to information or assets. A **digital certificate** is an attachment to an e-mail message or data embedded in a Web site that verifies the identity of a sender or Web site. A **certificate authority (CA)** is a trusted third-party organization or company that issues digital certificates. The CA is responsible for guaranteeing that the people or organizations granted these unique certificates are, in fact, who they claim to be. Digital certificates thus create a trust chain throughout the transaction, verifying both purchaser and supplier identities.

Secure Sockets Layer

All online shoppers fear the theft of credit card numbers and banking information. To help prevent this type of identity theft, the **Secure Sockets Layer (SSL)** communications protocol is used to secure sensitive data. The SSL communications protocol includes a handshake stage, which authenticates the server (and the client, if needed), determines the encryption and hashing algorithms to be used, and exchanges encryption keys. Following the handshake stage, data might be transferred. The data is always encrypted, ensuring that your transactions are not subject to interception or "sniffing" by a third party. Although SSL handles the encryption part of a secure e-commerce transaction, a digital certificate is necessary to provide server identification.

Electronic Cash

Electronic cash is an amount of money that is computerized, stored, and used as cash for e-commerce transactions. Typically, consumers must open an account with an electronic cash service provider by providing identification information. When the consumers want to withdraw electronic cash to make a purchase, they access the service provider via the Internet and present proof of identity—a digital certificate issued by a certification authority or a username and password. After verifying a consumer's identity, the system debits the consumer's account and credits the seller's account with the amount of the purchase. PayPal, BillMeLater, MoneyZap, and TeleCheck are four popular forms of electronic cash.

digital certificate
An attachment to an e-mail message or data embedded in a Web site that verifies the identity of a sender or Web site.

certificate authority (CA)
A trusted third-party organization or company that issues digital certificates.

Secure Sockets Layer (SSL)
A communications protocol used to secure sensitive data during e-commerce.

electronic cash
An amount of money that is computerized, stored, and used as cash for e-commerce transactions.

The PayPal service of eBay enables any person or business with an e-mail address to securely, easily, and quickly send and receive payments online. To send money, you enter the recipient's e-mail address and the amount you want to send. You can pay with a credit card, debit card, or funds from a checking account. The recipient gets an e-mail that says, "You've Got Cash!" Recipients can then collect their money by clicking a link in the e-mail that takes them to *www.paypal.com*. To receive the money, the user also must have a credit card or checking account to accept fund transfers. To request money for an auction, invoice a customer, or send a personal bill, you enter the recipient's e-mail address and the amount you are requesting. The recipient gets an e-mail and instructions on how to pay you using PayPal. PayPal has 78 million active accounts in 190 markets and makes payments in 19 currencies around the world.[27]

Credit, Charge, Debit, p-, and Smart Cards

Many online shoppers use credit and charge cards for most of their Internet purchases. A credit card, such as Visa or MasterCard, has a preset spending limit based on the user's credit history, and each month the user can pay all or part of the amount owed. Interest is charged on the unpaid amount. A charge card, such as American Express, carries no preset spending limit, and the entire amount charged to the card is due at the end of the billing period. Charge cards do not involve lines of credit and do not accumulate interest charges. American Express became the first company to offer disposable credit card numbers in 2000. Other banks, such as Citibank, protect the consumer by providing a unique number for each transaction. Debit cards look like credit cards, but they operate like cash or a personal check. Credit, charge, and debit cards currently store limited information about you on a magnetic strip. This information is read each time the card is swiped to make a purchase. All credit card customers are protected by law from paying more than $50 for fraudulent transactions.

A **p-card** (procurement card or purchasing card) is a credit card used to streamline the traditional purchase order and invoice payment processes. The p-card is typically issued to selected employees who must follow company rules and guidelines that may include a single purchase limit, a monthly spending limit, or merchant category code restrictions. Due to an increased risk of unauthorized purchases, each p-card holder's spending activity is reviewed periodically by someone independent of the cardholder to ensure adherence to the guidelines.

The **smart card** is a credit card–sized device with an embedded microchip to provide electronic memory and processing capability. Smart cards can be used for a variety of purposes, including storing a user's financial facts, health insurance data, credit card numbers, and network identification codes and passwords. They can also store monetary values for spending.

Smart cards are better protected from misuse than conventional credit, charge, and debit cards because the smart-card information is encrypted. Conventional credit, charge, and debit cards clearly show your account number on the face of the card. The card number, along with a forged signature, is all that a thief needs to purchase items and charge them against your card. A smart card makes credit theft practically impossible because a key to unlock the encrypted information is required, and there is no external number that a thief can identify and no physical signature a thief can forge. Table 5.4 compares various types of payment systems.

p-card (procurement card or purchasing card)
A credit card used to streamline the traditional purchase order and invoice payment processes.

smart card
A credit card–sized device with an embedded microchip to provide electronic memory and processing capability.

Payment System	Description	Advantages	Disadvantages
Credit card	Carries preset spending limit based on the user's credit history	Each month the user can pay all or part of the amount owed	Unpaid balance accumulates interest charges—often at a high rate of interest
Charge card	Looks like a credit card but carries no preset spending limit	Does not involve lines of credit and does not accumulate interest charges	The entire amount charged to the card is due at the end of the billing period
Debit card	Looks like a credit card or automated teller machine (ATM) card	Operates like cash or a personal check	Money is immediately deducted from user's account balance
Smart card	Is a credit card device with embedded microchip capable of storing facts about cardholder	Better protected from misuse than conventional credit, charge, and debit cards because the smart-card information is encrypted	Not widely used in the U.S.

The Dragon Hotel is a four-star, 527-room facility located in a scenic and tourist-friendly portion of Hangzhou, China. The hotel invested US $150 million to provide its clientele with a unique, personalized experience in its effort to become the first five-star hotel in the province of Zhejiang. It installed a smart card system "to automatically register visitors upon their arrival, direct them to their rooms, customize temperature settings to established preferences, and even record their attendance at conference events."[28]

Payments Using Cell Phones

A number of companies are exploring more convenient ways to enable payments by cell phones by converting cell phones into virtual checkbooks or credit cards so that users can simply use the touch pad to send payments. Two options are available: payments linked to your bank account and payments added to your phone bill. The goals are to make the payment process as simple and secure as possible and for it to work on many different phones and through many different cell phone service providers—not simple tasks. Fortunately, the intelligence built into the iPhone, BlackBerry, and other smartphones can make this all possible.[29]

Obopay is developing a service that enables people to transmit money from one another via text messaging. The MasterCard MoneySend service (currently in use in India) builds upon the Obopay technology. With MoneySend, funds transferred to a Maestro or Master-Card card can be available within one or two banking days. Users get an immediate confirmation that the funds have been successfully transferred. The recipient can access the transferred funds by making a purchase anywhere Maestro or MasterCard cards are accepted or by withdrawing cash at a participating ATM that accepts MasterCard cards and offers cash.[30]

Boku is attempting to start up a cell phone payment system based on the use of cell phone numbers rather than credit card numbers. The advantage is that while most people know their cell phone numbers, few remember their credit card numbers. The system sends a text message to the buyers asking them to authorize the purchase with a texted response. If authorized, a charge for the purchase then appears on the buyer's mobile phone bill.

AN OVERVIEW OF TRANSACTION PROCESSING SYSTEMS

Every organization has many *transaction processing systems (TPSs)*, which capture and process the detailed data necessary to update records about the fundamental business operations of the organization. These systems include order entry, inventory control, payroll, accounts payable, accounts receivable, and the general ledger, to name just a few. The input to these systems includes basic business transactions, such as customer orders, purchase orders, receipts, time cards, invoices, and customer payments. The processing activities include data collection, data editing, data correction, data manipulation, data storage, and document production. The result of processing business transactions is that the organization's records are updated to reflect the status of the operation at the time of the last processed transaction.

A TPS also provides employees involved in other business processes—via management information system/decision support system (MIS/DSS) and the special-purpose information systems—with data to help them achieve their goals. (MIS/DSS systems are discussed in Chapter 6.) A transaction processing system serves as the foundation for these other systems. See Figure 5.4.

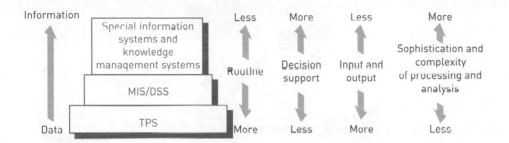

Figure 5.4

TPS, MIS/DSS, and Special Information Systems in Perspective

Transaction processing systems support routine operations associated with customer ordering and billing, employee payroll, purchasing, and accounts payable. The amount of support for decision making that a TPS directly provides managers and workers is low.

TPSs work with a large amount of input and output data and use this data to update the official records of the company about such things as orders, sales, and customers. As systems move from transaction processing to management information/decision support and special-purpose information systems, they involve less routine, more decision support, less input and output, and more sophisticated and complex analysis. These higher-level systems require the basic business transaction data captured by the TPS.

Newcastle Permanent Building Association is an Australian financial services organization that provides personal and business banking services and whose total assets exceed $6 billion. It operates a widespread network of branch offices and ATMs and also provides electronic banking services to its customers. Its transaction processing banking application manages all member account information and transactions and provides a master reference for other downstream systems, including ATMs and electronic banking services. The TPS also captures key data that produces daily and monthly reports to provide managers with information on the financial position of the firm.[31]

Traditional Transaction Processing Methods and Objectives

batch processing system
A form of data processing whereby business transactions are accumulated over a period of time and prepared for processing as a single unit or batch.

With **batch processing systems**, business transactions are accumulated over a period of time and prepared for processing as a single unit or batch. See Figure 5.5a. Transactions are accumulated for the length of time needed to meet the needs of the users of that system. For example, it might be important to process invoices and customer payments for the accounts receivable system daily. On the other hand, the payroll system might receive time cards and process them biweekly to create checks, update employee earnings records, and distribute labor costs. The essential characteristic of a batch processing system is that there is some delay between an event and the eventual processing of the related transaction to update the organization's records.

Figure 5.5

Batch versus Online Transaction Processing

(a) Batch processing inputs and processes data in groups. (b) In online processing, transactions are completed as they occur.

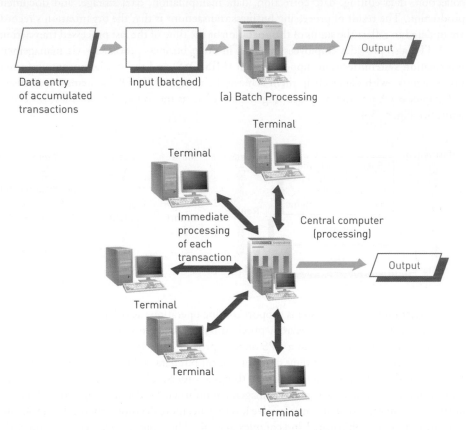

(a) Batch Processing

(b) Online Transaction Processing

Dean Health Plan (DHP) is a health maintenance organization offering its customers a network of more than 2,000 practitioners, 80 clinic sites, and 26 hospitals. DHP recently implemented a healthcare claims batch processing system to help it process claims more efficiently while meeting strict HIPAA (Health Insurance Portability and Accountability Act) standards for data privacy and portability.[32]

online transaction processing (OLTP)
A form of data processing where each transaction is processed immediately, without the delay of accumulating transactions into a batch.

With **online transaction processing (OLTP)**, each transaction is processed immediately, without the delay of accumulating transactions into a batch. See Figure 5.5b. Consequently, at any time, the data in an online system reflects the current status. This type of processing is essential for businesses that require access to current data such as airlines, ticket agencies, and stock investment firms. Many companies find that OLTP helps them provide faster, more efficient service—one way to add value to their activities in the eyes of the customer. Increasingly, companies are using the Internet to capture and process transaction data such as customer orders and shipping information from e-commerce applications.

Trimac Corporation is a trucking firm and uses an OLTP system to perform all tasks associated with order entry, dispatching, trip planning, and driver payment.

(Source: *www.trimac.com*.)

Trimac Corporation located in Calgary, Alberta in Canada is a bulk hauling carrier with 140 branch offices and some 3,000 tractors and 6,000 trailers driving 220 million miles per year. The firm uses an OLTP system to perform all tasks associated with order entry, dispatching, trip planning, and driver payment.[33]

TPS applications do not always run using online processing. For many applications, batch processing is more appropriate and cost effective. Payroll transactions and billing are typically done via batch processing. Specific goals of the organization define the method of transaction processing best suited for the various applications of the company.

Figure 5.6 shows the flow of key pieces of information from one TPS to another for a typical manufacturing organization. TPSs can be designed so that the flow of information from one system to another is automatic and requires no manual intervention or reentering of data. Such a set of systems is called an *integrated information system*. Many organizations have limited or no integration among their TPSs. In this case, data input to one TPS must be printed out and manually reentered into other systems. Of course, this increases the amount of effort required and introduces the likelihood of processing delays and errors.

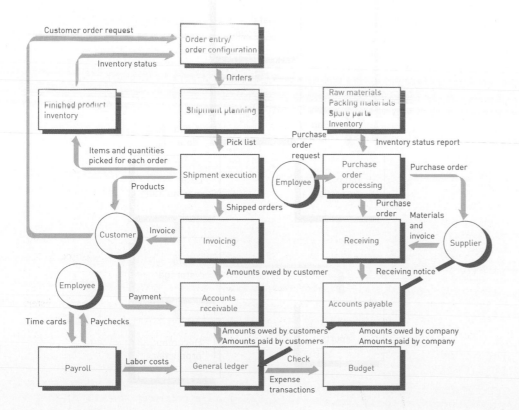

Figure 5.6

Integration of a Firm's TPS

Because of the importance of transaction processing, organizations expect their TPSs to accomplish a number of specific objectives including:

- Process data generated by and about transactions
- Maintain a high degree of accuracy and integrity
- Avoid processing fraudulent transactions
- Produce timely user responses and reports
- Increase labor efficiency
- Help improve customer service and/or loyalty

Depending on the specific nature and goals of the organization, any of these objectives might be more important than others. By meeting these objectives, TPSs can support corporate goals such as reducing costs; increasing productivity, quality, and customer satisfaction; and running more efficient and effective operations. For example, overnight delivery companies such as FedEx expect their TPSs to increase customer service. These systems can locate a client's package at any time—from initial pickup to final delivery. This improved customer information allows companies to produce timely information and be more responsive to customer needs and queries.

TRANSACTION PROCESSING ACTIVITIES

transaction processing cycle
The process of data collection, data editing, data correction, data manipulation, data storage, and document production.

Along with having common characteristics, all TPSs perform a common set of basic data-processing activities. TPSs capture and process data that describes fundamental business transactions. This data is used to update databases and to produce a variety of reports for people both within and outside the enterprise. The business data goes through a **transaction processing cycle** that includes data collection, data editing, data correction, data manipulation, data storage, and document production. See Figure 5.7.

Figure 5.7

Data Processing Activities Common to Transaction Processing Systems

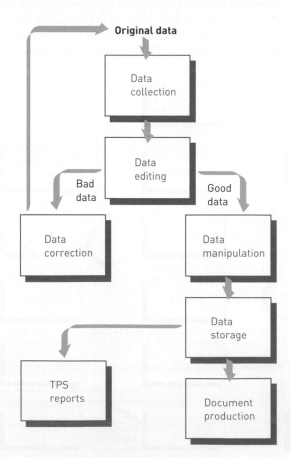

Data Collection

Capturing and gathering all data necessary to complete the processing of transactions is called **data collection**. In some cases it can be done manually, such as by collecting handwritten sales orders or changes to inventory. In other cases, data collection is automated via special input devices such as scanners, point-of-sale devices, and terminals.

Data collection begins with a transaction (e.g., taking a customer order) and results in data that serves as input to the TPS. Data should be captured at its source and recorded accurately in a timely fashion, with minimal manual effort, and in an electronic or digital form that can be directly entered into the computer. This approach is called *source data automation*. An example of source data automation is an automated device at a retail store that speeds the checkout process—either UPC codes read by a scanner or RFID signals picked up when the items approach the checkout stand. Using both UPC bar codes and RFID tags is quicker and more accurate than having a clerk enter codes manually at the cash register. The product ID for each item is determined automatically, and its price retrieved from the item database. The point-of-sale TPS uses the price data to determine the customer's bill. The store's inventory and purchase databases record the number of units of an item purchased, the date, the time, and the price. The inventory database generates a management report notifying the store manager to reorder items that have fallen below the reorder quantity. The detailed purchases database can be used by the store or sold to marketing research firms or manufacturers for detailed sales analysis. See Figure 5.8.

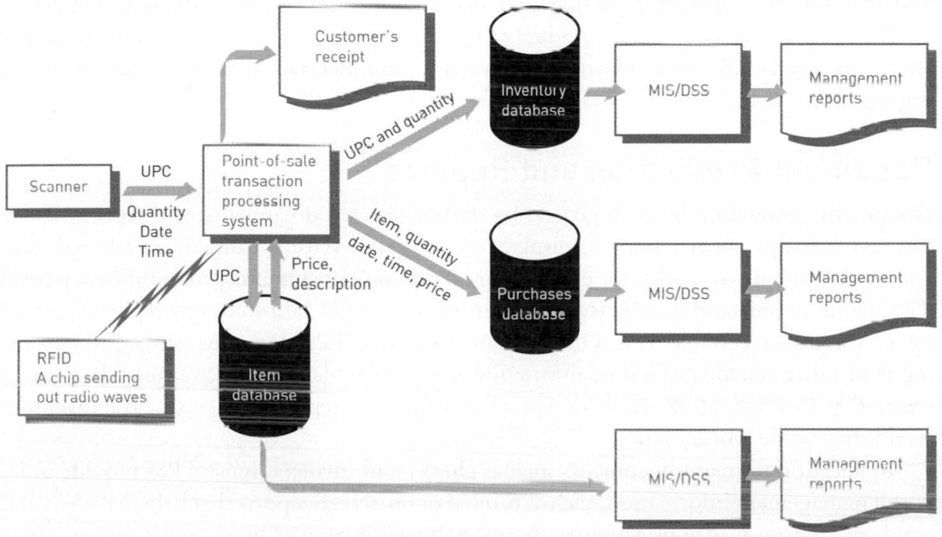

data collection
Capturing and gathering all data necessary to complete the processing of transactions.

Figure 5.8

Point-of-Sale Transaction Processing System

The purchase of items at the checkout stand updates a store's inventory database and its database of purchases.

Many grocery stores combine point-of-sale scanners and coupon printers. The systems are programmed so that each time a specific product—for example, a box of cereal—crosses a checkout scanner, an appropriate coupon—perhaps a milk coupon—is printed. Companies can pay to be promoted through the system, which is then reprogrammed to print those companies' coupons if the customer buys a competitive brand. These TPSs help grocery stores increase profits by improving their repeat sales and bringing in revenue from other businesses.

Data Editing

An important step in processing transaction data is to perform **data editing** for validity and completeness to detect any problems. For example, quantity and cost data must be numeric, and names must be alphabetic; otherwise, the data is not valid. Often, the codes associated with an individual transaction are edited against a database containing valid codes. If any code entered (or scanned) is not present in the database, the transaction is rejected.

data editing
The process of checking data for validity and completeness.

Data Correction

It is not enough simply to reject invalid data. The system should also provide error messages that alert those responsible for editing the data. Error messages must specify the problem so proper corrections can be made. A **data correction** involves reentering data that was not typed or scanned properly. For example, a scanned UPC code must match a code in a master table of valid UPCs. If the code is misread or does not exist in the table, the checkout clerk is given an instruction to rescan the item or type the information manually.

data correction
The process of reentering data that was not typed or scanned properly.

Data Manipulation

Another major activity of a TPS is **data manipulation**, the process of performing calculations and other data transformations related to business transactions. Data manipulation can include classifying data, sorting data into categories, performing calculations, summarizing results, and storing data in the organization's database for further processing. In a payroll TPS, for example, data manipulation includes multiplying an employee's hours worked by the hourly pay rate. Overtime pay, federal and state tax withholdings, and deductions are also calculated.

data manipulation
The process of performing calculations and other data transformations related to business transactions.

Data Storage

Data storage involves updating one or more databases with new transactions. After being updated, this data can be further processed and manipulated by other systems so that it is available for management reporting and decision making. Thus, although transaction databases can be considered a by-product of transaction processing, they have a pronounced effect on nearly all other information systems and decision-making processes in an organization.

data storage
The process of updating one or more databases with new transactions.

Document Production and Reports

Document production involves generating output records, documents, and reports. These can be hard-copy paper reports or displays on computer screens (sometimes referred to as *soft copy*). Printed paychecks, for example, are hard-copy documents produced by a payroll TPS, while an outstanding balance report for invoices might be a soft-copy report displayed by an accounts receivable TPS. Often, results from one TPS flow downstream to become input to other systems, which might use the results of updating the inventory database to create the stock exception report (a type of management report) of items whose inventory level is below the reorder point.

document production
The process of generating output records and reports.

In addition to major documents such as checks and invoices, most TPSs provide other useful management information, such as printed or on-screen reports that help managers and employees perform various activities. A report showing current inventory is one example; another might be a document listing items ordered from a supplier to help a receiving clerk check the order for completeness when it arrives. A TPS can also produce reports required by local, state, and federal agencies, such as statements of tax withholding and quarterly income statements.

Google Pulls Out Of China

Companies that serve customers around the world often need to make adjustments so that the products and services that they provide adhere to local laws. Conforming to the local laws of the countries in which you do business is not typically a major issue, unless those laws contradict the company's ethical values. Such was the case when Google decided to pull out of China.

The story begins in December of 2009 when Google and dozens of other companies and government organizations were the targets of a cyberattack based out of China. The purpose of the attacks was to gain access to the accounts of Chinese dissidents and journalists. For Google, the attack served as the final straw to building tensions between Google and the Chinese government. For years the Chinese government had required Google to filter search results served to Chinese citizen—a requirement that Google regards as unethical. China also occasionally required Google's China office to provide account information of Chinese bloggers that had criticized the government. In some cases, the information provided by Google reportedly resulted in arrests, and torture.

After it was clear to Google that the December attack could not have occurred without government sponsorship, Google laid down an ultimatum: Google would continue operations in China only if it was allowed to provide unfiltered search results. Google used the hacking incident as a lever to raise the ethical concerns it has with Chinese laws. China responded to the hacking allegation by downplaying the incident and reiterating that all businesses in China are bound to uphold China laws.

After months of closed-door negotiations, with China holding firm to its stance, Google closed the doors on its search engine in China, following through on its promise. But, rather than eliminating its filters on google.cn and risking the arrest of its China-based employees, Google redirected requests for google.cn to its Hong Kong search engine, google.com.hk, where it maintains unfiltered Chinese-language search results. Shortly after the switch, China was quick to apply its own censoring filters to the Internet DNS servers that feed its country.

Google's decision to close google.cn was shocking because of the large monetary sacrifice that Google has made. China has the world's largest population and has one of the most rapidly growing economies. Google has given up a large competitive advantage in exchange for its cleaner conscience. Google's move is causing many companies that do business in China to re-evaluate their own motivations and convictions.

So far, while many have applauded Google's decision, only a few have followed suit. Popular Web hosting company GoDaddy stopped registering domain names for the .cn domain. That decision came after the Chinese government demanded personal information about people who had purchased domain names from GoDaddy. Microsoft has stated that it intends to continue growing its business in China. Even Google continues other operations in China and looks forward to robust sales of Google Android phones in China in coming years.

China isn't the only country where Google censors content based on government-imposed policies. Google has posted the site www.google.com/governmentrequests that lists governments that require Google to censor content. Brazil, Germany, India, the UK, South Korea, and the U.S. rank high on the list. Granted, censorship is sometimes justified, such as in cases where it protects populations from physical harm. However, many feel that it is not justified when it is used to silence dissident opinions such as in China. Increasingly, technology companies such as Google and Internet service providers are assuming responsibility for policing Internet content. Google's stance against China's censorship has shown that the company is clearly uncomfortable with its role as a censor and causes some to wonder if it may not follow up with changes in policy elsewhere. Google's chief legal officer wrote that the China issue "goes to the heart of a much bigger global debate about freedom of speech."

Google's experience in China provides an extreme example of the considerations faced by technology companies providing services abroad. Similar considerations are faced by all kinds of international companies, at varying levels of complexity every day. While most companies comply with local laws and customs in the countries in which they operate without complaint, Google chose to use its financial power and influence to make a statement about its company's ethical position on political censorship.

Discussion Questions

1. What constraints imposed by China combined to cause Google to close its Chinese search engine?
2. What pressures might cause businesses to set aside ethical considerations to do business in China?

Critical Thinking Questions

1. Some have argued that Google's exit from China might have actually harmed the growth of democracy around the world. How might Google have positively influenced democracy if it had stayed in China?
2. What constraints might China place on a business like Wal-Mart, or India place on a restaurant like McDonalds? How do those constraints differ from those placed on Google in China?

SOURCES: Perez, Juan Carlos, "Google stops censoring in China," *Computerworld*, March 22, 2010, *www.computerworld.com*; Gross, Grant, "GoDaddy to stop registering .cn domain names," *Computerworld*, March 24, 2010, *www.computerworld.com*; Naone, Erica, "Google and Censorship," *Technology Review*, March 26, 2010, *www.technologyreview.com*; Wills, Ken, "China state media accuses Google of political agenda," Reuters, March 21, 2010, *www.reuters.com*.

TRADITIONAL TRANSACTION PROCESSING APPLICATIONS

A TPS typically includes the following types of systems:

- *Order processing systems.* Running these systems efficiently and reliably is so critical that the order processing systems are sometimes referred to as the lifeblood of the organization. The processing flow begins with the receipt of a customer order. The finished product inventory is checked to see if sufficient inventory is on hand to fill the order. If sufficient inventory is available, the customer shipment is planned to meet the customer's desired receipt date. A product pick list is printed at the warehouse from which the order is to be filled on the day the order is planned to be shipped. At the warehouse, workers gather the items needed to fill the order, and enter the item identifier and quantity for each item to update the finished product inventory. When the order is complete and sent on its way, a customer invoice is created with a copy included in the customer shipment.
- *Accounting systems.* The accounting systems must track the flow of data related to all the cash flows that affect the organization. As mentioned earlier, the order processing system generates an invoice for customer orders to include with the shipment. This information is also sent to the accounts receivable system to update the customer's account. When the customer pays the invoice, the payment information is also used to update the customer's account. The necessary accounting transactions are sent to the general ledger system to keep track of amounts owed and amounts paid. Similarly, as the purchasing systems generate purchase orders and those items are received, information is sent to the accounts payable system to manage the amounts owed by the company. Data about amounts owed and paid by customers to the company and from the company to vendors and others are sent to the general ledger system that records and reports all financial transactions for the company.
- *Purchasing systems.* The traditional transaction processing systems that support the purchasing business function include inventory control, purchase order processing, receiving, and accounts payable. Employees place purchase order requests in response to shortages identified in inventory control reports. Purchase order information flows to the receiving system and accounts payable systems. A record of receipt is created upon receipt of the items ordered. When the invoice arrives from the supplier, it is matched to the original order and the receiving report and a check is generated if all data is complete and consistent.

TRANSACTION PROCESSING SYSTEMS FOR SMALL AND MEDIUM-SIZE ENTERPRISES (SMES)

Many software packages provide integrated transaction processing system solutions for small and medium-size enterprises (SMEs), wherein small is an enterprise with fewer than 50 employees and medium is one with fewer than 250 employees. These systems are typically easy to install, easy to operate, and have a low total cost of ownership with an initial cost of a few hundred to a few thousand dollars. Such solutions are highly attractive to firms that have outgrown their current software but cannot afford a complex, high-end integrated system solution. Table 5.5 presents some of the dozens of such software solutions available.

Vendor	Software	Type of TPS Offered	Target Customers
AccuFund	AccuFund	Financial reporting and accounting	Nonprofit, municipal, and government organizations
OpenPro	OpenPro	Complete ERP solution, including financials, supply chain management, e-commerce, customer relationship management, and retail POS system	Manufacturers, distributors, and retailers
Intuit	QuickBooks	Financial reporting and accounting	Manufacturers, professional services, contractors, nonprofits, and retailers
Sage	Timberline	Financial reporting, accounting, and operations	Contractors, real estate developers, and residential builders
Redwing	TurningPoint	Financial reporting and accounting	Professional services, banks, and retailers

Table 5.5

Sample of Integrated TPS Solutions for SMEs

ENTERPRISE RESOURCE PLANNING

An **enterprise system** is central to an organization and ensures information can be shared across all business functions and all levels of management to support the running and managing of a business. Enterprise systems employ a database of key operational and planning data that can be shared by all (Figure 5.9). This eliminates the problems of lack of information and inconsistent information caused by multiple transaction processing systems that support only one business function or one department in an organization. Examples of enterprise systems include enterprise resource planning systems that support supply-chain processes, such as order processing, inventory management, and purchasing, and customer relationship management systems that support sales, marketing, and customer service-related processes.

Businesses rely on such systems to perform many of their daily activities in areas such as product supply, distribution, sales, marketing, human resources, manufacturing, accounting, and taxation so that work is performed quickly, while avoiding waste and mistakes. Without such systems, recording and processing business transactions would consume huge amounts of an organization's resources. This collection of processed transactions also forms a storehouse of data invaluable to decision making. The ultimate goal is to satisfy customers and provide a competitive advantage by reducing costs and improving service. See Figure 5.9.

enterprise system
A system central to the organization that ensures information can be shared across all business functions and all levels of management to support the running and managing of a business.

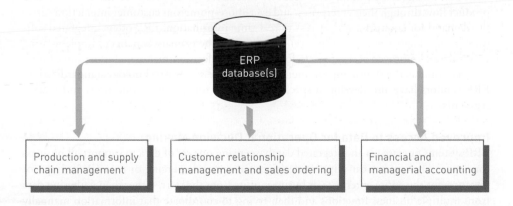

Figure 5.9

Enterprise Resource Planning System

An ERP integrates business processes and the ERP database.

Microsoft Dynamics is an ERP solution that is very popular among small businesses.

(Source: Courtesy of Microsoft Corporation)

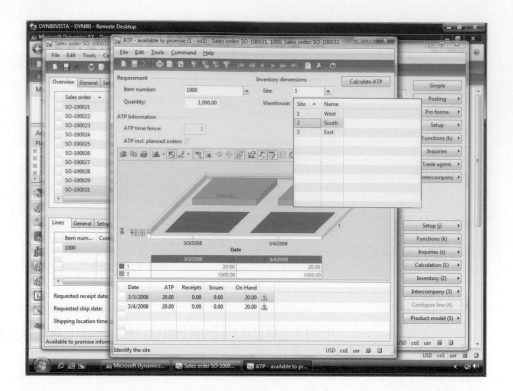

An Overview of Enterprise Resource Planning

ERP systems evolved from materials requirement planning systems (MRP) that tied together the production planning, inventory control, and purchasing business functions for manufacturing organizations. Many organizations recognized that their legacy transaction processing systems lacked the integration needed to coordinate activities and share valuable information across all the business functions of the firm. As a result, costs were higher and customer service poorer than desired. As a result, firms are scrapping large parts of their existing information systems and converting to new ERP systems. Large organizations, specifically members of the *Fortune* 1000, were the first to take on the challenge of implementing ERP. As they did, they uncovered many advantages as well as some disadvantages summarized in the following sections.

Advantages of ERP

Increased global competition, new needs of executives for control over the total cost and product flow through their enterprises, and ever-more-numerous customer interactions drive the demand for enterprise-wide access to real-time information. ERP offers integrated software from a single vendor to help meet those needs. The primary benefits of implementing ERP include improved access to data for operational decision making, elimination of inefficient or outdated systems, improvement of work processes, and technology standardization. ERP vendors have also developed specialized systems for specific applications and market segments.

Improved Access to Data for Operational Decision Making

ERP systems operate via an integrated database, using one set of data to support all business functions. The systems can support decisions on optimal sourcing or cost accounting, for instance, for the entire enterprise or business units from the start, rather than gathering data from multiple business functions and then trying to coordinate that information manually or reconciling data with another application. The result is an organization that looks seamless, not only to the outside world but also to the decision makers who are deploying resources within the organization. The data is integrated to facilitate operational decision making and allows companies to provide greater customer service and support, strengthen customer and supplier relationships, and generate new business opportunities.

Flambeau produces a wide range of plastic products and employs thousands of workers in eight manufacturing locations worldwide. It has grown through acquisition, and out of necessity was running multiple, disparate legacy information systems that drew data from multiple databases. The firm had to resort to the use of spreadsheets to manually track critical business information used for cost and inventory control. This inevitably led to errors and poor decision making. Finally the company implemented an ERP system to deliver timely, consistent data for both production and financial management purposes. Flambeau has used the system to lower its inventory costs, better manage its production operations, and provide access to a single set of data used to run the business.[34]

Elimination of Costly, Inflexible Legacy Systems

Adoption of an ERP system enables an organization to eliminate dozens or even hundreds of separate systems and replace them with a single, integrated set of applications for the entire enterprise. In many cases, these systems are decades old, the original developers are long gone, and the systems are poorly documented. As a result, the systems are extremely difficult to fix when they break, and adapting them to meet new business needs takes too long. They become an anchor around the organization that keeps it from moving ahead and remaining competitive. An ERP system helps match the capabilities of an organization's information systems to its business needs—even as these needs evolve.

Marin Municipal Water District provides drinking water to nearly 200,000 people residing in the 147 square mile area of Marin County, California, just north of San Francisco. The district relied on a combination of manual processes and antiquated information systems to manage the operation and maintenance of 925 miles of pipeline, 140 storage tanks, 94 pumping stations, and 3 water treatment plants. The Water District upgraded to an ERP system to meet this need as well as to integrate customer billing and Human Resources with its work order and maintenance operations. Now the Water District can perform maintenance on a preventative basis forecasting the work needed to keep everything operating smoothly without interruption. The ERP system has even been linked to a Geographic Information System (GIS) so that when maintenance is required, workers can generate maps to identify repair areas and get details on all Water District assets in the area.[35]

Improvement of Work Processes

Competition requires companies to structure their business processes to be as effective and customer oriented as possible. ERP vendors do considerable research to define the best business processes. They gather requirements of leading companies within the same industry and combine them with research findings from research institutions and consultants. The individual application modules included in the ERP system are then designed to support these best practices, the most efficient and effective ways to complete a business process. Thus, implementation of an ERP system ensures good work processes based on best practices. For example, for managing customer payments, the ERP system's finance module can be configured to reflect the most efficient practices of leading companies in an industry. This increased efficiency ensures that everyday business operations follow the optimal chain of activities, with all users supplied the information and tools they need to complete each step.

With 22,000 employees serving 4.7 million customers and generating revenue of 14 billion Euros, Achmea is the largest insurance company in the Netherlands. The company had grown rapidly through acquisition and had evolved to using a mix of manual data collection and reporting processes. The company converted to an ERP system to standardize on a set of industry best practices, streamlined work processes, and sophisticated data analysis tools across all divisions and operating companies. As a result, the company could reduce staffing levels in some areas of the business by as much as 30 percent, thus improving productivity and cutting costs. In addition, the time required to complete month-end financial reporting was reduced by 30 percent, with an increase in the accuracy and reliability of the data.[36]

Upgrade of Technology Infrastructure

When implementing an ERP system, an organization has an opportunity to upgrade the information technology (hardware, operating systems, databases, etc.) that it uses. While centralizing and formalizing these decisions, the organization can eliminate the hodgepodge of multiple hardware platforms, operating systems, and databases it is currently using—most likely from a variety of vendors. Standardizing on fewer technologies and vendors reduces ongoing maintenance and support costs as well as the training load for those who must support the infrastructure.

Barloworld Handling UK is the United Kingdom distributor of Hyster forklifts. It also provides parts and service through 26 service locations that field customer service calls, schedule and dispatch field techs, and manage the ordering and delivery of parts. This highly decentralized service operation resulted in inefficient work processes, high costs, and inconsistent service levels. Barloworld reengineered its service operations to squeeze out waste and inefficiency. Service techs were issued handheld computers programmed to follow the new work processes. The handheld devices could also access work orders, equipment information, and inventory data held in the firm's ERP database. By integrating mobile devices with improved work processes and access to ERP data, the firm achieved "paperless, real-time data entry; immediate parts lookup and availability checks with overnight delivery; time sheets completed as work progresses; and automatic dispatch of work orders," according to Robert S. Tennant, the firm's CIO. The number of service locations was reduced from 26 to 6, service tech efficiency was increased by 10 percent, and annual revenue increased by more than €500,000.[37]

Disadvantages of ERP Systems

Unfortunately, implementing ERP systems can be difficult and can disrupt current business practices. Some of the major disadvantages of ERP systems are the expense and time required for implementation, the difficulty in implementing the many business process changes that accompany the ERP system, the problems with integrating the ERP system with other systems, difficulty in loading data into the new system, the risks associated with making a major commitment to a single vendor, and the risk of implementation failure.

Expense and Time in Implementation

Getting the full benefits of ERP takes time and money. Although ERP offers many strategic advantages by streamlining a company's TPSs, large firms typically need three to five years and spend tens of millions of dollars to implement a successful ERP system.

Difficulty Implementing Change

In some cases, a company has to radically change how it operates to conform to the ERP's work processes—its best practices. These changes can be so drastic to long-time employees that they retire or quit rather than go through the change. This exodus can leave a firm short of experienced workers. Sometimes, the best practices simply are not appropriate for the firm and cause great work disruptions.

Difficulty Integrating with Other Systems

Most companies have other systems that must be integrated with the ERP system, such as financial analysis programs, e-commerce operations, and other applications. Many companies have experienced difficulties making these other systems operate with their ERP system. Other companies need additional software to create these links.

Difficulty in Loading Data into New ERP System

A major amount of work is required to load existing data from various sources into the new ERP database. The new ERP system may have the capability to store hundreds or even thousands of data items (e.g. customer name, bill to address, product description, etc.). The data items that will be required depend on the scope of ERP implementation. If certain processes or transactions are not included within the scope of implementation, there will be less data to load.

Data mapping is the examination of each data item required for the new ERP system and determining where that data item will come from. While most of the data for the new system will come from the files of existing legacy systems, some data items may need to be pulled from manual systems or may even need to be created for the new system.

Data cleanup is required because the legacy systems are likely to contain data that is inaccurate, incomplete, or inconsistent. For example, the same customer may be listed multiple times in existing customer files with varying bill to addresses or products may appear in the existing inventory files that have not been produced for years.

Data loading can be performed either by using data conversion software that reads the old data and converts it into a format for loading it into the database or by end users entering data via the input screens of the new system.

Risks in Using One Vendor

The high cost to switch to another vendor's ERP system makes it extremely unlikely that a firm will do so. After a company has adopted an ERP system, the vendor has less incentive to listen and respond to customer concerns. The high cost to switch also increases risk—in the event the ERP vendor allows its product to become outdated or goes out of business. Selecting an ERP system involves not only choosing the best software product, but also the right long-term business partner. It was unsettling for many companies that had implemented PeopleSoft, J.D. Edwards, or Siebel Systems enterprise software when these firms were acquired by Oracle.

Risk of Implementation Failure

Implementing an ERP system for a large organization is extremely challenging and requires tremendous amounts of resources, the best IS and businesspeople, and plenty of management support. Unfortunately, large ERP installations occasionally fail, and problems with an ERP implementation can require expensive solutions.

The following list provides tips for avoiding many common causes for failed ERP implementations:

- Assign a full-time executive to manage the project.
- Appoint an experienced, independent resource to provide project oversight and to verify and validate system performance.
- Allow sufficient time for transition from the old way of doing things to the new system and new processes.
- Plan to spend a lot of time and money training people; many project managers recommend that $10,000-$20,000 per employee be budgeted for training of personnel.
- Define metrics to assess project progress and to identify project-related risks.
- Keep the scope of the project well defined and contained to essential business processes.
- Be wary of modifying the ERP software to conform to your firm's business practices

ERP for Small and Medium-Size Enterprises [SMEs]

It is not only large *Fortune* 1000 companies that are successful in implementing ERP. SMEs (both for-profit and not-for-profit) can achieve real business benefits from their ERP efforts. Many of the SMEs elected to implement open-source ERP systems. With open-source software, anyone can see and modify the source code to customize it to meet their needs. Such systems are much less costly to acquire and are relatively easy to modify to meet business needs. A wide range of organizations can perform the system development and maintenance. Table 5.6 lists some of the open-source ERP systems geared for SMEs.

The following sections outline how an ERP system can support supply chain management and accounting, two major business processes.

Vendor	ERP Solutions
Apache	Open For Business ERP
Compiere	Compiere Open Source ERP
Openbravo	Openbravo Open Source ERP
WebERP	WebERP

supply chain management (SCM)

A system that includes planning, executing, and controlling all activities involved in raw material sourcing and procurement, converting raw materials to finished products, and warehousing and delivering finished product to customers.

Supply Chain Management (SCM)

Supply chain management (SCM) includes the planning, execution, and control of all activities involved in raw material sourcing and procurement, conversion of raw materials to finished products, and the warehousing and delivery of finished product to customers. The goal of SCM is to reduce costs and improve customer service, while at the same time reducing the overall investment in inventory in the supply chain.

The ERP system for a manufacturing organization typically encompasses SCM activities and manages the flow of materials, information and finances. Manufacturing ERP systems follow a systematic process for developing a production plan that draws on the information available in the ERP system database.

The process starts with *sales forecasting* to develop an estimate of future customer demand. This initial forecast is at a fairly high level, with estimates made by product group rather than by each individual product item. The sales forecast extends for months into the future. The sales forecast might be developed using an ERP software module or it might be produced by other means using specialized software and techniques. Many organizations are moving to a collaborative process with major customers to plan future inventory levels and production rather than relying on an internally generated sales forecast.

The *sales and operations plan* takes demand and current inventory levels into account and determines the specific product items that need to be produced and when to meet the forecast future demand. Production capacity and any seasonal variability in demand must also be considered. The result is a high-level production plan that balances market demand and production capacity.

Demand management refines the production plan by determining the amount of weekly or daily production needed to meet the demand for individual products. The output of the demand management process is the master production schedule which is a production plan for all finished goods.

Detailed scheduling uses the production plan defined by the demand management process to develop a detailed production schedule specifying details such as which item to produce first and when production should be switched from one item to another. A key decision is how long to make the production runs for each product. Longer production runs reduce the number of machine setups required, thus reducing production costs. Shorter production runs generate less finished product inventory and reduce inventory holding costs.

Materials requirement planning determines the amount and timing for placing raw material orders with suppliers. The types and amounts of raw materials required to support the planned production schedule are determined based on the existing raw material inventory and the bill of materials, or BOM, a sort of "recipe" of ingredients needed to make each product item. The quantity of raw materials to order also depends on the lead time and lot sizing. Lead time is the time it takes from the time a purchase order is placed until the raw materials arrive at the production facility. Lot size has to do with discrete quantities that the supplier will ship and the amount that is economical for the producer to receive and/or store. For example, a supplier might ship a certain raw material in units of 80,000-pound rail cars. The producer might need 95,000 pounds of the raw material. A decision must be made to order one or two rail cars of the raw material.

Purchasing uses the information from materials requirement planning to place purchase orders for raw materials and transmit them to qualified suppliers. Typically, the release of these purchase orders is timed so that raw materials arrive just in time to be used in production and minimize warehouse and storage costs. Often, producers will allow suppliers to tap into data via an extranet that enables them to determine what raw materials the supplier needs thus minimizing the effort and lead time to place and fill purchase orders.

Production uses the detailed schedule to plan the details of running and staffing the production operation.

Financial and Managerial Accounting

The general ledger is the main accounting record of a business. It is often divided into different categories, including assets, liabilities, revenue, expenses, and equity. These categories, in turn, are subdivided into subledgers to capture details such as cash, accounts payable, accounts receivable, and so on. The business processes required to capture and report these accounting details are essential to the operation of any organization and are frequently included within the scope of an organization's ERP system. In the ERP system, input to the general ledger occurs simultaneously with the input of a business transaction to a specific module. Here are several examples of how this occurs:

- An order clerk records a sale and the ERP system automatically creates an accounts receivable entry indicating that a customer owes money for goods received.
- A buyer enters a purchase order and the ERP system automatically creates an accounts payable entry in the general ledger registering that the company has an obligation to pay for goods that will be received at some time in the future.
- A dock worker enters a receipt of purchased materials from a supplier, and the ERP system automatically creates a general ledger entry to increase the value of inventory on hand.
- A production worker withdraws raw materials from inventory to support production, and the ERP system generates a record to reduce the value of inventory on hand.

Thus, the ERP system captures transactions entered by workers in all functional areas of the business. The ERP system then creates the associated general ledger record to track the financial impact of the transaction. This set of records is an extremely valuable resource that companies can use to support financial accounting and managerial accounting.

Financial accounting consists of capturing and recording all the transactions that affect a company's financial state and then using these documented transactions to prepare financial statements to external decision makers, such as stockholders, suppliers, banks, and government agencies. These financial statements include the profit and loss statement, balance sheet, and cash flow statement. They must be prepared in strict accordance to rules and guidelines of agencies such as the Securities and Exchange Commission, the Internal Revenue Service, and the Financial Accounting Standards Board. Data gathered for financial accounting can also form the basis for tax accounting because this involves external reporting of a firm's activities to the local, state, and federal tax agencies.

Managerial accounting involves using "both historical and estimated data in providing information that management uses in conducting daily operations, in planning future operations, and in developing overall business strategies."[38] Managerial accounting provides data to enable the firm's managers to assess the profitability of a given product line or specific product, identify underperforming sales regions, establish budgets, make profit forecasts, and measure the effectiveness of marketing campaigns.

All transactions that affect the financial state of the firm are captured and recorded in the database of the ERP system. This data is used in the financial accounting module of the ERP system to prepare the statements required by various constituencies. The data can also be used in the managerial accounting module of the ERP system along with various assumptions and forecasts to perform various analyses such as generating a forecasted profit and loss statement to assess the firm's future profitability.

Using an ERP with financial and managerial accounting systems can contribute significantly to a company's success. Spartan Foods of America produces and distributes breakfast sandwiches, pancakes, and pizzas to retail grocers and food service suppliers. The firm implemented a complete end-to-end ERP system including accounting modules to provide the vital accounting data it needed to accurately price and manage the inventory of its perishable products. Now the firm can make improved pricing decisions, which have enhanced its profitability and improved earnings.[39]

We will now cover customer relationship management, another form of enterprise system.

Customer Relationship Management

Customer relationship management (CRM) software automates and integrates the functions of sales, marketing, and service in an organization. The objective is to capture data about every contact a company has with a customer through every channel and store it in the CRM system so the company can truly understand customer actions. CRM software helps an organization build a database about its customers that describes relationships in sufficient detail so that management, salespeople, customer service providers—and even customers—can access information to match customer needs with product plans and offerings, remind them of service requirements, and know what other products they have purchased.

The key features of a CRM system include the following:

- *Contact management*: The ability to track data on individual customers and sales leads and access that data from any part of the organization.
- *Sales management*: The ability to organize data about customers and sales leads and then to prioritize the potential sales opportunities and identify appropriate next steps.
- *Customer support*: The ability to support customer service reps so that they can quickly, thoroughly, and appropriately address customer requests and resolve customers' issues while at the same time collecting and storing data about those interactions.
- *Marketing automation*: The ability to capture and analyze all customer interactions, generate appropriate responses, and gather data to create and build effective and efficient marketing campaigns.
- *Analysis*: The ability to analyze customer data to identify ways to increase revenue and decrease costs, identify the firm's "best customers," and determine how to retain them and find even more of them.
- *Social networking*: The ability to create and join groups like Facebook where salespeople can make contacts with potential customers.
- *Access by smartphones*: The ability to access Web-based customer relationship management software by devices such as the BlackBerry or Apple iPhone.
- *Import contact data*: The ability for users to import contact data from various data service providers such as Jigsaw, which offers company-level contact data that can be downloaded for free directly into the CRM application.

Figure 5.10 shows the SAP Contact Manager.

ACME is a flatbed trucking company with a fleet of 1,400 trucks handling some 4,000 loads per week. Data about each shipment for each customer is captured and entered into the CRM to provide a consolidated view of all customer activities. Every employee involved with customers, including sales staff, dispatchers, customer service, and the credit/collections department can use this data to enhance their interactions with customers. For example, before a national accounts manager visits a customer, the manager can obtain up-to-the-minute information about issues that may have occurred at any of the customers' multiple sites served by ACME, and be prepared to address them directly with the customer. The CRM also enables sales managers to track and analyze each phase of the field sales effort and share data with the national account management staff. According to Mike Coatney, president of ACME, "By giving our managers, sales staff, and dispatchers the information they need to address all customer issues, this solution is letting us streamline operational overhead and strengthen customer relationships with outstanding service."[40]

Organizations choose to implement CRM for a variety of reasons depending on their needs. ITSM Academy is an accredited provider of IT Service Management education, instructing companies on how to deliver IT services at reasonable costs. The Academy implemented a CRM to effectively manage its multistep, multiperson process to deliver training by tracking, linking, and coordinating everyone's activities.[41] Sovereign Bank is a large financial institution with $40 billion in assets, 525 offices, and 1,000 ATMs. The bank

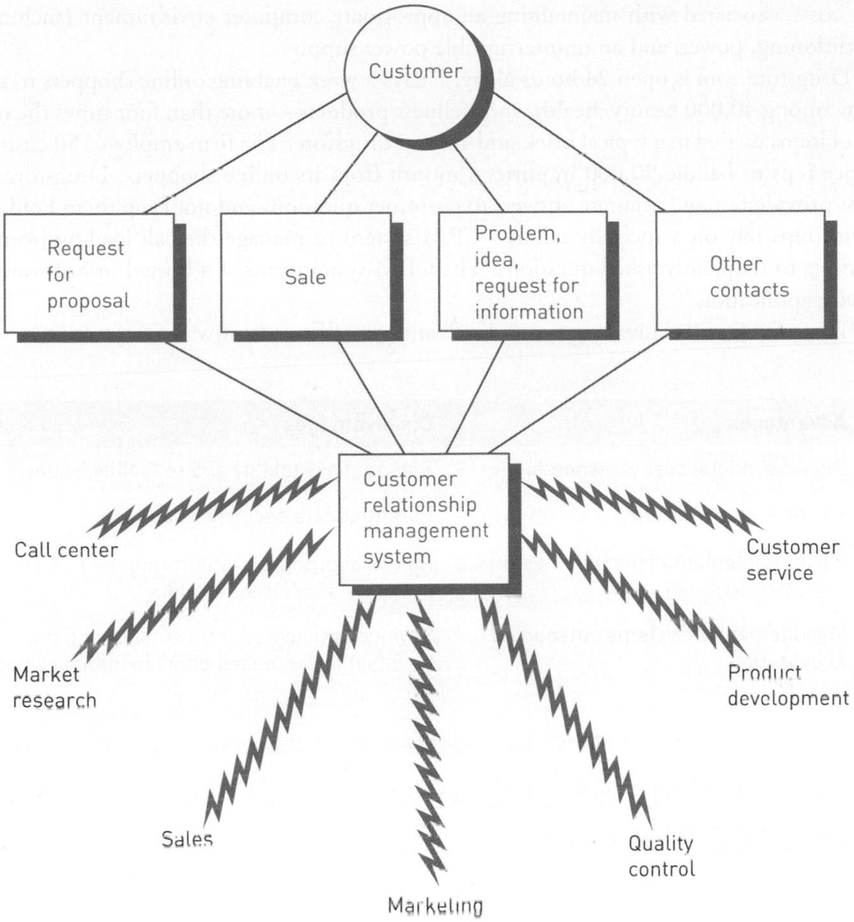

Figure 5.10

SAP Contact Manager

(Source: Copyright © by SAP AG.)

implemented a CRM to standardize and streamline its processes for finding, assigning, and managing new prospects, thus greatly increasing the efficiency of its Relationship Managers.[42] Jubilations Cheesecake makes gourmet cheesecakes for mail-order consumer gifts, business gifts, and fundraising. The firm implemented a CRM system to automate its 29-step process so its salesforce can maintain contact with prospects and customers as well as send and track outbound e-mails that are an essential part of its sales and marketing efforts.[43] Goodwill is a nonprofit organization that "enhances the dignity and quality of life of individuals, families and communities by eliminating barriers to opportunity and helping people in need reach their fullest potential through the power of work."[44] The organization implemented a CRM system to capture and use data about its donors and shoppers for use in cross-selling to customers, soliciting donors, and planning corporate donation drives.[45]

Hosted Software Model for Enterprise Software

Many business application software vendors are pushing the use of the hosted software model for SMEs. The goal is to help customers acquire, use, and benefit from the new technology while avoiding much of the associated complexity and high start-up costs. SAP, Microsoft, NetSuite, Intacct, Oracle, BizAutomation.com, Salesforce.com, NetBooks, and Workday are among the software vendors who offer hosted versions of their ERP or CRM software at a cost of $50-$200 per month per user.

This pay-as-you-go approach is appealing to SMEs because they can experiment with powerful software capabilities without making a major financial investment. Organizations can then dispose of the software without large investments if the software fails to provide value or otherwise misses expectations. Also, using the hosted software model means the small business firm does not need to employ a full-time IT person to maintain key business applications. The small business firm can expect additional savings from reduced hardware costs and costs associated with maintaining an appropriate computer environment (such as air conditioning, power, and an uninterruptible power supply).

Drugstore.com is open 24 hours a day, 7 days a week enabling online shoppers to select from among 40,000 beauty, health, and wellness products—more than four times the number of items carried in a typical brick-and-mortar drugstore. The firm employs 150 customer service reps to handle 90,000 inquiries a month from its online shoppers. Drugstore.com must provide fast and accurate answers to customer questions and not keep them hold. The service reps rely on a specially tailored CRM system to manage the call load and provide prompts to frequently asked questions. The CRM system runs as a hosted or Software as a Service application.[46]

Table 5.7 lists the advantages and disadvantages of hosted software.

Table 5.7

Advantages and Disadvantages of Hosted Software Model

Advantages	Disadvantages
Decreased total cost of ownership	Potential availability and reliability issues
Faster system startup	Potential data security issues
Lower implementation risk	Potential problems integrating the hosted products of different vendors
Management of systems outsourced to experts	Savings anticipated from outsourcing may be offset by increased effort to manage vendor

INTERNATIONAL ISSUES ASSOCIATED WITH ENTERPRISE SYSTEMS

Enterprise systems must support businesses that interoperate with customers, suppliers, business partners, shareholders, and government agencies in multiple countries. Different languages and cultures, disparities in IS infrastructure, varying laws and customs rules, and multiple currencies are among the challenges that must be met by an enterprise system of a multinational company.

SUMMARY

Principle

Electronic and mobile commerce are evolving, providing new ways of conducting business that present both potential benefits and problems.

E-commerce is the conducting of business activities electronically over networks. Business-to-business (B2B) e-commerce allows manufacturers to buy at a low cost worldwide, and it offers enterprises the chance to sell to a global market. B2B e-commerce is currently the largest type of e-commerce. Business-to-consumer (B2C) e-commerce enables organizations to sell directly to consumers, eliminating intermediaries. In many cases, this squeezes costs and inefficiencies out of the supply chain and can lead to higher profits and lower prices for consumers. Consumer-to-consumer (C2C) e-commerce involves consumers selling directly to other consumers. Online auctions are the chief method by which C2C e-commerce is currently conducted. e-Government is the use of information and communications technology to simplify the sharing of information, speed formerly paper-based processes, and improve the relationship between citizens and government.

Mobile commerce is the use of wireless devices such as cell phones and smartphones to facilitate the sale of goods or services—anytime, anywhere. The market for m-commerce in North America is expected to mature much later than in Western Europe and Japan. Although some industry experts predict great growth in this arena, several hurdles must be overcome, including improving the ease of use of wireless devices, addressing the security of wireless transactions, and improving network speed.

Electronic retailing (e-tailing) is the direct sale from a business to consumers through electronic storefronts designed around an electronic catalog and shopping cart model.

A cybermall is a single Web site that offers many products and services at one Internet location.

Manufacturers are joining electronic exchanges, where they can work with competitors and suppliers to use computers and Web sites to buy and sell goods, trade market information, and run back-office operations such as inventory control. They are also using e-commerce to improve the efficiency of the selling process by moving customer queries about product availability and prices online.

The Web allows firms to gather much more information about customer behavior and preferences than they could using other marketing approaches. This new technology has greatly enhanced the practice of market segmentation and enabled companies to establish closer relationships with their customers. Detailed information about a customer's behavior, preferences, needs, and buying patterns allows companies to set prices, negotiate terms, tailor promotions, add product features, and otherwise customize a relationship with a customer.

The Internet has also revolutionized the world of investment and finance, especially online stock trading and online banking.

The Internet has also created many options for electronic auctions, where geographically dispersed buyers and sellers can come together.

M-commerce provides a unique opportunity to establish one-on-one marketing relationships and support communications anytime and anywhere. M-commerce transactions are being used in many application arenas, including mobile banking, mobile price comparison, mobile advertising, and mobile coupons.

Businesses and people use e-commerce to reduce transaction costs, speed the flow of goods and information, improve the level of customer service, and enable the close coordination of actions among manufacturers, suppliers, and customers. E-commerce also enables consumers and companies to gain access to worldwide markets. E-commerce offers great promise for developing countries, helping them to enter the prosperous global marketplace, and hence helping to reduce the gap between rich and poor countries.

Principle

E-commerce and m-commerce require the careful planning and integration of a number of technology infrastructure components.

A number of infrastructure components must be chosen and integrated to support a large volume of transactions with customers, suppliers, and other business partners worldwide. These components include hardware, Web server software, security and identification services, Web site development tools, e-commerce software, and Web services.

The Wireless Application Protocol (WAP) is a standard set of specifications to enable development of m-commerce software for wireless devices. The development of the Wireless Application Protocol (WAP) and its derivatives addresses many m-commerce issues.

Electronic payment systems are a key component of the e-commerce infrastructure. A digital certificate is an attachment to an e-mail message or data embedded in a Web page that verifies the identity of a sender or a Web site. To help prevent the theft of credit card numbers and banking information, the Secure Sockets Layer (SSL) communications protocol is used to secure all sensitive data. There are several electronic cash alternatives that require the purchaser to open an account with an electronic cash service provider and to present proof of identity whenever payments are to be made. Payments can also be made by credit, charge, debit, and smart cards.

Principle

An organization must have information systems that support the routine, day-to-day activities that occur in the normal course of business and help a company add value to its products and services.

Transaction processing systems (TPSs) are at the heart of most information systems in businesses today. ATPS is an organized collection of people, procedures, software, databases, and devices used to capture fundamental data about events that affect the organization (transactions) and use that data to update the official records of the organization.

The methods of transaction processing systems include batch and online. Batch processing involves the collection of transactions into batches, which are entered into the system at regular intervals as a group. Online transaction processing (OLTP) allows transactions to be entered as they occur.

Organizations expect TPSs to accomplish a number of specific objectives, including processing data generated by and about transactions, maintaining a high degree of accuracy and information integrity, compiling accurate and timely reports and documents, increasing labor efficiency, helping provide increased and enhanced service, and building and maintaining customer loyalty. In some situations, an effective TPS can help an organization gain a competitive advantage.

All TPSs perform the following basic activities: data collection, which involves the capture of source data to complete a set of transactions; data editing, which checks for data validity and completeness; data correction, which involves providing feedback of a potential problem and enabling users to change the data; data manipulation, which is the performance of calculations, sorting, categorizing, summarizing, and storing data for further processing; data storage, which involves placing transaction data into one or more databases; and document production, which involves outputting records and reports.

Traditional TPS systems include the following types of systems: order processing, accounting, and purchasing systems.

The traditional TPSs that support the purchasing function include inventory control, purchase order processing, accounts payable, and receiving.

Many software packages provide integrated transaction processing solutions for SMEs.

Principle

A company that implements an enterprise resource planning system is creating a highly integrated set of systems, which can lead to many business benefits.

Enterprise resource planning (ERP) software supports the efficient operation of business processes by integrating activities throughout a business, including sales, marketing, manufacturing, logistics, accounting, and staffing.

Implementation of an ERP system can provide many advantages, including providing access to data for operational decision making; elimination of costly, inflexible legacy systems; providing improved work processes; and creating the opportunity to upgrade technology infrastructure.

Some of the disadvantages associated with ERP systems are that they are expensive and time consuming to implement; they may require the organization to change radically the way it operates; it is difficult to implement them with other systems; there can be difficulty in loading data into them; there are risks associated with using one vendor; and there is a risk of implementation failure.

Many SMEs have found open-source ERP systems to be effective solutions to their transaction processing and management reporting needs.

The production and supply chain management process starts with sales forecasting to develop an estimate of future customer demand. This initial forecast is at a fairly high level with estimates made by product group rather than by each individual product item. The sales and operations plan takes demand and current inventory levels into account and determines the specific product items that need to be produced and when to meet the forecast future demand. Demand management refines the production plan by determining the amount of weekly or daily production needed to meet the demand for individual products. Detailed scheduling uses the production plan defined by the demand management process to develop a detailed production schedule specifying details such as which item to produce first and when production should be switched from one item to another. Materials requirement planning determines the amount and timing for placing raw material orders with suppliers. Purchasing uses the information from materials requirement planning to place purchase orders for raw materials and transmit them to qualified suppliers. Production uses the detailed schedule to plan the details of running and staffing the production operation.

The business processes required to capture and report accounting details are often included within the scope of an organization's ERP system.

Principle

A company that implements a customer relationship management system is building a source of information about customers that can improve sales, marketing, and customer service.

A CRM helps an organization build a database about its customers that describes relationships in sufficient detail so that manage management, salespeople, customer service providers, and even customers can access information to match customer needs.

Business application software vendors are experimenting with the hosted software model to see if the approach meets customer needs and is likely to generate significant revenue.

Principle

There are many potential international issues associated with the operation of enterprise systems.

Numerous complications arise that multinational corporations must address in planning, building, and operating their TPSs. These challenges include dealing with different languages and cultures, disparities in IS infrastructure, varying laws and customs rules, and multiple currencies.

CHAPTER 5: SELF-ASSESSMENT TEST

Electronic and mobile commerce are evolving, providing new ways of conducting business that present both potential benefits and problems.

1. The market for m-commerce in North America is maturing much later than in Western Europe and Japan. True or False?

2. A form of e-commerce in which customers deal directly with an organization and avoid intermediaries is called ____.

3. Which form of e-commerce is the largest in terms of dollar volume?

E-commerce and m-commerce require the careful planning and integration of a number of technology infrastructure components.

4. The practice of _____ divides the pool of potential customers into subgroups, which are usually defined in terms of demographic characteristics.

5. An attachment to an e-mail message or data embedded in a Web site that verifies the identity of a sender or Web site is called a(n) ____.

An organization must have information systems that support the routine, day-to-day activities that occur in the normal course of business and help a company add value to its products and services.

6. Identify the missing TPS basic activity: data collection, data editing, data ____ , data manipulation, data storage, and document production.

7. A form of TPS whereby business transactions are accumulated over a period of time and processed all at once is called ____.

A company that implements an enterprise resource planning system is creating a highly integrated set of systems, which can lead to many business benefits.

8. Which of the following is a primary benefit of implementing an ERP system?
    a. elimination of inefficient systems
    b. easing adoption of improved work processes
    c. improving access to data for operational decision making
    d. all of the above

9. Because it is so critical to the operation of an organization, most companies can implement an ERP system without major difficulty. True or False?

10. Only large, multinational companies can justify the implementation of ERP systems. True or False?

A company that implements a customer relationship management system is building a source of information about customers that can improve sales, marketing, and customer service.

11. The objective of a CRM is to capture data about every _____ a company has with its customers through every channel and store it.

12. CRM systems cannot provide support for the access of data via a BlackBerry or Apple iPhone device. True or False?

There are many potential international issues associated with the operation of enterprise systems.

13. Many multinational companies roll out standard IS applications for all to use. However, standard applications often don't account for all the differences among business partners and employees operating in other parts of the world. Which of the following is a frequent modification that is needed for standard software?
    a. Software might need to be designed with local language interfaces to ensure the successful implementation of a new IS.
    b. Customization might be needed to handle date fields correctly.
    c. Users might also have to implement manual processes and overrides to enable systems to function correctly.
    d. all of the above

CHAPTER 5: SELF-ASSESSMENT TEST ANSWERS

(1) True (2) B2C (3) B2B (4) market segmentation (5) digital certificate (6) correction (7) batch processing (8) d (9) False (10) False (11) contact (12) False (13) d.

REVIEW QUESTIONS

1. Define the term e-Government. Identify three forms of e-Government and give an example of each.
2. Identify and briefly describe three limitations that complicate the use of handheld devices used for m-commerce.
3. What role do digital certificates and certificate authorities play in e-commerce?
4. What is mobile commerce? How big is the mobile commerce market in the U.S.?
5. Briefly explain the differences among smart, credit, charge, and debit cards.
6. Identify the key elements of the technology infrastructure required to successfully implement e-commerce within an organization.
7. What is the Secure Sockets Layer and how does it support e-commerce?
8. What problems can arise when an organization's TPS systems are not integrated?
9. An ERP system follows a systematic process for developing a production plan that draws on the information available in the ERP system database. Outline this process and identify the software modules that are used to support it.
10. Identify four complications that multinational corporations must address in planning, building, and operating their ERP systems.
11. What are the business processes included within the scope of supply chain management?
12. What is the role of a CRM system? What sort of business benefits can such a system produce?
13. Identify three characteristics of integrated processing software package solutions that make them attractive to SMEs.
14. What is the difference between managerial and financial accounting?

DISCUSSION QUESTIONS

1. You are a member of the organization's finance organization. The firm is considering the implementation of an ERP system. Make a convincing argument for finance and accounting to be included within the scope of the ERP implementation.
2. What do you think are the biggest barriers to wide-scale adoption of m-commerce by consumers? Who do you think is working on solutions to these problems and what might the solutions entail?
3. Identify and briefly describe three m-commerce applications you have used.
4. Discuss the use of e-commerce to improve spending on manufacturing, repair, and operations (MRO) of goods and services.
5. Identify three kinds of business organizations that would have difficulty in becoming a successful e-commerce organization.
6. Assume that you are the owner of a landscaping firm serving hundreds of customers in your area. Identify the kinds of customer information you would like your firm's CRM system to capture. How might this information be used to provide better service or increase revenue? Identify where or how you might capture this data.
7. Do you think that an SME would have less or more difficulty implementing an ERP system than a large, multinational corporation? Defend your position.
8. Your friend has been appointed the project manager of your firm's ERP implementation system. What advice would you offer to help ensure the success of the project?
9. What sort of benefits should the suppliers and customers of a firm that has successfully implemented an ERP system see? What sort of issues might arise for suppliers and customers during an ERP implementation?
10. Many organizations are moving to a collaborative process with their major customers to get their input on planning future inventory levels and production rather than relying on an internally generated demand forecast. Explain how such a process might work. What issues and concerns might a customer have in entering into an agreement to do this?

PROBLEM-SOLVING EXERCISES

1. Imagine that you are the new IS manager for a *Fortune* 1000 company. Surprisingly, the firm still operates with a hodgepodge of transaction processing systems—some are software packages from various vendors and some are in-house developed systems. Use a graphics package (e.g. PowerPoint) to prepare a slide presentation you will make to

senior company managers to convince them that it is time to implement a comprehensive ERP system. What sort of resistance and objections do you expect to encounter? How would you overcome these? Include appropriate slides to cover this as well.

2. Research the growth of B2B and B2C e-commerce and retail sales for the period 2000 to present. Use the graphics capability of your spreadsheet software to plot the growth of all three. Using current growth rates, predict the year that B2C e-commerce will exceed 10 percent of retail sales.

3. Your washing machine just gave out and must be replaced within the week! Use your Web-enabled smartphone (or borrow a friend's) to perform a price and product comparison to identify the manufacturer and model that best meets your needs and the retailer with the lowest delivered cost. Obtain peer input to validate your choice. Write a brief summary of your experience and identify the Web sites you found most useful.

TEAM ACTIVITIES

1. Imagine that your team has been hired as consultants to provide recommendations to boost the traffic to a Web site that sells environmentally friendly ("green") household cleaning products. Identify as many ideas as possible for how you can increase traffic to this Web site. Next, rank your ideas from best to worst.

2. Your team members should interview several business managers at a firm that has implemented a CRM system. Try to define the scope and a schedule for the major tasks of the overall project. Make a list of what they see as the primary benefits of the implementation. What were the biggest hurdles they had to overcome? Did the firm need to retrain its employees to place greater emphasis on putting the customer first?

WEB EXERCISES

1. Do research on the Web to find out more about the American Consumer Satisfaction Index methodology developed by the University of Michigan. Write a brief report about this methodology and explain how it was applied to rate B2C Web sites. Using this information, develop a list of at least six recommendations for someone developing a B2C Web site.

2. Do research on the Web and find a Web site that offers a demo of an ERP or CRM system. View the demo, perhaps more than once. Write a review of the software based on the demo. What are its strengths and weaknesses? What additional questions about the software do you have? E-mail your questions to the vendor and document their response to your questions.

CAREER EXERCISES

1. For your chosen career field, describe how you might use or be involved with e-commerce. If you have not chosen a career yet, answer this question for someone in marketing, finance, or human resources.

2. Imagine that you are a prescription drug salesperson for a large pharmaceutical firm and that you make frequent sales calls on physicians and other primary care people in doctor's offices. The purpose of these sales calls is to acquaint them with your firm's products and get them to begin prescribing your products to their patients. Describe the basic functionality you would want in your organization's CRM system for it to support you in preparing and making presentations to these people.

CASE STUDIES

Case One

Wrangler Sells Direct Online

The Wrangler Jeans story begins in 1904, when C.C. Hudson bought several sewing machines from his previous employer, leased space over a grocery store, and started the Hudson Overall Company. Fifteen years later, the name of the business was changed to Blue Bell Overall Company. Over decades of growth and acquisitions, Blue Bell become Wrangler and expanded to produce and sell western style clothing for men, women, and children around the world. Today, one in every five pairs of jeans sold globally is made by Wrangler.

In 2009, Wrangler decided to open its first B2C e-commerce site. Wrangler had several goals for the new site. It needed to provide visitors with a view into the spirit behind the Wrangler brand. Wrangler wanted to use the site to feature and promote various products as the market dictated. They also wanted the site to use the latest technologies to show that Wrangler is tech savvy and to empower Wrangler's marketing department with new ways to present its products.

One Wrangler slogan is "Enduring American freedom; it's in the spirit of people who work hard, have fun, and recognize courageous individuality." Wrangler's Web site design expresses this concept with large, bold photos of down-home American scenes featuring country music and rodeo stars alongside hard-working individuals and people just hanging out.

Wrangler's new site uses background image fade-ins with rotating product image overlays that allow Wrangler to feature different products as needed. Behind the scenes, the site is delivered in a Flash version and a pure HTML version, both fully integrated to support all browser requirements. Although this approach requires additional development time, it allows the site to "garner better natural search rankings."

Wrangler included additional database functionality to provide different views of products. For example, a visitor can click Jeans to go directly to the Jeans page or browse through product collections listed under themes that include Rodeo & Riding, Workwear & Safety, Music & Dancing, and Hanging Out. Each theme opens with a large splash screen that Wrangler Marketing uses to promote its brand in the selected environment. For example, clicking Music & Dancing opens a large photo of country music legend George Strait, with a paragraph on Wrangler's importance to those involved in country music and dancing.

The site uses the latest technologies to provide interesting page and page element presentations and site navigation. The latest technologies are also used on the back end to make sure that pages are loaded and refreshed extremely quickly. It features a single-page checkout system that makes the checkout process quick and efficient. It also features a powerful search utility that provides an alternative to the traditional drill-down approach to navigation.

Wrangler's new site took four months to develop and was completed on time and on budget. Wrangler has benefited from additional sales provided by its new B2C e-commerce site. Equally as valuable is the insight the company gains about customer interests and the platform it provides for Wrangler marketing to experiment with product and brand promotions.

Discussion Questions

1. How does Wrangler's new B2C e-commerce site assist Wrangler's brand recognition and marketing efforts?
2. What goals did Wrangler set for its e-commerce site? Visit *www.wrangler.com*. Do you think its new site meets those goals?

Critical Thinking Questions

1. Wrangler targets a very specific type of person with its marketing and Web site. How would you describe that group? What risks and benefits do companies assume when they target specific types of individuals? Do you think it pays off for Wrangler? Why?
2. The Wrangler site incorporates a lot of dynamic visual elements. How do these elements affect a shopper? What types of products are best suited for this type of marketing approach?

SOURCES: Todé, Chantel, "Wrangler launches first e-commerce portal," DMNews, September 10, 2009, *www.dmnews.com/wrangler-launches-first-e-commerce-portal/article/148482*; Zobrist Staff, "Wrangler," Zobrist Success Story, *www.zobristinc.com/our_clients/success_stories/Wrangler*, accessed April 10, 2010; Wrangler Web site, *www.wrangler.com*, accessed April 10, 2010.

Case Two

Dubai Bank Improves Customer Satisfaction with CRM

Dubai Bank is one of the top Islamic banks based in Dubai, United Arab Emirates, with 25 branches across the UAE and total assets of AED 14.4 billion. Banking is a highly competitive business in Dubai, a city of glass-and-steel skyscrapers and state-of-the-art massive engineering projects, rooted in oil money, and growing in leaps and bounds. Banks work hard to win over customers with lavish lobbies, financial incentives, and impeccable customer service.

As Dubai Bank grew over time, adding services and customers, it became apparent that the bank needed a way to easily gather customer information. Dubai Bank customers often had data spread across three separate databases—one for account information, another for credit card information, and yet another for investments and loans. If customers held multiple bank accounts, separate database records were created and even more information was duplicated.

The complexity of customer information systems caused Dubai Bank agents frustration in finding information and setting up new accounts. Even worse was the aggravation it caused valuable customers. If a customer phoned customer service with a question, the agent might have had to access eight systems to collect customer information, leaving the customer waiting on the line. If the customer followed up the next day with a visit to the bank, the teller would have had no knowledge of the previous discussion with customer service, further aggravating the customer.

When customers set up a new account, agents were required to fill out multiple applications. Credit checks needed to be performed manually. With this level of customer service, Dubai Bank was having a difficult time keeping customers.

Faizal Eledath, chief information officer at Dubai Bank, recognized the need for an enterprise-wide customer relationship management (CRM) system. He sought a system that would integrate all customer records into one cohesive system, providing agents and managers detailed information about each customer through a single interface.

Dubai Bank is recognized as an Islamic institution that strictly adheres to the Shari'a principles—that is, the sacred law of Islam. These principles include conducting business with the highest level of transparency, integrity, fairness, respect, and care. Dubai Bank's information systems would need to support its Islamic principles.

Dubai Bank hired information systems consultant Veripark to assist in building the ideal CRM for the bank. Dubai made its choice based on positive past experiences it shared with Veripark and the Microsoft products it represents. Veripark selected Microsoft Dynamics CRM package that integrates all critical operational banking systems, including credit cards, data warehouse, wealth management, and risk systems, into the CRM.

The new CRM provides bank representatives with a 360 degree view of the customer, whereby information can be entered and accessed through a single interface. Veripark customized the package to comply with Islamic banking regulations. Business process automation is programmed into the system to further assist the bank in its strict adherence to Shari'a principles.

The new CRM system records information from all customer interactions with the bank. If a customer makes a withdrawal from an ATM, it is recorded in the CRM. When a customer phones customer service, notes are recorded in the system as well. Whenever a customer has a need, a bank representative can easily review the customer's history and quickly recommend a course of action.

Since the installation of its new CRM system, both customers and bank agents are much happier. Customer service agents can now provide speedy service because information is provided in one interface rather than eight. New accounts are created in a quarter of the time previously

required. Compliance with Islamic banking requirements is assured and automated, without involving additional effort on the part of bank officials or agents. Most importantly, Dubai Bank knows and understands its customers more deeply and can use that information to provide services and implement programs that increase customer satisfaction.

Discussion Questions

1. What conditions brought Dubai Bank to the realization that it could benefit from a CRM system?
2. How did the CRM system make life easier for Dubai Bank agents and customers?

Critical Thinking Questions

1. Dubai Bank had regulation imposed on it by the Islamic faith that affected its information systems. What types of regulations are imposed on U.S. banks that have a similar impact?
2. How can the information collected by a CRM system be used to gain insight and boost a business' profits? Provide some examples.

SOURCES: Microsoft Staff, "Dubai Bank," Microsoft Case Studies, March 26, 2010, *www.microsoft.com/casestudies/Case_Study_Detail.aspx?casestudyid=4000006766*; Dubai Bank Web site, *www.dubaibank.ae*, accessed May 2, 2010.

Questions for Web Cases

See the Web site for this book to read about the Altitude Online case for this chapter. The following are questions concerning this Web case.

Altitude Online: E-Commerce Considerations

Discussion Questions

1. How does Altitude Online's Web site contribute to the company's commerce?
2. How will the new ERP system impact Altitude Online's Web presence?

Critical Thinking Questions

1. How can companies like Altitude Online, which sell services rather than physical products, use e-commerce to attract customers and streamline operations?
2. Consider a company like Fluid by reviewing its site: *www.fluid.com*. Fluid is similar to Altitude Online in the services it offers. What site features do you think are effective for e-commerce? How might you design the site differently?

Altitude Online: Enterprise System Considerations

Discussion Questions

1. Judging from the ERP features, how important is an ERP to the functioning of a business? Explain.
2. What consideration do you think led Altitude Online to decide to host the ERP on its own servers rather than using SaaS? What are the benefits and drawbacks of both approaches?

Critical Thinking Questions

1. What challenges lay ahead for Altitude Online as it rolls out its new ERP system?
2. How might the ERP affect Altitude Online's future growth and success?

NOTES

Sources for the opening vignette: ATG Staff, "ShopTommy.com Dresses for Online Success – OnDemand," ATG Case Studies, *www.atg.com/resource-library/case-studies/CS-Tommy-Hilfiger.pdf*, accessed March 31, 2010; Tommy Hilfiger Web site, *www.tommy.com*, accessed March 31, 2010; ATG Web site, I, accessed March 31, 2010.

1 Thibodeau, Patrick, "Study: Internet Economy Has Created 1.2 Million Jobs," *Computerworld*, June 10, 2009.
2 B2B Computer Products Web site, *www.b2bcomp.com*. accessed March 7, 2010.
3 O'Reilly, Joseph, "P&G Explores B2C," *Inbound Logistics*, February 10, 2010, *www.inboundlogistics.com/articles/trends/trends0210.shtml*, accessed March 6, 2010.
4 "Etsy Crafts a Recession Success," *eMarketer Digital Intelligence*, May 1, 2009, *www.emarketer.com/Article.aspx?R=1007066*, accessed March 13, 2010.
5 Mitchell, Robert L., "Recovery.com Relaunch Puts Stimulus Spending on the Map," *Computerworld*, September 28, 2009.
6 "Better Service for Those Who Need It Most," *Government Technology*, www.govtech.com/gt/case_study/736529January 2010.
7 *http://georgewbush-whitehouse.archives.gov/omb/egov/c-2-4-evital.html*, accessed April 30, 2010.
8 *http://gos2.geodata.gov/wps/portal/gov*, accessed April 20, 2010.
9 Shah, Agam, "IDC: 1 Billion Mobile Devices Will Go Online by 2013," *Computerworld*, December 9, 2009.
10 "2010 Top 10 Mobile Shopping Sites," *www.catalogs.com/blog*, accessed March 24, 2010.
11 Mobile Mammoth Web site, *www.mobilemammoth.com*, accessed March 24, 2010.
12 Burns, Enid, "Study: Consumer Satisfaction in E-Commerce Slips," *ClickZ*, May 11, 2009, *www.clickz.com/3633686*, accessed March 14, 2010.
13 Avendra, LLC Profile, Hoovers, *www.hoovers.com/company/Avendra_Llc/ryxskci-1.html*, accessed March 26, 2010.
14 Mancini, Michael, "B2B Discovers Market Segmentation," *NielsenWire*, November 2, 2009, *blog.nielsen.com/nielsenwire/consumer/b2b-discovers-market-segmentation*, accessed March 3, 2010.
15 "mobiThinking Guide to Mobile Advertising Networks (2010)," *mobilethinking.com/mobile-ad-network-guide*, accessed March 29, 2010.
16 Weintraub, Seth, "Google to Acquire AdMob for $750 Million in Stock," *Computerworld*, November 9, 2009.
17 "Motorola 2009 Retail Holiday Season Shopper Study," January 11, 2010, *http://mediacenter.motorola.com/imagelibrary/detail.aspx?MediaDetailsID=861*, accessed March 29, 2010.
18 Purdy, Kevin, "Google Product Search Goes Mobile," April 24, 2009, *http://lifehacker.com/5225858/google-product-search-goes-mobile*, accessed March 27, 2010.
19 Frucall Web site, *www.frucall.com*, accessed March 27, 2010.
20 Wortham, Jenna, "Coupons You Don't Clip, Sent to Your Cellphone," *The New York Times*, August 29, 2009.
21 "Target Launches First Ever Scannable Mobile Coupon Program, *SuperMarket Industry News*, March 11, 2010.
22 "Valpak Mobile Coupons 2010," March 19, 2010, *webfloss.com/Valpak-mobile-coupons-2010*, accessed March 28, 2010.
23 "Mobile Applications," Bank of America Web site, *www.bankofamerica.com/onlinebanking/index.cfm?template=mobile_banking*, accessed March 27, 2010.
24 Muir, Courtney, "Gilt Groupe iPad App Comprimses 2 Percent of Sales in First 3 Days," *Mobile Commerce Daily*, April 12, 2010.
25 Miller, Claire Cain, "On the iPad, Gilt is for All Kinds of Shoppers," *The New York Times*, April 2, 2010.
26 McDonough, Michele, "Microsoft Internet Information Services Basics," Bright Hub Web site, August 22, 2009, *www.brighthub.com/computing/hardware/articles/16160.aspx*, accessed March 26, 2010.
27 "About Us," PayPal Web site, *https://www.paypal-media.com/aboutus.cfm,* accessed March 21, 2010.
28 "The Dragon Hotel Uses Smart Card Technology to Deliver a World-Class, Personalized Experience, IBM Web site, *www-01.ibm.com/software/success/cssdb.nsf/CS/LMCM-7XBP7B?OpenDocument&Site=corp&cty=en_us*, accessed March 22, 2010.
29 Miller, Claire Cain and Richtel, Matt, "Investors Bet on Payments via Cellphone," *New York Times*, June 22, 2009.
30 MasterCard MoneySend, *www.mastercard.com/in/personal/en/moneysend/index.html*, accessed March 22, 2010.
31 "Newcastle Permanent Accelerates Transaction Processing," *www.integra-pc.com/io.nsf/html/WEBB7ZLSXA/$FILE/ATTPEPIX.pdf*, accessed May 15, 2010.
32 McDougall, Paul, "Informatica Launches Healthcare Claims System," *Information Week*, October 26, 2009.
33 "Trimac Corporation," *www.sybase.com/detail?id=1033799*, accessed April 5, 2010.
34 "Customer Success Stories: Plastic Manufacturer Sets Higher Standards of Performance with Rapid ERP Implementation of ERP from IQMS," *www.iqms.com/company/flambeau/index.html*, accessed April 9, 2010.
35 "Marin Municipal Water Optimizing Assets Through Integration and Visualization," *http://204.154.71.138/pdf/MarinWaterSuccess.pdf*, accessed April 9, 2010.
36 "Achmea Transforms Financial Reporting with SAP and IBM," March 11, 2010 *www-01.ibm.com/software/success/cssdb.nsf/CS/STRD-83ELTK?OpenDocument&Site=corp&cty=en_us*, accessed April 9, 2010.
37 "Barloworld Handling UK Driving Optimal Performance with Mobile Technology," *www.google.com/search?q=Barloworld+Handling+UK+Driving+Optimal+Performance+&rls=com.microsoft:en-us:IE-SearchBox&ie=UTF-8&oe=UTF-8&sourceid=ie7&rlz=1I7ADBF_en*, accessed April 9, 2010.

38 Glossary of terms at *www.crfonline.org/oc/glossary/m.html*, accessed April 24, 2010.

39 Success Stories Spartan Foods of America *www.sagenorthamerica.com/company/success_stories/details?*, accessed April 25, 2010.

40 "ACME Truck Line Selects SageCRM to Drive Its Customer Service Network," SageCRM Customer Success Stories, *www.sagecrmsolutions.com/company/newsroom/successstories*, accessed March 3, 2010.

41 "Success Stories: ITSM Academy," *www.salesnet.com/customer-success-stories-itsm.aspx*, accessed April 23, 2010.

42 "Success Stories: Sovereign Bank," *www.salesnet.com/customer-success-stories-itsm.aspx*, accessed April 23, 2010.

43 "Success Stories: Jubilations Cheesecake," *www.salesnet.com/customer-success-stories-itsm.aspx*, accessed April 23, 2010.

44 "Goodwill: Our Mission," *www.goodwill.org/about-us/our-mission*, accessed April 23, 2010.

45 "Success Stories: Goodwill," *www.salesnet.com/customer-success-stories-itsm.aspx*, accessed April 23, 2010.

46 "drugstore.com," *www.rightnow.com/customers-drugstore.php*, accessed April 24, 2010.

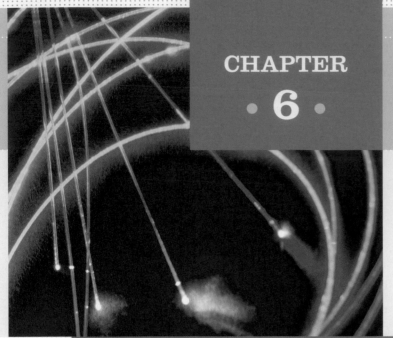

CHAPTER · 6 ·

Information and Decision Support Systems

PRINCIPLES

- Good decision-making and problem-solving skills are the key to developing effective information and decision support systems.

- The management information system (MIS) must provide the right information to the right person in the right format at the right time.

- Decision support systems (DSSs) are used when the problems are unstructured.

- Specialized support systems, such as group support systems (GSSs) and executive support systems (ESSs), use the overall approach of a DSS in situations such as group and executive decision making.

LEARNING OBJECTIVES

- Define the stages of decision making.
- Discuss the importance of implementation and monitoring in problem solving.

- Explain the uses of MISs and describe their inputs and outputs.
- Discuss information systems in the functional areas of business organizations.

- List and discuss important characteristics of DSSs that give them the potential to be effective management support tools.
- Identify and describe the basic components of a DSS.

- State the goals of a GSS and identify the characteristics that distinguish it from a DSS.
- Identify the fundamental uses of an ESS and list the characteristics of such a system.
- List and discuss other special-purpose systems.

Information Systems in the Global Economy
Tru-Test, New Zealand

MIS Reports Guide Decision Makers at Tru-Test

Tru-Test, the world's leading manufacturer of livestock scales, milk-metering equipment, and electric fencing, specializes in livestock and pasture management. Based on extensive research and development, Tru-Test has created technologies that allow their products to stand head and horns above the competition. Tru-Test livestock scales have a patented algorithm that makes it so livestock can be weighed while they're moving. This makes it possible for ranchers to weigh 500 sheep per hour or 150 cattle per hour—far more than with traditional scales. Tru-Test milk meters use an advanced technology and design for precise and accurate milk sampling and testing during the milking process. The company's electric fencing uses a patented Cyclic Wave technology that supports a safer and more economic use of electricity from AC, DC, or solar sources. Like business information systems, Tru-Test products allow ranchers to work more efficiently and maximize the return on their livestock investment.

Tru-Test is headquartered in Auckland, New Zealand, with offices and operations in Australia, the United States, and Mexico. Tru-Test has more than 400 staff worldwide and customers spread across 70 countries. During its lifetime, Tru-Test has expanded and diversified into four business units that house eight brands. The business units are Livestock Management, Pasture Management and Animal Containment, Shearing Equipment, and Contract Manufacturing. As with all successful businesses, Tru-Test has experienced growth pains trying to keep up with its ever-expanding amount of information. Recently Tru-Test took steps to gain control of its information and to leverage it to support wise strategic business planning.

Through a period of sustained growth and multiple acquisitions, Tru-Test found itself supporting nine separate information systems. It was clear that the first step towards information management was to merge those nine systems into one central system. Tru-Test chose an ERP system from JD Edwards to manage finances, procurement, customer service, warehousing, and manufacturing. The ERP system is housed on servers at Tru-Test headquarters, and limited access is made available to Tru-Test subsidiaries around the world through virtual desktop software from Citrix.

The impact of the new ERP system was immediately felt across Tru-Test. CFO Ian Hadwin states that "the standardization of business processes across the group on a single centralized architecture is easier to manage." The new system performs basic financial tasks much more quickly and efficiently than the previous system. This allows employees to be more productive while operating costs decrease. But the largest benefit is in the power that the new system provides to corporate decision makers.

The management information systems that make up the ERP provide Tru-Test decision makers with a variety of reports that contain detailed information about the state of the company. Scheduled reports on sales and inventory provide production and warehouse managers with the information they need to make sure sales are fulfilled in a timely manner. Tru-Test can now accurately forecast demand for its products and implement a just-in-time production model, reducing its inventory to precisely match demand. Eliminating surplus stock means an increase in profits for the company.

Tru-Test extended its demand planning to all areas of its business using Demantra, specialized software from Oracle. Demantra is a group support system designed to support demand planning across an enterprise. Using enterprise data from the ERP and models designed to represent various business activities, Demantra assists managers in sales, operations, manufacturing, marketing, and other business areas with planning for the future.

Tru-Test's database of customers and customer interactions supports a powerful CRM system. Reports from the CRM system provide sales information that allows decision makers in marketing to launch more effective marketing campaigns. The finance department has designed reports that allow for reliable revenue forecasting.

Reports providing Tru-Test top management with important corporate facts, called key indicators, support high-level decisions that steer the company toward meeting its goals. Gaining control of its information, collecting it into a centralized database, and mining it for valuable information depend on Tru-Test's ability to meet and exceed its goals to maximize profits with a minimum investment.

As you read this chapter, consider the following:

- How is an MIS used in the various functional areas of a business?
- How do an MIS and DSS affect a company's business practices and its ability to compete in a market?

Why Learn About Information and Decision Support Systems?

You have seen throughout this book how information systems can make you more efficient and effective through the use of database systems, the Internet, e-commerce, enterprise systems, and many other technologies. The true potential of information systems, however, is in helping you and your coworkers make more informed decisions. This chapter shows you how to slash costs, increase profits, and uncover new opportunities for your company using management information and decision support systems. Transportation coordinators can use management information reports to find the least expensive way to ship products to market and to solve bottlenecks. A loan committee at a bank or credit union can use a group support system to help them determine who should receive loans. Store managers can use decision support systems to help them decide what and how much inventory to order to meet customer needs and increase profits. An entrepreneur who owns and operates a temporary storage company can use vacancy reports to help determine what price to charge for new storage units. Everyone can be a better problem solver and decision maker. This chapter shows you how information systems can help. It begins with an overview of decision making and problem solving.

As shown in the opening vignette, information and decision support are the lifeblood of today's organizations. Thanks to information and decision support systems, managers and employees can obtain useful information in real time. As discussed in Chapter 5, TPS and ERP systems capture a wealth of data. When this data is filtered and manipulated, it can provide powerful support for managers and employees. The ultimate goal of management information and decision support systems is to help managers and executives at all levels make better decisions and solve important problems. The result can be increased revenues, reduced costs, and the realization of corporate goals. Many companies, for example, are using Internet video sharing and social networking sites such as YouTube, MySpace, and Facebook to reduce advertising costs.[1] No matter what type of information and decision support system you use, its primary goal should be to help you and others become better decision makers and problem solvers.

DECISION MAKING AND PROBLEM SOLVING

In most cases, strategic planning and the overall goals of the organization set the course for decision making, helping employees and business units achieve their objectives and goals. Often, information systems also assist with problem solving, helping people make better decisions and save lives. For example, an information system at Hackensack University Medical Center (*www.humed.com*) in New Jersey analyzes possible drug interactions. In one case, an AIDS patient taking drugs for depression was advised not to take a therapeutic AIDS medication that could have dangerously interacted with the depression medication.

Decision Making as a Component of Problem Solving

In business, one of the highest compliments you can receive is to be recognized by your colleagues and peers as a "real problem solver." Problem solving is a critical activity for any business organization. After identifying a problem, the process of solving the problem begins with decision making. A well-known model developed by Herbert Simon divides the **decision-making phase** of the problem-solving process into three stages: intelligence, design, and choice. This model was later incorporated by George Huber into an expanded model of the entire problem-solving process. See Figure 6.1.

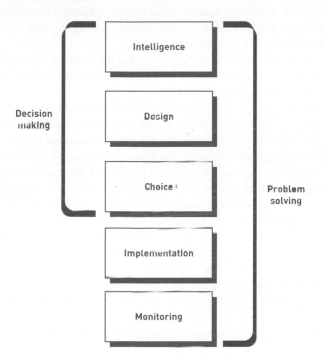

decision-making phase
The first part of problem solving, including three stages: intelligence, design, and choice.

How Decision Making Relates to Problem Solving

The three stages of decision making—intelligence, design, and choice—are augmented by implementation and monitoring to result in problem solving.

The first stage in the problem-solving process is the **intelligence stage**. During this stage, you identify and define potential problems or opportunities. You also investigate resource and environmental constraints. For example, if you were a Hawaiian farmer, during the intelligence stage, you would explore the possibilities of shipping tropical fruit from your farm in Hawaii to stores in Michigan. The perishability of the fruit and the maximum price that consumers in Michigan are willing to pay for the fruit are problem constraints. As another example, researchers are gathering information during the intelligence stage about a new encryption approach to protect sensitive data that was introduced in a Ph.D. dissertation.[2] The new approach would allow calculations on encrypted data instead of removing the encryption to make the calculations, which could make the data vulnerable to theft or privacy concerns. According to one researcher, "It's like one of those boxes with the gloves that are used to handle toxic chemicals. All the manipulation happens inside the box." If successful, this new approach could have a dramatic impact on data security techniques.

In the **design stage**, you develop alternative solutions to the problem and evaluate their feasibility. In the tropical fruit example, you would consider the alternative methods of shipment, including the transportation times and costs associated with each. During this stage, you might determine that shipment by freighter to California and then by truck to Michigan is not feasible because the fruit would spoil. In another example, many hedge funds and investment firms are developing high-frequency trading systems during the design stage that increase speed and potential profitability.[3] High-frequency trading, also called flash trading, allows individuals and companies to make money by placing trades on very fast computers

intelligence stage
The first stage of decision making, in which you identify and define potential problems or opportunities.

design stage
The second stage of decision making, in which you develop alternative solutions to the problem and evaluate their feasibility.

and high-speed telecommunications lines whenever prices fluctuate slightly during the trading day.[4] These profits would not be available using normal computers and trading approaches. Some regulatory agencies, such as the Securities and Exchange Commission (SEC), are considering limiting or banning some types of high-frequency trading because it may give some large, sophisticated firms a trading and profitability advantage.[5]

The last stage of the decision-making phase, the **choice stage**, requires selecting a course of action. In the tropical fruit example, you might select the method of shipping fruit by air from your Hawaiian farm to Michigan as the solution. The choice stage would then conclude with selection of an air carrier. As you will see later, various factors influence choice; the act of choosing is not as simple as it might first appear.

choice stage

The third stage of decision making, which requires selecting a course of action.

Problem solving includes and goes beyond decision making. It also includes the **implementation stage**, when the solution is put into effect. For example, if your decision is to ship tropical fruit to Michigan as air freight using a specific carrier, implementation involves informing your field staff of the new activity, getting the fruit to the airport, and actually shipping the product to Michigan.

problem solving

A process that goes beyond decision making to include the implementation stage.

The final stage of the problem-solving process is the **monitoring stage**. In this stage, decision makers evaluate the implementation to determine whether the anticipated results were achieved and to modify the process in light of new information. Monitoring can involve feedback and adjustment. For example, after the first shipment of fruit from Hawaii to Michigan, you might learn that the flight of your chosen air freight firm routinely stops in Phoenix, Arizona, where the plane sits on the runway for a number of hours while loading additional cargo. If this unforeseen fluctuation in temperature and humidity adversely affects the fruit, you might have to readjust your solution to include a new carrier that does not make such a stop, or perhaps you would consider a change in fruit packaging. In another example, a children's hospital in Pittsburgh monitored the number of deaths and decided it could improve its information system and reduce the death rate for its patients.[6] With the old system, one doctor might have to treat a child, while another doctor could spend 10 minutes or more trying to order needed drugs. As a result of these inefficiencies, the hospital decided to develop a new system that streamlined all operations, including ordering critical drugs. The new information system helped reduce death rates by more than 30 percent. In another case, a popular GPS company monitored the performance of one of its marine navigation systems and determined that the GPS device incorrectly computed water depths.[7] As a result of the monitoring stage, the company issued a worldwide recall of all affected devices.

implementation stage

A stage of problem solving in which a solution is put into effect.

monitoring stage

The final stage of the problem-solving process, in which decision makers evaluate the implementation.

Programmed versus Nonprogrammed Decisions

In the choice stage, various factors influence the decision maker's selection of a solution. One factor is whether the decision can be programmed. **Programmed decisions** are made using a rule, procedure, or quantitative method. For example, to say that inventory should be ordered when inventory levels drop to 100 units is a programmed decision because it adheres to a rule. Programmed decisions are easy to computerize using traditional information systems. For example, you can easily program a computer to order more inventory when levels for a certain item reach 100 units or less. Cisco Systems, a large computer equipment and server manufacturing company, controls its inventory and production levels using programmed decisions embedded into its computer systems.[8] The programmed decision-making process has improved forecasting accuracy and reduced the possibility of manufacturing the wrong types of inventory, which has saved money and preserved cash reserves. Management information systems can also reach programmed decisions by providing reports on problems that are routine and in which the relationships are well defined (in other words, they are structured problems).

programmed decision

A decision made using a rule, procedure, or quantitative method.

Nonprogrammed decisions deal with unusual or exceptional situations. In many cases, these decisions are difficult to quantify. Determining the appropriate training program for a new employee, deciding whether to develop a new type of product line, and weighing the benefits and drawbacks of installing an upgraded pollution control system are examples. Each of these decisions contains unique characteristics, and standard rules or procedures might not apply to them. Today, decision support systems help solve many nonprogrammed decisions, in which the problem is not routine and rules and relationships are not well defined

nonprogrammed decision

A decision that deals with unusual or exceptional situations.

(unstructured or ill-structured problems). These problems can include deciding the best location for a manufacturing plant or whether to rebuild a hospital that was severely damaged from a hurricane or tornado.

Optimization, Satisficing, and Heuristic Approaches

In general, computerized decision support systems can either optimize or satisfice. An **optimization model** finds the best solution, usually the one that will best help the organization meet its goals.[9] For example, an optimization model can find the best route to ship products to markets, given certain conditions and assumptions. StatoilHydro, a Norwegian oil and natural gas organization, used optimization software called GassOpt that not only helped the organization minimize shipping costs for natural gas, but also helped executives determine new shipping routes.[10] The optimization application has saved the organization, which is partially owned by the country of Norway, about $2 billion since its development. Laps Care is an information system that used optimization to assign medical personnel to home healthcare patients in Sweden while minimizing healthcare costs. The optimization system has improved the quality of medical care delivered to the elderly.[11] The system has also improved healthcare efficiency by 10 to 15 percent and lowered costs by more than 20 million euros. In another case, Xerox has developed an optimization system to increase printer productivity and reduce costs, called Lean Document Production (LDP) solutions.[12] The optimization routine was able to reduce labor costs by 20 to 40 percent in some cases. Marriott International used optimization to help determine the optimal price of a group or block of rooms. The optimization routine helped Marriott increase revenues and profits as a result.[13]

Optimization models use problem constraints. A limit on the number of available work hours in a manufacturing facility is an example of a problem constraint. Shermag, a Canadian furniture manufacturing company, used an optimization program to reduce raw materials costs, including wood, in its manufacturing operations.[14] The optimization program, which used the C++ programming language and CPLEX optimization software, helped the company reduce total costs by more than 20 percent. Some spreadsheet programs, such as Excel, have optimizing features, as shown in Figure 6.2. Optimization software also allows decision makers to explore various alternatives.[15]

A **satisficing model** is one that finds a good—but not necessarily the best—solution to a problem. Satisficing is used when modeling the problem properly to get an optimal decision would be too difficult, complex, or costly. Satisficing normally does not look at all possible solutions but only at those likely to give good results. Consider a decision to select a location

optimization model
A process to find the best solution, usually the one that will best help the organization meet its goals.

satisficing model
A model that will find a good—but not necessarily the best—solution to a problem.

for a new manufacturing plant. To find the optimal (best) location, you must consider all cities in the United States or the world. A satisficing approach is to consider only five or ten cities that might satisfy the company's requirements. Limiting the options might not result in the best decision, but it will likely result in a good decision, without spending the time and effort to investigate all cities. Satisficing is a good alternative modeling method because it is sometimes too expensive to analyze every alternative to find the best solution.

Figure 6.2

Optimization Software

Some spreadsheet programs, such as Microsoft Excel, have optimizing routines. This figure shows Solver, which can find an optimal solution given certain constraints.

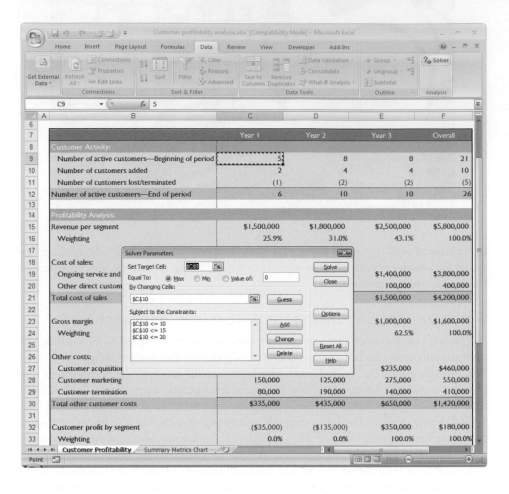

heuristics

"Rules of thumb," or commonly accepted guidelines or procedures that usually find a good solution.

Heuristics, also known as "rules of thumb," are commonly accepted guidelines or procedures that usually find a good solution. A heuristic that baseball team managers use is to place batters most likely to get on base at the top of the lineup, followed by the power hitters who can drive them in to score. An example of a heuristic used in business is to order four months' supply of inventory for a particular item when the inventory level drops to 20 units or less; although this heuristic might not minimize total inventory costs, it can serve as a good rule of thumb to avoid stockouts without maintaining excess inventory. Symantec, a provider of antivirus software, has developed an antivirus product that is based on heuristics.[16] The software can detect viruses that are difficult to detect using traditional antivirus software techniques.

The Benefits of Information and Decision Support Systems

The information and decision support systems covered in this chapter and the next help individuals, groups, and organizations make better decisions, solve problems, and achieve their goals. These systems include management information systems, decision support systems, group support systems, executive support systems, knowledge management systems, and a variety of special-purpose systems. As shown in Figure 6.3, the benefits are a measure of increased performance of these systems versus the cost to deliver them. The plus sign (+) by the arrow from *performance* to *benefits* indicates that increased performance has a positive impact on benefits. The minus sign (-) from *cost* to *benefits* indicates that increased cost has a negative impact on benefits.

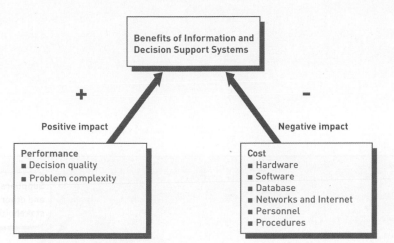

Figure 6.3

The Benefits of Information and
Decision Support Systems

The performance of these systems is typically a function of decision quality and problem complexity. Decision quality can result in increased effectiveness, increased efficiency, higher productivity, and many other measures first introduced in Chapter 1.[17] Problem complexity depends on how hard the problem is to solve and implement. The cost of delivering these systems are the expenditures of the information technology components covered in Part II of this book, including hardware, software, databases, networks and the Internet, people, and procedures. But how do these systems actually deliver benefits to the individuals, groups, and organizations that use them? It depends on the system. We begin our discussion with traditional management information systems.

AN OVERVIEW OF MANAGEMENT INFORMATION SYSTEMS

A management information system (MIS) is an integrated collection of people, procedures, databases, and devices that provides managers and decision makers with information to help achieve organizational goals. MISs can often give companies and other organizations a competitive advantage by providing the right information to the right people in the right format and at the right time.

Management Information Systems in Perspective

The primary purpose of an MIS is to help an organization achieve its goals by providing managers with insight into the regular operations of the organization so that they can control, organize, and plan more effectively. One important role of the MIS is to provide the right information to the right person in the right format at the right time. In short, an MIS provides managers with information, typically in reports, that supports effective decision making and provides feedback on daily operations. Figure 6.4 shows the role of MISs within the flow of an organization's information. Note that business transactions can enter the organization through traditional methods, or via the Internet, or via an extranet connecting customers and suppliers to the firm's ERP or transaction processing systems. The use of MISs spans all levels of management. That is, they provide support to and are used by employees throughout the organization.

Figure 6.4

Sources of Managerial
Information

The MIS is just one of many sources
of managerial information. Decision
support systems, executive support
systems, and specialized systems
also assist in decision making.

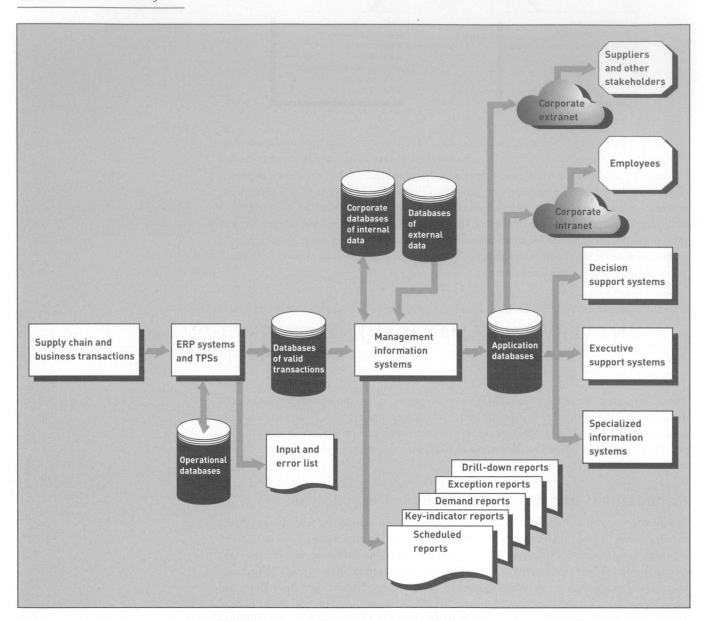

ETHICAL AND SOCIETAL ISSUES

Southwest Airlines Applies MIS to Customer Service

Perhaps one of the most frustrating parts of long distance travel is dealing with unexpected flight delays, which can be caused by weather, aircraft mechanical problems, overbooked flights, or other unexpected circumstances. Delays strand travelers away from their destinations for hours or sometimes even for days.

Some airlines make attempts to notify passengers about flight delays as soon as they become evident in hopes of saving the passenger needless waiting. Southwest Airlines has been doing this for years. Southwest prides itself on clever, unique approaches to customer service that it calls "The Southwest Way." The airline makes special efforts to satisfy customers that have been inconvenienced. Fred Taylor is Southwest's Senior Manager of Proactive Customer Service Communications. The *New York Times* has nicknamed Fred Southwest's "Chief Apology Officer." It's Fred's job to make sure customers who experience difficulties are left with options that leave them feeling satisfied. When Fred found out that Southwest was given a low score for its flight notification service in a *Wall Street Journal* poll, he took action.

Although information systems are typically credited for providing business managers and decision makers with information they can use, they also supply information that customers can use. Information generated by MISs and DSSs can act as a service to a company's customers, such as with Google Search, or as a value added, such as UPS package tracking. Another example is Amazon, which uses information about purchases customers have made on the Amazon Web site to recommend other products that might interest them. Fred Taylor of Southwest Airlines wanted to empower his customers with up-to-the-minute reports on flight information. Not only would his system notify travelers of flight delays, but also of gate changes, opportunities to upgrade, cancellations and rescheduling, and other information that can assist travelers with their preparations.

Fred decided that the best way to inform passengers of preflight information was by phone. Southwest does not have access to all passengers' e-mail addresses or cell phone numbers, so it cannot depend on e-mail or text messaging. It does, however, have access to all customers' phone numbers. Fred and his team reviewed a dozen services that offer automated phone messaging systems. It settled on Varolii Corporation, which specializes in helping companies stay in touch with employees and customers through personalized automated phone messages.

Fred and his team had to develop an MIS that pulled information from its reservation database provided by Sabre reservations systems and from Southwest's own database that manages flight information. The output of the MIS needed to be a report containing passenger names and phone numbers of those who had reservations on flights that were changed, along with details about the disruption. This information had to be delivered in a format that Varolii's system could take as input. The reports also had to be generated quickly so that passengers received the notification as it was delivered to Southwest personnel. In the early testing of the system, Fred received a call from a flustered gate attendant who had a line of passengers asking about a gate change notification. The passengers had received notification before the attendant! After that, a slight delay was programmed into the system to make sure that the information was disseminated to the right people at the right time.

As Fred and his crew rolled out the new system, they included a failsafe to make sure no misinformation was mistakenly sent to passengers. Each notification requires a human review prior to sending the information to Varolii for phoning. Once a notification gets approved, customers' phones ring within seconds. Upon answering, the customer hears a chime, indicating an automated message, and a personable voice that proclaims, "This is Southwest Airlines calling." The message continues with details on flight information and advice on how to proceed and ends with the offer to connect the customer to a service representative.

Southwest is now collecting travelers' cell phone numbers and e-mail addresses so it can extend its service to other forms of communication. Its Chief Apology Officer continues to explore new ways to apply information systems to making customers happy.

Discussion Questions

1. What were the unique aspects and requirements of Southwest's customer notification MIS?
2. How does this information assist Southwest in gaining a competitive advantage?

Critical Thinking Questions

1. In what other ways could airline information systems assist with increasing customer satisfaction?
2. Many banks use Varolii's phone service for notifying customers who are late making loan payments. What are the benefits and drawbacks of using automated communications systems compared to human communication?

SOURCES: Carr, David, "Southwest Upgrades Customer Service," CIO Insight, August 31, 2009, *www.cioinsight.com/index2.php? option=content&task=view&id=882809&pop=1&hide_ads=1&page=0 &hide_js=1*; Southwest Airlines Web site, *www.southwest.com*, accessed March 24, 2010; Varolii Web site *www.varolii.com*, accessed March 24, 2010.

Inputs to a Management Information System

As shown in Figure 6.4, data that enters an MIS originates from both internal and external sources, including a company's supply chain, first discussed in Chapter 1. The most significant internal data sources for an MIS are the organization's various TPS and ERP systems and related databases. External sources of data can include customers, suppliers, competitors, and stockholders whose data is not already captured by TPS and ERP systems, as well as other sources, such as the Internet. As discussed in Chapter 3, companies also use data warehouses and data marts to store valuable business information. Business intelligence, also discussed in Chapter 3, can be used to turn a database into useful information throughout the organization. According to one study, almost 70 percent of consumer goods companies surveyed said they are going to make business intelligence available to more employees in the future.[18]

Outputs of a Management Information System

The output of most MISs is a collection of reports that are distributed to managers. Many MIS reports come from an organization's databases, first discussed in Chapter 3. These reports can be tailored for each user and can be delivered in a timely fashion. Today, many hospitals and healthcare facilities use their health-related databases to streamline MIS reports, reduce recordkeeping costs, and save lives by avoiding medical errors in diagnoses, treatments, and adverse drug interactions.[19] Providence Washington Insurance Company used Report-Net from Cognos (*www.cognos.com*), an IBM company, to reduce the number of paper reports they produce and the associated costs.[20] The new reporting system creates an "executive dashboard" that shows current data, graphs, and tables to help managers make better real-time decisions. Executives from Dunkin' Donuts use a dashboard to see the status of new stores. The dashboard displays geographic areas and the new stores that are being developed. By clicking a store icon, executives can see the details of how new stores are being constructed and if any stores are being delayed. The company hopes to grow to 15,000 franchises around the globe in the next several years.[21] The city of Atlanta, Georgia also used Cognos to measure the performance of its various departments and to keep track of its expenditures and budgets. See Figure 6.5. Microsoft makes a reporting system called Business Scorecard Manager to give decision makers timely information about sales and customer information.[22] The software, which competes with Business Objects and Cognos, can integrate with other Microsoft software products, including Microsoft Office Excel. Hewlett-Packard's OpenView Dashboard is another MIS package that can quickly and efficiently render pictures, graphs, and tables that show how a business is functioning.

Figure 6.5

An Executive Dashboard

OpenView Dashboard provides a graphic overview of how a business is functioning.

(Source: Courtesy of Hewlett-Packard.)

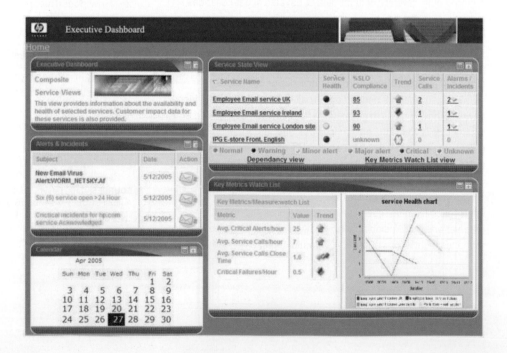

Management reports can come from various company databases, data warehouses, and other sources. These reports include scheduled reports, key-indicator reports, demand reports, exception reports, and drill-down reports. See Figure 6.6.

(a) Scheduled Report

Daily Sales Detail Report

Prepared: 08/10/08

Order #	Customer ID	Salesperson ID	Planned Ship Date	Quantity	Item #	Amount
P12453	C89321	CAR	08/12/08	144	P1234	$3,214
P12453	C89321	CAR	08/12/08	288	P3214	$5,660
P12454	C03214	GWA	08/13/08	12	P4902	$1,224
P12455	C52313	SAK	08/12/08	24	P4012	$2,448
P12456	C34123	JMW	08/13/08	144	P3214	$720
....	....	....	....	....	....	....

(b) Key-Indicator Report

Daily Sales Key-Indicator Report

	This Month	Last Month	Last Year
Total Orders Month to Date	$1,808	$1,694	$1,914
Forecasted Sales for the Month	$2,406	$2,224	$2,608

(c) Demand Report

Daily Sales by Salesperson Summary Report

Prepared: 08/10/00

Salesperson ID	Amount
CAR	$42,345
GWA	$38,950
SAK	$22,100
JWN	$12,350
....	....

(d) Exception Report

Daily Sales Exception Report—Orders Over $10,000

Prepared: 08/10/08

Order #	Customer ID	Salesperson ID	Planned Ship Date	Quantity	Item #	Amount
P12345	C89321	GWA	08/12/08	576	P1234	$12,856
P22153	C00453	CAR	08/12/08	288	P2314	$28,800
P23023	C32832	JMN	08/11/08	144	P2323	$14,400
....	....	....	....	....	....	....

(e) First-Level Drill-Down Report

Earnings by Quarter (Millions)

		Actual	Forecast	Variance
2nd Qtr.	2008	$12.6	$11.8	6.8%
1st Qtr.	2008	$10.8	$10.7	0.9%
4th Qtr.	2008	$14.3	$14.5	-1.4%
3rd Qtr.	2008	$12.8	$13.3	-3.8%

(f) Second-Level Drill-Down Report

Sales and Expenses (Millions)

Qtr: 2nd Qtr. 2008	Actual	Forecast	Variance
Gross Sales	$110.9	$108.3	2.4%
Expenses	$ 98.3	$ 96.5	1.9%
Profit	$ 12.6	$ 11.8	6.8%

(g) Third-Level Drill-Down Report

Sales by Division (Millions)

Qtr: 2nd Qtr. 2008	Actual	Forecast	Variance
Beauty Care	$ 34.5	$ 33.9	1.8%
Health Care	$ 30.0	$ 28.0	7.1%
Soap	$ 22.8	$ 23.0	-0.9%
Snacks	$ 12.1	$ 12.5	-3.2%
Electronics	$ 11.5	$ 10.9	5.5%
Total	$110.9	$108.3	2.4%

(h) Fourth-Level Drill-Down Report

Sales by Product Category (Millions)

Qtr: 2nd Qtr. 2008 Division: Health Care	Actual	Forecast	Variance
Toothpaste	$12.4	$10.5	18.1%
Mouthwash	$ 8.6	$ 8.8	-2.3%
Over-the-Counter Drugs	$ 5.8	$ 5.3	9.4%
Skin Care Products	$ 3.2	$ 3.4	-5.9%
Total	$30.0	$28.0	7.1%

Scheduled Reports

Scheduled reports are produced periodically, such as daily, weekly, or monthly. For example, a production manager could use a weekly summary report that lists total payroll costs to monitor and control labor and job costs. Monthly bills are also examples of scheduled reports. Using its monthly bills, the Sacramento Municipal Utility District compares how people use energy, trying to encourage better energy usage.[23] Other scheduled reports can help managers control customer credit, performance of sales representatives, inventory levels, and more.

A **key-indicator report** summarizes the previous day's critical activities and is typically available at the beginning of each workday. These reports can summarize inventory levels, production activity, sales volume, and the like. Key-indicator reports are used by managers and executives to take quick, corrective action on significant aspects of the business. Some believe that the U.S. economy will recover in the next several years as a result of important key-indicator reports, including production rates and sales figures.[24]

Demand Reports

Demand reports are developed to provide certain information upon request. In other words, these reports are produced on demand rather than on a schedule. Like other reports discussed in this section, they often come from an organization's database system. For example, an executive might want to know the production status of a particular item—a demand report can be generated to provide the requested information by querying the company's database. Suppliers and customers can also use demand reports. FedEx, for example, provides demand reports on its Web site to allow customers to track packages from their source to their final destination. Other examples of demand reports include reports requested by executives to

Figure 6.6

Reports Generated by an MIS

The types of reports are (a) scheduled, (b) key indicator, (c) demand, (d) exception, and (e–h) drill down.

(Source: George W. Reynolds, *Information Systems for Managers*, Third Edition. St. Paul, MN: West Publishing Co., 1995.)

scheduled report
A report produced periodically, such as daily, weekly, or monthly.

key-indicator report
A summary of the previous day's critical activities, typically available at the beginning of each workday.

demand report
A report developed to give certain information at someone's request rather than on a schedule.

show the hours worked by a particular employee, total sales to date for a product, and so on. Today, many demand reports are generated from the Internet or by using cloud computing.[25] Jive Software, for example, allows employees to search large databases and get important demand reports using keywords and natural-language questions. The software can also generate useful graphics, including pie charts and bar graphs.

Exception Reports

exception report
A report automatically produced when a situation is unusual or requires management action.

Exception reports are reports that are automatically produced when a situation is unusual or requires management action. For example, a manager might set a parameter that generates a report of all inventory items with fewer than the equivalent of five days of sales on hand. This unusual situation requires prompt action to avoid running out of stock on the item. The exception report generated by this parameter would contain only items with fewer than five days of sales in inventory.

As with key-indicator reports, exception reports are most often used to monitor aspects important to an organization's success. In general, when an exception report is produced, a manager or executive takes action. Parameters, or *trigger points*, should be set carefully for an exception report. Trigger points that are set too low might result in too many exception reports; trigger points that are too high could mean that problems requiring action are overlooked. For example, if a manager at a large company wants a report that contains all projects over budget by $100 or more, the system might retrieve almost every company project. The $100 trigger point is probably too low. A trigger point of $10,000 might be more appropriate.

Drill-Down Reports

drill-down report
A report providing increasingly detailed data about a situation.

Drill-down reports provide increasingly detailed data about a situation. Using these reports, analysts can see data at a high level first (such as sales for the entire company), then at a more detailed level (such as the sales for one department of the company), and then at a very detailed level (such as sales for one sales representative). Boehringer Ingelheim (*www.boehringer-ingelheim.com/corporate/home/home.asp*), a large German drug company with more than $7 billion in revenues and thousands of employees in 60 countries, uses a variety of drill-down reports so it can respond rapidly to changing market conditions. Managers can drill down into more levels of detail to individual transactions if they want. Companies and organizations of all sizes and types use drill-down reports.

Characteristics of a Management Information System

Scheduled, key-indicator, demand, exception, and drill-down reports have all helped managers and executives make better, more timely decisions. In general, MISs perform the following functions:

- Provide reports with fixed and standard formats
- Produce hard-copy and soft-copy reports
- Use internal data stored in the computer system
- Allow users to develop custom reports
- Require user requests for reports developed by systems personnel

FUNCTIONAL ASPECTS OF THE MIS

Most organizations are structured along functional areas. This functional structure is usually apparent from an organization chart, which typically shows a hierarchy in roles or positions. Some traditional functional areas include finance, manufacturing, marketing, human resources, and other specialized information systems. The MIS can also be divided along those functional lines to produce reports tailored to individual functions. See Figure 6.7.

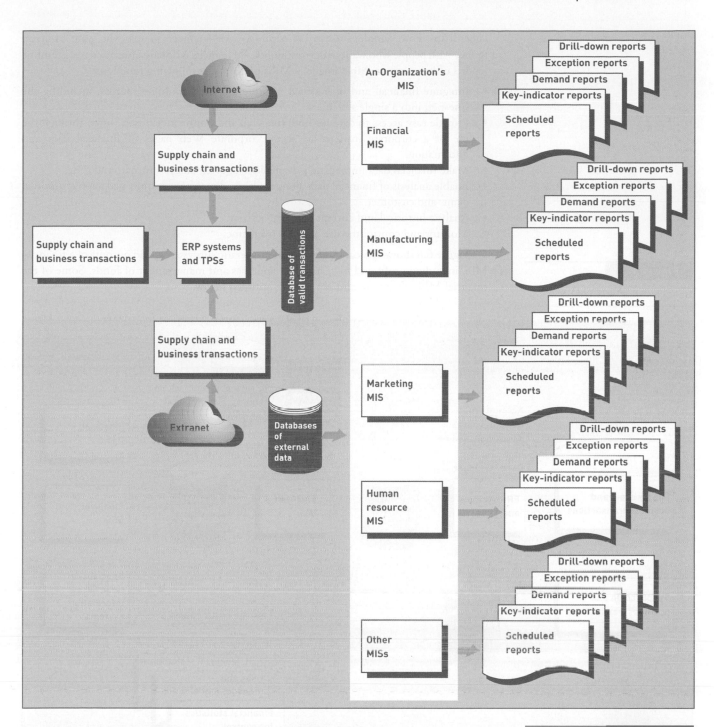

Financial Management Information Systems

A **financial MIS** provides financial information not only for executives, but also for a broader set of people who need to make better decisions on a daily basis. Reuters, for example, has developed an automated reporting system that scans articles about companies for its stock traders to determine if the news is favorable or unfavorable. The reports can result in buy orders if the news is positive or sell orders if the news is negative. Eventually, the system will be tied into machine trading that doesn't require trade orders generated by people.[26] Web sites can also provide financial information. For example, Web sites such as Kiva (*www.kiva.com*) provide information for people seeking small loans, called microloans. Many microloans are for $100 or less and are made for a period of several months. People in Europe, Japan, and other countries are starting to use their smartphones and mobile devices to transfer money directly to other people to make loans or repay them.[27] The Internet is also used for

Figure 6.7

An Organization's MIS

The MIS is an integrated collection of functional information systems, each supporting particular functional areas.

financial MIS

An information system that provides financial information for executives and for a broader set of people who need to make better decisions on a daily basis.

larger loans. For example, Prosper (*www.prosper.com*) connects people who want to invest money with people who want to borrow money.[28] Financial MISs are often used to streamline reports of transactions. Most financial MISs perform the following functions:

- Integrate financial and operational information from multiple sources, including the Internet, into a single system.
- Provide easy access to data for both financial and nonfinancial users, often through the use of a corporate intranet to access corporate Web pages of financial data and information.
- Make financial data immediately available to shorten analysis turnaround time.
- Enable analysis of financial data along multiple dimensions—time, geography, product, plant, and customer.
- Analyze historical and current financial activity.
- Monitor and control the use of funds over time.

Figure 6.8 shows typical inputs, function-specific subsystems, and outputs of a financial MIS, including profit and loss, auditing, and uses and management of funds. Some of the financial MIS subsystems and outputs are outlined below.

Figure 6.8

Overview of a Financial MIS

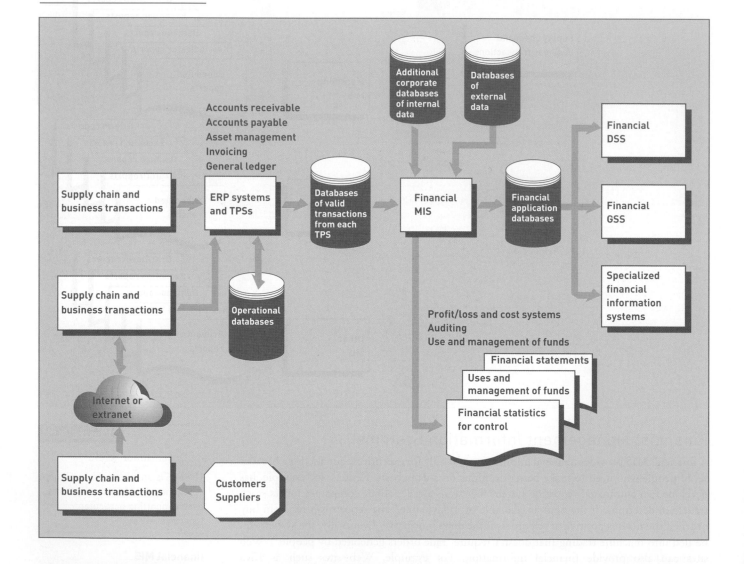

profit center
A department within an organization that focuses on generating profits.

revenue center
A division within a company that generates sales or revenues.

cost center
A division within a company that does not directly generate revenue.

Figure 6.9

Income Statement

An income statement that shows a corporation's business results, including all revenues, earnings, expenses, costs and taxes.

- *Profit/loss and cost systems.* Many departments within an organization are **profit centers**, which means that they focus on generating profits. An investment division of a large insurance or credit card company is an example of a profit center. See Figure 6.9. Other departments can be **revenue centers**, which are divisions within the company that focus primarily on sales or revenues, such as a marketing or sales department. Still other departments can be **cost centers**, which are divisions within a company that do not directly generate revenue, such as manufacturing or research and development. In most cases, information systems are used to compute revenues, costs, and profits.

Projected Five-Year Income Statement

Upland International

Recreational Products Division
Projected Five-Year Income Statement

Prepared: 4/29/2013

| | | | Tax Rate | | 33% |
| | | | Cost of Goods | | 75% |

	Year 1	Year 2	Year 3	Year 4	Year 5
Revenues	$3,200,000	$3,541,382	$3,919,184	$4,337,290	$4,800,000
Cost of Sales	$2,400,000	$2,656,037	$2,939,388	$3,252,967	$3,600,000
Gross Profit	$800,000	$885,346	$979,796	$1,084,322	$1,200,000
Accounting	$12,600	$14,112	$15,805	$17,702	$19,826
Advertising & Promotion	$37,800	$42,336	$47,416	$53,106	$59,479
Insurance	$3,600	$4,032	$4,516	$5,058	$5,665
Maintenance	$8,640	$9,677	$10,838	$12,139	$13,595
Utilities	$13,680	$15,322	$17,160	$19,219	$21,526
Miscellaneous	$4,300	$4,816	$5,394	$6,041	$6,766
Total General Expenses	$80,620	$90,294	$101,130	$113,265	$126,857
Earnings before Interest, Depr. & Tax	$719,380	$795,051	$878,666	$971,057	$1,073,143
Depreciation Expense	$141,550	$120,459	$102,511	$87,237	$74,238
Operating Profit	$577,830	$674,592	$776,156	$883,821	$998,905
Interest Expense	$65,733	$53,102	$39,490	$24,822	$9,016
Earnings Before Taxes	$512,097	$621,490	$736,665	$858,998	$989,889
Estimated Tax	$168,992	$205,092	$243,099	$283,469	$326,663
Net Income	$343,105	$416,398	$493,566	$575,529	$663,226

auditing
Analyzing the financial condition of
an organization and determining
whether financial statements and
reports produced by the financial
MIS are accurate.

internal auditing
Auditing performed by individuals
within the organization.

external auditing
Auditing performed by an outside
group.

- *Auditing.* **Auditing** involves analyzing the financial condition of an organization and determining whether financial statements and reports produced by the financial MIS are accurate.[29] **Internal auditing** is performed by individuals within the organization. For example, the finance department of a corporation might use a team of employees to perform an audit. **External auditing** is performed by an outside group, usually an accounting or consulting firm such as PricewaterhouseCoopers, Deloitte & Touche, or one of the other major, international accounting firms. Computer systems are used in all aspects of internal and external auditing. Even with internal and external audits, some companies can mislead investors and others with fraudulent accounting statements. It was reported that one large company paid $50 million in fines to settle a U.S. fraud case.[30] According to one investigator, the company bent accounting rules beyond the breaking point.

- *Uses and management of funds.* Internal uses of funds include purchasing additional inventory, updating plants and equipment, hiring new employees, acquiring other companies, buying new computer systems, increasing marketing and advertising, purchasing raw materials or land, investing in new products, and increasing research and development. External uses of funds are typically investment related. Companies often use financial MISs to invest excess funds in such external revenue generators as bank accounts, stocks, bonds, bills, notes, futures, options, and foreign currency. A number of powerful personal finance applications also help people manage and use their money.[31] Mint Software, for example, can help people budget their expenditures.[32] Brokerage companies and some specialized Internet sites such as Economic Security Planning can be used for financial planning. Yahoo Finance (*http://finance.yahoo.com*) and Portfolio Monkey (*www.portfoliomonkey.com*) are just a couple of Internet sites that can be used to monitor personal investments.

Manufacturing Management Information Systems

Without question, advances in information systems have revolutionized manufacturing. As a result, many manufacturing operations have been dramatically improved over the last decade. The use of computerized systems is emphasized at all levels of manufacturing—from the shop floor to the executive suite. Dell Computer has used both optimization and heuristic software to help it manufacture a larger variety of products.[33] Dell was able to double its product variety while saving about $1 million annually in manufacturing costs. Figure 6.10 gives an overview of some of the manufacturing MIS inputs, subsystems, and outputs.

The manufacturing MIS subsystems and outputs are used to monitor and control the flow of materials, products, and services through the organization. As raw materials are converted to finished goods, the manufacturing MIS monitors the process at almost every stage. New technology could make this process easier. Using specialized computer chips and tiny radio transmitters, companies can monitor materials and products through the entire manufacturing process. Procter & Gamble, Wal-Mart, and Target have funded research into this manufacturing MIS. According to a survey of manufacturing companies, increased efficiency, lower costs, and new products and services are important manufacturing MIS features that can increase profitability and success.[34]

The success of an organization can depend on the manufacturing function. Some common information subsystems and outputs used in manufacturing are provided in the following list:

- *Design and engineering.* Manufacturing companies often use computer-aided design (CAD) with new or existing products.[35] For example, Boeing (*www.boeing.com*) uses a CAD system to develop a complete digital blueprint of an aircraft before it begins the manufacturing process. As mock-ups are built and tested, the digital blueprint is constantly revised to reflect the most current design. Using such technology helps Boeing reduce manufacturing costs and the time to design a new aircraft.

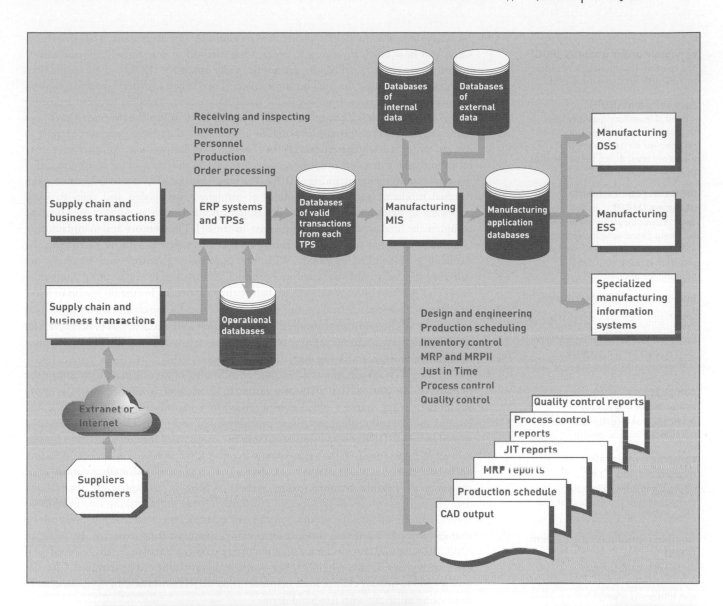

Figure 6.10

Overview of a Manufacturing MIS

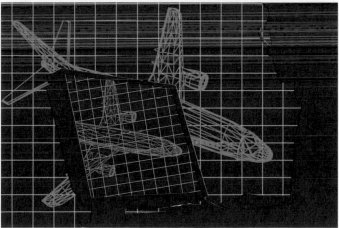

Computer-aided design (CAD) is used in the development and design of complex products or structures.

(Source: © Dennis Hallinan/Alamy.)

- *Master production scheduling.* Scheduling production and controlling inventory are critical for any manufacturing company. The overall objective of master production scheduling is to provide detailed plans for both short-term and long-range scheduling of manufacturing facilities. Some companies hire outside companies to help them with inventory control.

economic order quantity (EOQ)
The quantity that should be reordered to minimize total inventory costs.

reorder point (ROP)
A critical inventory quantity level that calls for more inventory to be ordered for an item when the inventory level drops to the reorder point or critical level.

material requirements planning (MRP)
A set of inventory-control techniques that help coordinate thousands of inventory items when the demand of one item is dependent on the demand for another.

just-in-time (JIT) inventory
An inventory management approach in which inventory and materials are delivered just before they are used in manufacturing a product.

computer-assisted manufacturing (CAM)
A system that directly controls manufacturing equipment.

computer-integrated manufacturing (CIM)
Using computers to link the components of the production process into an effective system.

flexible manufacturing system (FMS)
An approach that allows manufacturing facilities to rapidly and efficiently change from making one product to making another.

Computer-assisted manufacturing systems control complex processes on the assembly line and provide users with instant access to information.

(Source: © Phototake/Alamy.)

• *Inventory control.* Most techniques are used to minimize inventory costs. They determine when and how much inventory to order. One method of determining the amount of inventory to order is called the **economic order quantity** (**EOQ**). This quantity is calculated to minimize the total inventory costs. The "When to order?" question is based on inventory usage over time. Typically, the question is answered in terms of a **reorder point** (**ROP**), which is a critical inventory quantity level. When the inventory level for a particular item falls to the reorder point, or critical level, the system generates a report so that an order is immediately placed for the EOQ of the product. Another inventory technique used when demand for one item depends on the demand for another is called **material requirements planning** (**MRP**). The basic goals of MRP are to determine when finished products, such as automobiles or airplanes, are needed and then to work backward to determine deadlines and resources needed, such as engines and tires, to complete the final product on schedule. **Just-in-time** (**JIT**) **inventory** and manufacturing is an approach that maintains inventory at the lowest levels without sacrificing the availability of finished products. With this approach, inventory and materials are delivered just before they are used in a product. Many drug manufacturing companies use JIT to produce flu vaccinations just before they are needed for the flu season to make sure the vaccinations are freshest and the most effective.[36] One chemical company used JIT to deliver chemicals to customers just before they were needed.[37] JIT, however, can result in some organizations running out of inventory when demand exceeds expectations or there are problems with the manufacturing process.

• *Process control.* Managers can use a number of technologies to control and streamline the manufacturing process. For example, computers can directly control manufacturing equipment, using systems called **computer-assisted manufacturing** (**CAM**).[38] CAM systems can control drilling machines, assembly lines, and more. **Computer-integrated manufacturing** (**CIM**) uses computers to link the components of the production process into an effective system. CIM's goal is to tie together all aspects of production, including order processing, product design, manufacturing, inspection and quality control, and shipping. A **flexible manufacturing system** (**FMS**) is an approach that allows manufacturing facilities to rapidly and efficiently change from making one product to another. In the middle of a production run, for example, the production process can be changed to make a different product or to switch manufacturing materials. By using an FMS, the time and cost to change manufacturing jobs can be substantially reduced, and companies can react quickly to market needs and competition. For example, Chrysler used a FMS to quickly change from manufacturing diesel minivans with right-hand drive to gasoline minivans with left-hand drive.[39]

- *Quality control and testing.* With increased pressure from consumers and a general concern for productivity and high quality, today's manufacturing organizations are placing more emphasis on **quality control**, a process that ensures that the finished product meets the customers' needs. Information systems are used to monitor quality and take corrective steps to eliminate possible quality problems.

Marketing Management Information Systems

A **marketing MIS** supports managerial activities in product development, distribution, pricing decisions, promotional effectiveness, and sales forecasting.[40] Marketing functions are increasingly being performed on the Internet. Estee Lauder and Sony have developed a sitcom called Sufie's Diary on an Internet site to entertain young viewers and advertise Clinique, a popular Estee Lauder cosmetic, along with various Sony products.[41] Viewers can vote on an Internet blog and have an impact on the plot and future episodes. Some companies make gifts to people who write reviews on blogs or the Internet, which can result in favorable, biased reviews.[42] The U.S. Federal Trade Commission has issued a guideline that bloggers report or identify reviews for which they received gifts or other compensation. Companies such as Salesforce (*www.salesforce.com*) offer a range of applications over the Internet, including customer relationship management (CRM), first introduced in Chapter 1, and many other programs that help a company's marketing efforts.[43] Salesforce generates about $1 billion in revenues annually. Newer marketing companies, such as AdMob, Inc. (*www.admob.com*), are placing ads on cell phones and mobile devices with Internet access. Unilever (*www.unilever.com*) is trying a marketing campaign that sends coupons to mobile phones. Customers can use the coupons in stores to buy products.[44] According to a Unilever executive, "This has been a Holy Grail thing that people have been trying to figure out. I think this is on target for where consumers' heads are at right now." One marketing research firm estimates that some companies spend millions of dollars annually on social networking Internet sites to promote their products and services online.[45]

Some marketing departments are actively using the Internet to advertise their products and services and keep customers happy using services such as Facebook (*www.facebook.com*). YouTube (*www.youtube.com*), the video-sharing Internet site, sells video ads on its site to companies, including Ford, BMW, Time Warner, and others. After about 10 seconds, the video ad disappears unless the user clicks it. Corporate marketing departments also use social networking sites, such as Second Life (*www.secondlife.com*), to advertise their products and perform marketing research. Companies such as JetBlue are using blogs and Twitter to listen to customers and perform basic marketing research.[46] According to JetBlue's director of corporate communications, "Our primary use is just watching and listening. Twitter can easily be one of the best marketing research tools available." When the founder of TweetPhoto decided to launch his new photo-sharing company, he used Twitter to advertise the new company and its services.[47] According to an Internet consultant, "Twitter is a digital handshake. It's one of the fastest ways you can reach out to people." Figure 6.11 shows the inputs, subsystems, and outputs of a typical marketing MIS.

quality control
A process that ensures that the finished product meets the customers' needs.

marketing MIS
An information system that supports managerial activities in product development, distribution, pricing decisions, promotional effectiveness, and sales forecasting.

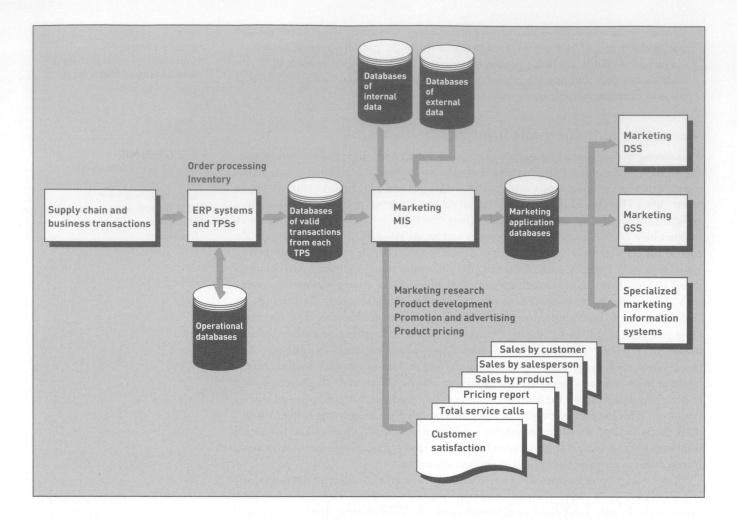

Figure 6.11

Overview of a Marketing MIS

Subsystems for the marketing MIS include marketing research, product development, promotion and advertising, and product pricing. These subsystems and their outputs help marketing managers and executives increase sales, reduce marketing expenses, and develop plans for future products and services to meet the changing needs of customers.

Corporate marketing departments use social networking sites, such as Second Life (*www.secondlife.com*), to advertise their products and perform marketing research.

(Source: *www.secondlife.com*.)

- *Marketing research.* The purpose of marketing research is to conduct a formal study of the market and customer preferences. Computer systems are used to help conduct and analyze the results of surveys, questionnaires, pilot studies, and interviews. BMW, the German luxury car maker, performs marketing research using search engines to determine customer preferences and to target ads to people who might want to buy one of its cars.[48] Hyundai Motor America asked a group of men and women to look at certain parts of a new Hyundai car while wearing head caps with wires attached to their skulls to try to determine customer preferences.[49] According to one company executive, "We want to know what consumers think about a car before we start manufacturing thousands of them." In addition to knowing what you buy, marketing research can determine where you buy. This can help in developing new products and services and tailoring ads and promotions. With the use of GPS, marketing firms can promote products to phones and other mobile devices by knowing your location. Internet sites such as Loopt (*www.loopt.com*), BrightKite (*www.brightkite.com*), and Latitude by Google (*www.google.com/latitude/intro.html*) allow you to locate people, stores, restaurants, and other businesses and landmarks that are close to you.[50] Using GPS and location analysis from cell phone towers, advertisers will be able to promote products and services in stores and shops that are close to people with cell phones.[51] In other words, you could receive ads on your cell phone for a burger restaurant as you walk or drive close to it. Other marketing research companies, such as Betawave (*www.betawave.com*), are performing marketing research on customer engagement and attentiveness in responding to ads.[52] If successful, this marketing research might result in advertisers charging more for higher levels of customer engagement and attentiveness, instead of charging for the number of viewers of a particular advertisement. ATG is developing predictive marketing software that attempts to determine buying behaviors, such as which days and times women are more likely to be shopping for men.[53]

Marketing research data yields valuable information for the development and marketing of new products.

(Source: © Michael Newman/PhotoEdit.)

- *Product development.* Product development involves the conversion of raw materials into finished goods and services and focuses primarily on the various attributes of the product and its supply chain first introduced in Chapter 1. Sysco, a Texas-based food distribution company, uses software and databases to prepare and ship more than 20 million tons of meat, produce, and other food items to restaurants and other outlets every year.[54] The huge company supplies one out of three cafeterias, sports stadiums, restaurants, and other food outlets. Many factors, including plant capacity, labor skills, engineering issues, and materials are important in product development decisions. In many cases, a computer program analyzes these various factors and selects the appropriate mix of labor, materials, plant and equipment, and engineering designs. Make-or-buy decisions can also be made with the assistance of computer programs.
- *Promotion and advertising.* One of the most important functions of any marketing effort is promotion and advertising. Product success is a direct function of the types of advertising and sales promotion done. Increasingly, organizations are using the Internet to advertise and sell products and services. Johnson & Johnson used Internet cartoons instead of extensive TV advertising to promote a popular baby lotion. The goal is to target ads to a specific group of people who are likely to purchase the advertised goods and services. Today, about $35 billion is spent annually on newspaper ads, about $29 billion on local TV ads, and about $23 billion on Internet ads.[55] Companies are also trying to

measure the effectiveness of different advertising approaches, such as TV and Internet advertising. Toyota, for example, used IAG Research to help it measure the effectiveness of TV and Internet advertising. Several companies, including ScanScout (*www.scanscout.com*) and YuMeNetworks (*www.yumenetworks.com*), are trying to match video content on the Internet with specific ads that cater to those watching the videos.[56] Cell phones are becoming very popular advertising outlets.[57] Burger King, Lions Gate Entertainment, and other companies are now promoting their products on games and other applications for Apple's iPhone and other devices.[58] In some cases, the advertising is hidden within free gaming applications. When you download and start playing the game, the advertising pops up on the phone. Other companies, like Honda Motors and Marriott International, are placing ads on mobile search engines available on phones and other mobile devices.[59] Increased use of the Internet to place advertisements, however, has made some computers and mobile devices vulnerable to virus attacks.[60]

Cell phones are becoming increasingly popular advertising outlets.

(Source: © Courtesy of PRNewsFoto/State Farm Insurance Companies)

- *Product pricing.* Product pricing is another important and complex marketing function. Retail price, wholesale price, and price discounts must be set. Most companies try to develop pricing policies that will maximize total sales revenues. Computers are often used to analyze the relationship between prices and total revenues. Traditionally, executives used costs to determine prices. They simply added a profit margin to total costs to guarantee a decent profit. Today, however, more executives look at the marketplace to determine product pricing. In one case, a company increased its revenue by about $200 million, in part by using a more aggressive pricing policy based on what the market was willing to pay.
- *Sales analysis.* Computerized sales analysis is important to identify products, sales personnel, and customers that contribute to profits and those that do not. This analysis can be done for sales and ads that help generate sales. Engagement ratings, for example, show how ads convert to sales.[61] According to Ford's marketing chief, "It's almost like we're getting price guarantees on how well our ad is doing, and that's close to the Holy Grail for me." IBM used the OnTarget sales analysis tool to identify new sales opportunities with existing customers.[62] The sales analysis tool helps IBM assign sales

personnel to sales opportunities to improve both sales and profitability. Several reports can be generated to help marketing managers make good sales decisions. See Figure 6.12. The sales-by-product report lists all major products and their sales for a specified period of time. This report shows which products are doing well and which need improvement or should be discarded altogether. The sales-by-salesperson report lists total sales for each salesperson for each week or month. This report can also be subdivided by product to show which products are being sold by each salesperson. The sales-by-customer report is a tool that can be used to identify high- and low-volume customers.

(a) Sales by Product

Product	August	September	October	November	December	Total
Product 1	34	32	32	21	33	152
Product 2	156	162	177	163	122	780
Product 3	202	145	122	98	66	633
Product 4	345	365	352	341	288	1,691

(b) Sales by Salesperson

Salesperson	August	September	October	November	December	Total
Jones	24	42	42	11	43	162
Kline	166	155	156	122	133	732
Lane	166	155	104	99	106	630
Miller	245	225	305	291	301	1,367

(c) Sales by Customer

Customer	August	September	October	November	December	Total
Ang	234	334	432	411	301	1,712
Braswell	56	62	77	61	21	277
Celec	1,202	1,445	1,322	998	667	5,634
Jung	45	65	55	34	88	287

Figure 6.12

Reports Generated to Help Marketing Managers Make Good Decisions

(a) This sales-by-product report lists all major products and their sales for the period from August to December. (b) This sales-by-salesperson report lists total sales for each salesperson for the same time period. (c) This sales-by-customer report lists sales for each customer for the period. Like all MIS reports, totals are provided automatically by the system to show managers at a glance the information they need to make good decisions.

Human Resource Management Information Systems

A **human resource MIS (HRMIS)**, also called the *personnel MIS*, is concerned with activities related to previous, current, and potential employees of the organization. Increasingly, the HRMIS is being used to oversee and manage part-time, virtual work teams and job sharing, in additional to traditional job titles and duties.[63] Because the personnel function relates to all other functional areas in the business, the HRMIS plays a valuable role in ensuring organizational success. Some of the activities performed by this important MIS include workforce analysis and planning, hiring, training, job and task assignment, and many other personnel-related issues. An effective HRMIS allows a company to keep personnel costs at a minimum while serving the required business processes needed to achieve corporate goals. Although human resource information systems focus on cost reduction, many of today's HR systems concentrate on hiring and managing existing employees to get the total potential of the human talent in the organization. Figure 6.13 shows some of the inputs, subsystems, and outputs of the human resource MIS.

human resource MIS (HRMIS)
An information system that is concerned with activities related to employees and potential employees of an organization, also called a personnel MIS.

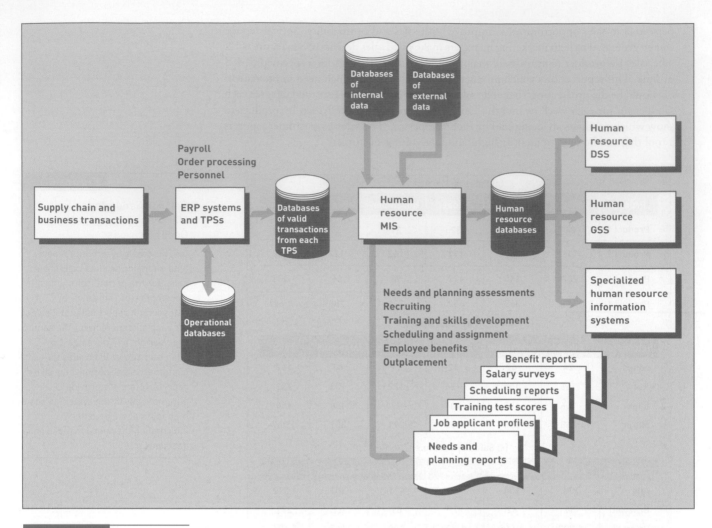

Figure 6.13

Overview of a Human Resource
MIS

Human resource MIS subsystems
help to determine personnel needs
and match employees to jobs.

(Source: © Josh Hodge/
iStockphoto.)

- *Human resource planning.* One of the first aspects of any human resource MIS is determining personnel and human needs. The overall purpose of this MIS subsystem is to put the right number and types of employees in the right jobs when they are needed, including internal employees that work exclusively for the organization and outside workers that are hired when they are needed. Some experts believe that workers should be managed like a supply chain, using supply chain management (SCM) and just-in-time techniques, first discussed in Chapter 1.

- *Personnel selection and recruiting.* If the human resource plan reveals that additional personnel are required, the next logical step is recruiting and selecting personnel. Companies seeking new employees often use computers to schedule recruiting efforts and trips and to test potential employees' skills. Many companies now use the Internet to screen for job applicants. Applicants use a template to load their résumé onto the Internet site. HR managers can then access these résumés and identify applicants they are interested in interviewing.

- *Training and skills inventory.* Some jobs, such as programming, equipment repair, and tax preparation, require very specific training for new employees. Other jobs may require general training about the organizational culture, orientation, dress standards, and expectations of the organization. When training is complete, employees often take computer-scored tests to evaluate their mastery of skills and new material.

- *Scheduling and job placement.* Employee schedules are developed for each employee, showing job assignments over the next week or month. Job placements are often determined based on skills inventory reports showing which employee might be best suited to a particular job. Sophisticated scheduling programs are often used in the airline industry, the military, and many other areas to get the right people assigned to the right jobs at the right time.

- *Wage and salary administration.* Another human resource MIS subsystem involves determining wages, salaries, and benefits, including medical payments, savings plans, and retirement accounts. Wage data, such as industry averages for positions, can be taken from the corporate database and manipulated by the human resource MIS to provide wage information and reports to higher levels of management.

- *Outplacement.* Employees leave a company for a variety of reasons. Outplacement services are offered by many companies to help employees make the transition. *Outplacement* can include job counseling and training, job and executive search, retirement and financial planning, and a variety of severance packages and options. Many employees use the Internet to plan their future retirement or to find new jobs, using job sites such as *www.monster.com*.

Other Management Information Systems

In addition to finance, manufacturing, marketing, and human resource MISs, some companies have other functional management information systems. For example, most successful companies have well-developed accounting functions and a supporting accounting MIS. Also, many companies use geographic information systems for presenting data in a useful form.

Accounting MISs

In some cases, accounting works closely with financial management. An **accounting MIS** performs a number of important activities, providing aggregate information on accounts payable, accounts receivable, payroll, and many other applications. The organization's enterprise resource planning and transaction processing system captures accounting data, which is also used by most other functional information systems.

Some smaller companies hire outside accounting firms to assist them with their accounting functions. These outside companies produce reports for the firm using raw accounting data. In addition, many excellent integrated accounting programs are available for personal computers in small companies. Depending on the needs of the small organization and its staff's computer experience, using these computerized accounting systems can be a very cost-effective approach to managing information.

accounting MIS
An information system that provides aggregate information on accounts payable, accounts receivable, payroll, and many other applications.

geographic information system (GIS)

A computer system capable of assembling, storing, manipulating, and displaying geographic information, that is, data identified according to its location.

Geographic Information Systems

Increasingly, managers want to see data presented in graphical form. A **geographic information system (GIS)** is a computer system capable of assembling, storing, manipulating, and displaying geographically referenced information; that is, data identified according to its location. Google, for example, has developed a GIS that combines geothermal information with its mapping applications to help identify carbon emissions.[64] Another Google GIS has been developed to map flu trends around the world by tracking flu inquiries on its Web site. Staples Inc., the large office supply store chain, uses a GIS to select new store locations. Finding the best location is critical. It can cost up to $1 million for a failed store because of a poor location. Staples also uses a GIS tool from Tactician Corporation (*www.tactician.com*) along with software from SAS. It is possible to use GIS to analyze customer preferences and shopping patterns in various locations using GPS and location analysis from cell phone towers.[65] In addition, a number of applications follow the location of people and places using GPS, including Loopt (*www.loopt.com*), Where (*www.where.com*), and Google Latitude.[66] Some people are concerned about privacy concerns with these GPS applications.[67]

Google Latitude tracks the physical location of people who give their permission to be tracked.

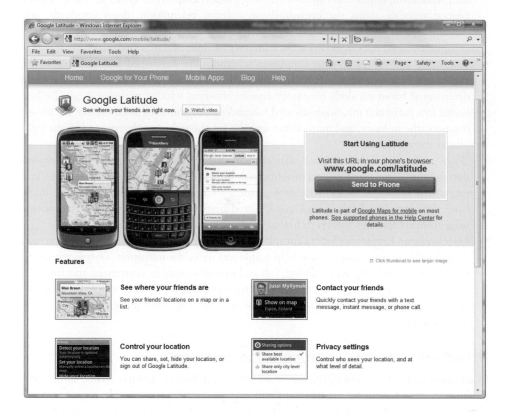

We saw earlier in this chapter that management information systems (MISs) provide useful summary reports to help solve structured and semistructured business problems. Decision support systems (DSSs) offer the potential to assist in solving both semistructured and unstructured problems. These systems are discussed next.

AN OVERVIEW OF DECISION SUPPORT SYSTEMS

A DSS is an organized collection of people, procedures, software, databases, and devices used to help make decisions that solve problems. The focus of a DSS is on decision-making effectiveness when faced with unstructured or semistructured business problems. Decision support systems offer the potential to generate higher profits, lower costs, and better products and services. TurboRouter, for example, is a decision support system developed in Norway to reduce shipping costs and cut emissions of merchant ships.[68] Operating a single ship can cost more than $10,000 every day. TurboRouter schedules and manages the use of ships to transport oil and other products to locations around the world. Billerud, a paper mill company in Sweden, used a DSS optimization program to control a bleaching process that reduced the need for certain chemicals by 10 percent and saved the company about 2 million euros annually.[69]

Decision support systems, although skewed somewhat toward the top levels of management, can be used at all levels. The CEO of Serena, a California software company, uses his cell phone to get rapid feedback on the performance of each of his key executives.[70] He calls the approach the "victory plan" and said, "The fact that everybody can see how we're doing, good, or bad, builds trust and forces accountability." DSSs are also used in government, law enforcement, and nonprofit organizations. See Figure 6.14. They also support optimization, satisficing, and heuristic approaches, as shown in Figure 6.15.

Figure 6.14

Decision support systems are used by healthcare professionals in many settings.

(Source: Image copyright 2010, Konstantin Sutyagin. Used under license from Shutterstock.com.)

Capabilities of a Decision Support System

Developers of decision support systems strive to make them more flexible than management information systems and to give them the potential to assist decision makers in a variety of situations. DSSs can assist with all or most problem-solving phases, decision frequencies, and varying degrees of problem structure. DSS approaches can also help at all levels of the decision-making process. A single DSS might provide only a few of these capabilities, depending on its uses and scope.

Support for Problem-Solving Phases

The objective of most decision support systems is to assist decision makers with the phases of problem solving. As previously discussed, these phases include intelligence, design, choice, implementation, and monitoring. A specific DSS might support only one or a few phases. By supporting all types of decision-making approaches, a DSS gives the decision maker a great deal of flexibility in getting computer support for decision-making activities.

ad hoc DSS

A DSS concerned with situations or decisions that come up only a few times during the life of the organization.

institutional DSS

A DSS that handles situations or decisions that occur more than once, usually several times per year or more. An institutional DSS is used repeatedly and refined over the years.

highly structured problems

Problems that are straightforward and require known facts and relationships.

semistructured or unstructured problems

More complex problems in which the relationships among the pieces of data are not always clear, the data might be in a variety of formats, and the data is often difficult to manipulate or obtain.

Support for Various Decision Frequencies

Decisions can range on a continuum from one-of-a-kind to repetitive decisions. One-of-a-kind decisions are typically handled by an **ad hoc DSS**. An ad hoc DSS is concerned with situations or decisions that come up only a few times during the life of the organization; in small businesses, they might happen only once. For example, a company might need to decide whether to build a new manufacturing facility in another area of the country. Repetitive decisions are addressed by an institutional DSS. An **institutional DSS** handles situations or decisions that occur more than once, usually several times per year or more. An institutional DSS is used repeatedly and refined over the years. Examples of institutional DSSs include systems that support portfolio and investment decisions and production scheduling. These decisions might require decision support numerous times during the year. Between these two extremes are decisions that managers make several times, but not routinely.

Support for Various Problem Structures

As discussed previously, decisions can range from highly structured and programmed to unstructured and nonprogrammed. **Highly structured problems** are straightforward, requiring known facts and relationships. **Semistructured** or **unstructured problems**, on the other hand, are more complex. The relationships among the pieces of data are not always clear, the data might be in a variety of formats, and it might be difficult to manipulate or obtain. In addition, the decision maker might not know the information requirements of the decision in advance.

Amenities Inc. Gets a Grip on Pachinko Information

Just as Las Vegas has its casinos and slot machines, Japan has pachinko parlors and pachinko machines. Pachinko is a game played on vertical boards where small metal balls are flung to the top and cascade down through numerous metal pins. If a ball happens to land in one of the cups, points are scored and more balls are released. Winning scores collect prizes.

Amenities Inc. owns 16 pachinko parlors in Japan, as well as cinemas, karaoke booths, bowling alleys, and restaurants. In 2005, Amenities became the first in the pachinko industry to adopt an ERP system to manage its data and provide useful information for business decision making. In 2008, Amenities installed a business intelligence platform to further benefit from the information it gathered about its businesses and customers. More recently, Amenities began a quest to organize and display information in a manner that better supports strategic decision-making.

Until recently, Amenities used a decision-making process referred to as the plan-do-check-act (PDCA) approach. This is a model based on the scientific method that begins with a hypothesis (plan), tested by an experiment (do), followed by an examination of results (check), and adjustments applied based on observations (act), which cycles back to the planning stage. Amenities realized that the PDCA approach was not ideal for the strategic decisions it needed to make in the pachinko industry. It thought that a more appropriate model was the observe, orient, decide, and act (OODA) approach. OODA is used by the military in combat operations and in business to gain advantages in tough competitive markets.

The OODA model is designed to defeat an adversary and survive. It depends on ongoing observation and analysis of a situation. It requires an information system that can supply up-to-the minute information about the business, its competitors, and the environment in which it operates. Over the years, Amenities had built such an information system. Now it needed to focus on getting relevant data to its decision makers.

Amenities purchased three software packages intended to deliver the right information to the right people at the right time in the right format. SAP BusinessObjects Web Intelligence software is designed to allow decision makers to access business reports through a Web browser interface. SAP BusinessObjects Live Office software is designed to obtain and analyze data directly in Microsoft Excel and Word. Xcelsius software is designed to provide a dashboard that graphically displays key indicators. This was combined with Crystal Reports software that provides additional visual reporting tools.

Armed with this arsenal of MIS tools, Amenities CIO Kazuo Yoshida set out to deliver useful information to its decision makers.

For corporate executives, a dashboard was designed that included three key indicator charts and meters: profits and losses, sales amounts, and customer amounts. The charts were integrated with the company's groupware so that every time executives check e-mail or collaborate online, they get a quick glimpse of the company's current health. If the graphs indicate anything unusual, an executive can click the graph to drill down into more detailed information.

Lower-ranked business managers within Amenities have a similar dashboard showing key indicators that include bar graphs for sales, profits, and expenses with goals and current status listed side by side. The layout allows managers to easily see their current degree of success in achieving key indicators and corporate goals. Managers also have access to interactive forms that allow them to submit daily reports to headquarters.

Amenities' new information delivery system allows its decision makers with up-to-the-minute corporate information. If a competitor launches a one-day promotion, managers can react before midday. Executives can quickly react to significant sales declines by launching ad-hoc promotional events. Unexpected sales increases allow executives to recognize new opportunities and quickly take advantage of them. With its ERP system, business intelligence system, and reporting tools in place, Amenities will continue to refine the manner in which it delivers valuable information to individuals who are able to act upon it and improve its position in the market.

Discussion Questions

1. What information system components did Amenities use to get the right information to the right people at the right time in the right format?
2. Why did Amenities decide to combine its dashboard charts and meters with its groupware?

Critical Thinking Questions

1. What advantages did Amenities think it could gain by adopting an OODA approach to decision making?
2. How might real-time and continuous access to key-indicator data influence a business' decision making?

SOURCES: SAP Staff, "Amenities - Sap Software Optimizes Decision, Making And Information Quality," SAP Success Stories, *www.sap.com/solutions/sap-businessobjects/sme/reporting-dashboarding/customers/index.epx*, accessed March 23, 2010.

Support for Various Decision-Making Levels

Decision support systems can provide help for managers at various levels within the organization. Operational managers can get assistance with daily and routine decision making. Tactical decision makers can use analysis tools to ensure proper planning and control. At the strategic level, DSSs can help managers by providing analysis for long-term decisions requiring both internal and external information. See Figure 6.16.

Figure 6.16

Decision-Making Level

Strategic managers are involved with long-term decisions, which are often made infrequently. Operational managers are involved with decisions that are made more frequently.

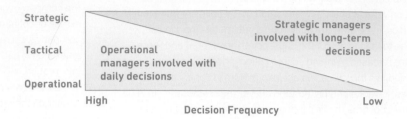

A Comparison of DSS and MIS

A DSS differs from an MIS in numerous ways, including the type of problems solved, the support given to users, the decision emphasis and approach, and the type, speed, output, and development of the system used. Table 6.1 lists brief descriptions of these differences.

Table 6.1

Comparison of DSSs and MISs

Factor	DSS	MIS
Problem Type	Can handle unstructured problems that cannot be easily programmed.	Normally used only with structured problems.
Users	Supports individuals, small groups, and the entire organization. In the short run, users typically have more control over a DSS.	Supports primarily the organization. In the short run, users have less control over an MIS.
Support	Supports all aspects and phases of decision making; it does not replace the decision maker—people still make the decisions.	In some cases, makes automatic decisions and replaces the decision maker.
Emphasis	Emphasizes actual decisions and decision-making styles.	Usually emphasizes information only.
Approach	Serves as a direct support system that provides interactive reports on computer screens.	Typically serves as an indirect support system that uses regularly produced reports.
System	Uses computer equipment that is usually online (directly connected to the computer system) and related to real time (providing immediate results). Computer terminals and display screens are examples—these devices can provide immediate information and answers to questions.	Uses printed reports that might be delivered to managers once per week, so it cannot provide immediate results.
Speed	Is flexible and can be implemented by users, so it usually takes less time to develop and is better able to respond to user requests.	Provides response time usually longer than a DSS.
Output	Produces reports that are usually screen oriented, with the ability to generate reports on a printer.	Is oriented toward printed reports and documents.
Development	Has users who are usually more directly involved in its development. User involvement usually means better systems that provide superior support. For all systems, user involvement is the most important factor for the development of a successful system.	Is frequently several years old and often was developed for people who are no longer performing the work supported by the MIS.

COMPONENTS OF A DECISION SUPPORT SYSTEM

dialogue manager

A user interface that allows decision makers to easily access and manipulate the DSS and to use common business terms and phrases.

At the core of a DSS are a database and a model base. In addition, a typical DSS contains a user interface, also called **dialogue manager**, which allows decision makers to easily access and manipulate the DSS and to use common business terms and phrases. Finally, access to the Internet, networks, and other computer-based systems permits the DSS to tie into other powerful systems, including enterprise systems or function-specific subsystems. Figure 6.17 shows a conceptual model of a DSS. Specific DSSs might not have all the components shown in Figure 6.17.

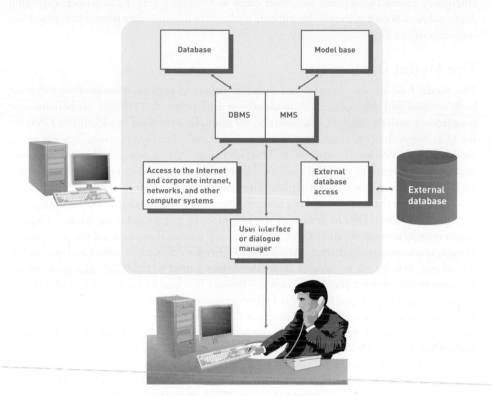

Figure 6.17

Conceptual Model of a DSS

DSS components include a model base; database; external database access; access to the Internet and corporate intranet, networks, and other computer systems; and a user interface or dialogue manager.

The Database

The database management system allows managers and decision makers to perform *qualitative analysis* on the company's vast stores of data in databases, data warehouses, and data marts, discussed in Chapter 3. A *data-driven DSS* primarily performs qualitative analysis based on the company's databases.[71] Data-driven DSSs tap into vast stores of information contained in the corporate database, retrieving information on inventory, sales, personnel, production, finance, accounting, and other areas. TD Securities used a pilot program from IBM called "stream computing" that allows the trading company to analyze 5 million pieces of data per second before the data is placed into corporate databases.[72] The data-driven DSS helps TD Securities sort through an immense amount of data and place profitable security trades automatically. According to a company executive, "In this business, quicker decisions are better decisions. If you fall behind, you're dealing with stale data and that puts you at a disadvantage." Exeros, which was acquired by IBM, provides technology and software to help companies sift through a mountain of data and return valuable information and decision support.[73] As another example, Tween Brands specialty retail store used the Oracle database to provide decision support to reduce inventory costs. Data mining and business intelligence, introduced in Chapter 3, are often used in a data-driven DSS.

Data-driven DSSs can be used in emergency medical situations to make split-second, life-or-death treatment decisions. Data-driven medical DSSs allow doctors to access the complete medical records of a patient.[74] Some medical record systems allow patients to enter their own health information into the database, such as medicines, allergies, and family health histories. WebMD, iHealthRecord, Walgreens, and PersonalHealthKey allow people to put their medical records online for rapid access.

A database management system can also connect to external databases to give managers and decision makers even more information and decision support. External databases can include the Internet, libraries, government databases, and more. Access to a combination of internal and external database can give key decision makers a better understanding of the company and its environment. Schumacher Group, for example, used software to mash up, or tie together, information from TV reports, maps, computerized phone books, and other sources to analyze the impact of hurricanes on how doctors are to be scheduled at different emergency rooms in Lafayette and other cities in Louisiana. Other companies, including Audi and AccuWeather, use similar software packages to integrate data from different sources into data-driven DSSs.

The Model Base

model base

Part of a DSS that allows managers and decision makers to perform quantitative analysis on both internal and external data.

The **model base** allows managers and decision makers to perform *quantitative analysis* on both internal and external data. A *model-driven DSS* primarily performs mathematical or quantitative analysis. Applied Mathematics, for example, developed model-driven DSSs for the U.S. Navy, the U.S. Coast Guard, and a number of other organizations.[75] Some of the model-driven DSSs have been developed for submarine warfare, search and rescue missions, healthcare information, and crop analysis. The model base gives decision makers access to a variety of models so that they can explore different scenarios and see their effects. Ultimately, it assists them in the decision-making process. Organic, Chrysler's digital marketing agency, used a model-driven DSS to determine the best place to invest marketing funds.[76] Organic used a team of economists and statisticians to develop models that predicted the effectiveness of various advertising alternatives. According to Chrysler's director of media and events, "As a marketer, it helps me be smarter about the dollars I need to reach the sales goals we are responsible for. It gives you some science." Procter & Gamble, maker of Pringles potato chips, Pampers diapers, and hundreds of other consumer products, used a model-driven DSS to streamline how raw materials and products flow from suppliers to customers. The model-driven DSS has saved the company hundreds of millions of dollars in supply chain-related costs. Model-driven DSSs are excellent at predicting customer behaviors. Some stock trading and investment firms use sophisticated model-driven DSSs to make trading decisions and huge profits. Some experts believe that a slight time advantage in computerized trading programs can result in millions of dollars of trading profits. Not all model-driven DSSs, however, provide superior results. In 2009, for example, many stock trading companies lost millions of dollars with models that didn't accurately predict stock market trends.[77] According to one chief investment officer of a large hedge fund, "We have been wrong more than usual." To help traders effectively use model-driven DSSs, a number of programs and Internet sites, such as *www.kaChing.com*, let people try quantitative trading strategies before they make real investments.[78]

model management software (MMS)

Software that coordinates the use of models in a DSS.

Model management software (MMS) can coordinate the use of models in a DSS, including financial, statistical analysis, graphical, and project-management models. Depending on the needs of the decision maker, one or more of these models can be used (see Table 6.2). What is important is how the mathematical models are used, not the number of models that an organization has available. In fact, too many model-based tools can be a disadvantage. According to the vice president of technology solutions at Allstate Insurance Company, "Usually people have more tools than they need, and that can be distracting."[79] MMS can often help managers effectively use multiple models in a DSS.

Table 6.2

Model Management Software

Model Type	Description	Software
Financial	Provides cash flow, internal rate of return, and other investment analysis	Spreadsheet, such as Microsoft Excel
Statistical	Provides summary statistics, trend projections, hypothesis testing, and more	Statistical programs, such as SPSS or SAS
Graphical	Assists decision makers in designing, developing, and using graphic displays of data and information	Graphics programs, such as Microsoft PowerPoint
Project Management	Handles and coordinates large projects; also used to identify critical activities and tasks that could delay or jeopardize an entire project if they are not completed in a timely and cost-effective fashion	Project management software, such as Microsoft Project

The User Interface or Dialogue Manager

The user interface or dialogue manager allows users to interact with the DSS to obtain information. It assists with all aspects of communications between the user and the hardware and software that constitute the DSS. In a practical sense, to most DSS users, the user interface is the DSS. Upper-level decision makers are often less interested in where the information came from or how it was gathered than that the information is both understandable and accessible.

GROUP SUPPORT SYSTEMS

The DSS approach has resulted in better decision making for all levels of individual users. However, many DSS approaches and techniques are not suitable for a group decision-making environment. A **group support system** (GSS), also called a *group decision support system* and a *computerized collaborative work system*, consists of most of the elements in a DSS, plus software to provide effective support in group decision-making settings (see Figure 6.18).

group support system (GSS)
Software application that consists of most elements in a DSS, plus software to provide effective support in group decision making; also called *group support system* or *computerized collaborative work system*.

Figure 6.18

Configuration of a GSS

A GSS contains most of the elements found in a DSS, plus software to facilitate group member communications.

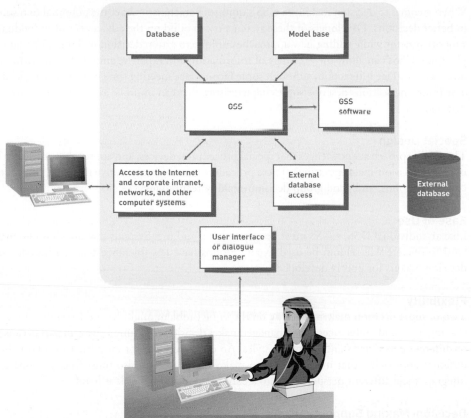

Group support systems are used in most industries, governments, and the military. Kraft, for example, used iPhones and other mobile devices to help managers and workers stay connected and work together on important projects.[80] Engineers at MWH, a Colorado firm, used social mapping software to show which employees effectively work together and collaborate on projects.[81] The software keeps track of employees that work together and then draws maps showing the relationships. The social mapping software helped MWH identify key employees that collaborate on projects and foster innovation and creative thinking. Engineers can use Mathcad Enterprise, another GSS. The software allows engineers to create, share, and reuse calculations. Collaboration using GSS can also save money. TransUnion, a large credit reporting company, reduced computer-related costs by about $2.5 million annually by investing $50,000 in a corporate social networking Internet site to improve collaboration.[82] According to the chief technology officer, "The savings mostly come out of teams that would have historically said 'Buy me more hardware' or 'I need a new software tool' who figured out how to solve their problems without asking for those things." Social networking Internet sites, such as Facebook and MySpace, can be used to support group decision making.[83] Serena, a software company in California, used the Facebook social networking site to collaborate on projects and exchange documents. The company believes this type of collaboration is so important that it has instituted "Facebook Fridays" to encourage its employees to use the social networking site to collaborate and make group decisions. Facebook Fridays also helped the company work with clients and recruit new employees. Many other organizations have used Facebook or have developed their own social networking sites to help their employees collaborate on important projects. With Ning (*www.ning.com*), for example, companies and individuals can create their own social networking sites. Many companies are using Ning and similar social networking sites to stay in contact with their customers.[84] Some executives, however, believe that social networking Internet sites are a waste of time and corporate resources.

Characteristics of a GSS That Enhance Decision Making

When it comes to decision making, a GSS's unique characteristics have the potential to result in better decisions. Developers of these systems try to build on the advantages of individual support systems while adding new approaches unique to group decision making. For example, some GSSs can allow the exchange of information and expertise among people without direct face-to-face interaction, although some face-to-face meeting time is usually beneficial. The following sections describe some characteristics that can improve and enhance decision making.

Special Design

The GSS approach acknowledges that special procedures, devices, and approaches are needed in group decision-making settings. These procedures must foster creative thinking, effective communications, and good group decision-making techniques.

Ease of Use

Like an individual DSS, a GSS must be easy to learn and use. Systems that are complex and hard to operate will seldom be used. Many groups have less tolerance than do individual decision makers for poorly developed systems.

Flexibility

Two or more decision makers working on the same problem might have different decision-making styles and preferences. Each manager makes decisions in a unique way, in part because of different experiences and cognitive styles. An effective GSS not only has to support the different approaches that managers use to make decisions, but also must find a means to integrate their different perspectives into a common view of the task at hand.

Decision-Making Support

delphi approach

A decision-making approach in which group decision makers are geographically dispersed; this approach encourages diversity among group members and fosters creativity and original thinking in decision making.

brainstorming

A decision-making approach that consists of members offering ideas "off the top of their heads."

A GSS can support different decision-making approaches, including the **delphi approach**, in which group decision makers are geographically dispersed throughout the country or the world. This approach encourages diversity among group members and fosters creativity and original thinking in decision making. In another approach, called **brainstorming**, members offer ideas "off the top of their heads," fostering creativity and free thinking. The $80 billion consumer products company Procter & Gamble used the Internet, video conferencing, and other technologies to brainstorm and think creatively about new products and services.[85] According to the director of global business services for the company, "It is an absolute necessity to be able to collaborate every day. We have a mandate to brainstorm, to listen, to innovate where competition is fierce." The **group consensus approach** forces members in the group to reach a unanimous decision. The Shuttle Project Engineering Office at the Kennedy Space Center has used the Consensus-Ranking Organizational-Support System (CROSS) to evaluate space projects in a group setting. The group consensus approach analyzes the benefits of various projects and their probabilities of success. CROSS is used to evaluate and prioritize advanced space projects. With the **nominal group technique**, each decision maker can participate; this technique encourages feedback from individual group members, and the final decision is made by voting, similar to a system for electing public officials.

Anonymous Input

Many GSSs allow anonymous input, whereby the person giving the input is not known to other group members. For example, some organizations use a GSS to help rank the performance of managers. Anonymous input allows the group decision makers to concentrate on the merits of the input without considering who gave it. In other words, input given by a top-level manager is given the same consideration as input from employees or other members of the group. Some studies have shown that groups using anonymous input can make better decisions and have superior results compared with groups that do not use anonymous input.

Reduction of Negative Group Behavior

One key characteristic of any GSS is the ability to suppress or eliminate group behavior that is counterproductive or harmful to effective decision making. In some group settings, dominant individuals can take over the discussion, which can prevent other members of the group from participating. In other cases, one or two group members can sidetrack or subvert the group into areas that are nonproductive and do not help solve the problem at hand. Other times, members of a group might assume they have made the right decision without examining alternatives—a phenomenon called *groupthink*. If group sessions are poorly planned and executed, the result can be a tremendous waste of time. Today, many GSS designers are developing software and hardware systems to reduce these types of problems. Procedures for effectively planning and managing group meetings can be incorporated into the GSS approach. A trained meeting facilitator is often employed to help lead the group decision-making process and to avoid groupthink. See Figure 6.19.

group consensus approach

A decision-making approach that forces members in the group to reach a unanimous decision.

nominal group technique

A decision-making approach that encourages feedback from individual group members, and the final decision is made by voting, similar to the way public officials are elected.

Figure 6.19

Using the GSS Approach

A trained meeting facilitator can help lead the group decision-making process and avoid groupthink.

(Source: © Alexander Hafemann/ iStockphoto.)

Parallel and Unified Communication

With traditional group meetings, people must take turns addressing various issues. One person normally talks at a time. With a GSS, every group member can address issues or make comments at the same time by entering them into a PC or workstation. These comments and issues are displayed on every group member's PC or workstation immediately. *Parallel communication* can speed meeting times and result in better decisions. Organizations are using unified communications to support group decision making. *Unified communications* ties together and integrates various communication systems, including traditional phones, cell phones, e-mail, text messages, the Internet, and more. With unified communications, members of a group decision-making team use a wide range of communications methods to help them collaborate and make better decisions.

Automated Recordkeeping

Most GSSs can keep detailed records of a meeting automatically. Each comment that is entered into a group member's PC or workstation can be anonymously recorded. In some cases, literally hundreds of comments can be stored for future review and analysis. In addition, most GSS packages have automatic voting and ranking features. After group members vote, the GSS records each vote and makes the appropriate rankings.

GSS Software

GSS software, often called *groupware* or *workgroup software*, helps with joint work group scheduling, communication, and management. Software from Autodesk, for example, has GSS capabilities that allow group members to work together on the design of manufacturing facilities, buildings, and other structures. Designers, for example, can use Autodesk's Buzzsaw Professional Online Collaboration Service, which works with AutoCAD, a design and engineering software product from Autodesk.

One popular package, IBM's Lotus Notes, can capture, store, manipulate, and distribute memos and communications that are developed during group projects. Some companies standardize on messaging and collaboration software, such as Lotus Notes. Lotus Connections (*www-01.ibm.com/software/lotus/products/connections/features.html*) is a feature of Lotus Notes that allows people to post documents and information on the Internet. The feature is similar to popular social networking sites such as Facebook, Linkedin, and MySpace but is designed for business use. Microsoft has invested billions of dollars in GSS software to incorporate collaborative features into its Office suite and related products. Office Communicator (*www.microsoft.com/unifiedcommunications*), for example, is a Microsoft product developed to allow better and faster collaboration. Other GSS software packages include Collabnet, OpenMind, and TeamWare. All of these tools can aid in group decision making. *Shared electronic calendars* can be used to coordinate meetings and schedules for decision-making teams. Using electronic calendars, team leaders can block out time for all members of the decision-making team.

A number of additional collaborative tools are available on the Internet. Sharepoint (*www.microsoft.com*), WebOffice (*www.weboffice.com*), and BaseCamp (*www.basecamphq.com*) are just a few examples. Fuze Meeting (*www.fuzemeeting.com*) provides video collaboration tools on the Internet.[86] The service can automatically bring participants into a live chat, allow workers to share information on their computer screens, and broadcast video content in high definition. Twitter (*www.twitter.com*) and Jaiku (*www.jaiku.com*) are Internet sites that some organizations use to help people and groups stay connected and coordinate work schedules. Yammer (*www.yammer.com*) is an Internet site that helps companies provide short answers to frequently asked questions. Managers and employees must first log into their private company network on Yammer to get their questions answered. Sermo (*www.sermo.com*) is a social networking site used by doctors to collaborate with other doctors, share their medical experiences, and even help make diagnoses. Teamspace (*www.teamspace.com*) is yet another collaborative software package that assists teams to successfully complete projects. Many of these Internet packages embrace the use of Web 2.0 technologies. Some executives, however, worry about security and corporate compliance issues with any new technology.

In addition to stand-alone products, GSS software is increasingly being incorporated into existing software packages. Today, some transaction processing and enterprise resource planning packages include collaboration software. Some ERP producers (see Chapter 5), for example, have developed groupware to facilitate collaboration and to allow users to integrate applications from other vendors into the ERP system of programs. In addition to groupware, GSSs use a number of tools discussed previously, including the following:

- E-mail, instant messaging (IM), and text messaging (TM)
- Video conferencing
- Group scheduling
- Project management
- Document sharing

WebEx WebOffice is a collaboration suite that lets colleagues and other team members work together from anywhere in the world.

(Source: *www.weboffice.com*.)

GSS Alternatives

Group support systems can take on a number of network configurations, depending on the needs of the group, the decision to be supported, and the geographic location of group members. GSS alternatives include a combination of decision rooms, local area networks, teleconferencing, and wide area networks.

- The **decision room** is a room that supports decision making, with the decision makers in the same building, combining face-to-face verbal interaction with technology to make the meeting more effective and efficient. Hewlett-Packard, for example, has developed the Halo system (*www.hp.com/halo/introducing.html*), which uses identical rooms containing a number of high-resolution screens and other devices to facilitate group decision making.[87] In some cases, the decision room might have a few computers and a projector for presentations. In other cases, the decision room can be fully equipped with a network of computers and sophisticated GSS software. A typical decision room is shown in Figure 6.20.

decision room
A room that supports decision making, with the decision makers in the same building, combining face-to-face verbal interaction with technology to make the meeting more effective and efficient.

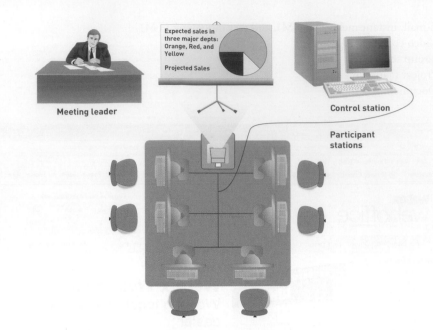

- The *local area decision network* can be used when group members are located in the same building or geographic area and under conditions in which group decision making is frequent. In these cases, the technology and equipment for the GSS approach is placed directly into the offices of the group members.
- *Teleconferencing* is used when the decision frequency is low and the location of group members is distant. These distant and occasional group meetings can tie together multiple GSS decision-making rooms across the country or around the world.
- The *wide area decision network* is used when the decision frequency is high and the location of group members is distant. In this case, the decision makers require frequent or constant use of the GSS approach. This GSS alternative allows people to work in **virtual workgroups**, in which teams of people located around the world can work on common problems.

virtual workgroups
Teams of people located around the world working on common problems.

EXECUTIVE SUPPORT SYSTEMS

executive support system (ESS)
Specialized DSS that includes all hardware, software, data, procedures, and people used to assist senior-level executives within the organization.

Because top-level executives often require specialized support when making strategic decisions, many companies have developed systems to assist executive decision making. This type of system, called an **executive support system (ESS)**, is a specialized DSS that includes all the hardware, software, data, procedures, and people used to assist senior-level executives within the organization. In some cases, an ESS, also called an *executive information system (EIS)*, supports decision making of members of the board of directors, who are responsible to stockholders. These top-level decision-making strata are shown in Figure 6.21. A nonprofit hospital in Boston developed an ESS for hospital administrators that will help hospital executives determine what critical drugs need to be immediately available to patients, spot possible safety issues, and help provide critical information for important staffing decisions.[88] The system should cost about $70 million and might be funded by federal stimulus spending. Valero Energy used an EIS to monitor plant performance.[89] According to the CIO of the company, "They review how each plant and unit is performing compared to the plan."

An ESS can also be used by individuals at middle levels in the organizational structure. Once targeted at the top-level executive decision makers, ESSs are now marketed to—and used by—employees at other levels in the organization.

Figure 6.21

The Layers of Executive
Decision Making

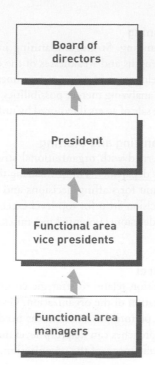

Executive Support Systems in Perspective

An ESS is a special type of DSS, and, like a DSS, is designed to support higher-level decision making in the organization. The two systems are, however, different in important ways. DSSs provide a variety of modeling and analysis tools to enable users to thoroughly analyze problems—that is, they allow users to *answer* questions. ESSs present structured information about aspects of the organization that executives consider important. The characteristics of an ESS are summarized in the following list:

- Are tailored to individual executives
- Are easy to use
- Have drill down abilities
- Support the need for external data
- Can help with situations that have a high degree of uncertainty
- Have a future orientation
- Are linked with value-added business processes

Capabilities of Executive Support Systems

The responsibility given to top-level executives and decision makers brings unique problems and pressures to their jobs. This section discusses some of the characteristics of executive decision making that are supported through the ESS approach. ESSs take full advantage of data mining, the Internet, blogs, podcasts, executive dashboards, social networking sites, and many other technological innovations. As you will note, most of these decisions are related to an organization's overall profitability and direction. An effective ESS should have the capability to support executive decisions with components such as strategic planning and organizing, crisis management, and more.

Support for Defining an Overall Vision

One of the key roles of senior executives is to provide a broad vision for the entire organization. This vision includes the organization's major product lines and services, the types of businesses it supports today and in the future, and its overriding goals.

strategic planning
Determining long-term objectives
by analyzing the strengths and
weaknesses of the organization,
predicting future trends, and
projecting the development of new
product lines.

Support for Strategic Planning

ESSs also support strategic planning. **Strategic planning** involves determining long-term objectives by analyzing the strengths and weaknesses of the organization, predicting future trends, and projecting the development of new product lines. It also involves planning the acquisition of new equipment, analyzing merger possibilities, and making difficult decisions concerning downsizing and the sale of assets if required by unfavorable economic conditions.

Support for Strategic Organizing and Staffing

Top-level executives are concerned with organizational structure. For example, decisions concerning the creation of new departments or downsizing the labor force are made by top-level managers. Overall direction for staffing decisions and effective communication with labor unions are also major decision areas for top-level executives. ESSs can help executives analyze the impact of staffing decisions, potential pay raises, changes in employee benefits, and new work rules.

Support for Strategic Control

Another type of executive decision relates to strategic control, which involves monitoring and managing the overall operation of the organization. Goal seeking can be done for each major area to determine what performance these areas need to achieve to reach corporate expectations. Effective ESS approaches can help top-level managers make the most of their existing resources and control all aspects of the organization.

Support for Crisis Management

Even with careful strategic planning, a crisis can occur. Major incidents, including natural disasters, fires, and terrorist activities, can totally shut down major parts of the organization. Handling these emergencies is another responsibility for top-level executives. In many cases, strategic emergency plans can be put into place with the help of an ESS. These contingency plans help organizations recover quickly if an emergency or crisis occurs.

Decision making is a vital part of managing businesses strategically. IS systems such as information and decision support, group support, and executive support systems help employees by tapping existing databases and providing them with current, accurate information. The increasing integration of all business information systems—from enterprise systems to MISs to DSSs—can help organizations monitor their competitive environment and make better-informed decisions. Organizations can also use specialized business information systems, discussed in the next chapter, to achieve their goals.

SUMMARY

Principle:

Good decision-making and problem-solving skills are the key to developing effective information and decision support systems.

Every organization needs effective decision making and problem solving to reach its objectives and goals. Problem solving begins with decision making. A well-known model developed by Herbert Simon divides the decision-making phase of the problem-solving process into three stages: intelligence, design, and choice. During the intelligence stage, potential problems or opportunities are identified and defined. Information is gathered that relates to the cause and scope of the problem. Constraints on the possible solution and the problem environment are investigated. In the design stage, alternative solutions to the problem are developed and explored. In addition, the feasibility and implications of these alternatives are evaluated. Finally, the choice stage involves selecting the best course of action. In this stage, the decision makers evaluate the implementation of the solution to determine whether the anticipated results were achieved and to modify the process in light of new information learned during the implementation stage.

Decision making is a component of problem solving. In addition to the intelligence, design, and choice steps of decision making, problem solving also includes implementation and monitoring. Implementation places the solution into effect. After a decision has been implemented, it is monitored and modified if necessary.

Decisions can be programmed or nonprogrammed. Programmed decisions are made using a rule, procedure, or quantitative method. Ordering more inventory when the level drops to 100 units or fewer is an example of a programmed decision. A nonprogrammed decision deals with unusual or exceptional situations. Determining the best training program for a new employee is an example of a nonprogrammed decision.

Decisions can use optimization, satisficing, or heuristic approaches. Optimization finds the best solution. Optimization problems often have an objective such as maximizing profits given production and material constraints. When a problem is too complex for optimization, satisficing is often used. Satisficing finds a good, but not necessarily the best, decision. Finally, a heuristic is a "rule of thumb" or commonly used guideline or procedure used to find a good decision.

Principle:

The management information system (MIS) must provide the right information to the right person in the right format at the right time.

A management information system is an integrated collection of people, procedures, databases, and devices that provides managers and decision makers with information to help achieve organizational goals. An MIS can help an organization achieve its goals by providing managers with insight into the regular operations of the organization so that they can control, organize, and plan more effectively and efficiently. The primary difference between the reports generated by the TPS and those generated by the MIS is that MIS reports support managerial decision making at the higher levels of management.

Data that enters the MIS originates from both internal and external sources. The most significant internal sources of data for the MIS are the organization's various TPSs and ERP systems. Data warehouses and data marts also provide important input data for the MIS. External sources of data for the MIS include extranets, customers, suppliers, competitors, and stockholders.

The output of most MISs is a collection of reports that are distributed to managers. These reports include scheduled reports, key-indicator reports, demand reports, exception reports, and drill-down reports. Scheduled reports are produced periodically, such as daily, weekly, or monthly. A key-indicator report is a special type of scheduled report. Demand reports are developed to provide certain information at a manager's request. Exception reports are automatically produced when a situation is unusual or requires management action. Drill-down reports provide increasingly detailed data about situations.

Management information systems have a number of common characteristics, including producing scheduled, demand, exception, and drill-down reports; producing reports with fixed and standard formats; producing hard-copy and soft-copy reports; using internal data stored in organizational computerized databases; and having reports developed and implemented by IS personnel or end users. Increasingly, MIS reports are being delivered over the Internet and through mobile devices, such as cell phones.

Most MISs are organized along the functional lines of an organization. Typical functional management information systems include financial, manufacturing, marketing, human

Altitude Online: Information and Decision Support System Considerations

Discussion Questions

1. What functional areas of Altitude Online are supported by MISs?
2. How do MISs and DSSs provide a value add to Altitude Online's products?

Critical Thinking Questions

1. How do you think MISs and DSSs assist Altitude Online's top executives in guiding the direction of the company?
2. How can the quality of information systems affect Altitude Online's ability to compete in the online marketing industry?

NOTES

Sources for the opening vignette: Oracle Staff, "Tru-Test Consolidates Applications, Integrates Business Units, and Increases Margins," Oracle Case Studies, *www.oracle.com/customers/solutions/bi.html*, accessed March 23, 2010; Tru-Test Web site, *www.tru- test.com*, accessed March 23, 2010; Citrix Web site, *www.citrix.com*, accessed March 23, 2010; Demantra Web page, *www.oracle.com/demantra*, accessed March 23, 2010.

1   Steel, Emily, "Marketers Take Search Ads Beyond Search Engines," *The Wall Street Journal,* January 20, 2009, p. B4.
2   Greenberg, Andy, "The Blindfolded Calculator," *Forbes,* July 13, 2009, p. 40.
3   Patterson, Scott, and Rogow, Geoffrey, "What's Behind High-Frequency Trading," *The Wall Street Journal,* August 1, 2009, p. B1.
4   Smith, Randall, "The Flash Trading Thorn in NYSE's Side," *The Wall Street Journal,* August 31, 2009, p.. C1.
5   Patterson, Scott, et al, "Ban on Flash Orders Is Considered by SEC," *The Wall Street Journal,* August 5, 2009, p. C1.
6   Langreth, Robert, "Wiring Medicine," *Forbes,* May 11, 2009, p. 40.
7   Ricknas, Mikael, "GPS Algorithm Error Prompts Garmin Recall," *PC World,* August 2009, p. 30.
8   Brandel, William, "Free Up Cash," *Computerworld,* August 17, 2009, p. 28.
9   Gnanlet, A., et al, "Sequential and Simultaneous Decision Making for Optimizing Health Care," *Decision Sciences,* May 2009, p. 295.
10 Romo, Frode, et al, "Optimizing the Norwegian Natural Gas Production and Transport," *Interfaces,* January-February 2009, p. 46.
11 Eveborn, P., et al, "Operations Research Improves Quality and Efficiency in Home Care," *Interfaces,* January 2009, p. 18.
12 Rai, S., et al, "LDP—O.R. Enhanced Productivity Improvements for the Printing Industry," *Interfaces,* January 2009, p. 69.
13 Hormby, S., et al, "Marriott International Increases Revenue by Implementing a Group Pricing Optimizer," *Interfaces,* January-February, 2010, p. 47.
14 D'Amours, S., et al, "Optimization Helps Shermag Gain Competitive Advantage," *Interfaces,* July-August, 2009, p. 329.
15 *www.cmis.csiro.au*, accessed July 7, 2009.
16 Staff, "Norton Internet Security 2010, *The New Zealand Herald,* October 5, 2009.
17 Weier, Mary Hayes, "Collaboration Is Key to Increased Efficiency," *Information Week,* September 14, 2009, p. 90.
18 Claburn, Thomas, "BI and the Web Are Front and Center," *Information Week,* September 14, 2009, p. 85.
19 Ioffe, Julia, "Tech Rx For Health Care," *Fortune,* March 16, 2009, p. 36.
20 *www.cognos.com,* accessed July 07, 2009.
21 *www.cognos.com/news/releases/2007/1112.html?mc=-web_hp,* accessed July 07, 2009.
22 *www.microsoft.com/dynamics/product/ business_scorecard_manager.mspx* , July 07, 2009.

23 Sanserino, Michael, "Peer Pressure And Other Pitches," *The Wall Street Journal,* September 14, 2009, p. B6.
24 Haynes, Dion V., "Retail Sales Rise for a Change," *The Washington Post,* October 9, 2009, p. A17.
25 Jackson-Higgins, Kelly, "Jive Wikis Meet SAP Analytics," *Information Week,* July 6, 2009, p. 15.
26 Cui, Carolyn, "Computer-Trading Models Meet Match," *The Wall Street Journal,* April 20, 2009, p. C3.
27 Feldman, Amy, "Buddy, Can You E-Mail Me 100 Bucks?" *BusinessWeek,* November 23, 2009, p. 68.
28 *www.prosper.com,* accessed July 07, 2009.
29 Kelly, C., "Turning an F Into Fun," *Computerworld,* May 26, 2008, p. 32.
30 Glader, Paul and Scannel, Kara, "GE Settles Civil-Fraud Charges," *The Wall Street Journal,* August 5, 2009, p. B2.
31 Banjo, Shelly, "The Best Online Tools for Personal Finance," *The Wall Street Journal,* June 8, 2009, p. R1.
32 Young, Lauren, "Big Banks Take Hint from Mint.com," *BusinessWeek,* October 12, 2009, p. 62.
33 *www.dell.com,* accessed October 12, 2009.
34 Weier, Mary Hayes, "Collaboration Is Key to Increased Efficiency," *Information Week,* September 14, 2009, p. 90.
35 Staff, "CimatronE to Power Live Cutting," *Drug Week,* October 9, 2009, p. 1550.
36 Linblad, C. and Maurer, H., "A Drumbeat of Deals," *BusinessWeek,* October 12, 2009, p. 4.
37 Dror, M., et al, "Deux Chemicals Goes Just-in-Time," *Interfaces,* November-December, 2009, p. 503.
38 Staff, "Education In Paper Magazine," *The Weekly Times,* July 8, 2009, p. 16.
39 Van Alphen, Tony, "Chrysler Adds Right-Hand Drive Minivan," *The Toronto Star,* September 9, 2009, p. B03.
40 Sandler, Kathy, "Web Ad Sales in Britain Overtake TV.," *The Wall Street Journal,* September 30, 2009, p. B7.
41 Fong, Mei, "Clinique, Sony Star in Web Sitcom," *The Wall Street Journal,* March 27, 2009, p. B4.
42 Schatz, Amy, "U.S. Seeks To Restrict Gifts Made To Bloggers," *The Wall Street Journal,* October 6, 2009, p. A1.
43 Hempel, Jessi, "Salesforce Hits Its Stride," *Fortune,* March 2, 2009, p. 29.
44 LaVallee, Andrew, "Unilever to Test Mobile Coupons," *The Wall Street Journal,* May 29, 2009, p. B8.
45 Staff, "Can You Believe What You Read on the Web," *Parade,* June, 21, 2009, p. 8.
46 Wagner, Mitch, "Opportunity Tweets," *Information Week,* June 1, 2009, p. 24.
47 Richmond, Riva, "A Start-Up's Tale: Tweet by Tweet," *The Wall Street Journal,* September 28, 2009, p. R10.
48 Hof, Robert, "Google's New Ad Weapon," *BusinessWeek,* June 22, 2009, p. 52.

49 Dornan, David and Jules, "Battle for the Brain," *Forbes,* November 16, 2009, p. 76.

50 Staff, "Loopt Now Available in the Android Market," *Drug Week,* January 2, 2009, p. 1852.

51 Baker, Stephen, "The Next Net," *BusinessWeek,* March 9, 2009, p. 42.

52 Helm, Burt, "Online Ads: Beyond Counting Clicks," *BusinessWeek,* March 9, 2009, p. 56.

53 Baker, Stephen, "The Web Knows What You Want," *BusinessWeek,* July 27, 2009, p. 48.

54 Yang, Jia L., "Veggie Tales," *Fortune,* June 8, 2009, p. 25.

55 Vascellaro, Jessica, "Radio Tunes Out Google," *The Wall Street Journal,* May 12, 2009, p. A1.

56 Vascellaro, Jessica, "Radio Tunes Out Google," *The Wall Street Journal,* May 12, 2009, p. A1.

57 Lowry, Tom, "Pandora: Unleashing Mobile Phone Ads," *BusinessWeek,* June 1, 2009, p. 52.

58 Kane, Y., "iPhone Gets Bigger as Ad Medium," *The Wall Street Journal,* May 12, 2009, p. B6.

59 Vascellaro, Jessica and Steel, Emily, "Something New Gains With Something Borrowed," *The Wall Street Journal,* June 5, 2009, p. B6.

60 Steel, Emily, "Web Ad Sales Open Door to Viruses," *The Wall Street Journal,* June 15, 2009, p. B7.

61 Kiley, David, "Paying for Viewers Who Pay Attention," *BusinessWeek,* May 18, 2009, p. 56.

62 Lawrence, Rick, et al, "Operations Research Improves Sales Force Productivity at IBM," *Interfaces,* January-February 2010, p. 33.

63 Moy, Patsy, "Sharing Just the Job For Women," *The Standard,* October 12, 2009.

64 Hardy, Quentin, "High Network Individual," *Forbes,* March 16, 2009, p. 20.

65 Baker, Stephen, "The Next Net," *BusinessWeek,* March 9, 2009, p. 42.

66 Mossberg, Walter, "Tracking Friends the Google Way," *The Wall Street Journal,* February 4, 2009, p. D2.

67 Wingfield, Nick, "Sharing Where You Are When You Care to Share," *The Wall Street Journal,* May 21, 2009, p. D1.

68 Christiansen, et al, "An Ocean of Opportunities," *OR/MS Today,* April 2009, p. 26.

69 Flisberg, Patrik, et al, "Billerud Optimizes Its Bleaching Process," *Interfaces,* March-April 2009, p. 119.

70 Staff, "How Clouds Can Change Management," *BusinessWeek,* June 15, 2009, p. 45.

71 Zweig, Jason, "Data Mining Isn't a Good Bet for Stock-Market Predictions," *The Wall Street Journal,* August 8, 2009, p. B1.

72 Hamm, Steve, "Big Blue Goes into Analysis," *BusinessWeek,* April 27, 2009, p. 16.

73 McDougall, Paul, "IBM Takes Aim at Data Deluge," *Information Week,* May 11, 2009, p. 24.

74 Ilie, V., et al, "Paper Versus Electronic Medical Records," *Decision Sciences,* May 2009, p. 213.

75 Browning, William, "Applied Mathematics Offers Wide Range of Services," *OR/MS Today,* April 2009, p. 22.

76 Steel, Emily, "Modeling Tool Stretches Ad Dollars," *The Wall Street Journal,* May 18, 2009, p. B7.

77 Cui, Carolyn, "Computer-Trading Models Meet Match," *The Wall Street Journal,* April 20, 2009, p. C3.

78 Lobb, Annelena, "Aspiring Quant Traders Get Taste of the Fantasy Behind Equations," *The Wall Street Journal,* June 17, 2009, p. C5.

79 Mitchell, Robert, "BI On A Budget," *Computerworld,* September 14, 2009, p. 23.

80 Weier, Mary Hayes, "Business Gone Mobile," *Information Week,* March 30, 2009, p. 23.

81 Dvorak, Phred, "Engineering Firm Charts Ties," *The Wall Street Journal,* January 26, 2009, p. B7.

82 Murphy, Chris, "TransUnion Finds Cost Savings, Seeks More," *Information Week,* March 23, 2009, p. 24.

83 Hempel, Jessi, "How Facebook Is Taking Over Our Lives," *Fortune,* March 2, 2009, p. 49.

84 Stern, Jack, "Build a Social Network for Your Business," *PC World,* December 2009, p. 31.

85 Hamblen, Matt, "Finding The Stars With Bright Ideas," *Computerworld,* September 14, 2009, p. 10.

86 Stern, Zack, "Collaboration Online with Fuze Meeting Service," *PC World,* March 2009, p. 34.

87 Scheck, J. and White, B., "Telepresence Is Taking Hold," *The Wall Street Journal,* May 6, 2009, p. B6.

88 Ioffe, Julia, "Tech RX for Health Care," *Fortune,* March 16, 2009, p. 36.

89 Henschen, Doug, "From Gut to Facts," *Information Week,* November 16, 2009, p. 6.

As you read this chapter, consider the following:

- What steps can a business take to retain corporate knowledge within the business?
- How can computer intelligence be harnessed to serve corporate needs in various industries?
- How can people and businesses make the best use of specialized systems?

Why Learn About Knowledge Management and Specialized Information Systems?

Knowledge management and specialized information systems are used in almost every industry. If you are a manager, you might use a knowledge management system to support decisive action to help you correct a problem. If you are a production manager at an automotive company, you might oversee robots that attach windshields to cars or paint body panels. As a young stock trader, you might use a special system called a *neural network* to uncover patterns and make millions of dollars trading stocks and stock options. As a marketing manager for a PC manufacturer, you might use virtual reality on a Web site to show customers your latest laptop and desktop computers. If you are in the military, you might use computer simulation as a training tool to prepare you for combat. In a petroleum company, you might use an expert system to determine where to drill for oil and gas. You will see many additional examples of using these specialized information systems throughout this chapter. Learning about these systems will help you discover new ways to use information systems in your day-to-day work.

Like other aspects of an information system, the overall goal of knowledge management and the specialized systems discussed in this chapter is to help people and organizations achieve their goals. In this chapter, we explore knowledge management, artificial intelligence, and many other specialized information systems, including expert systems, robotics, vision systems, natural language processing, learning systems, neural networks, genetic algorithms, intelligent agents, multimedia, and virtual reality.

KNOWLEDGE MANAGEMENT SYSTEMS

Chapter 1 defines and discusses data, information, and knowledge. Recall that *data* consists of raw facts, such as an employee number, number of hours worked in a week, inventory part numbers, or sales orders. A list of the quantity available for all items in inventory is an example of data. When these facts are organized or arranged in a meaningful manner, they become information. *Information* is a collection of facts organized so that they have additional value beyond the value of the facts themselves. An exception report of inventory items that might be out of stock in a week because of high demand is an example of information. *Knowledge* is the awareness and understanding of a set of information and the ways that information can be made useful to support a specific task or reach a decision. Knowing the procedures for ordering more inventory to avoid running out is an example of knowledge. In a sense, information (low inventory levels for some items), tells you what has to be done while knowledge tells you how to do it (make two important phone calls to the right people to get the needed inventory shipped overnight). See Figure 7.1.

Figure 7.1

The Differences Among Data, Information, and Knowledge

Data	There are 20 PCs in stock at the retail store.
Information	The store will run out of inventory in a week unless more is ordered today.
Knowledge	Call 800-555-2222 to order more inventory.

A *knowledge management system (KMS)* is an organized collection of people, procedures, software, databases, and devices used to create, store, share, and use the organization's knowledge and experience.[1] KMSs cover a wide range of systems, from software that contains some KMS components to dedicated systems designed specifically to capture, store, and use knowledge.

Overview of Knowledge Management Systems

Like the other systems discussed throughout the book, including information and decision support systems, knowledge management systems attempt to help organizations achieve their goals. For businesses, this usually means increasing profits or reducing costs. One study of a large information systems consulting firm found an $18.60 return on every dollar invested in its knowledge management system, representing a greater than 1,000 percent return on investment (ROI).[2] This outstanding ROI was a result of time and cost savings from superior knowledge retrieval and usage. Advent, a San Francisco company that develops investment applications for hedge funds and financial services companies, used a KMS to help its employees locate and use critical information.[3] According to a company executive, "Advent has been doing KM for over a decade—we're pretty experienced, and it's integrated into our culture." For nonprofit organizations, KM can mean providing better customer service or providing special needs to people and groups.

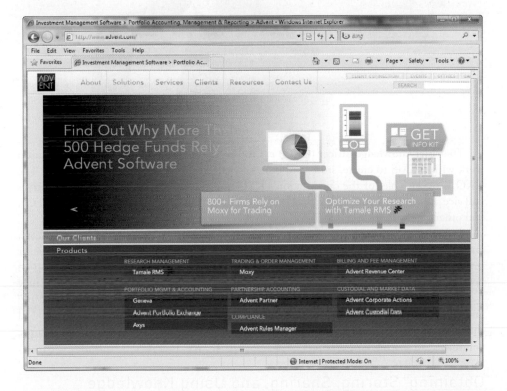

Advent Software used a knowledge management system to help its employees find critical investment information.

(Source: *www.advent.com*.)

A KMS can involve different types of knowledge.[4] *Explicit knowledge* is objective and can be measured and documented in reports, papers, and rules. For example, knowing the best road to take to minimize drive time from home to the office when a major highway is closed is explicit knowledge. It can be documented in a report or a rule, as in "If I-70 is closed, take Highway 6 to the office." *Tacit knowledge*, on the other hand, is hard to measure and document and typically is not objective or formalized. Knowing the best way to negotiate with a foreign government about nuclear disarmament or a volatile hostage situation often requires a lifetime of experience and a high level of skill. These are examples of tacit knowledge. It is difficult to write a detailed report or a set of rules that would always work in every hostage situation. Many organizations actively attempt to convert tacit knowledge to explicit knowledge to make the knowledge easier to measure, document, and share with others.[5]

Data and Knowledge Management Workers and Communities of Practice

The personnel involved in a KMS include data workers and knowledge workers. Secretaries, administrative assistants, bookkeepers, and similar data-entry personnel are often called *data workers*. As mentioned in Chapter 1, *knowledge workers* are people who create, use, and disseminate knowledge.[6] They are usually professionals in science, engineering, or business and they usually work in offices and belong to professional organizations. Other examples of knowledge workers include writers, researchers, educators, and corporate designers. See Figure 7.2.

Figure 7.2

Knowledge Workers

Knowledge workers are people who create, use, and disseminate knowledge, including professionals in science, engineering, business, and other areas.

(Source: © Josh Hodge/ iStockphoto.)

chief knowledge officer (CKO)
A top-level executive who helps the organization use a KMS to create, store, and use knowledge to achieve organizational goals.

The **chief knowledge officer (CKO)** is a top-level executive who helps the organization work with a KMS to create, store, and use knowledge to achieve organizational goals. The CKO is responsible for the organization's KMS and typically works with other executives and vice presidents, including the chief executive officer (CEO), chief financial officer (CFO), and others.

Some organizations and professions use *communities of practice (COP)* to create, store, and share knowledge. A COP is a group of people dedicated to a common discipline or practice, such as open-source software, auditing, medicine, or engineering. A group of oceanographers investigating climate change or a team of medical researchers looking for new ways to treat lung cancer are examples of COPs. COPs excel at obtaining, storing, sharing, and using knowledge. A study of knowledge workers in a large insurance company showed that a COP shares information better, solves problems more collaboratively, and is more committed to sharing best practices.[7]

Obtaining, Storing, Sharing, and Using Knowledge

Obtaining, storing, sharing, and using knowledge is the key to any KMS. MWH Global (*www.mwhglobal.com*), located in Colorado, uses a KMS to create, disseminate, and use knowledge specializing in environmental engineering, construction, and management activities worldwide. The company has about 7,000 employees and 170 offices around the world. Using a KMS often leads to additional knowledge creation, storage, sharing, and usage. Drug companies and medical researchers invest billions of dollars in creating knowledge on cures for diseases. Knowledge management systems can also diminish the reliance on paper reports, which reduces costs and helps protect the environment.[8] According to one expert, "Going green has become a topic of increased attention lately, but it's nothing new to knowledge management." Although knowledge workers can act alone, they often work in teams to create or obtain knowledge. See Figure 7.3.

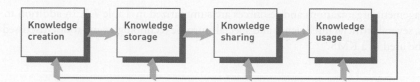

Figure 7.3

Knowledge Management System

Obtaining, storing, sharing, and using knowledge is the key to any KMS.

After knowledge is created, it is often stored in a *knowledge repository* that includes documents, reports, files, and databases.[9] The knowledge repository can be located both inside the organization and outside. Some types of software can store and share knowledge contained in documents and reports. Adobe Acrobat PDF files, for example, allow you to store corporate reports, tax returns, and other documents and send them to others over the Internet.[10] This publisher and the authors of this book used PDF files to store, share, and edit each chapter. Traditional databases, data warehouses, and data marts, discussed in Chapter 3, often store the organization's knowledge. Specialized knowledge bases in expert systems, discussed later in this chapter, can also be used.

Because knowledge workers often work in groups or teams, they can use collaborative work software and group support systems (discussed in Chapter 6) to share knowledge, such as groupware, meeting software, and collaboration tools. Intranets and password protected Internet sites also provide ways to share knowledge. Many businesses, however, use patents, copyrights, trade secrets, Internet firewalls, and other measures to keep prying eyes from seeing important knowledge that is expensive and hard to create.

Using a knowledge management system begins with locating the organization's knowledge. This is often done using a *knowledge map* or directory that points the knowledge worker to the needed knowledge. Drug companies have sophisticated knowledge maps that include database and file systems to allow scientists and drug researchers to locate previous medical studies. Medical researchers, university professors, and even textbook authors use Lexis-Nexis to locate important knowledge. Corporations often use the Internet or corporate Web portals to help their knowledge workers find knowledge stored in documents and reports.

Technology to Support Knowledge Management

KMSs use a number of tools discussed throughout the book. In Chapter 1, for example, we explored the importance of *organizational learning* and *organizational change*. An effective KMS is based on learning new knowledge and changing procedures and approaches as a result. A manufacturing company, for example, might learn new ways to program robots on the factory floor to improve accuracy and reduce defective parts. The new knowledge will likely cause the manufacturing company to change how it programs and uses its robots. In Chapter 3, we investigated the use of *data mining* and *business intelligence*. These powerful tools can be important in capturing and using knowledge. Enterprise resource planning tools, such as SAP, include knowledge management features. In Chapter 6, we showed how *groupware* can improve group decision making and collaboration. Groupware can also be used to help capture, store, and use knowledge. Of course, hardware, software, databases, telecommunications, and the Internet, discussed in Part II, are important technologies used to support most knowledge management systems.

Hundreds of organizations provide specific KM products and services.[11] See Figure 7.4. In addition, researchers at colleges and universities have developed tools and technologies to support knowledge management. American companies pay billions of dollars on knowledge management technology every year. Companies such as IBM have many knowledge management tools in a variety of products, including Lotus Notes, discussed in Chapter 6. Microsoft offers a number of knowledge management tools, including Digital Dashboard, which is based on the Microsoft Office suite. Digital Dashboard integrates information from a variety of sources, including personal, group, enterprise, and external information and documents. Other tools from Microsoft include Web Store Technology, which uses wireless technology to deliver knowledge to any location at any time; Access Workflow Designer, which helps database developers create effective systems to process transactions and keep work flowing through the organization; and related products. Some additional knowledge management organizations and resources are summarized in Table 7.1. In addition to these tools, several artificial intelligence and special-purpose technologies and tools, discussed next, can be used in a KMS.

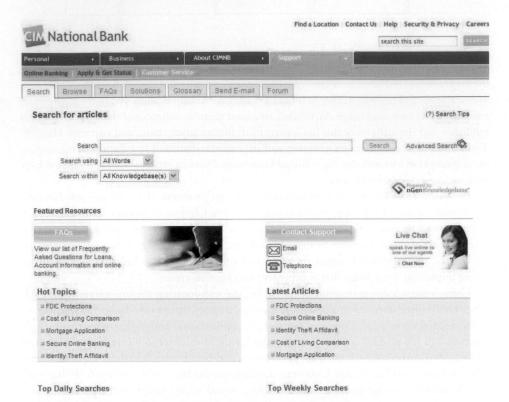

Company	Description	Web Site
CortexPro	Knowledge management collaboration tools	*www.cortexpro.com*[12]
Delphi Group	A knowledge management consulting company	*www.delphigroup.com*[13]
Knowledge Management Resource Center	Knowledge management sites, products and services, magazines, and case studies	*www.kmresource.com*[14]
Knowledge Management Solutions, Inc.	Tools to create, capture, classify, share, and manage knowledge	*www.kmsi.us*[15]
Knowledge Management Web Directory	A directory of knowledge management Web sites	*www.knowledge-manage.com*[16]
KnowledgeBase	Content creation and management	*www.knowledgebase.net*[17]
Law Clip Knowledge Manager	A service that collects and organizes text, Web links, and more from law-related Web sites	*www.lawclip.com*[18]
Knowledge Management Consortium International	Offers knowledge management training and support	*www.kmci.org/index.html*[19]

AN OVERVIEW OF ARTIFICIAL INTELLIGENCE

At a Dartmouth College conference in 1956, John McCarthy proposed the use of the term *artificial intelligence (AI)* to describe computers with the ability to mimic or duplicate the functions of the human brain. For example, advances in AI have led to systems that can recognize complex patterns.

Many AI pioneers attended this first conference; a few predicted that computers would be as "smart" as people by the 1960s. The prediction has not yet been realized, but the benefits of artificial intelligence in business and research can be seen today, and research continues.

Artificial Intelligence in Perspective

Artificial intelligence systems include the people, procedures, hardware, software, data, and knowledge needed to develop computer systems and machines that demonstrate characteristics of human intelligence. Artificial intelligence can be used by most industries and applications. Researchers, scientists, and experts on how human beings think are often involved in developing these systems.

artificial intelligence systems
People, procedures, hardware, software, data, and knowledge needed to develop computer systems and machines that demonstrate the characteristics of intelligence.

The Nature of Intelligence

From the early AI pioneering stage, the research emphasis has been on developing machines with the ability to "learn" from experiences and apply knowledge acquired from those experiences, handle complex situations, solve problems when important information is missing, determine what is important, react quickly and correctly to a new situation, understand visual images, process and manipulate symbols, be creative and imaginative, and use heuristics, which together is considered **intelligent behavior**.[20] In a book called *The Singularity Is Near* and in articles by and about him, Ray Kurzweil predicts computers will have human-like intelligence in 20 years.[21] The author also foresees that, by 2045, human and machine intelligence might merge. Machine intelligence, however, is hard to achieve.

The *Turing Test* attempts to determine whether the responses from a computer with intelligent behavior are indistinguishable from responses from a human being. No computer has passed the Turing Test, developed by Alan Turing, a British mathematician. The Loebner Prize offers money and a gold medal for anyone developing a computer that can pass the Turing Test (see *www.loebner.net*). Some of the specific characteristics of intelligent behavior include the ability to do the following:

intelligent behavior
The ability to learn from experiences and apply knowledge acquired from experience, handle complex situations, solve problems when important information is missing, determine what is important, react quickly and correctly to a new situation, understand visual images, process and manipulate symbols, be creative and imaginative, and use heuristics.

- *Learn from experience and apply the knowledge acquired from experience.* Learning from past situations and events is a key component of intelligent behavior and is a natural ability of humans, who learn by trial and error. This ability, however, must be carefully programmed into a computer system. Today, researchers are developing systems that can "learn" from experience.[22] For instance, computerized AI chess software can "learn" to improve while playing human competitors. In one match, Garry Kasparov competed against a personal computer with AI software developed in Israel, called Deep Junior. This match was a 3-3 tie, but Kasparov picked up something the machine would have no interest in—$700,000. The 20 questions (20q) Web site, *www.20q.net*, is another example of a system that learns.[23] The Web site is an artificial intelligence game that learns as people play.

- *Handle complex situations.* People are often involved in complex situations. In a business setting, top-level managers and executives must handle a complex market, challenging competitors, intricate government regulations, and a demanding workforce. Even human experts make mistakes in dealing with these situations. Very careful planning and elaborate computer programming are necessary to develop systems that can handle complex situations.

- *Solve problems when important information is missing.* An integral part of decision making is dealing with uncertainty. Often, decisions must be made with little information or inaccurate information because obtaining complete information is too costly or

20Q is an online game in which users play the popular game, Twenty Questions, against an artificial intelligence foe.

(Source: *www.20q.net*.)

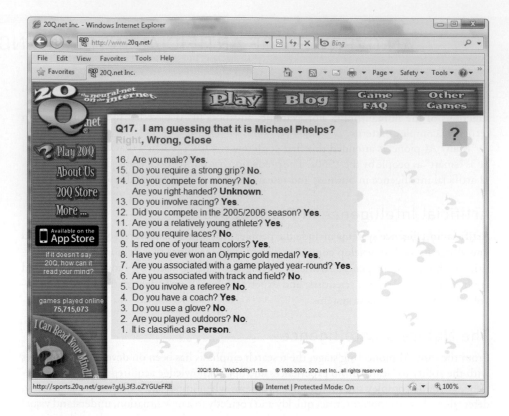

impossible. Today, AI systems can make important calculations, comparisons, and decisions even when information is missing.

- *Determine what is important.* Knowing what is truly important is the mark of a good decision maker. Developing programs and approaches to allow computer systems and machines to identify important information is not a simple task.

- *React quickly and correctly to a new situation.* A small child, for example, can look over an edge and know not to venture too close. The child reacts quickly and correctly to a new situation. Computers, on the other hand, do not have this ability without complex programming.

- *Understand visual images.* Interpreting visual images can be extremely difficult, even for sophisticated computers. Moving through a room of chairs, tables, and other objects can be trivial for people but extremely complex for machines, robots, and computers. Such machines require an extension of understanding visual images, called a **perceptive system**. Having a perceptive system allows a machine to approximate the way a person sees, hears, and feels objects. Military robots, for example, use cameras and perceptive systems to conduct reconnaissance missions to detect enemy weapons and soldiers.[24] Detecting and destroying them can save lives.

- *Process and manipulate symbols.* People see, manipulate, and process symbols every day. Visual images provide a constant stream of information to our brains. By contrast, computers have difficulty handling symbolic processing and reasoning. Although computers excel at numerical calculations, they aren't as good at dealing with symbols and three-dimensional objects. Recent developments in machine-vision hardware and software, however, allow some computers to process and manipulate some symbols.

- *Be creative and imaginative.* Throughout history, some people have turned difficult situations into advantages by being creative and imaginative. For instance, when defective mints with holes in the middle arrived at a candy factory, an enterprising entrepreneur decided to market these new mints as LifeSavers instead of returning them to the manufacturer. Ice cream cones were invented at the St. Louis World's Fair when an imaginative store owner decided to wrap ice cream with a waffle from his grill for portability. Developing new products and services from an existing (perhaps negative)

perceptive system
A system that approximates the way a person sees, hears, and feels objects.

situation is a human characteristic. Although software has been developed to enable a computer to write short stories, few computers can be imaginative or creative in this way.

• *Use heuristics.* For some decisions, people use heuristics (rules of thumb arising from experience) or even guesses. In searching for a job, you might rank the companies you are considering according to profits per employee. Today, some computer systems, given the right programs, obtain good solutions that use approximations instead of trying to search for an optimal solution, which would be technically difficult or too time consuming.

This list of traits only partially defines intelligence. Unlike the terminology used in virtually every other field of IS research, in which the objectives can be clearly defined, the term *intelligence* is a formidable stumbling block. Another challenge is linking a human brain to a computer.[25]

The Brain Computer Interface

Developing a link between the human brain and the computer is another exciting area that touches all aspects of artificial intelligence. Called *Brain Computer Interface (BCI)*, the idea is to directly connect the human brain to a computer and have human thought control computer activities.[26] Two exciting studies conducted at the Massachusetts General Hospital in Boston will attempt to use a chip called BrainGate to connect a human brain to a computer. If successful, the BCI experiment will allow people to control computers and artificial arms and legs through thought alone. The objective is to give people without the ability to speak or move (called Locked-in Syndrome) the ability to communicate and move artificial limbs using advanced BCI technologies. Honda Motors has developed a BCI system that allows a person to complete certain operations, like bending a leg, with 90 percent accuracy.[27] The new system uses a special helmet that can measure and transmit brain activity to a computer.

The Major Branches of Artificial Intelligence

AI is a broad field that includes several specialty areas, such as expert systems, robotics, vision systems, natural language processing, learning systems, and neural networks (see Figure 7.5). Many of these areas are related; advances in one can occur simultaneously with or result in advances in others.

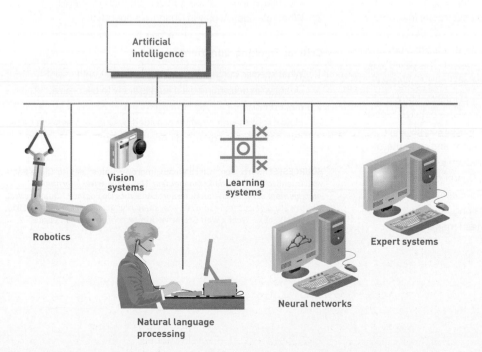

Figure 7.5

A Conceptual Model of Artificial Intelligence

Artificial intelligence

Robotics

Vision systems

Natural language processing

Learning systems

Neural networks

Expert systems

WebEx Uses AI-Powered Analytics to Focus Salesforce

WebEx is an online conferencing service that allows participants to view a common computer screen broadcasted over an Internet connection while converseing over the phone. With WebEx, a salesperson or academic researcher could present a PowerPoint presentation from his or her office to interested people scattered around the world. Software developers use WebEx to demonstrate products to clients. Corporations use WebEx to hold online meetings and train new employees. Other people use WebEx for so-called webinars (seminars on the Web) that launch new ideas or products and for help desk operations and customer support.

WebEx is not the only company with an online conferencing product. It fights for market share like most commercial enterprises. WebEx has a large salesforce that searches for potential clients, or leads, and then works to turn leads into customers. In sales, it is often difficult to predict which leads will pan out and which will dry up. Because of this, sales representatives sometimes waste time working with leads who have no intention of becoming paying customers. Salesforces use various lead management systems to help ensure that leads will eventually close a deal. The rate at which leads decide to close a deal and become a customer is referred to as the lead-to-close rate. A successful salesforce works hard to improve its lead-to-close rate.

To determine which leads will turn into sales, you must predict the future. You can make accurate predictions if you understand history, current facts, and trends. What types of behavior and conditions have led to sales in the past? Do those behaviors and conditions exist in a current scenario? Are past trends the same as current trends?

To accurately predict the future of a sale, it helps to study and understand customer behavior, past and present. This practice is sometimes referred to as behavioral analytics. Some businesses specialize in behavioral analytics. They collect information about customers, study it thoroughly, and then provide a detailed report on the characteristics of likely customers. More recently, new behavioral analytics software can analyze leads and customers and then advise a sales representative on which leads to pursue.

WebEx selected a SaaS solution from Quantivo Corporation to assist its salesforce in improving its lead-to-close rate. The system collects data from WebEx's CRM system along with financial records, Web site analytics, and WebEx's lead management system. It provides this information to sophisticated AI algorithms. Using data-mining techniques, the Quantivo software "learns" from past sales successes and failures to determine which lead characteristics make for likely sales. Using that information, the system creates lead profiles that the salesforce uses to identify current prospects that are ready to buy. Customers and potential customers are also evaluated over time to determine critical touch points, which are opportunities when they may be open to further investment.

Quantivo is easy to use so that sales representatives unfamiliar with the details of behavioral analytics can understand the customer qualities to identify. WebEx also uses the software to identify market segments on which to concentrate the efforts of its entire salesforce so they can win more market share with less investment. Sales managers can also evaluate sales patterns relative to individual sales representatives, competitors, sales regions, and other filters.

The amount of information that Quantivo processes in a few seconds or minutes would take a human analyst days or weeks to calculate. Because Quantivo is programmed with machine intelligence to identify profile attributes that lead to sales, it carries out this otherwise human activity continuously and tirelessly, providing new valuable lead data every day. All the sales representative needs to do is enter data collected about each new lead, and Quantivo takes it from there.

Discussion Questions

1. How does Quantivo assist a salesforce in being more productive and a company in gaining market share?
2. What role does AI play in Quantivo's operations?

Critical Thinking Questions

1. What specific customer data do you think would assist Quantivo in determining if a lead is likely to become a customer?
2. How might the information provided by Quantivo assist a sales rep in seeking out new and unknown leads?

SOURCES: Schwartz, Joe, "Cisco WebEx Improves Sales Lead-to-Close Rates with Quantivo," *Information Management*, May 2009, *www.information-management.com*; "What Can Behavioral Analytics Do For Me?," Quantivo Web site, *www.quantivo.com/resources/me_sales.php*, accessed February 2, 2010; "What is WebEx," WebEx Web site, *www.webex.com*, accessed February 2, 2010.

Expert Systems

An **expert system** consists of hardware and software that stores knowledge and makes inferences, similar to those of a human expert.[28] Because of their many business applications, expert systems are discussed in more detail in the next several sections of the chapter.

Robotics

Robotics involves developing mechanical or computer devices that can paint cars, make precision welds, and perform other tasks that require a high degree of precision or are tedious or hazardous for human beings. The word "robot" comes from a play by Karel Capek in the 1920s, when he used the word "robota" to describe factory machines that do drudgery work and revolt.[29] The use of robots has expanded and is likely to increase in the future. For many businesses, robots are used to do the three Ds—dull, dirty, and dangerous jobs. Manufacturers use robots to locate, assemble, and paint products.[30] Some robots, such as the ER series by Intelitek (*www.intelitek.com*), can be used for training or entertainment. Contemporary robotics combine both high-precision machine capabilities and sophisticated controlling software. The controlling software in robots is what is most important in terms of AI.

The field of robotics has many applications, and research into these unique devices continues. The following are a few examples:

- The Robot Learning Laboratory, part of the computer science department and the Robotics Institute at Carnegie Mellon University (*www.ri.cmu.edu*), conducts research into the development and use of robotics.[31]
- IRobot *(www.irobot.com)* is a company that builds a number of robots, including the Roomba Floorvac for cleaning floors and the PackBot, an unmanned vehicle used to assist and protect soldiers.
- The DEKA arm *(www.dekaresearch.com/deka_arm.shtml)*, named "Luke" after Luke Skywalker in the Star Wars movies, is a robotic artificial arm under development that developers are hoping can be controlled by human thought in the future.
- Robots are used in a variety of ways in medicine. The Porter Adventist Hospital (*www.porterhospital.org*) in Denver, Colorado, uses a $1.2 million Da Vinci Surgical System to perform surgery on prostate cancer patients.[32] The robot has multiple arms that hold surgical tools. According to one doctor at Porter, "The biggest advantage is it improves recovery time. Instead of having an eight-inch incision, the patient has a 'band-aid' incision. It's much quicker." The Heart-Lander is a very small robot that is inserted below the rib cage and used to perform delicate heart surgery. Cameron Riviere at the Carnegie Mellon Robotics Institute (*www.ri.cmu.edu*) developed the robot with help from Johns Hopkins University.
- DARPA (The Defense Advanced Research Project Agency) sponsors the DARPA Grand Challenge (*www.darpagrandchallenge.com*), a 132-mile race over rugged terrain for computer-controlled cars. The agency also sponsors other races and challenges.[33]
- The Hybrid Assistive Limb (HAL) lab is developing a robotic suit to help paraplegics and stroke victims move so they can perform basic functions. The suit helps with lifting heavy objects, walking long distances, or performing other basic movements that can't be done otherwise. HAL was also the name of an artificial-intelligence computer in the classic movie *2001: A Space Odyssey*. The letters in HAL are each one letter up from the letters in IBM.
- In the military, robots are moving beyond movie plots to become real weapons. The Air Force is developing a smart robotic jet fighter. Often called *unmanned combat air vehicles (UCAVs)*, these robotic war machines, such as the X-45A, will be able to identify and destroy targets without human pilots. UCAVs send pictures and information to a central command center and can be directed to strike military targets. These machines extend the current Predator and Global Hawk technologies the military used in Afghanistan and Iraq after the September 11 terrorist attacks. Big Dog, made by Boston Dynamics (*www.bostondynamics.com*), is a robot that can carry up to 200 pounds of military gear in field conditions.

robotics
The development of mechanical or computer devices that perform tasks requiring a high degree of precision or that are tedious or hazardous for humans.

Big Dog, manufactured by Boston Dynamics, is a robot that can carry up to 200 pounds of military gear in field conditions.

(Source: Courtesy of Boston Dynamics.)

Although most of today's robots are limited in their capabilities, future robots will find wider applications in banks, restaurants, homes, doctors' offices, and hazardous working environments such as nuclear stations. The Repliee Q1 and Q2 robots from Japan are ultra-human-like robots or androids that can blink, gesture, speak, and even appear to breathe (*www.ed.ams.eng.osaka-u.ac.jp/development/Android_ReplieeQ2_e.html*). See Figure 7.6. Microrobotics, also called *micro-electro-mechanical systems (MEMS)*, are also being developed (*www.memsnet.org/mems/what-is.html*). MEMS can be used in a person's bloodstream to monitor the body and in air bags, cell phones, refrigerators, and more.

Figure 7.6

The Repliee Q2 Robot from Japan

(Source: AP Photo/Katsumi Kasahara.)

Vision Systems

Another area of AI involves vision systems. **Vision systems** include hardware and software that permit computers to capture, store, and manipulate visual images.[34] Vision systems are effective at identifying people based on facial features. Nvidia's GeForce 3D is software that can display images on a computer screen that are three-dimensional when viewed using special glasses.[35]

vision systems
The hardware and software that permit computers to capture, store, and manipulate visual images.

Natural Language Processing and Voice Recognition

As discussed in Chapter 2, **natural language processing** allows a computer to understand and react to statements and commands made in a "natural" language, such as English.[36] Google, for example, has a service called Google Voice Local Search that allows you to dial a toll-free number and search for local businesses using voice commands and statements. Many companies provide natural language processing help over the phone. When you call the help phone number, you are typically given a menu of options and asked to speak your responses. Many people, however, are frustrated talking to a machine instead of a human.

In some cases, voice recognition is used with natural language processing. *Voice recognition* involves converting sound waves into words. After converting sounds into words, natural language processing systems react to the words or commands by performing a variety of tasks. Brokerage services are a perfect fit for voice recognition and natural language processing technology to replace the existing "press 1 to buy or sell a stock" touchpad telephone menu system. Using voice recognition to convert recordings into text is also possible. Some companies claim that voice recognition and natural language processing software is so good that customers forget they are talking to a computer and start discussing the weather or sports scores.

natural language processing
Processing that allows the computer to understand and react to statements and commands made in a "natural" language, such as English.

Dragon Systems' NaturallySpeaking 10 uses continuous voice recognition, or natural speech, allowing the user to speak to the computer at a normal pace without pausing between words. The spoken words are transcribed immediately onto the computer screen.

(Source: Courtesy of Nuance Communications, Inc.)

Learning Systems

Another part of AI deals with **learning systems**, a combination of software and hardware that allows a computer to change how it functions or reacts to situations based on feedback it receives. For example, some computerized games have learning abilities. If the computer does not win a game, it remembers not to make the same moves under the same conditions again.[37] IBM, for example, has developed a computer called Watson after one of its founders that will challenge humans in the popular TV program Jeopardy.[38] The objective is to demonstrate how a sophisticated computer system can challenge humans and provide fast, accurate answers to difficult questions. *Reinforcement Learning* is a learning system involving sequential decisions with learning taking place in between each decision.[39] Reinforcement learning often involves sophisticated computer programming and optimization techniques, first discussed in Chapter 6. The computer makes a decision, analyzes the results, and then makes a better decision based on the analysis. The process, often called *dynamic programming*, is repeated until it is impossible to make improvements in the decision.

Learning systems software requires feedback on the results of actions or decisions. At a minimum, the feedback needs to indicate whether the results are desirable (winning a game) or undesirable (losing a game). The feedback is then used to alter what the system will do in the future.

learning systems
A combination of software and hardware that allows the computer to change how it functions or reacts to situations based on feedback it receives.

Neural Networks

An increasingly important aspect of AI involves neural networks, also called neural nets. A **neural network** is a computer system that can act like or simulate the functioning of a human brain. The systems can use massively parallel processors in an architecture that is based on the human brain's own mesh-like structure. In addition, neural network software simulates

neural network
A computer system that can act like or simulate the functioning of a human brain.

a neural network using standard computers. Neural networks can process many pieces of data at the same time and learn to recognize patterns.[40] A neural network, for example, was used to appraise property values of more than 40,000 properties with lower pricing errors compared to traditional appraisal methods.[41] The use of neural networks and artificial intelligence was also used to help predict fuel wood prices in Greece.[42] (Fuel wood is a renewable energy source that can replace imported natural gas and oil products.) Some oil and gas exploration companies use a program called the Rate of Penetration based on neural networks to monitor and control drilling operations. The neural network program helps engineers slow or speed drilling operations to help increase drilling accuracy and reduce costs.

More businesses are firing up neural nets to help them navigate ever-thicker forests of data and make sense of a myriad of customer traits and buying habits. Computer Associates has developed Neugents (*www.neugents.com*), neural intelligence agents that "learn" patterns and behaviors and predict what will happen next. For example, Neugents can track the habits of insurance customers and predict which ones will not renew an automobile policy. They can then suggest to an insurance agent what changes to make in the policy to persuade the consumer to renew it.

Other Artificial Intelligence Applications

A few other artificial intelligence applications exist in addition to those just discussed. A **genetic algorithm**, also called a genetic program, is an approach to solving large, complex problems in which many repeated operations or models change and evolve until the best one emerges.[43] The approach is based on the theory of evolution that requires (1) variation and (2) natural selection. The first step is to change or vary competing solutions to the problem. This can be done by changing the parts of a program or by combining different program segments into a new program, mimicking the evolution of species, in which the genetic makeup of a plant or animal mutates or changes over time. The second step is to select only the best models or algorithms, which continue to evolve. Programs or program segments that are not as good as others are discarded, a process similar to what happens in natural selection, in which only the best species survive and continue to evolve. This process of variation and selection continues until the genetic algorithm yields the best possible solution to the original problem. A genetic algorithm can be used to help schedule airline crews to meet flight requirements while minimizing total costs.[44] Natural Selection, a San Diego company, originally developed a genetic algorithm that attempted to analyze past inventions and suggest future ones.[45] Although the original genetic algorithm was not an immediate success, the approach has been used by General Electric, the U.S. Air Force, and others to cut costs and streamline delivery routes of products.

An **intelligent agent** (also called an *intelligent robot* or *bot*) consists of programs and a knowledge base used to perform a specific task for a person, a process, or another program. Like a sports agent who searches for the best endorsement deals for a top athlete, an intelligent agent often searches to find the best price, schedule, or solution to a problem. The programs used by an intelligent agent can search large amounts of data as the knowledge base refines the search or accommodates user preferences. Often used to search the vast resources of the Internet, intelligent agents can help people find information on any topic, such as the best price for a new digital camera. An intelligent agent can also help determine how land and other natural resources can be best used when different people and groups have different interests, such as forest management, recreational uses, and tree harvesting.[46]

genetic algorithm
An approach to solving large, complex problems in which many related operations or models change and evolve until the best one emerges.

intelligent agent
Programs and a knowledge base used to perform a specific task for a person, a process, or another program; also called *intelligent robot* or *bot*.

NASA's Jet Propulsion Laboratory has an agent that monitors inventory, planning, and scheduling equipment ordering to keep costs down, as well as food storage facilities. These agents usually monitor complex computer networks that can keep track of the configuration of each computer connected to the network.

(Source: Courtesy of Alan Mak.)

AN OVERVIEW OF EXPERT SYSTEMS

As mentioned earlier, an expert system behaves similarly to a human expert in a particular field. Like human experts, computerized expert systems use heuristics, or rules of thumb, to arrive at conclusions or make suggestions. One company uses the Lantek expert system to cut and fabricate metal into finished products for the automotive, construction, and mining industries. The expert system helped reduce raw material waste and increase profits. The U.S. Army uses the Knowledge and Information Fusion Exchange (KnIFE) expert system to help soldiers in the field make better military decisions based on successful decisions made in previous military engagements.

Expert systems are used in metal fabrication plants to aid in decision making.

(Source: © H. Mark Weidman Photography/Alamy.)

When to Use Expert Systems

Sophisticated expert systems can be difficult, expensive, and time consuming to develop. This is especially true for large expert systems implemented on mainframes. The following is a list of factors that normally make expert systems worth the expenditure of time and money. People and organizations should develop an expert system if it can do any of the following:

- Provide a high potential payoff or significantly reduce downside risk
- Capture and preserve irreplaceable human expertise
- Solve a problem that is not easily solved using traditional programming techniques
- Develop a system more consistent than human experts
- Provide expertise needed at a number of locations at the same time or in a hostile environment that is dangerous to human health
- Provide expertise that is expensive or rare
- Develop a solution faster than human experts can
- Provide expertise needed for training and development to share the wisdom and experience of human experts with many people

Components of Expert Systems

An expert system consists of a collection of integrated and related components, including a knowledge base, an inference engine, an explanation facility, a knowledge base acquisition facility, and a user interface. A diagram of a typical expert system is shown in Figure 7.7. In this figure, the user interacts with the interface, which interacts with the inference engine. The inference engine interacts with the other expert system components. These components must work together to provide expertise. This figure shows the inference engine coordinating the flow of knowledge to other components of the expert system. Note that different knowledge flows can exist, depending on what the expert system is doing and on the specific expert system involved.

Figure 7.7

Components of an Expert System

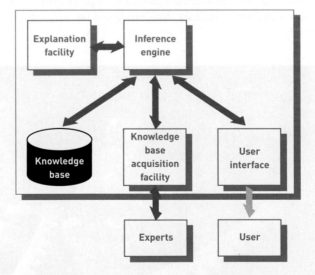

The Knowledge Base

The **knowledge base** stores all relevant information, data, rules, cases, and relationships that the expert system uses. As shown in Figure 7.8, a knowledge base is a natural extension of a database (presented in Chapter 3) and an information and decision support system (presented in Chapter 6). A knowledge base must be developed for each unique application. For example, a medical expert system contains facts about diseases and symptoms. The following are some tools and techniques that can be used to create a knowledge base. [47]

- *Using rules.* A **rule** is a conditional statement that links conditions to actions or outcomes. In many instances, these rules are stored as **IF-THEN statements**, which are rules that suggest certain conclusions. For example: "If a certain set of network conditions exists,

rule
A conditional statement that links conditions to actions or outcomes.

IF-THEN statements
Rules that suggest certain conclusions.

then a certain network problem diagnosis is appropriate." In an expert system for a weather forecasting operation, for example, the rules could state that if certain temperature patterns exist with a given barometric pressure and certain previous weather patterns over the last 24 hours, then a specific forecast will be made, including temperatures, cloud coverage, and wind-chill factor. IBM has used a rule-based expert system to help detect execution errors in its large mainframe computers.[48] Figure 7.9 shows how to use expert system rules in determining whether a person should receive a mortgage loan from a bank. These rules can be placed in almost any standard programming language (discussed in Chapter 2) using "IF-THEN" statements or into special expert system shells and products, discussed later in the chapter. In general, as the number of rules that an expert system knows increases, the precision of the expert system also increases.

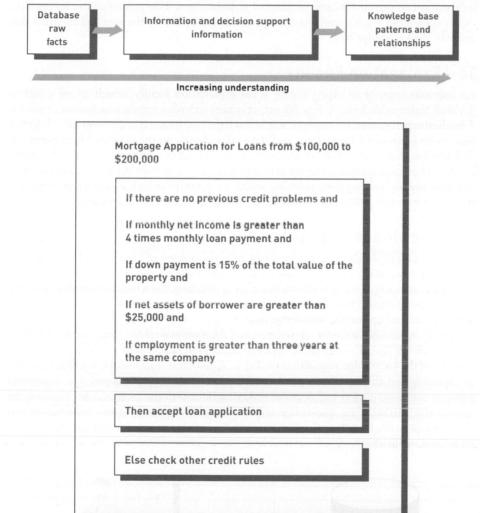

| Database raw facts | → | Information and decision support information | → | Knowledge base patterns and relationships |

Increasing understanding

The Relationships Among Data, Information, and Knowledge

Mortgage Application for Loans from $100,000 to $200,000

If there are no previous credit problems and

If monthly net income is greater than 4 times monthly loan payment and

If down payment is 15% of the total value of the property and

If net assets of borrower are greater than $25,000 and

If employment is greater than three years at the same company

Then accept loan application

Else check other credit rules

Figure 7.9

Rules for a Credit Application

- *Using cases.* An expert system can use cases in developing a solution to a current problem or situation. This process involves (1) finding cases stored in the knowledge base that are similar to the problem or situation at hand and (2) modifying the solutions to the cases to fit or accommodate the current problem or situation. For example, a company might use an expert system to determine the best location for a new service facility in the state of New Mexico. The expert system might identify two previous cases involving the location of a service facility where labor and transportation costs were also important— one in the state of Colorado and the other in the state of Nevada. The expert system can modify the solution to these two cases to determine the best location for a new facility in New Mexico. In another situation, a case-based expert system was used to help determine how jobs were assigned in a chip fabrication factory in Taiwan.[49]

The Inference Engine

The overall purpose of an **inference engine** is to seek information and relationships from the knowledge base and to provide answers, predictions, and suggestions similar to the way a human expert would. In other words, the inference engine is the component that delivers the expert advice. Consider the expert system that forecasts future sales for a product. One approach is to start with a fact such as "The demand for the product last month was 20,000 units." The expert system searches for rules that contain a reference to product demand. For example, "IF product demand is over 15,000 units, THEN check the demand for competing products." As a result of this process, the expert system might use information on the demand for competitive products. Next, after searching additional rules, the expert system might use information on personal income or national inflation rates. This process continues until the expert system can reach a conclusion using the data supplied by the user and the rules that apply in the knowledge base.

The Explanation Facility

An important part of an expert system is the **explanation facility**, which allows a user or decision maker to understand how the expert system arrived at certain conclusions or results. A medical expert system, for example, might reach the conclusion that a patient has a defective heart valve given certain symptoms and the results of tests on the patient. The explanation facility allows a doctor to find out the logic or rationale of the diagnosis made by the expert system. The expert system, using the explanation facility, can indicate all the facts and rules that were used in reaching the conclusion, which the doctors can look at to determine whether the expert system is processing the data and information correctly and logically.

The Knowledge Acquisition Facility

A difficult task in developing an expert system is the process of creating and updating the knowledge base. In the past, when more traditional programming languages were used, developing a knowledge base was tedious and time consuming. Each fact, relationship, and rule had to be programmed into the knowledge base. In most cases, an experienced programmer had to create and update the knowledge base.

Today, specialized software allows users and decision makers to create and modify their own knowledge bases through the knowledge acquisition facility shown in Figure 7.10. The purpose of the **knowledge acquisition facility** is to provide a convenient and efficient means of capturing and storing all components of the knowledge base.[50] Knowledge acquisition software can present users and decision makers with easy-to-use menus. After entering the appropriate attributes, the knowledge acquisition facility correctly stores information and relationships in the knowledge base, making the knowledge base easier and less expensive to set up and maintain.

Figure 7.10

Knowledge Acquisition Facility

The knowledge acquisition facility acts as an interface between experts and the knowledge base.

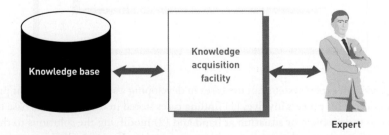

The User Interface

The main purpose of the user interface is to make an expert system easier for users and decision makers to develop and use. At one time, skilled computer personnel created and operated most expert systems; today, the user interface permits decision makers to develop and use their own expert systems. A typical user interface between a user and the expert system for the credit application shown in Figure 7.9 follows.

EXPERT SYSTEM: Are there any previous credit problems?

USER: No.

EXPERT SYSTEM: Is the applicant's net income greater than four times the monthly loan payment?

USER: Yes.

EXPERT SYSTEM: Is the down payment greater than 15% of the property value?

USER: Yes.

EXPERT SYSTEM: Are the net assets of the borrower greater than $25,000?

USER: Yes.

EXPERT SYSTEM: Has the borrower worked more than three years at the same company?

USER: Yes.

EXPERT SYSTEM: **Accept the loan application.**

Participants in Developing and Using Expert Systems

Typically, several people are involved in developing and using an expert system. (See Figure 7.11.)

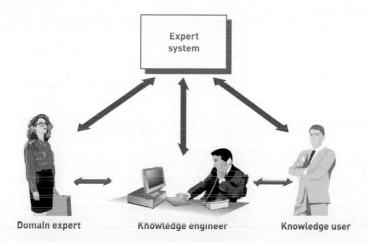

Domain expert   **Knowledge engineer**   **Knowledge user**

Figure 7.11

Participants in Expert Systems Development and Use

There are a number of participants in the development and use of an expert system. The **domain expert** is the person or group with the expertise or knowledge the expert system is trying to capture (domain). In most cases, the domain expert is a group of human experts. Research has shown that good domain experts can increase the overall quality of an expert system.[51] A **knowledge engineer** is a person who has training or experience in the design, development, implementation, and maintenance of an expert system, including training or experience with expert system shells. Knowledge engineers can help transfer the knowledge from the expert system to the knowledge user.[52] The **knowledge user** is the person or group who uses and benefits from the expert system. Knowledge users do not need any previous training in computers or expert systems.

domain expert
The person or group with the expertise or knowledge the expert system is trying to capture (domain).

knowledge engineer
A person who has training or experience in the design, development, implementation, and maintenance of an expert system.

knowledge user
The person or group who uses and benefits from the expert system.

Austin Energy First to Implement a Smart Grid

The power generators that provide homes and businesses in the United States with electricity are responsible for 40 percent of the country's energy consumption. Transportation is second, consuming around 29 percent of the country's energy. Creating efficiencies in energy consumption in these two sectors could greatly decrease our dependency on nonrenewable resources. The solution would also reduce carbon emissions, resulting in economic, environmental, and social benefits.

One solution that has become popular for saving electricity involves special-purpose systems referred to as Intelligent Utility Networks (IUNs), or smart grid technologies. In 2009, the U.S. Department of Energy distributed $3.4 billion in financial stimulus funding to launch 100 smart grid projects across the country. By that time, Austin Energy in Austin, Texas had already rolled out its smart grid and was designing what it calls Smart Grid 2.0. The lessons Austin Energy learned and its pioneering spirit have served as guidelines for other companies now embarking on smart grid projects.

In 2003, Austin Energy worked with IBM and Ascendant Technology (an IT consulting company) to launch its IUM initiative, which it calls Smart Grid 1.0. So how can a utility network be intelligent or smart? As you are learning in this chapter, computers can be programmed to carry out tasks the way a domain expert does, even making judgment calls based on past experience. However, to act intelligently, the computers that control a utility network need access to consumption and production information. That means the first steps in transforming a traditional electric grid into a smart grid are replacing traditional electricity meters with digital meters and deploying sensors across the network to collect production and flow information.

Digital meters, also called smart meters, collect detailed information about electric consumption and wirelessly report it to the electric company using standard Internet protocols. It took Austin Energy five years to install 410,000 smart meters for its customers. Additionally, Austin Energy installed 86,000 smart thermostats that also contribute information to the smart grid. They installed 2,500 sensors on the grid, along with 3,000 computers, additional servers, and network gear to collect and process the information. By October 2008, Austin Energy was processing hundreds of terabytes of data flowing in from meters and sensors, which provided detailed consumption data every 15 minutes.

With access to detailed utility network data, Austin Energy can now determine where energy is most required on its network at any time of the day. Using a distributed system, rather than one central power distribution point, Austin Energy automates the supply of electricity to various points on its grid as needed. The intelligent network can foresee power outages and make adjustments to prevent them.

Austin Energy is also experimenting with many forms of alternative energy sources. Its smart grid makes it possible for energy consumers to produce energy (through solar, wind, or other generation methods) and provide it back to the network for credit. Tracking energy consumption and production and efficiently routing electricity to where it is most urgently needed becomes too laborious for human operators. Instead, Austin Energy programmed its knowledge into its computer systems, which are better equipped to manage the huge amounts of continuously streaming data.

Having access to detailed energy information empowers both consumers and policy makers so they can make wiser decisions. Consumers using a smart grid can go online to view and evaluate their energy consumption patterns, perhaps finding that minor changes in lifestyle could create major savings in energy consumption. Policy makers can determine whether the country's energy supply is being consumed wisely and, in times of crisis, can make decisions that maintain essential systems.

Austin Energy is sold on the idea that information empowers. The second phase of its initiative, Smart Grid 2.0, extends its reach into the homes and businesses of its consumers. Using smart appliances and smart thermostats, Austin Energy can assist consumers in learning how to best consume energy within their domain. Austin Energy is even looking at ways to use electric car batteries as a household backup power supply; an electric car charging in the garage could reverse its current to provide five hours of electricity to the home during power failures.

By analyzing energy consumption in a highly detailed fashion, smart grids can assist the country in squeezing the most usefulness out of every watt produced. Austin Energy found that it saved 660 megawatts of electricity in its first month using America's first smart grid.

Discussion Questions

1. How can smart grids provide economic, environmental, and social benefits?
2. What are the key components in a smart grid system?

Critical Thinking Questions

1. Why do you think the U.S. government decided to invest billions of dollars to jump start smart grids across the country?
2. What privacy concerns, if any, might arise from Austin Energy's Smart Grid 2.0 project?

SOURCES: Fehrenbacher, Katie, "Smart Grid Stimulus Funding Revealed!" Earth2Tech, October 27, 2009, *http://earth2tech.com*; Carvallo, Andres, "LIGHTSON: Austin Energy Delivers First Smart Grid in the US," Electric Energy Online, *www.electricenergyonline.com*, accessed February 2, 2010; Carvallo, Andres, "Austin Energy Plans Its Smart Grid 2.0," CIO Master, April 18, 2009, *www.ciomaster.com*; LaMonica, Martin, "Will anyone pay for the 'smart' power grid," *Cnet News*, May 16, 2007, *news.cnet.com*; "Austin Energy Smart Grid Program," Austin Energy Web site, *www.austinenergy.com/About%20Us/ Company%20Profile/smartGrid*, accessed February 2, 2010.

Expert Systems Development Tools and Techniques

Theoretically, expert systems can be developed from any programming language. Since the introduction of computer systems, programming languages have become easier to use, more powerful, and better able to handle specialized requirements. In the early days of expert systems development, traditional high-level languages, including Pascal, FORTRAN, and COBOL, were used, as shown in Figure 7.12. LISP was one of the first special languages developed and used for artificial intelligence applications. PROLOG was also developed for AI applications. Since the 1990s, however, other expert system products (such as shells) have become available that remove the burden of programming, allowing nonprogrammers to develop and benefit from the use of expert systems.

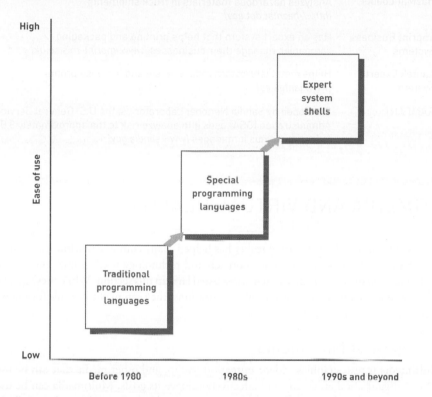

Figure 7.12

Expert Systems Development

Software for expert systems development has evolved greatly since 1980, from traditional programming languages to expert system shells.

Expert System Shells and Products

An *expert system shell* is a collection of software packages and tools used to design, develop, implement, and maintain expert systems. Expert system shells are available for both personal computers and mainframe systems. Some shells are inexpensive, costing less than $500. In addition, off-the-shelf expert system shells are complete and ready to run. The user enters the appropriate data or parameters, and the expert system provides output to the problem or situation. Table 7.2 lists a few expert system products.

Name of Product	Application and Capabilities
Exsys Corvid	An expert system tool that simulates a conversation with a human expert from Exsys (www.exsys.com)[53]
EZ-Xpert	A rule-based expert system that results in complete applications in the C++ or Visual Basic programming languages by EZ-Xpert (www.ez-xpert.com)[54]
G2	Assists in oil and gas operations; Transco, a British company, uses it to help in the transport of gas to more than 20 million commercial and domestic customers
HazMat Loader	Analyzes hazardous materials in truck shipments (http://hazmat.dot.gov)
Imprint Business Systems	Has an expert system that helps printing and packaging companies manage their businesses (www.imprint-mis.co.uk)
Lantek Expert System	Helps metal fabricators reduce waste and increase profits (www.lantek.es)
RAMPART	Developed by Sandia National Laboratories, the U.S. General Services Administration (GSA) uses it to analyze risk to the approximately 8,000 federal buildings it manages (www.sandia.gov)

MULTIMEDIA AND VIRTUAL REALITY

The use of multimedia and virtual reality has helped many companies achieve a competitive advantage and increase profits. The approach and technology used in multimedia is often the foundation of virtual reality systems, discussed later in this section. While these specialized information systems are not used by all organizations, they can play a key role for many. We begin with a discussion of multimedia.

Overview of Multimedia

multimedia
Text, graphics, video, animation, audio, and other media that can be used to help an organization efficiently and effectively achieve its goals.

Multimedia is text, graphics, video, animation, audio, and other media that can be used to help an organization efficiently and effectively achieve its goals. Multimedia can be used to create stunning brochures, presentations, reports, and documents. Although not all organizations use the full capabilities of multimedia, most use text and graphics capabilities.

Text and Graphics

All large organizations and most small and medium-sized ones use text and graphics to develop reports, financial statements, advertising pieces, and other documents used internally and externally. Internally, organizations use text and graphics to communicate policies, guidelines, and much more to managers and employees. Externally, organizations use text and graphics to communicate to suppliers, customers, federal and state organizations, and a variety of other stakeholders. Text can have different sizes, fonts, and colors. Graphics can include photographs, illustrations, drawings, a variety of charts, and other still images. Graphic images can be stored in a variety of formats, including JPEG (Joint Photographic Experts Group format) and GIF (Graphics Interchange Format).

While standard word-processing programs are an inexpensive and simple way to develop documents and reports that require text and graphics, most organizations use specialized software. Adobe Illustrator, for example, can be used to create attractive and informative charts, illustrations, and brochures. The software can also be used to develop digital art, reference manuals, profit and loss statements, and a variety of reports required by state and federal governments. Adobe Photoshop is a sophisticated and popular software package that can be used to edit photographs and other visual images. Once created, these documents and reports can be saved in an Adobe PDF (portable document format) file and sent over the

Internet or saved on a CD or similar storage device. Microsoft Silverlight can be used to add high-definition video and animation to Internet sites and other programming.[55] The National Football League (NFL) plans to use Silverlight in many of its broadcasts. PowerPoint, also by Microsoft, can be used to develop a presentation that is displayed on a large viewing screen with sound and animation. There are many other graphics programs, including Paint and PhotoDraw by Microsoft, CorelDraw, and others. Many graphics programs can create 3D images. James Cameron's movie *Avatar* used sophisticated computers and 3D imaging to create one of the most profitable movies in history.[56] Once used primarily in movies, 3D technology can be used by companies to design products, such as motorcycles, jet engines, bridges, and more.[57] Autodesk, for example, makes exciting 3D software that companies can use to design large skyscrapers and other buildings.[58] The software can also be used by Hollywood animators to develop exciting movies. The technology used to produce 3D movies will also be available with some TV programs.[59] British Sky Broadcasting, for example, will offer 3D programming on some of its channels in the United Kingdom. Graphics software and systems can be used to help scientists understand our universe, doctors perform delicate surgeries, military specialists guide unmanned drones used to locate and destroy enemy targets, geologists locate oil and gas, and many more applications. See Figure 7.13.

Figure 7.13

Digital Graphics

Businesses create graphics such as charts, illustrations, and brochures using software such as Adobe Photoshop or Adobe Illustrator.

(Source: Image © Viktor Gmyria, 2010. Used under license from Shutterstock.com.)

Audio

Audio includes music, human voices, recorded sounds, and a variety of computer-generated sounds. Audio can be stored in a variety of file formats, including MP3 (Motion Picture Experts Group Audio Layer 3), WAV (wave format), MIDI (Musical Instrument Digital Interface), and other formats. When audio files are played while being downloaded from the Internet, it is called *streaming audio*.

Input to audio software includes audio-recording devices like microphones, imported music or sound from CDs or audio files, MIDI instruments that can create music and sounds directly, and other audio sources. Once stored, audio files can be edited and augmented using audio software, including Apple's QuickTime, Microsoft's Sound Recorder, Adobe's Audition, and other software. Audio files are used by scientists to monitor ocean life, by submarines for navigation and enemy ship detection, by radio stations to broadcast news and music, and by law enforcement to catch criminals. Once edited, audio files can also be used to enhance presentations, create music, broadcast satellite radio signals, develop audio books, record podcasts for iPods and other audio players, provide realism to movies, and enhance video and animation (discussed next).

Apple Soundtrack Pro provides editing tools for editing and producing audio files in a variety of formats.

(Source: Courtesy of Apple Inc.)

Video and Animation

The moving images of video and animation are typically created by rapidly displaying one still image after another. Video and animation can be stored in AVI (Audio Video Interleave) files used with many Microsoft applications, MPEG (Motion Picture Experts Group format) files, and MOV (QuickTime format) files used with many Apple applications. When video files are played while being downloaded from the Internet, it is called *streaming video*. On the Internet, Java applets (small downloadable programs) and animated GIF files can be used to animate or create "moving" images.

A number of video and animation software products can be used to create and/or edit video and animation files. Many video and animation programs can create realistic 3D moving images. Adobe's Premiere and After Effects and Apple's Final Cut Pro can be used to edit video images taken from cameras and other sources. Final Cut Pro, for example, has been used to edit and produce full-length motion pictures shown in movie theaters. Adobe Flash and LiveMotion can be used to add motion and animation to Web pages.

There are many business uses of video and animation. Companies that develop computer-based or Internet training materials often use video and audio software. An information kiosk at an airport or shopping mall can use animation to help customers check into to a flight or get information. Of course, movie studios use audio and video techniques to make standard and animated movies. Pixar, for example, used sophisticated animation software to create dazzling 3D movies. The exact process can be seen at Pixar's Web site.[60] See Figure 7.14.

File Conversion and Compression

Most multimedia applications are created, edited, and distributed in a digital file format, such as the ones discussed above. Older inputs to these applications, however, can be in an analog format from old home movies, magnetic tape, vinyl records, or similar sources. In addition, there are older digital formats that are no longer popular or used. In these cases, the analog and older digital formats must be converted into a newer digital format before they can be edited and processed by today's multimedia software. This can be done with a conversion program or specialized hardware. Some of the multimedia software discussed above, such as Adobe Premium, Adobe Audition, and many others, have this analog-to-digital conversion capability. Standalone software and specialized hardware can also be used.[61] Grass Valley, for example, is a hardware device that can be used to convert analog video to digital video or digital video to analog video. With this device, you can convert old VHS tapes to digital video files or digital video files to an analog format.

Figure 7.14

Creating Animation for Pixar

Pixar uses sophisticated animation software called iClone4 to create cutting-edge 3D movies.

(Source: Courtesy of Reallusion, Inc.)

Because multimedia files can be large, file compression can be important. Many of the multimedia software programs discussed above can be used to compress files to make them easier to download from the Internet or send as e-mail attachments. In addition, standalone file conversion programs, such as WinZip, can be used to compress multimedia files in many file formats.

Designing a Multimedia Application

Designing multimedia applications requires careful thought and a systematic approach. The overall approach to modifying any existing application or developing a new one is discussed in the next chapter on systems development. There are, however, some additional considerations in developing a multimedia application. Multimedia applications can be printed on beautiful brochures, placed into attractive corporate reports, uploaded to the Internet, or displayed on large screens for viewing. Because these applications are typically more expensive than preparing documents and files in a word-processing program, it is important to spend time designing the best possible multimedia application. Designing a multimedia application requires that the end use of the document or file be carefully considered. For example, some text styles and fonts are designed for Internet display. Because different computers and Web browsers display information differently, it is a good idea to select styles, fonts, and presentations based on computers and browsers that are likely to display the multimedia application. Because large files can take much longer to load into a Web page, smaller files are usually preferred for Web-based multimedia applications.

Overview of Virtual Reality

The term *virtual reality* was initially coined by Jaron Lanier, founder of VPL Research, in 1989. Originally, the term referred to *immersive virtual reality* in which the user becomes fully immersed in an artificial, 3D world that is completely generated by a computer. Immersive virtual reality can represent any three-dimensional setting, real or abstract, such as a building, an archaeological excavation site, the human anatomy, a sculpture, or a crime scene reconstruction. Through immersion, the user can gain a deeper understanding of the virtual world's behavior and functionality. The Media Grid at Boston College has a number of initiatives in the use of immersive virtual reality in education.[62]

A **virtual reality system** enables one or more users to move and react in a computer-simulated environment. Virtual reality simulations require special interface devices that transmit the sights, sounds, and sensations of the simulated world to the user. These devices can also record and send the speech and movements of the participants to the simulation program, enabling users to sense and manipulate virtual objects much as they would real objects. This natural style of interaction gives the participants the feeling that they are immersed in the simulated world. For example, an auto manufacturer can use virtual reality to help it simulate and design factories.

virtual reality system
A system that enables one or more users to move and react in a computer-simulated environment.

Interface Devices

To see in a virtual world, often the user wears a head-mounted display (HMD) with screens directed at each eye. The HMD also contains a position tracker to monitor the location of the user's head and the direction in which the user is looking. Using this information, a computer generates images of the virtual world—a slightly different view for each eye—to match the direction that the user is looking, and displays these images on the HMD. Many companies sell or rent virtual reality interface devices, including Virtual Realities (*www.vrealities.com*), Amusitronix (*www.amusitronix.com*), I-O Display Systems (*www.i-glassesstore.com*), and others.

The Electronic Visualization Laboratory at the University of Illinois at Chicago introduced a room constructed of large screens on three walls and the floor on which the graphics are projected. The CAVE, as this room is called, provides the illusion of immersion by projecting stereo images on the walls and floor of a room-sized cube (*http://cave.ncsa.uiuc.edu*). Several persons wearing lightweight stereo glasses can enter and walk freely inside the CAVE. A head-tracking system continuously adjusts the stereo projection to the current position of the leading viewer.

Military personnel train in an immersive CAVE system.

(Source: Courtesy of Fakespace Systems, Inc.)

Users hear sounds in the virtual world through earphones. The information reported by the position tracker is also used to update audio signals. When a sound source in virtual space is not directly in front of or behind the user, the computer transmits sounds to arrive at one ear a little earlier or later than at the other and to be a little louder or softer and slightly different in pitch.

The *haptic* interface, which relays the sense of touch and other physical sensations in the virtual world, is the least developed and perhaps the most challenging to create.[63] A Japanese virtual reality company has developed a haptic interface device that can be placed on a person's fingertips to give an accurate feel for game players, surgeons, and others.[64] Currently, with the use of a glove and position tracker, the computer locates the user's hand and measures finger movements. The user can reach into the virtual world and handle objects; however, it is difficult to generate the sensations of a person tapping a hard surface, picking up an object, or running a finger across a textured surface. Touch sensations also have to be synchronized with the sights and sounds users experience. Today, some virtual reality developers are even trying to incorporate taste and smell into virtual reality applications.[65] According to a virtual reality researcher at the University of Warwick Digital Lab, "The crucial thing for virtual reality is that it will hit all five senses in a highly realistic manner ... We need to have smell, we need to have taste."

Forms of Virtual Reality

Aside from immersive virtual reality, virtual reality can also refer to applications that are not fully immersive, such as mouse-controlled navigation through a 3D environment on a graphics monitor, stereo viewing from the monitor via stereo glasses, stereo projection systems, and others. *Augmented reality,* a newer form of virtual reality, has the potential to superimpose digital data over real photos or images.[66] GPS maps, for example, can be combined with real pictures of stores and streets to help you locate your position or find your way to a new destination. Using augmented reality, you can point a smartphone camera at a historic landmark such as a castle, museum, or other building and have information about the landmark appear on your screen, including a brief description of the landmark, admission price, and hours of operation. Although still in its early phases of implementation, augmented reality has the potential to become an important feature of tomorrow's smartphones and similar mobile devices.

Some virtual reality applications allow views of real environments with superimposed virtual objects. Telepresence systems (such as telemedicine and telerobotics) immerse a viewer in a real world that is captured by video cameras at a distant location and allow for the remote manipulation of real objects via robot arms and manipulators. Many believe that virtual reality will reshape the interface between people and information technology by offering new ways to communicate information, visualize processes, and express ideas creatively.

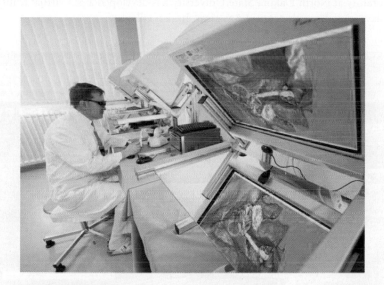

Using virtual reality technology, health professionals can perform remote medical procedures or examinations, even virtual surgery.

(Source: AP Photo/Jens Meyer.)

Virtual Reality Applications

You can find thousands of applications of virtual reality, with more being developed as the cost of hardware and software declines and people's imaginations are opened to the potential of virtual reality. The following are a few virtual reality applications in medicine, education and training, business, and entertainment. See Figure 7.15.

Figure 7.15

Virtual Reality Applications

Virtual reality has been used to increase real estate sales in several powerful ways. RealSpace Vision Communication, for example, helps real estate developers showcase their properties with virtual reality tours.

(Source: Courtesy of RealSpace Vision Communication Inc.)

Medicine

Barbara Rothbaum, the director of the Trauma and Recovery Program at Emory University School of Medicine and cofounder of Virtually Better, uses an immersive virtual reality system to help in the treatment of anxiety disorders.[67] One VR program, called SnowWorld, helps treat burn patients.[68] Using VR, the patients can navigate through icy terrain and frigid waterfalls. VR helps because it gets a patient's mind off the pain.

Education and Training

Virtual environments are used in education to bring exciting new resources into the classroom. According to the founder of Mantis Development Corporation, a software company that specialized in digital media and virtual reality, "In order to learn, you need to engage the mind, and immersive education is engaging."[69] In development for more than 10 years, *3D Rewind Rome* is a virtual reality show developed at a virtual reality lab at University of California Los Angeles (UCLA).[70] The show is historically accurate with more than 7,000 reconstructed buildings on a background of realistic landscape. The Archaeology Technologies Laboratory at North Dakota State University has developed a 3D virtual reality system that displays an eighteenth-century American Indian village.

3D Rewind Rome is based on more than 10 years of research by archaeologists and historians coordinated by the University of California, Los Angeles.

(Source: *www.3drewind.com*.)

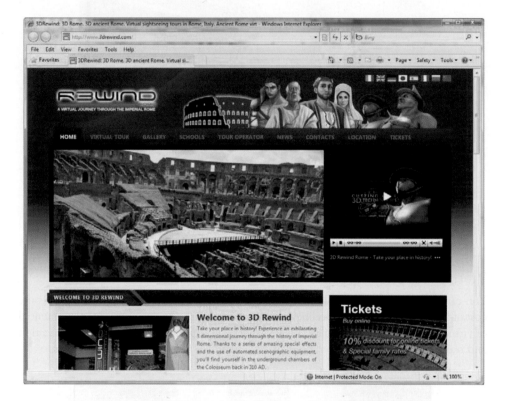

Virtual technology has also been applied by the military. To help with aircraft maintenance, a virtual reality system has been developed to simulate an aircraft and give a user a sense of touch, while computer graphics provide a sense of sight and sound. The user sees, touches, and manipulates the various parts of the virtual aircraft during training. The Virtual Aircraft Maintenance System simulates real-world maintenance tasks that are routinely performed on the AV8B vertical takeoff and landing aircraft used by the U.S. Marines. Also, the Pentagon is using a virtual reality training lab to prepare for a military crisis. The virtual reality system simulates various war scenarios.

Business and Commerce

Virtual reality has been used in all areas of business. Boeing used virtual reality to help it design and manufacture airplane parts and new planes, including the 787 Dreamliner. Boeing used 3D PLM from Dassault Systems.[71] One healthcare institution used Second Life to create a virtual hospital when it started construction of a real multimillion dollar hospital. The purpose of the Second Life virtual hospital was to show clients and staff the layout and capabilities of the new hospital. Second Life has also been used in business and recruiting. Second Life (*www.secondlife.com*) also allows people to play games, interact with avatars, and build structures, such as homes. A number of companies are using VR in advertising.[72] Pizza chain Papa John's used VR as an advertising tool. It placed a VR image on many of its pizza boxes. When the image is viewed by a Web camera on a computer, a standard keyboard can be used to manipulate images of a Chevrolet Camaro on the computer screen. It is a moving image of the Camaro that the founder of Papa John's sold to start his pizza company.

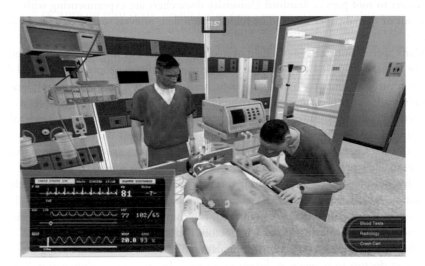

Although it looks like a video game, Pulse!! is a serious training tool for nurses and physicians developed by Breakaway.

(Source: Pulse!! is a research project of Texas A&M University-Corpus Christi in collaboration with the Office of Naval Research and Breakaway Ltd.)

Entertainment

Computer-generated image technology, or CGI, has been around since the 1970s. Many movies use this technology to bring realism to the silver screen, including *Avatar, Finding Nemo, Spider-Man II,* and *Star Wars Episode II—Attack of the Clones.* A team of artists rendered the roiling seas and crashing waves of *Perfect Storm* almost entirely on computers using weather reports, scientific formulas, and their imagination. Other films that have used CGI technology include *Dinosaur* with its realistic talking reptiles, *Titan A.E.*'s beautiful 3D spacescapes, and the casts of computer-generated crowds and battles in *Gladiator* and *The Patriot.* CGI can also be used for sports simulation to enhance the viewers' knowledge and enjoyment of a game. SimCity (*http://simcity.ea.com/*), a virtual reality game, allows people to experiment with decisions related to urban planning.

SimCity simulates urban planning by letting you create, build, and run a virtual city.

(Source: *simcity.ea.com.*)

The explanation facility of an expert system allows the user to understand what rules were used in arriving at a decision. The knowledge acquisition facility helps the user add or update knowledge in the knowledge base. The user interface makes it easier to develop and use the expert system.

The people involved in the development of an expert system include the domain expert, the knowledge engineer, and the knowledge users. The domain expert is the person or group who has the expertise or knowledge being captured for the system. The knowledge engineer is the developer whose job is to extract the expertise from the domain expert. The knowledge user is the person who benefits from the use of the developed system.

Expert systems can be implemented in several ways. Previously, traditional high-level languages, including Pascal, FORTRAN, and COBOL, were used. LISP and PROLOG are two languages specifically developed for creating expert systems from scratch. A faster and less-expensive way to acquire an expert system is to purchase an expert system shell or existing package. The shell program is a collection of software packages and tools used to design, develop, implement, and maintain expert systems.

The benefits of using an expert system go beyond the typical reasons for using a computerized processing solution. Expert systems display "intelligent" behavior, manipulate symbolic information and draw conclusions, provide portable knowledge, and can deal with uncertainty. Expert systems can be used to solve problems in many fields or disciplines and can assist in all stages of the problem-solving process. Past successes have shown that expert systems are good at strategic goal setting, planning, design, decision making, quality control and monitoring, and diagnosis.

Applications of expert systems and artificial intelligence include credit granting and loan analysis, catching cheats and terrorists, budgeting, games, information management and retrieval, AI and expert systems embedded in products, plant layout, hospitals and medical facilities, help desks and assistance, employee performance evaluation, virus detection, repair and maintenance, shipping, and warehouse optimization.

Principle

Multimedia and virtual reality systems can reshape the interface between people and information technology by offering new ways to communicate information, visualize processes, and express ideas creatively.

Multimedia is text, graphics, video, animation, audio, and other media that can be used to help an organization efficiently and effectively achieve its goals. Multimedia can be used to create stunning brochures, presentations, reports,

and documents. Although not all organizations use the full capabilities of multimedia, most use text and graphics capabilities. Other applications of multimedia include audio, video, and animation. File compression and conversion are often needed in multimedia applications to import or export analog files and to reduce file size when storing multimedia files and sending them to others. Designing a multimedia application requires careful thought to get the best results and achieve corporate goals.

A virtual reality system enables one or more users to move and react in a computer-simulated environment. Virtual reality simulations require special interface devices that transmit the sights, sounds, and sensations of the simulated world to the user. These devices can also record and send the speech and movements of the participants to the simulation program. Thus, users can sense and manipulate virtual objects much as they would real objects. This natural style of interaction gives the participants the feeling that they are immersed in the simulated world.

Virtual reality can also refer to applications that are not fully immersive, such as mouse-controlled navigation through a three-dimensional environment on a graphics monitor, stereo viewing from the monitor via stereo glasses, stereo projection systems, and others. Some virtual reality applications allow views of real environments with superimposed virtual objects. Augmented reality, a newer form of virtual reality, has the potential to superimpose digital data over real photos or images. Virtual reality applications are found in medicine, education and training, real estate and tourism, and entertainment.

Principle

Specialized systems can help organizations and individuals achieve their goals.

A number of specialized systems have recently appeared to assist organizations and individuals in new and exciting ways. Segway, for example, is an electric scooter that uses sophisticated software, sensors, and gyro motors to transport people through warehouses, offices, downtown sidewalks, and other spaces. Originally designed to transport people around a factory or around town, more recent versions are being tested by the military for gathering intelligence and transporting wounded soldiers to safety. Radio Frequency Identification (RFID) tags are used in a variety of settings. Game theory involves the use of information systems to develop competitive strategies for people, organizations, and even countries. Informatics combines traditional disciplines, such as science and medicine, with computer science. Bioinformatics and medical informatics are examples. A number of special-purpose telecommunications systems can be placed in products for varied uses.

CHAPTER 7: SELF-ASSESSMENT TEST

Knowledge management allows organizations to share knowledge and experience among their managers and employees.

1. _____ knowledge is objective and can be measured and documented in reports, papers, and rules.

2. What type of person creates, uses, and disseminates knowledge?
   a. knowledge worker
   b. information worker
   c. domain expert
   d. knowledge engineer

3. A community of practice (COP) is a group of people or community dedicated to a common discipline or practice, such as open-source software, auditing, medicine, engineering, and other areas. True or False?

Artificial intelligence systems form a broad and diverse set of systems that can replicate human decision making for certain types of well-defined problems.

4. The Turing Test attempts to determine whether the responses from a computer with intelligent behavior are indistinguishable from responses from a human. True or False?

5. _____ are rules of thumb arising from experience or even guesses.

6. What is not an important attribute for artificial intelligence?
   a. the ability to use sensors
   b. the ability to learn from experience
   c. the ability to be creative
   d. the ability to make complex calculations

7. _____ involves mechanical or computer devices that can paint cars, make precision welds, and perform other tasks that require a high degree of precision or are tedious or hazardous for human beings.

8. What branch of artificial intelligence involves a computer understanding and reacting to statements in English or another language?
   a. expert systems
   b. neural networks
   c. natural language processing
   d. vision systems

9. A(n) _____ is a combination of software and hardware that allows the computer to change how it functions or reacts to situations based on feedback it receives.

Expert systems can enable a novice to perform at the level of an expert but must be developed and maintained very carefully.

10. What is a disadvantage of an expert system?
    a. the inability to solve complex problems
    b. the inability to deal with uncertainty
    c. limitations to relatively narrow problems
    d. the inability to draw conclusions from complex relationships

11. A(n) _____ is a collection of software packages and tools used to develop expert systems that can be implemented on most popular PC platforms to reduce development time and costs.

12. A heuristic consists of a collection of software and tools used to develop an expert system to reduce development time and costs. True or False?

13. What stores all relevant information, data, rules, cases, and relationships used by the expert system?
    a. the knowledge base
    b. the data interface
    c. the database
    d. the acquisition facility

14. A disadvantage of an expert system is the inability to provide expertise needed at a number of locations at the same time or in a hostile environment that is dangerous to human health. True or False?

15. What allows a user or decision maker to understand how the expert system arrived at a certain conclusion or result?
    a. the domain expert
    b. the inference engine
    c. the knowledge base
    d. the explanation facility

16. An important part of an expert system is the _____, which allows a user or decision maker to understand how the expert system arrived at certain conclusions or results.

17. In an expert system, the domain expert is the individual or group who has the expertise or knowledge one is trying to capture in the expert system. True or False?

Multimedia and virtual reality systems can reshape the interface between people and information technology by offering new ways to communicate information, visualize processes, and express ideas creatively.

18. _____ can be used to create stunning brochures, presentations, reports, and documents.

19. What type of virtual reality is used to make human beings feel as though they are in a three-dimensional setting, such as a building, an archaeological excavation site, the human anatomy, a sculpture, or a crime scene reconstruction?
    a. cloud
    b. relative
    c. immersive
    d. visual

Specialized systems can help organizations and individuals achieve their goals.

20. _____ involves the use of information systems to develop competitive strategies for people, organizations, or even countries.

CHAPTER 7: SELF-ASSESSMENT TEST ANSWERS

(1) Explicit (2) a (3) True (4) True (5) Heuristics (6) d (7) Robotics (8) c (9) learning system (10) c (11) expert system shell (12) False (13) a (14) False (15) d (16) explanation facility (17) True (18) multimedia (19) c (20) Game theory

REVIEW QUESTIONS

1. What is a *knowledge management system?*
2. What is a *community of practice?*
3. What is a *chief knowledge officer?* What are his or her duties?
4. What is a vision system? Discuss two applications of such a system.
5. What is natural language processing? What are the three levels of voice recognition?
6. Describe three examples of the use of robotics. How can a microrobot be used?
7. What is a learning system? Give a practical example of such a system.
8. What is a neural network? Describe two applications of neural networks.
9. Under what conditions is the development of an expert system likely to be worth the effort?
10. Identify the basic components of an expert system and describe the role of each.
11. Describe several business uses of multimedia.
12. What is virtual reality? Give three examples of its use.
13. Expert systems can be built based on rules or cases. What is the difference between the two?
14. Describe the roles of the domain expert, the knowledge engineer, and the knowledge user in expert systems.
15. What is informatics? Give three examples.
16. Describe game theory and its use.
17. Identify three special interface devices developed for use with virtual reality systems.
18. Identify and briefly describe three specific virtual reality applications.
19. What is informatics? How is it used?
20. Give three examples of other specialized systems.

DISCUSSION QUESTIONS

1. What are the requirements for a computer to exhibit human-level intelligence? How long will it be before we have the technology to design such computers? Do you think we should push to accelerate such a development? Why or why not?
2. You work for an insurance company as an entry-level manager. The company contains both explicit and tacit knowledge. Describe the types of explicit and tacit knowledge that might exist in your insurance company. How would you capture each type of knowledge?
3. Describe the duties of a chief knowledge officer.
4. What are some of the tasks at which robots excel? Which human tasks are difficult for robots to master? What fields of AI are required to develop a truly perceptive robot?
5. Describe how natural language processing could be used in a university setting.
6. Discuss how learning systems can be used in a military war simulation to train future officers and field commanders.
7. You have been hired to develop an expert system for a university career placement center. Develop five rules a student could use in selecting a career.
8. What is the relationship between a database and a knowledge base?
9. Imagine that you are developing the rules for an expert system to select the strongest candidates for a medical school. What rules or heuristics would you include?
10. Describe how game theory can be used in a business setting.
11. Describe how a university might use multimedia.
12. What application of virtual reality has the most potential to generate increased profits in the future?
13. Describe a situation for which RFID could be used in a business setting.

PROBLEM-SOLVING EXERCISES

1. You are a senior vice president of a company that manufactures kitchen appliances. You are considering using robots to replace up to ten of your skilled workers on the factory floor. Using a spreadsheet, analyze the costs of acquiring several robots to paint and assemble some of your products versus the cost savings in labor. How many years would it take to pay for the robots from the savings in fewer employees? Assume that the skilled workers make $20 per hour, including benefits.
2. Assume that you have just won a lottery worth $100,000. You have decided to invest half the amount in the stock market. Develop a simple expert system to pick ten stocks to consider. Using your word-processing program, create seven or more rules that could be used in such an expert system. Create five cases and use the rules you developed to determine the best stocks to pick.
3. Use a graphics program, such as PowerPoint, to develop a brochure for a small restaurant. Contrast your brochure to one that could have been developed using a multimedia application. Write a report using a word-processing application on the advantages of a multimedia application compared to a graphics program.

TEAM ACTIVITIES

1. Do research with your team to identify KMSs in three different businesses or nonprofit organizations. Describe the types of tacit and explicit knowledge that would be needed by each organization or business.
2. Have your team develop a game between two contestants in which the winning contestant receives fake money. At random, pick two team members to play the game against each other. Have your other team members write a report on the winning strategy and how the game could be improved.
3. Have your team members explore the use of a special-purpose system in an industry of your choice. Describe the advantages and disadvantages of this special-purpose system.

WEB EXERCISES

1. Use the Internet to find information about the use of multimedia in a business setting. Describe what you found.
2. This chapter discussed several examples of expert systems. Search the Internet for two examples of the use of expert systems. Which one has the greatest potential to increase profits for a medium-sized firm? Explain your choice.
3. Use the Internet to get information about the application of game theory in business or the military. Write a report about what you found.

CAREER EXERCISES

1. Describe how a COP can be used to help advance your career.
2. Describe the roles and salaries of two people involved in multimedia applications, such as movie production, sound production, or another multimedia application.

CASE STUDIES

Case One

MITRE Taps the Brain Trust of Top U.S. Experts

MITRE Corporation is responsible for managing the Research and Development (R&D) Centers for the U.S. Department of Defense, the Federal Aviation Administration, the Internal Revenue Service, U.S. Department of Veterans Affairs, and the Department of Homeland Security. MITRE also researches new technologies that may assist in solving its clients' problems.

More than 7,000 scientists, engineers, and support specialists work in labs managed by MITRE, and most have master's or doctoral degrees. Staff members are engaged in hundreds of different projects across the company. Each staff member possesses valuable technical, operational, and domain knowledge that MITRE wants to tap to its full value and potential. When knowledge management (KM) systems came on the scene in the mid-1990s, MITRE immediately saw the benefit for its researchers and has been tinkering with KM ever since.

With so many research specialists engaged across its labs, the value of tapping each other's knowledge and collaborating on projects is immense. However, it's a challenge to interact efficiently with low overhead while researchers are simultaneously working on hundreds of separate projects. For knowledge management, MITRE takes a gradual learn-while-you-go approach.

MITRE's first step in providing knowledge management was to simply track its research staff. A people locator was developed as part of the larger MITRE Information Infrastructure (MII). The people locator works like an electronic phone book, identifying which employees worked on which assignments over time. The system drew information from the existing project management systems and human resource systems. Using the people locator, staff could find colleagues with useful knowledge based on previous work or the sponsoring organization.

As MITRE researchers used the people finder, developers refined the system based on user feedback. Over time, they introduced additional capabilities. For example, they added an Expertise Finder to help find researchers with expertise in special areas. MITRE also included a library of best practices for systems engineering and project management in the system. MITRE experimented with technology exchange meetings and an annual Innovation Exchange, which allowed researchers to share their successes with colleagues. When they found new technologies and ideas useful, developers added them to the KM system. More recently, MITRE has experimented with Web 2.0 technologies similar to Facebook and Wikipedia for its KM system.

MITRE's approach to KM has been evolutionary. New ideas are piloted, and those proven valuable and viable are kept in the system. The success of MITRE's KM system is in its unique approach to KM as a journey with continuous improvements.

Discussion Questions

1. Why is KM extremely valuable in areas of research and development?
2. How do the different components of MITRE's KM system assist in spreading knowledge throughout its labs and in storing knowledge for use in the future?

Critical Thinking Questions

1. What unique challenges do research and development labs provide for KM implementation?
2. What is the benefit of MITRE's evolutionary approach to KM?

SOURCES: Swanborg, Rick, "Mitre's Knowledge Management Journey," *CIO*, February 27, 2009, *www.cio.com*; "About MITRE," MITRE Web site, *www.mitre.org/about*, accessed January 31, 2010.

Case Two

JEA Uses AI to Optimize Water Delivery

JEA supplies much of the Jacksonville, Florida area with electricity, water, and sewer services. The JEA water system uses 150 artesian wells to tap the Florida Aquifer, distributing water to 44 water treatment plants and then down 3,480 miles of underground water pipes to businesses and residences.

Making sure that its wells are producing enough water to meet customer demand is a tricky and sometimes wasteful process. Reservoirs accommodate anticipated demand, often overcompensating to make sure they have enough water to accommodate an unusually high demand. When too much water is kept on hand, the quality of the water decreases (due to salt intrusion), and the health of the well decreases (due to inactivity).

Recently, JEA decided to put artificial intelligence to work assisting its wells in pumping just enough water to meet customer demand. It purchased AI-based software from Gensym and created a neural network to predict hourly consumption of water over a given time frame. The neural network is trained using previous water-usage data. The consumption forecasts created by the neural net are fed into an expert system. The expert system allocates the total anticipated demand to the 44 water treatment plants and artesian well pumps.

The schedule for hourly water production is fed into automated systems in the production and treatment plant to control production and delivery of the water. The Gensym software allows JEA engineers to define inputs (including reservoir sensors and meter-reading) and outputs (such as controls on equipment) to optimize the production and flow of water through the JEA system.

It took JEA six months to develop and implement the optimized expert system and automation. The system has resulted in better use of reservoir capacity by filling reservoirs only to levels required with each hour of operation. The needs-based production has minimized costs while maximizing water quality. Salt intrusion has been reduced, and the health of the wells is increasing. As JEA rolls out its new system to all of its wells and treatment plants, it will enjoy considerable benefits and savings, including lower energy costs and reduced equipment failures from smarter pumping.

Discussion Questions

1. What problem did JEA face that required the use of an expert system and automation?
2. What benefits were provided by JEA's new automated system?

Critical Thinking Questions

1. What are the components of JEA's new system, and what tasks does each component accomplish?
2. What other industries might benefit from a Gensym optimization and automation system like JEA's? Why?

SOURCES: Gensym Staff, "Success Story: JEA," Gensym Web site, *www.gensym.com*, accessed January 31, 2010; "About JEA," JEA Web site, *www.jea.com/about*, accessed January 31, 2010.

Questions for Web Case

See the Web site for this book to read about the Altitude Online case for this chapter. Following are questions concerning this Web case.

Altitude Online: Knowledge Management and Other Considerations

Discussion Questions

1. Why do you think it is a good idea for Altitude Online to maintain records of all advertising projects?
2. How can social networks and blogs serve as knowledge management systems?

Critical Thinking Questions

1. What challenges lie in filling a wiki with information provided by employees?
2. What other tools could Altitude Online use to capture employee knowledge, build community, and reward productive employees?

NOTES

Sources for the opening vignette: Gruman, Galen, "Capgemini adopts social networking tools for knowledge management," *Computerworld*, June 1, 2009, *www.computerworld.com*; "Capgemini – Who We Are," Capgemini Web site, *www.us.capgemini.com/about*, accessed January 31, 2010; Fitzgerald, Michael, "Why Social Computing Aids Knowledge Management," *CIO*, June 13, 2008, *www.cio.com*.

1   Staff, "Knowledge Management in Practice," *Information Today*," June 2009, p. 48.
2   Aaron, Bruce, "Determining the Business Impact of Knowledge Management," *Performance Improvement*, April 2009, p. 35.
3   Larger, Marshall, "Investing in Knowledge Management," *Customer Relationship Management*," June 2009, p. 46.
4   Nguyen, Le, et al, "Acquiring Tacit and Explicit Marketing Knowledge from Foreign Partners in IJVs," *Journal of Business Research*, November 2007, p. 1152.
5   Gerard, J. et al, "Empirically Testing Explicit and Tacit Knowledge Assumptions," *The Business Review*, Summer 2009, p. 1.
6   Holtshouse, Dan, "The Future of Knowledge Workers," *KM World*, September 2009, p. 2.
7   Hemmasi, M. and Csanda C., "The Effectiveness of Communities of Practice," *Journal of Managerial Issues*," Summer 2009, p. 262.
8   Lamont, Judith, "Knowledge Management: Naturally Green," *KM World*, February 2009, p. 5.
9   Subramenian, A. and Soh, P., "Contributing Knowledge to Knowledge Repositories," *Information Resources Management Journal*, January 2009, p. 45.
10 "Adobe Creative Suite 3," *www.adobe.com/products/creativesuite/*, accessed July 20, 2009.
11 McKellar, Hugh, "100 Companies That Matter in Knowledge Management," *KM World*, March 2009, p. 18.
12 *www.cortexpro.com*, accessed July 19, 2009.
13 *www.delphigroup.com*, accessed July 19, 2009.
14 *www.kmresource.com*, accessed July 19, 2009.
15 *www.kmsi.us*, accessed July 19, 2009.
16 *www.knowledge-manage.com*, accessed July 19, 2009.
17 *www.knowledgebase.net*, accessed on October 26, 2009.
18 *www.lawclip.com*, accessed July 19, 2009.
19 *www.kmci.org/index.html*, accessed July 19, 2009.
20 Gomes, Lee, "When Smart is Dumb," *Forbes*, April 13, 2009, p. 42.
21 O'Keefe, Brian, "The Smartest, the Nuttiest Futurist on Earth," *Fortune*, May 14, 2007, p. 60.
22 Markoff, John, "IBM Computer Program to Take on Jeopardy," *The New York Times*, April 27, 2009, p. 11.
23 *www.20q.net*, accessed July 20, 2009.
24 Harris, Mark, "Stand Aside, Soldier, We Robots Are in Command," *The Sunday Times*, May 31, 2009, p. 4.
25 Fahey, Jonathan, "Reconnecting the Brain," *Forbes*, December 28, 2009, p. 48.

26 Staff, "Mind-Machine Meld: Brain-Computer Interfaces for ALS, Paralysis," *Alzheimer Research Forum,* June 22, 2009, *www.alzforum.org/new/detail.asp?id=2173,* accessed July 21, 2009.

27 Rowley, Ian, "Drive, He Thought," *Businessweek,* April 20, 2009, p. 10.

28 Zebda, A. and McEacham, M., "Accounting Expert Systems," *The Business Review,* December 2008, p. 11.

29 Abate, Tom, "Future Moving From I, Robot to My Robot," *Rocky Mountain News,* February 26, 2007, p. 8.

30 *www.irobot.com,* accessed July 20, 2009.

31 *www.cs.cmu.edu/~rll,* accessed July 20, 2009.

32 *www.porterhospital.org,* accessed July 20, 2009.

33 *www.darpagrandchallenge.com,* accessed July 20, 2009.

34 Thryft, A., "Vision Systems Enables Zero Defects," *Test & Measurement,* October 2009, p. 48.

35 Wildstrom, Stephen, "Coming at You: 3D On Your PC," *Businessweek,* January 19, 2009, p. 65.

36 Koit, M., et al, "Towards Computer-Human Interaction in Natural Language," *International Journal of Computer Applicatioins in Technology,* Vol. 34, 2009, p. 291.

37 Young, Peyton, "Learning by Trial and Error," *Games and Economic Behavior,* March 2009, p. 626.

38 Staff, "IBM Developing Computing System to Challenge Humans on America's Favorite Quiz Show, Jeopardy," IBM Press Room, *www-03.ibm.com/press/us/en/pressrelease/27324.wss,* accessed July 22, 2009.

39 Gosavi, Abhijit, "Reinforcement Learning," *Informs,* Spring 2009, p. 178.

40 Dengiz, B., et al, "Optimization of Manufacturing Systems Using a Neural Network," *The Journal of the Operational Research Society,* September 2009, p. 1191.

41 Peterson, S., et al, "Neural Network Hedonic Pricing Models in Mass Real Estate Appraisal, *The Journal of Real Estate Research,* April-June 2009, p. 147.

42 Koutroumanidis, T., et al, "Predicting Fuelwood Prices in Greece," *Energy Policy,* September 2009, p. 3627.

43 He, J., et al, "A Hybrid Parallel Genetic Algorithm for Yard Crane Scheduling," *Transportation Research,* January 2009, p. 136.

44 Souai, N., et al, "Genetic Algorithm Based Approach for the Integrated Airline Crew-Pairing and Rostering Problem," *European Journal of Operations Research,* December 16, 2009, p. 674.

45 Reena, J., "Dusting Off a Big Idea in Hard Times," *Businessweek,* June 22, 2009, p. 44.

46 Bone, C. and Dragicevic, S., "GIS and Intelligent Agents," *Transactions in GIS,* June 2009, p. 253.

47 Ahmed, M., et al, "Handling Imprecision and Uncertainty in Software Development," *Information and Software Technology,* March 2009, p. 640.

48 Sinz, C., et al, "Detection of Dynamic Execution Errors in IBM System Automation Rule-Based Expert System," *Information and Software Technology,* November 1, 2002, pg 857.

49 Chiu, C., et al, "A Case-Based Expert Support System for Due-Date Assignment in Wafer Fabrication," *Journal of Intelligent Manufacturing,* June-August, 2003, p. 14.

50 Wagner, W., "Knowledge Acquisition for Marketing Expert Systems," *Marketing Intelligence & Planning,* Vol. 23, 2005, p. 403.

51 Guimareas, T., et al, "Empirically Testing Some Important Factors for Expert System Quality," *The Quality Management Journal,* Vol. 13, 2006, p. 7.

52 Feng, W., et al, "Understanding Expert Systems Applications from a Knowledge Transfer Perspective," *Knowledge Management Research & Practices,* June 2009, p. 131.

53 EXSYS, *www.exsys.com/,* accessed July 21, 2009.

54 EZ-Xpert Expert System, *www.ez-xpert.com,* accessed July 21, 2009.

55 Wingfield, Nick, "Silverlight Is Still Racing Flash," *The Wall Street Journal,* September 15, 2009, p. B4.

56 Betts, Mitch, "Data Center Plays Supporting Role in Avatar," *Computerworld,* January 18, 2010, p. 4.

57 Copeland, Michael, "3-D Gets Down to Business," *Fortune,* March 30, 2009, p. 32.

58 Foust, Dean, "Top Performing Companies," *Business Week,* April 6, 2009, p. 40.

59 Sonne, P. and Scheichner, S., "After Conquering The Movies, 3-D Viewing Makes Its Way Toward Home TV," *The Wall Street Journal,* August 17, 2009, p. B1.

60 Staff, "How We Do It," *www.pixar.com/howwedoit/index.html#,* accessed July 18, 2009.

61 West, Jackson, "Digitize All of Your Old Analog Media, Easily," *PC World,* August 2009, p. 104.

62 *www.mediagrid.org,* accessed July 20, 2009.

63 *http://osl-www.colorado.edu/Research/haptic/ hapticInterface.shtml,* accessed July 20, 2009.

64 Staff, "AIST Brings Feel of Reality into Virtual Reality," *June 11, 2007, p. 1.*

65 Madrigal, Alexis, "Researchers Want to Add Touch, Taste, Smell to Virtual Reality," *Wired Science,* March 04, 2009, p. 1.

66 Wildstrom, Stephen, "Augmented Reality," *Businessweek,* November 20, 2009, p. 75.

67 *www.emory.edu/EMORY_MAGAZINE/winter96/rothbaum.html,* accessed July 20, 2009.

68 *www.temple.edu/ispr/examples/ex03_07_23.html,* accessed July 20, 2009.

69 Walsh, Aaron, "Dossier," *Computerworld,* August 17, 2009, p. 12.

70 Vorais, Richard, "Ancient Rome 2.0," *Forbes,* March 16, 2009, p. 68.

71 *www.3ds.com/home,* accessed July 20, 2009.

72 Vranica, Suzanne, "Madison Avenue Flirts with 3-D," *The Wall Street Journal,* May 26, 2009, p. B10.

73 Corcoran, Elizabeth, "Fab Labs," *Forbes,* August 24, 2009, p. 32.

74 Gaudin, Sharon, "Nanotech Creates Batteries Out of Paper," *Computerworld,* December 21, 2009, p. 6.

75 Staff, "Eagle Eyes Project, *www.bc.edu/schools/csom/eagleeyes/,* accessed June 22, 2009.

76 Mossberg, Walter, "Intel Makes Leap In Device To Aid Impaired Readers," *The Wall Street Journal,* November 19, 2009, p. D1.

77 Staff, "Innovation Awards," *The Wall Street Journal,* September 14, 2009, p. R3.

78 Clark, Don, "Take Two Digital Pills and Call Me in The Morning," *The Wall Street Journal,* August 4, 2009, p. A6.

79 Worthen, Ben, "Doctor, can you see me now?" *The Wall Street Journal,* October 20, 2009, p. D1.

80 Choi, K. "GM Daewood Auto," *The Wall Street Journal,* April 8, 2009, p. B3.

81 Greenburg, Z., "Segway Owners Are Challenging The Notion That Polo Should Be Played On Horseback," *Forbes,* October 19, 2009, p. 242.

82 Naik, Gautam, "To Sketch A Thief," *The Wall Street Journal,* March 27, 2009, p. A9.

83 Staff, "FBI Scanning Driver Photos For Fugitives," *The Tampa Tribune,* p. 15.

84 Franklin, Curtis, "Pimp My Dash," *Information Week,* November 16, 2009, p. 22.

85 Hamm, Steve, "Brains in the Concrete and Steel," *Business Week,* March 2, 2009, p. 43.

86 Casselman, Ben, "Chevron Engineers Squeeze New Oil From Old Wells," *The Wall Street Journal,* October 9, 2009, p. B1.

87 Weier, Mary Hayes, "RFID-Based Dispensers," *Information Week,* June 8, 2009, p. 30.

88 Michaels, Daniel, "Airline Industry Gets Smarter With Bags," *The Wall Street Journal,* September 30, 2009, p. B5.

89 Pita, J., et al, "Using Game Theory for Los Angeles Airport Security," *AI Magazine,* Spring 2009, p. 43.

90 Flood, M., "Embracing Change: Financial Informatics and Risk Analysis," *Quantitative Finance,* April 2009, p. 243.

Systems Development and Social Issues

Chapter 8   Systems Development

CHAPTER
· 8 ·

Systems Development

PRINCIPLES	LEARNING OBJECTIVES
▪ Effective systems development requires a team effort of stakeholders, users, managers, systems development specialists, and various support personnel, and it starts with careful planning.	▪ Identify the key participants in the systems development process and discuss their roles. ▪ Define the term *information systems planning* and discuss the importance of planning a project.
▪ Systems development often uses different approaches and tools such as traditional development, prototyping, rapid application development, end-user development, computer-aided software engineering, and object-oriented development to select, implement, and monitor projects.	▪ Discuss the key features, advantages, and disadvantages of the traditional, prototyping, rapid application development, and end-user systems development life cycles. ▪ Discuss the use of computer-aided software engineering (CASE) tools and the object-oriented approach to systems development.
▪ Systems development starts with investigation and analysis of existing systems.	▪ State the purpose of systems investigation. ▪ Discuss the importance of performance and cost objecti ▪ State the purpose of systems analysis and discuss some of the tools and techniques used in this phase of systems development.
▪ Designing new systems or modifying existing ones should always be aimed at helping an organization achieve its goals.	▪ State the purpose of systems design and discuss the differences between logical and physical systems desigr ▪ Discuss the use of environmental design in the systems development process.
▪ The primary emphasis of systems implementation is to make sure that the right information is delivered to the right person in the right format at the right time.	▪ State the purpose of systems implementation and discuss the various activities associated with this phase of systems development.
▪ Maintenance and review add to the useful life of a system but can consume large amounts of resources, so they benefit from the same rigorous methods and project management techniques applied to systems development.	▪ State the importance of systems and software maintenance and discuss the activities involved. ▪ Describe the systems review process.

Information Systems in the Global Economy ›
LEGO, Denmark

LEGO Builds Information Systems from Modular Blocks

LEGO blocks are one of the best-known toys in the world. Founded in 1932 in Denmark by the Kirk Kristiansen family, which still owns the company, the LEGO Group has grown to 8,000 employees, providing fun building toys to children in more than 130 countries. The word LEGO is derived from an abbreviation of the two Danish words "leg godt," which translates to "play well." Children can "play well" with LEGO blocks because of the modular framework, which allows children to explore their creativity and to solve problems.

Recently LEGO has enjoyed a resurgence of popularity. The company has experienced growing net profits over the past decade, with 2009 annual net profits increasing by 63 percent to DKK 2,204 million. LEGO is building and riding its tidal wave of success by diversifying and growing its product line. Its popular LEGO and DUPLO blocks take advantage of the latest media fads by offering kits for popular titles such as Star Wars, Toy Story, SpongeBob, and Space Police. Its Bionicle line is popular with tweens, and its Mindstorms computer-driven robots appeal to technically and scientifically minded children and young adults. Adults enjoy LEGO's more complicated kits such as its Architecture line. Recently, LEGO expanded into software games that duplicate their physical block packages in virtual reality software. The company has also launched LEGO Universe—a massively multiplayer online game (MMOG). The LEGO Group has even opened Discovery Centers featuring educational LEGO activities and theme parks featuring more than 50 LEGO-themed rides, shows, and attractions in Denmark, the UK, the U.S., and Germany. By continuously reshaping itself, LEGO has reenergized its brand, leading to unprecedented growth for the company.

The rapid growth of the LEGO Group has provided substantial challenges for its information systems. Until recently, its mainframe-based enterprise system could not provide the flexibility to keep up with the rapid changes of the toy market. LEGO systems engineers were tasked with upgrading LEGO systems to handle the company's growth and its diverse business model. LEGO required a system that could support the needs of a large enterprise while being flexible and nimble enough to accommodate rapid change. Esben Viskum, Senior Director of the LEGO Service Center, defines rapid change as the ability to "respond to the market quickly, using short product development processes without losing control of cost and quality, and being able to manage both people and operations effectively and efficiently."

After considering the problem, LEGO systems analysts decided that the best system for the task would need to be as modular and standardized as the LEGO blocks themselves. A modular standardized model would make it possible for the company to quickly expand into new markets.

LEGO system analysts spent months evaluating LEGO's current systems and data to determine the exact needs for the new system. Next, the team performed numerous feasibility studies to confirm that they could build the new system within their economic, technical, operational, and time constraints.

Because the new system would be large and comprehensive, the LEGO team would require the assistance of information system market leaders. LEGO selected SAP as the software foundation of its enterprise-wide systems. This foundation included SAP ERP Human Capital Management software and SAP Product Lifecycle Management software. Esben Viskum defines these as "business-critical" solutions.

LEGO selected IBM for the system infrastructure, including servers and storage. Esben Viskum says that IBM provided the "best way to deliver robust operations, and provide a repeatable template for each new LEGO venture."

The resulting system clearly supports LEGO's corporate goals. According to Viskum, LEGO wants to become "a much larger business, with the products, sales, and infrastructure to become truly robust." Its new enterprise systems will allow it to do just that. LEGO plans to expand with new sales offices, manufacturing plants, and retail shops. The SAP/IBM system provides a cookie-cutter approach to stamping out new business extensions with information systems that support local operations integrated into the system.

LEGO's investment of €45 million in its new systems is estimated to produce business benefits of €150 million—a threefold payback. The savings result from improved information delivery, which provides managers with better control and allows executives to respond more quickly and effectively to opportunities and problems.

As you read this chapter, consider the following:

- What situations can arise within a business to trigger new systems development initiatives?
- What are the best methods for a business to use in approaching new systems development projects?

Why Learn About Systems Development?

Throughout this book, you have seen many examples of the use of information systems in a variety of careers. A manager at a hotel chain can use an information system to look up client preferences. An entrepreneur can use systems development to build a new information systems and a new business. An accountant at a manufacturing company can use an information system to analyze the costs of a new plant. A sales representative for a music store can use an information system to determine which CDs to order and which to discount because they are not selling. Information systems have been designed and implemented for almost every career and industry. An individual can use systems development to create applications for smartphones and other mobile devices for profit or enjoyment. But where do you start to acquire these systems or have them developed? How can you work with IS personnel, such as systems analysts and computer programmers, to get what you need to succeed on the job? This chapter gives you the answer. You will see how you can initiate the systems development process and analyze your needs with the help of IS personnel. In this chapter, you will learn how your project can be planned, aligned with corporate goals, rapidly developed, and much more. We start with an overview of the systems development process.

When an organization needs to accomplish a new task or change a work process, how does it do so? It develops a new system or modifies an existing one. Systems development is the activity of creating or modifying systems. It refers to all aspects of the process—from identifying problems to solve or opportunities to exploit to implementing and refining the chosen solution.

AN OVERVIEW OF SYSTEMS DEVELOPMENT

In today's businesses, managers and employees in all functional areas work together and use business information systems. As a result, they are helping with development and, in many cases, leading the way. Users might request that a systems development team determine whether they should purchase a few PCs or create an attractive Web site using the tools discussed in Chapter 4. In another case, an entrepreneur might use systems development to build an Internet site to compete with large corporations. A number of individuals, for example, have developed applications for Apple's iPhone that are sold on Apple's applications store (App Store).[1] According to Steve Jobs, one of the founders of Apple Computer, "The App Store is like nothing the industry has ever seen before in both scale and quality. With 1.5 billion apps downloaded, it is going to be very hard for others to catch up." This chapter provides you with a deeper appreciation of the systems development process for individuals and organizations.

Participants in Systems Development

Effective systems development requires a team effort. The team usually consists of stakeholders, users, managers, systems development specialists, and various support personnel. This team, called the *development team*, is responsible for determining the objectives of the information system and delivering a system that meets these objectives. Selecting the best IS team for a systems development project is critical to project success.[2] A *project* is a planned collection of activities that achieves a goal, such as constructing a new manufacturing plant or developing a new decision support system. Nevsun Resources, a Canadian mining operation, used a software package called Unifier to oversee its large African mining operations.[3] The company used the project-management software to obtain real time reviews of its mining projects in remote areas, such as Africa.

In the context of systems development, **stakeholders** are people who, either themselves or through the organization they represent, ultimately benefit from the systems development project. **Users** are people who will interact with the system regularly. They can be employees, managers, or suppliers. For large-scale systems development projects, in which the investment in and value of a system can be high, it is common for senior-level managers, including the functional vice presidents (of finance, marketing, and so on), to be part of the development team.

stakeholders
People who, either themselves or through the organization they represent, ultimately benefit from the systems development project.

users
People who will interact with the system regularly.

Because stakeholders ultimately benefit from the systems development project, they often work with others in developing a computer application.

(Source: © Jacob Wackerhausen/ iStockphoto.)

Depending on the nature of the systems project, the development team might include systems analysts and programmers, among others. A **systems analyst** is a professional who specializes in analyzing and designing business systems. One popular magazine rated a systems analyst job as one of the 50 best jobs in America.[4] Systems analysts play various roles while interacting with the stakeholders and users, management, vendors and suppliers, external companies, programmers, and other IS support personnel. See Figure 8.1. Like an architect developing blueprints for a new building, a systems analyst develops detailed plans for the new or modified system. The **programmer** is responsible for modifying or developing programs to satisfy user requirements. Like a contractor constructing a new building or renovating an existing one, the programmer takes the plans from the systems analyst and builds or modifies the necessary software.

systems analyst
A professional who specializes in analyzing and designing business systems.

programmer
A specialist responsible for modifying or developing programs to satisfy user requirements.

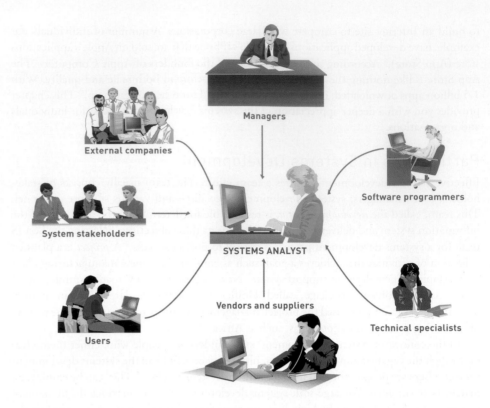

Individual Systems Developers and Users

For decades, systems development was oriented towards corporations and corporate teams or groups. The major participants were discussed above. While this continues to be an important part of systems development, we are seeing individual systems developers and users to a greater extent.

An *individual systems developer* is a person that performs all of the systems development roles, including systems analyst, programmer, technical specialist, and other roles described in the above section. While individual systems developers can create applications for a group or entire organization, many specialize in developing applications for individuals. A large number of these applications are available for smartphones and other handheld computing devices. Individual developers from around the world, for example, are using the steps of systems development to create unique applications for the iPhone.[5] In addition, Apple has special tools for iPhone application developers, including GPS capabilities, turn-by-turn directions, instant messaging, cut-and-paste features, and audio streaming, to make it easier for people to craft unique applications.[6] Apple is also allowing systems developers to charge users in a variety of ways, including fixed prices and subscription fees through Apple's App Store. Ted Sullivan, for example, has developed a free download he hopes to make available on Apple's App Store called Game Changer.[7] The application will collect baseball statistics from kids' baseball games and send them to the iPhones of parents, relatives, and others. Sullivan hopes to make millions of dollars from advertisers and monthly subscription fees for the service. Before an individual developer can have his or her application placed or sold on Apple's application store, however, Apple must approve or select the application.[8] Apple has tens of thousands of applications that can be downloaded and used. Other companies also have application stores. BlackBerry has an application store, called App World, and Google has Android Market store.[9] Google also has a systems development tool called Wave that lets individual developers collaborate and communicate with others in creating documents.[10] Other people can add text, multimedia, and a variety of applications to a document. According to a Wave cofounder, "We're banking on Wave having a very large impact, but a lot of it depends on our ability to explain this to users. That's part of the reason why we're putting this out early to developers." Some applications, such as Google's Secure Data Connector, allow data to be downloaded from secure corporate databases, including customer and supplier information.[11]

Individual users acquire applications for both personal and professional use.[12] Cisco, the large networking company, has developed an iPhone application to help individual security personnel respond to IS and computer-related threats.[13] The application, called Security Intelligence Operations To Go, can instantly notify security professionals of ongoing security attacks as they occur and help them recover if they occur. Another personal application turns an iPhone into a musical flute that can be played by blowing into the microphone and pushing keys on a virtual keyboard.[14] The program costs 99 cents but has generated about $1 million for the developer, an Assistant Professor of Music. Another application is a sophisticated patient monitoring system for doctors and other healthcare professionals.[15] Individual applications can be used to compare prices of products, analyze loans, locate organic food, find reliable repairmen, and locate an apartment in your area.[16] You can also turn a smartphone into a powerful scientific or financial calculator.[17] Other applications can synchronize with popular calendar and contact applications on laptop or desktop computers.[18] Applications can cost less than $1 or more than $100, including games, a compass and maps to show your direction and location, word-processing and spreadsheet programs, airline flight information, and much more. While most people purchase individual applications from authorized Web sites, unauthorized application stores that are not supported by the smartphone or cellular company can be used to purchase or acquire useful applications.[19]

It is also possible for one person to be both an individual developer and user. The term **end-user systems development** describes any systems development project in which business managers and users assume the primary effort. User-developed systems range from the very small (such as a software routine to merge form letters) to those of significant organizational value (such as customer contact databases for the Web).[20] Like any systems developer, individual developers and end users should follow the approach and techniques of the systems development process described in this chapter. Even if you develop your own applications, you will likely want to have an IS department develop applications for you that are too complex or time consuming to develop on your own. In this case, you will be involved in initiating systems development, discussed next.

end-user systems development
Any systems development project in which the primary effort is undertaken by a combination of business managers and users.

Many end users today are demonstrating their systems development capability by designing and implementing their own PC-based systems.

(Source: India Today Group/Getty Images.)

Information Systems Planning and Aligning Corporate and IS Goals

Information systems planning and aligning corporate and IS goals are important aspects of any systems development project. Achieving a competitive advantage is often the overall objective of systems development.

The term **information systems planning** refers to translating strategic and organizational goals into systems development initiatives. See Figure 8.2.[21] Proper IS planning ensures that specific systems development objectives support organizational goals. Long-range planning

information systems planning
Translating strategic and organizational goals into systems development initiatives.

can also be important and result in getting the most from a systems development effort. It can also align IS goals with corporate goals and culture, which is discussed next.

Figure 8.2

Information Systems Planning

Information systems planning transforms organizational goals outlined in the strategic plan into specific systems development activities.

Aligning organizational goals and IS goals is critical for any successful systems development effort.[22] Because information systems support other business activities, IS staff and people in other departments need to understand each other's responsibilities and tasks. Most corporations have profits and return on investment (ROI), first introduced in Chapter 1, as primary goals. According to the chief technology officer for the Financial Industry Regulatory Authority, "ROI is a key metric for technology initiatives, and the business case needs to include both initial development costs and subsequent maintenance costs."[23] Another difficult aspect of aligning corporate and IS goals is the changing nature of business goals and priorities.[24]

SYSTEMS DEVELOPMENT LIFE CYCLES

The systems development process is also called a *systems development life cycle (SDLC)* because the activities associated with it are ongoing. As each system is built, the project has timelines and deadlines, until at last the system is installed and accepted. The life of the system continues as it is maintained and reviewed. If the system needs significant improvement beyond the scope of maintenance, if it needs to be replaced because of a new generation of technology, or if the IS needs of the organization change significantly, a new project will be initiated and the cycle will start over.

Hess Information Systems Take the Long View

Hess Corporation is a global energy company engaged in the exploration for and production of crude oil and natural gas. Hess has offices in 18 countries across six continents, with key headquarters in Houston, London, Kuala Lumpur (Malaysia), and Woodbridge (New Jersey).

Like most smart, mature companies, Hess aligns its information systems with long-term organizational goals and strategies. The focus on long-term, however, was further emphasized with the arrival of a new CIO, Jeff Steinhorn. Steinhorn discovered that the Hess information systems group had historically taken a short-term approach to project planning. The company had its IS personnel focused on supporting near-term initiatives and the separate needs of each business division. No one was analyzing how these smaller short-term projects were assisting the company as a group at the highest level to meet its long-term objectives.

While this short-sighted approach served Hess adequately during the years that the company was focused solely on the oil business, it became more important as the company diversified for IS to assist in high-level long-term goals. In the past decade, Hess has expanded into natural gas and electricity. It was clear to Steinhorn that the company's IS initiatives needed to support and connect the organization's more diverse interests over the long haul.

Bobby Cameron at Forrester Research says that Steinhorn's predicament is not unique. The majority of new CIOs find themselves addressing the same problem. This is the result of the recent trend of organizations moving to use IS in ways important to the organization as a whole. Although businesses have traditionally called upon IS to support various initiatives, enterprise IS is a current force that drives business processes and spurs innovation. With this in mind, more and more IS initiatives are aligned with high-level, long-term organizational goals.

Steinhorn began by centralizing the data source that fed all information systems. Since IS projects had been conducted in each business division, the divisions weren't sharing important information such as customer records and market information. A central data store would eliminate data redundancy and improve data accuracy.

Next, Steinhorn set out to develop a five-year IS strategic plan. Once established, Steinhorn had to work to gain senior management approval. Senior management was resistant to change, and Steinhorn had not yet proved himself within the organization. Steinhorn brought some of the Hess' best regarded IS managers onto the planning team and, with their assistance, gained senior management support.

Steinhorn also had trouble gaining the support of division executives who were accustomed to viewing IS as a support facility rather than a contributor to organizational goals and objectives. Steinhorn finally won them over by selling them on the ROI of long-term planning. Steinhorn persuaded them by showing that the costs of his plan were one-tenth as much as the benefits would return.

Steinhorn's five-year plan was divided into three components:

Business or "B" projects that assisted in improving business processes, reducing costs, and increasing revenue.

Enabler or "E" projects that provided decision makers with information that enabled them to make better decisions, including business intelligence and analytical systems.

Process or "P" projects that assisted the IS group in organizing and standardizing its own work processes.

Within nine months of Steinhorn's five-year plan, 17 projects had been kicked off and seven were completed. This included a major upgrade of the company's SAP retail energy system.

Under its "P" project category, Steinhorn's group developed standardized systems for application development, project management, and IS governance. Steinhorn also implemented a performance tracking and scorecard system with which project success could be gauged. These advances in IS project management dramatically improved the quality and efficiency of the IS team's work.

With each successful project, Steinhorn's efforts are gaining increased support across Hess. The three divisions of Hess—oil, gas, and electricity—have become more interested in working together. Approaching IS from a long-term, enterprise level "has really elevated the decision-making to help decide where it's best to invest IT dollars to get the greatest returns," says William Hanna, vice president of electric operations at Hess and a program sponsor for the IT strategic planning initiative.

Discussion Questions

1. What challenges did Steinhorn face when he took the job as CIO of Hess?
2. Why are many businesses switching to a long-term, enterprise-wide emphasis for IS projects?

Critical Thinking Questions

1. How did Steinhorn organize IS projects under his five-year plan? What other categories of projects might you have included?
2. Why did Steinhorn face resistance from top-level managers and executives? What smart moves did he make to win them over?

SOURCES: "Amerada Hess Uses SAP Oil & Gas to Reduce Cost of Joint-Venture Activity," SAP Customer Implementation Success, accessed May 23, 2010, www.sap.com/industries/oil-gas/pdf/50026246.pdf; Hess Web site, accessed May 23, 2010, www.hess.com; Hoffman, Thomas, "Hess builds a project pipeline with long-term vision," Computerworld, April 7, 2008, www.computerworld.com/s/article/314711/ Building_an_IT_Project_Pipeline?taxonomyId=74& pageNumber=1.

The Traditional Systems Development Life Cycle

Traditional systems development efforts can range from a small project, such as purchasing an inexpensive computer program, to a major undertaking. The steps of traditional systems development might vary from one company to the next, but most approaches have five common phases: investigation, analysis, design, implementation, and maintenance and review. See Figure 8.3.

Figure 8.3

The Traditional Systems Development Life Cycle

Sometimes, information learned in a particular phase requires cycling back to a previous phase.

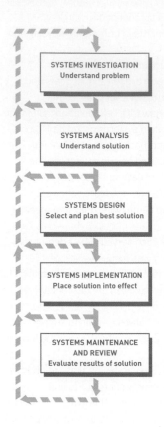

systems investigation
The systems development phase during which problems and opportunities are identified and considered in light of the goals of the business.

systems analysis
The systems development phase that attempts to answer the question "What must the information system do to solve the problem?"

systems design
The systems development phase that defines how the information system will do what it must do to obtain the problem solution.

systems implementation
The systems development phase involving the creation or acquisition of various system components detailed in the systems design, assembling them, and placing the new or modified system into operation.

In the **systems investigation** phase, potential problems and opportunities are identified and considered in light of the goals of the business. Systems investigation attempts to answer the questions "What is the problem, and is it worth solving?" The primary result of this phase is a defined development project for which business problems or opportunity statements have been created, to which some organizational resources have been committed, and for which systems analysis is recommended. **Systems analysis** attempts to answer the question "What must the information system do to solve the problem?" This phase involves studying existing systems and work processes to identify strengths, weaknesses, and opportunities for improvement. The major outcome of systems analysis is a list of requirements and priorities. **Systems design** seeks to answer the question "How will the information system do what it must do to obtain the problem solution?" The primary result of this phase is a technical design that either describes the new system or describes how existing systems will be modified. The system design details system outputs, inputs, and user interfaces; specifies hardware, software, database, telecommunications, personnel, and procedure components; and shows how these components are related. **Systems implementation** involves creating or acquiring the various system components detailed in the systems design, assembling them, and placing the new or modified system into operation. An important task during this phase is to train the users. Systems implementation results in an installed, operational information system that meets the business needs for which it was developed. It can also involve phasing out or removing old systems, which can be difficult for existing users, especially when the systems are free.

The purpose of **systems maintenance and review** is to ensure that the system operates as intended and to modify the system so that it continues to meet changing business needs. As shown in Figure 8.3, a system under development moves from one phase of the traditional SDLC to the next.

Prototyping

Prototyping takes an iterative approach to the systems development process.[25] During each iteration, requirements and alternative solutions to the problem are identified and analyzed, new solutions are designed, and a portion of the system is implemented. Users are then encouraged to try the prototype and provide feedback. See Figure 8.4. Prototyping begins with creating a preliminary model of a major subsystem or a scaled-down version of the entire system. For example, a prototype might show sample report formats and input screens. After they are developed and refined, the prototypical reports and input screens are used as models for the actual system, which can be developed using an end-user programming language such as Visual Basic. The first preliminary model is refined to form the second- and third-generation models, and so on until the complete system is developed. See Figure 8.5.

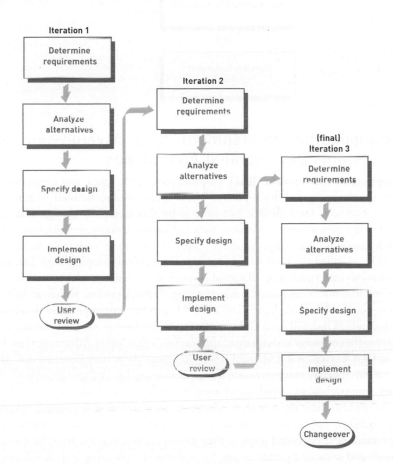

Figure 8.5

Refining during Prototyping

Each generation of prototype is a refinement of the previous generation based on user feedback.

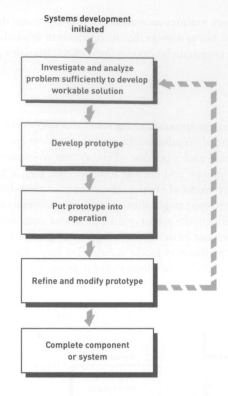

Rapid Application Development, Agile Development, and Other Systems Development Approaches

rapid application development (RAD)

A systems development approach that employs tools, techniques, and methodologies designed to speed application development.

Rapid application development (RAD) employs tools, techniques, and methodologies designed to speed application development. These tools can also be used to make systems development projects more flexible and agile to be able to rapidly change with changing conditions and environments.[26] Vendors such as Computer Associates International, IBM, and Oracle market products targeting the RAD market. Rational Software, a division of IBM, has a RAD tool called Rational Rapid Developer to make developing large Java programs and applications easier and faster. Rational allows both systems developers and users to collaborate on systems development projects using Team Concert, which is like a social networking site for IBM developers and users.[27] Locus Systems, a program developer, used a RAD tool called OptimalJ from Compuware (*www.compuware.com*) to generate more than 60 percent of the computer code for three applications it developed. Advantage Gen, formerly known as COOL:Gen, is a RAD tool from Computer Associates International. It can be used to rapidly generate computer code from business models and specifications.

Other approaches to rapid development, such as *agile development* or *extreme programming* (*XP*), allow the systems to change as they are being developed.[28] Agile development requires cooperation and frequent face-to-face meetings with all participants, including systems developers and users, as they modify, refine, and test how the system meets users' needs and what its capabilities are.[29] Organizations are using agile development to a greater extent today to improve the results of systems development, including global systems development projects requiring IS resources distributed in different locations.[30] Agile development can be a good approach when the requirements of a new or modified system aren't completely known in advance.[31] According to one IS professional, "Agile is perfect when you're not sure what you're getting into." There are many variations of agile development with different names, including Dynamic Systems Development Method, Crystal, Agile Modeling, and several other names. [32]

Extreme programming (XP) uses pairs of programmers who work together to design, test, and code parts of the systems they develop.[33] Research has shown that a pair of programmers usually outperforms the average individual programmer and on a level with the organization's best programmers.[34] The iterative nature of XP helps companies develop robust systems with fewer errors. Sabre Airline Solutions, a $2 billion computer company serving the airline travel industry, used XP to eliminate programming errors and shorten program development times.

In addition to the systems development approaches discussed previously, a number of other agile and innovative systems development approaches have been created by computer vendors and authors of systems development books. These approaches all attempt to deliver better systems in a shorter amount of time. A few agile development tools are listed below.

- *Adaptive Software Development.* Adaptive Software Development (ASD) grew out of rapid application development techniques and stresses an iterative process that involves analysis, design, and implementation at each cycle or iteration. The approach was primarily developed by James Highsmith.
- *Lean Software Development.* Lean Software Development came from a book with the same title by Mary and Tom Poppendieck. The approach comes from lean manufacturing practices used by Toyota and stresses the elimination of waste, continuous learning, just-in-time decision making, and empowering systems development teams.[35]
- *Rational Unified Process (RUP).* Rational Unified Process is an iterative systems development approach developed by IBM and includes a number of tools and techniques that are typically tailored to fit the needs of a specific company or organization. RUP uses an iterative approach to software development that stresses quality as the software is changed and updated over time.[36] Many companies have used RUP to their advantage.[37]
- *Feature-Driven Development (FDD).* Originally used to complete a systems development project at a large bank, Feature-Driven Development is an iterative systems development approach that stresses the features of the new or modified system and involves developing an overall model, creating a list of features, planning by features, designing by features, and building by features.[38]
- *Crystal Methodologies.* Crystal Methodologies is a family of systems development approaches developed by Alistair Cockburn that concentrates on effective team work and the reduction of paperwork and bureaucracy to make development projects faster and more efficient.

Outsourcing and On-Demand Computing

Many companies hire an outside consulting firm or computer company that specializes in systems development to take over some or all of its development and operations activities.[39] The drug company Pfizer, for example, used outsourcing to allow about 4,000 of its employees to outsource some of their jobs to other individuals or companies around the globe.[40] Vodafone used outsourcing to help it innovate and find creative solutions in providing wireless services.[41] According to its chief executive, "We were a bit naive thinking everything could be done in-house."

Small and medium-sized firms are using outsourcing to a greater extent today to cut costs and acquire needed technical expertise that would be difficult to afford with in-house personnel.[42] According to one outsourcing expert, "The downturn is making it harder for companies to tie outsourcing to broader goals than basic cost cutting." The market for outsourcing services for small and medium-sized firms is expected to increase by 15 percent annually through 2012 and beyond. Reducing costs, obtaining state-of-the-art technology, eliminating staffing and personnel problems, and increasing technological flexibility are reasons that companies have used the outsourcing and on-demand computing approaches.

A number of companies and nonprofit organizations offer outsourcing and on-demand computing services—from general systems development to specialized services. IBM's Global Services, for example, is one of the largest full-service outsourcing and consulting services.[43] IBM has consultants located in offices around the world and generates over $50 billion in revenues each year. Electronic Data Systems (EDS) is another large company that specializes in consulting and outsourcing.[44] EDS has approximately 140,000 employees in almost 60 countries and more than 9,000 clients worldwide. EDS, which was acquired by Hewlett-Packard, generates over $20 billion annually.[45] Accenture is another company that specializes in consulting and outsourcing.[46] The company has more than 75,000 employees in 47 countries with annual revenues that exceed $20 billion. Wipro Technologies, head-quartered in India, is another worldwide outsourcing company with more than $4 billion in annual revenues.[47] Amazon, the large online retailer of books and other products, will offer on-demand computing to individuals and other companies of all sizes, allowing them to use Amazon's computer expertise and database capacity. Individuals and companies will only pay for the computer services they use. See Figure 8.6.

FACTORS AFFECTING SYSTEMS DEVELOPMENT SUCCESS

Successful systems development means delivering a system that meets user and organizational needs—on time and within budget. Achieving a successful systems development project, however, can be difficult. The state of Florida, for example, had problems with its $15 billion Medicaid program, making it difficult or impossible for some beneficiaries to gain access to the system or receive payments.[48] According to a February statement made by one frustrated state senator, "I'm not looking for zero errors, but you're saying it's going to be into the summer before we can expect our constituents to be able to get through on the call lines?"

Getting users and stakeholders involved in systems development is critical for most systems development projects. Some researchers believe that how a systems development project is managed and run is one of the best indicators of systems development success.[49] Having the support of top-level managers is also important. In addition to user involvement and top management support, other factors can contribute to successful systems development efforts—at a reasonable cost. These factors are discussed next.

Degree of Change

A major factor that affects the quality of systems development is the degree of change associated with the project. The scope can vary from enhancing an existing system to major reengineering. The project team needs to recognize where they are on this spectrum of change.

Continuous Improvement versus Reengineering

As discussed in Chapter 1, continuous improvement projects do not require a lot of changes or retraining of people; thus, they have a high degree of success.[50] Typically, because continuous improvements involve minor improvements, these projects also have relatively modest benefits. On the other hand, reengineering involves fundamental changes in how the organization conducts business and completes tasks. The factors associated with successful reengineering are similar to those of any development effort, including top management support, clearly defined corporate goals and systems development objectives, and careful management of change. Major reengineering projects tend to have a high degree of risk but also a high potential for major business benefits. See Figure 8.7.

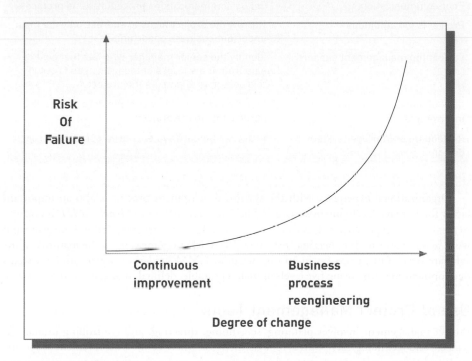

Figure 8.7

The degree of change can greatly affect the probability of a project's success.

Managing Change

The ability to manage change is critical to the success of systems development. New systems inevitably cause change. Unfortunately, not everyone adapts easily, and the increasing complexity of systems can multiply the problems. Some systems developers believe that system complexity is a major cause of systems development failures.[51] Managing change requires the ability to recognize existing or potential problems (particularly the concerns of users) and deal with them before they become a serious threat to the success of the new or modified system. Here are several of the most common problems that often need to be addressed as a result of new or modified systems:

- Fear that the employee will lose his or her job, power, or influence within the organization
- Belief that the proposed system will create more work than it eliminates
- Reluctance to work with "computer people"
- Anxiety that the proposed system will negatively alter the structure of the organization
- Belief that other problems are more pressing than those solved by the proposed system or that the system is being developed by people unfamiliar with "the way things need to get done"
- Unwillingness to learn new procedures or approaches

The Importance of Planning

The bigger the project, the more likely that poor planning will lead to significant problems. Many companies find that large systems projects fall behind schedule, go over budget, and do not meet expectations. Although proper planning cannot guarantee that these types of problems will be avoided, it can minimize the likelihood of their occurrence. Good systems development is not automatic. Certain factors contribute to the failure of systems development projects. These factors and the countermeasures to eliminate or alleviate the problem are summarized in Table 8.1.

Table 8.1

Project Planning Issues Frequently Contributing to Project Failure

Factor	Countermeasure
Solving the wrong problem	Establish a clear connection between the project and organizational goals
Poor problem definition and analysis	Follow a standard systems development approach
Poor communication	Set up communications procedures and protocols
Project is too ambitious	Narrow the project focus to address only the most important business opportunities
Lack of top management support	Identify the senior manager who has the most to gain from the success of the project and recruit this person to champion the project
Lack of management and user involvement	Identify and recruit key stakeholders to be active participants in the project
Inadequate or improper system design	Follow a standard systems development approach

Organizational experience with the systems development process is also an important factor for systems development success.[52] The *Capability Maturity Model (CMM)* is one way to measure this experience.[53] It is based on research done at Carnegie Mellon University and work by the Software Engineering Institute (SEI). CMM is a measure of the maturity of the software development process in an organization. CMM grades an organization's systems development maturity using five levels: initial, repeatable, defined, managed, and optimized.

Use of Project Management Tools

Project management involves planning, scheduling, directing, and controlling human, financial, and technological resources for a defined task whose result is achievement of specific goals and objectives. Corporations and nonprofit organizations use these important tools and techniques.

A **project schedule** is a detailed description of what is to be done. Each project activity, the use of personnel and other resources, and expected completion dates are described. A **project milestone** is a critical date for the completion of a major part of the project. The completion of program design, coding, testing, and release are examples of milestones for a programming project. The **project deadline** is the date the entire project is to be completed and operational— when the organization can expect to begin to reap the benefits of the project.

In systems development, each activity has an earliest start time, earliest finish time, and slack time, which is the amount of time an activity can be delayed without delaying the entire project. The **critical path** consists of all activities that, if delayed, would delay the entire project. These activities have zero slack time. Any problems with critical-path activities will cause problems for the entire project. To ensure that critical-path activities are completed in a timely fashion, formalized project management approaches have been developed. Tools such as Microsoft Project are available to help compute these critical project attributes.

Although the steps of systems development seem straightforward, larger projects can become complex, requiring hundreds or thousands of separate activities. For these systems development efforts, formal project management methods and tools become essential. A formalized approach called **Program Evaluation and Review Technique (PERT)** creates three time estimates for an activity: shortest possible time, most likely time, and longest possible time. A formula is then applied to determine a single PERT time estimate. A **Gantt chart** is a graphical tool used for planning, monitoring, and coordinating projects; it is essentially a grid that lists activities and deadlines. Each time a task is completed, a marker such as a darkened line is placed in the proper grid cell to indicate the completion of a task. (See Figure 8.8.)

project schedule
A detailed description of what is to be done.

project milestone
A critical date for the completion of a major part of the project.

project deadline
The date the entire project is to be completed and operational.

critical path
Activities that, if delayed, would delay the entire project.

Program Evaluation and Review Technique (PERT)
A formalized approach for developing a project schedule.

Gantt chart
A graphical tool used for planning, monitoring, and coordinating projects.

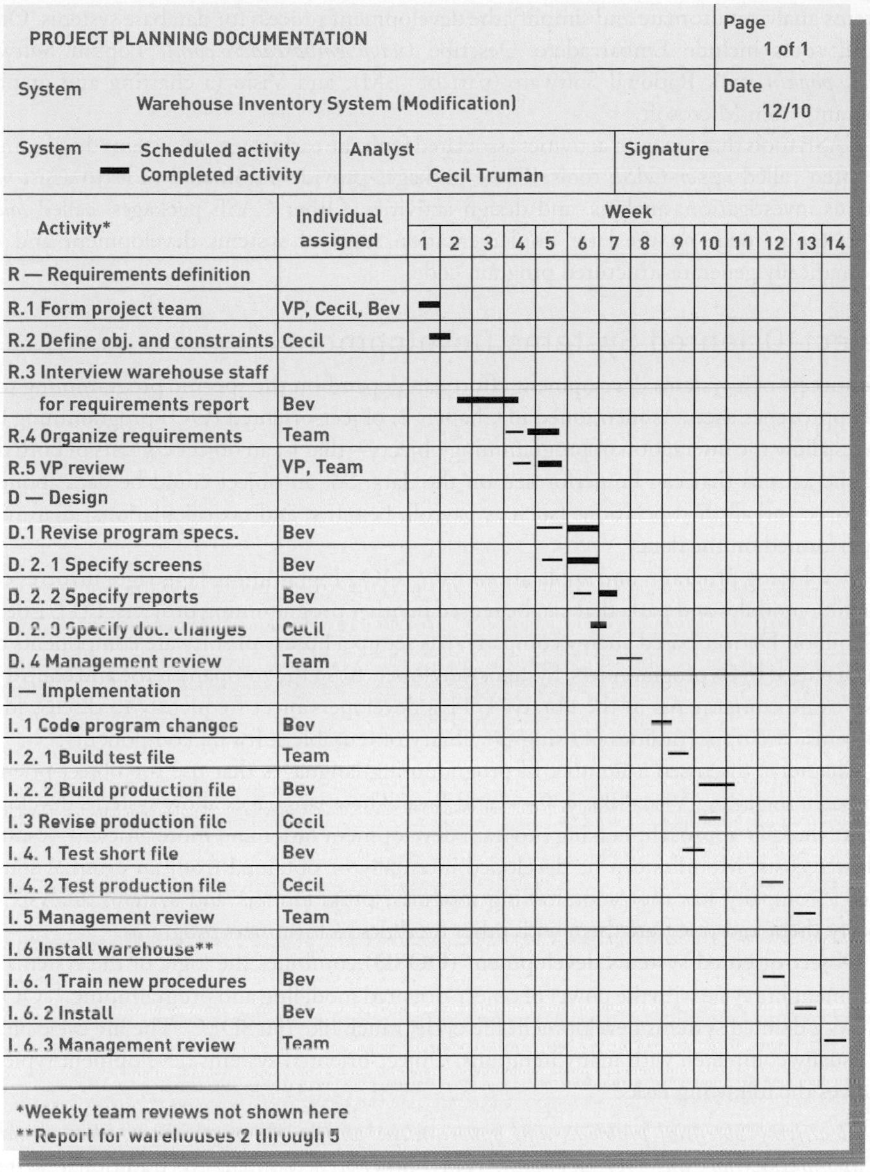

Figure 8.8

Sample Gantt Chart

A Gantt chart shows progress through systems development activities by putting a bar through appropriate cells.

PROJECT PLANNING DOCUMENTATION															Page 1 of 1
System	Warehouse Inventory System (Modification)														Date 12/10
System — Scheduled activity ▬ Completed activity	Analyst Cecil Truman											Signature			

Activity*	Individual assigned	Week 1	2	3	4	5	6	7	8	9	10	11	12	13	14
R — Requirements definition															
R.1 Form project team	VP, Cecil, Bev	▬													
R.2 Define obj. and constraints	Cecil	▬													
R.3 Interview warehouse staff for requirements report	Bev		▬												
R.4 Organize requirements	Team				▬										
R.5 VP review	VP, Team				▬										
D — Design															
D.1 Revise program specs.	Bev						▬								
D.2.1 Specify screens	Bev						▬								
D.2.2 Specify reports	Bev						▬								
D.2.3 Specify doc. changes	Cecil						▬								
D.4 Management review	Team							▬							
I — Implementation															
I.1 Code program changes	Bev								▬						
I.2.1 Build test file	Team								▬						
I.2.2 Build production file	Bev										▬				
I.3 Revise production file	Cecil										▬				
I.4.1 Test short file	Bev									▬					
I.4.2 Test production file	Cecil											▬			
I.5 Management review	Team												▬		
I.6 Install warehouse**															
I.6.1 Train new procedures	Bev											▬			
I.6.2 Install	Bev												▬		
I.6.3 Management review	Team														▬

*Weekly team reviews not shown here
**Report for warehouses 2 through 5

Both PERT and Gantt techniques can be automated using project management software. Project management software helps managers determine the best way to reduce project completion time at the least cost. Several project management software packages are identified in Table 8.2.

Table 8.2

Selected Project Management Software

Software	Vendor
OpenPlan	Welcom (www.welcom.com)
Microsoft Project	Microsoft (www.microsoft.com)
Unifier	Skire (www.skire.com)

Use of Computer-Aided Software Engineering (CASE) Tools

Computer-aided software engineering (CASE) tools automate many of the tasks required in a systems development effort and encourage adherence to the SDLC, thus instilling a high degree of rigor and standardization to the entire systems development process. Oracle Designer by Oracle (*www.oracle.com*) and Visible Analyst by Visible Systems Corporation (*www.visible.com*) are examples of CASE tools. Oracle Designer is a CASE tool that can help

computer-aided software engineering (CASE)

Tools that automate many of the tasks required in a systems development effort and encourage adherence to the SDLC.

systems analysts automate and simplify the development process for database systems. Other CASE tools include Embarcadero Describe (*www.embarcadero.com*), Popkin Software (*www.popkin.com*), Rational Software (part of IBM), and Visio (a charting and graphics program) from Microsoft.

CASE tools that focus on activities associated with the early stages of systems development are often called *upper-CASE* tools. These packages provide automated tools to assist with systems investigation, analysis, and design activities. Other CASE packages, called *lower-CASE* tools, focus on the later implementation stage of systems development and can automatically generate structured program code.

Object-Oriented Systems Development

The success of a systems development effort can depend on the specific programming tools and approaches used. As mentioned in Chapter 2, object-oriented (OO) programming languages allow the interaction of programming objects—that is, an object consists of both data and the actions that can be performed on the data. So, an object could be data about an employee and all the operations (such as payroll, benefits, and tax calculations) that might be performed on the data.

Developing programs and applications using OO programming languages involves constructing modules and parts that can be reused in other programming projects. DTE Energy, a $7 billion Detroit-based energy company, has set up a library of software components that can be reused by its programmers. Systems developers from the company reuse and contribute to software components in the library. DTE's developers meet frequently to discuss ideas, problems, and opportunities of using the library of reusable software components.

Chapter 2 discussed a number of programming languages that use the object-oriented approach, including Visual Basic, C++, and Java. These languages allow systems developers to take the OO approach, making program development faster and more efficient, resulting in lower costs. Modules can be developed internally or obtained from an external source. After a company has the programming modules, programmers and systems analysts can modify them and integrate them with other modules to form new programs.

Object-oriented systems development (OOSD) combines the logic of the systems development life cycle with the power of object-oriented modeling and programming.[54] OOSD follows a defined systems development life cycle, much like the SDLC. The life cycle phases are usually completed with many iterations. Object-oriented systems development typically involves the following tasks:

object-oriented systems development (OOSD)
An approach to systems development that combines the logic of the systems development life cycle with the power of object-oriented modeling and programming.

- *Identifying potential problems and opportunities within the organization that would be appropriate for the OO approach.* This process is similar to traditional systems investigation. Ideally, these problems or opportunities should lend themselves to the development of programs that can be built by modifying existing programming modules.
- *Defining what kind of system users require.* This analysis means defining all the objects that are part of the user's work environment (object-oriented analysis). The OO team must study the business and build a model of the objects that are part of the business (such as a customer, an order, or a payment). Many of the CASE tools discussed in the previous section can be used, starting with this step of OOSD.
- *Designing the system.* This process defines all the objects in the system and the ways they interact (object-oriented design). Design involves developing logical and physical models of the new system by adding details to the object model started in analysis.
- *Programming or modifying modules.* This implementation step takes the object model begun during analysis and completed during design and turns it into a set of interacting objects in a system. Object-oriented programming languages are designed to allow the programmer to create classes of objects in the computer system that correspond to the objects in the actual business process. Objects such as customer, order, and payment are redefined as computer system objects—a customer screen, an order entry menu, or a dollar sign icon. Programmers then write new modules or modify existing ones to produce the desired programs.

- *Evaluation by users.* The initial implementation is evaluated by users and improved. Additional scenarios and objects are added, and the cycle repeats. Finally, a complete, tested, and approved system is available for use.
- *Periodic review and modification.* The completed and operational system is reviewed at regular intervals and modified as necessary.

SYSTEMS INVESTIGATION

As discussed earlier in the chapter, systems investigation is the first phase in the traditional SDLC of a new or modified business information system. The purpose is to identify potential problems and opportunities and consider them in light of the goals of the company. Some stock trading companies that automatically place buy and sell orders based on news items are investigating their automated trading systems.[55] The automated stock trading programs caused the stock of a large airline company to fall after an erroneous news item surfaced on an online news report. After investigating its current information system, a nonprofit hospital in Boston decided to undergo a large systems development project. The objective of the project was to streamline healthcare delivery and centralize most of the healthcare company's records. It will also help hospital executives determine what critical drugs need to be immediately available to patients, spot possible safety issues, and help provide critical information for important staffing decisions.[56] The new system should cost about $70 million and might be possibly funded by federal stimulus spending. According to one hospital executive, "It's going to be one collective brain that encompasses all a patient's needs." In general, systems investigation attempts to uncover answers to the following questions:

- What primary problems might a new or enhanced system solve?
- What opportunities might a new or enhanced system provide?
- What new hardware, software, databases, telecommunications, personnel, or procedures will improve an existing system or are required in a new system?
- What are the potential costs (variable and fixed)?
- What are the associated risks?

Initiating Systems Investigation

Because systems development requests can require considerable time and effort to implement, many organizations have adopted a formal procedure for initiating systems development, beginning with systems investigation. The **systems request form** is a document that is filled out by someone who wants the IS department to initiate systems investigation. This form typically includes the following information:

- Problems in or opportunities for the system
- Objectives of systems investigation
- Overview of the proposed system
- Expected costs and benefits of the proposed system

The information in the systems request form helps to rationalize and prioritize the activities of the IS department. Based on the overall IS plan, the organization's needs and goals, and the estimated value and priority of the proposed projects, managers make decisions regarding the initiation of each systems investigation for such projects.

Feasibility Analysis

A key step of the systems investigation phase is **feasibility analysis**, which assesses technical, economic, legal, operational, and schedule feasibility. (See Figure 8.9.) **Technical feasibility** is concerned with whether the hardware, software, and other system components can be acquired or developed to solve the problem.

systems request form
A document filled out by someone who wants the IS department to initiate systems investigation.

feasibility analysis
Assessment of the technical, economic, legal, operational, and schedule feasibility of a project.

technical feasibility
Assessment of whether the hardware, software, and other system components can be acquired or developed to solve the problem.

Figure 8.9

Technical, Economic, Legal, Operational, and Schedule Feasibility

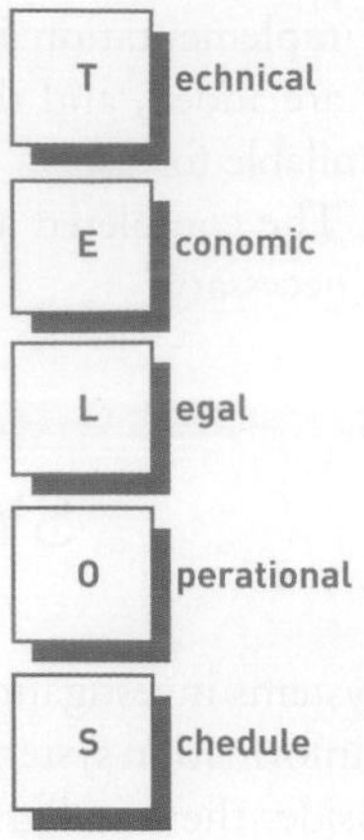

T   echnical

E   conomic

L   egal

O   perational

S   chedule

economic feasibility
The determination of whether the project makes financial sense and whether predicted benefits offset the cost and time needed to obtain them.

legal feasibility
The determination of whether laws or regulations may prevent or limit a systems development project.

operational feasibility
The measure of whether the project can be put into action or operation.

schedule feasibility
The determination of whether the project can be completed in a reasonable amount of time.

Economic feasibility determines whether the project makes financial sense and whether predicted benefits offset the cost and time needed to obtain them. Economic feasibility can involve cash flow analysis such as that done in internal rate of return (IRR) or total cost of ownership (TCO) calculations, first discussed in Chapter 1. Spreadsheet programs, such as Microsoft Excel, have built-in functions to compute internal rate of return and other cash flow measures.

Legal feasibility determines whether laws or regulations can prevent or limit a systems development project. Legal feasibility involves an analysis of existing and future laws to determine the likelihood of legal action against the systems development project and the possible consequences.

Operational feasibility is a measure of whether the project can be put into action or operation. It can include logistical and motivational (acceptance of change) considerations. Motivational considerations are important because new systems affect people and data flows and can have unintended consequences. As a result, power and politics might come into play, and some people might resist the new system.

Schedule feasibility determines whether the project can be completed in a reasonable amount of time—a process that involves balancing the time and resource requirements of the project with other projects.

Object-Oriented Systems Investigation

The object-oriented approach can be used during all phases of systems development, from investigation to maintenance and review. Consider a kayak rental business in Maui, Hawaii in which the owner wants to computerize its operations, including renting kayaks to customers and adding new kayaks into the rental program. (See Figure 8.10.) As you can see, the kayak rental clerk rents kayaks to customers and adds new kayaks to the current inventory available for rent. The stick figure is an example of an *actor*, and the ovals each represent an event, called a *use case*. In our example, the actor (the kayak rental clerk) interacts with two use cases (rent kayaks to customers and add new kayaks to inventory). The use case diagram is part of the Unified Modeling Language (UML) that is used in object-oriented systems development.

Figure 8.10

Use Case Diagram for a Kayak Rental Application

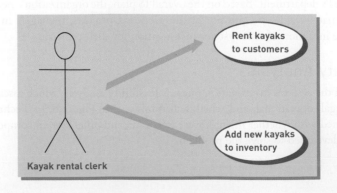

The Systems Investigation Report

The primary outcome of systems investigation is a **systems investigation report**, also called a *feasibility study*. This report summarizes the results of systems investigation and the process of feasibility analysis and recommends a course of action: Continue on into systems analysis, modify the project in some manner, or drop it. A typical table of contents for the systems investigation report is shown in Figure 8.11.

systems investigation report
A summary of the results of the systems investigation and the process of feasibility analysis and recommendation of a course of action.

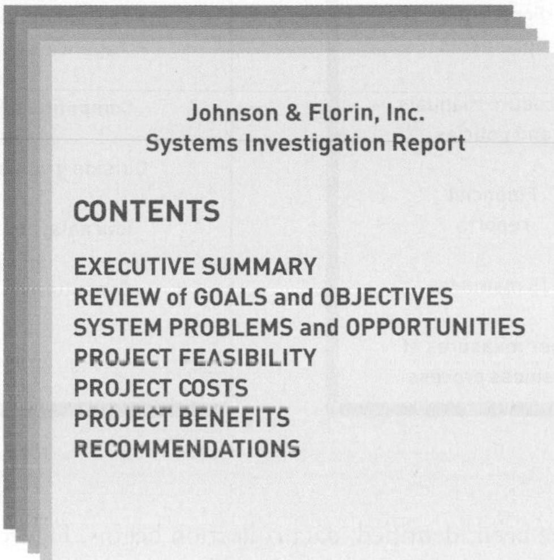

Figure 8.11

A Typical Table of Contents for a Systems Investigation Report

Johnson & Florin, Inc.
Systems Investigation Report

CONTENTS

EXECUTIVE SUMMARY
REVIEW of GOALS and OBJECTIVES
SYSTEM PROBLEMS and OPPORTUNITIES
PROJECT FEASIBILITY
PROJECT COSTS
PROJECT BENEFITS
RECOMMENDATIONS

The systems investigation report is reviewed by senior management, often organized as an advisory committee, or **steering committee**, consisting of senior management and users from the IS department and other functional areas. These people help IS personnel with their decisions about the use of information systems in the business and give authorization to pursue further systems development activities. After review, the steering committee might agree with the recommendation of the systems development team or suggest a change in project focus to concentrate more directly on meeting a specific company objective. Another alternative is that everyone might decide that the project is not feasible and cancel the project.

steering committee
An advisory group consisting of senior management and users from the IS department and other functional areas.

SYSTEMS ANALYSIS

After a project has been approved for further study, the next step is to answer the question "What must the information system do to solve the problem?" The overall emphasis of analysis is gathering data on the existing system, determining the requirements for the new system, considering alternatives within these constraints, and investigating the feasibility of the solutions. The primary outcome of systems analysis is a prioritized list of systems requirements.

Data Collection

The purpose of data collection is to seek additional information about the problems or needs identified in the systems investigation report. During this process, the strengths and weaknesses of the existing system are emphasized.

Identifying Sources of Data

Data collection begins by identifying and locating the various sources of data, including both internal and external sources. See Figure 8.12.

Figure 8.12

Internal and External Sources of
Data for Systems Analysis

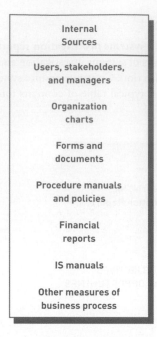

Collecting Data

After data sources have been identified, data collection begins. Figure 8.13 shows the steps involved. Data collection might require a number of tools and techniques, such as interviews, direct observation, and questionnaires.

Figure 8.13

The Steps in Data Collection

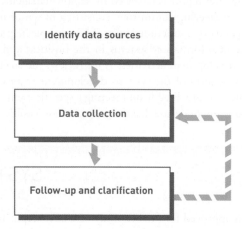

structured interview
An interview in which the questions are written in advance.

unstructured interview
An interview in which the questions are not written in advance.

direct observation
Directly observing the existing system in action by one or more members of the analysis team.

questionnaires
A method of gathering data when the data sources are spread over a wide geographic area.

Interviews can either be structured or unstructured. In a **structured interview**, the questions are written in advance. In an **unstructured interview**, the questions are not written in advance; the interviewer relies on experience in asking the best questions to uncover the inherent problems of the existing system.

With **direct observation**, one or more members of the analysis team directly observe the existing system in action.

When many data sources are spread over a wide geographic area, **questionnaires** might be the best method. Like interviews, questionnaires can be either structured or unstructured. In most cases, a pilot study is conducted to fine-tune the questionnaire. A follow-up questionnaire can also capture the opinions of those who do not respond to the original questionnaire.

Direct observation is a method of data collection. One or more members of the analysis team directly observes the existing system in action.

(Source: © Track5/iStockphoto.)

Data Analysis

The data collected in its raw form is usually not adequate to determine the effectiveness of the existing system or the requirements for the new system. The next step is to manipulate the collected data so that the development team members who are participating in systems analysis can use the data. This manipulation is called **data analysis**. Data and activity modeling and using data-flow diagrams and entity relationship diagrams are useful during data analysis to show data flows and the relationships among various objects, associations, and activities. Other common tools and techniques for data analysis include application flowcharts, grid charts, CASE tools, and the object oriented approach.

Data Modeling

Data modeling, first introduced in Chapter 3, is a commonly accepted approach to modeling organizational objects and associations that employ both text and graphics. How data modeling is employed, however, is governed by the specific systems development methodology.

Data modeling is most often accomplished through the use of entity-relationship (ER) diagrams. Recall from Chapter 3 that an entity is a generalized representation of an object type—such as a class of people (employee), events (sales), things (desks), or places (city)—and that entities possess certain attributes. Objects can be related to other objects in many ways. An entity-relationship diagram, such as the one shown in Figure 8.14a, describes a number of objects and the ways they are associated. An ER diagram (or any other modeling tool) cannot by itself fully describe a business problem or solution because it lacks descriptions of the related activities. It is, however, a good place to start because it describes object types and attributes about which data might need to be collected for processing.

Activity Modeling

To fully describe a business problem or solution, the related objects, associations, and activities must be described. Activities in this sense are events or items that are necessary to fulfill the business relationship or that can be associated with the business relationship in a meaningful way.

Activity modeling is often accomplished through the use of data-flow diagrams. A **data-flow diagram (DFD)** models objects, associations, and activities by describing how data can flow between and around various objects. DFDs work on the premise that every activity involves some communication, transference, or flow that can be described as a data element. DFDs describe the activities that fulfill a business relationship or accomplish a business task, not how these activities are to be performed. That is, DFDs show the logical sequence of associations and activities, not the physical processes. A system modeled with a DFD could operate manually or could be computer based; if computer based, the system could operate with a variety of technologies.

DFDs are easy to develop and easily understood by nontechnical people. Data-flow diagrams use four primary symbols, as illustrated in Figure 8.14b.

data analysis
The manipulation of collected data so that the development team members who are participating in systems analysis can use the data.

data-flow diagram (DFD)
A model of objects, associations, and activities that describes how data can flow between and around various objects.

Figure 8.14

Data and Activity Modeling

(a) An entity-relationship diagram.
(b) A data-flow diagram. (c) A
semantic description of the
business process.

(Source: G. Lawrence Sanders, *Data Modeling*, Boyd & Fraser Publishing, Danvers, MA: 1995.)

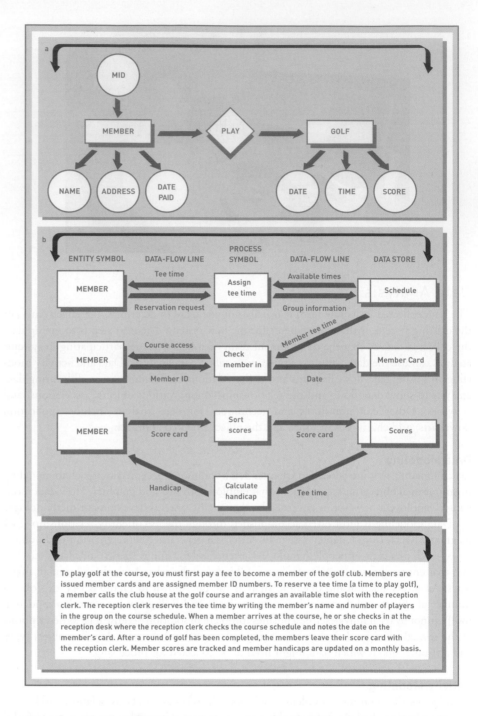

To play golf at the course, you must first pay a fee to become a member of the golf club. Members are issued member cards and are assigned member ID numbers. To reserve a tee time (a time to play golf), a member calls the club house at the golf course and arranges an available time slot with the reception clerk. The reception clerk reserves the tee time by writing the member's name and number of players in the group on the course schedule. When a member arrives at the course, he or she checks in at the reception desk where the reception clerk checks the course schedule and notes the date on the member's card. After a round of golf has been completed, the members leave their score card with the reception clerk. Member scores are tracked and member handicaps are updated on a monthly basis.

data-flow line
Arrows that show the direction of data element movement.

process symbol
Representation of a function that is performed.

entity symbol
Representation of either a source or destination of a data element.

data store
Representation of a storage location for data.

- *Data flow.* The **data-flow line** includes arrows that show the direction of data element movement.
- *Process symbol.* The **process symbol** reveals a function that is performed. Computing gross pay, entering a sales order, delivering merchandise, and printing a report are examples of functions that can be represented with a process symbol.
- *Entity symbol.* The **entity symbol** shows either the source or destination of the data element. An entity can be, for example, a customer who initiates a sales order, an employee who receives a paycheck, or a manager who receives a financial report.
- *Data store.* A **data store** reveals a storage location for data. A data store is any computerized or manual data storage location, including magnetic tape, disks, a filing cabinet, or a desk.

Comparing entity-relationship diagrams with data-flow diagrams provides insight into the concept of top-down design. Figures 8.14a and b show an entity-relationship diagram and a data-flow diagram for the same business relationship—namely, a member of a golf club playing golf. Figure 8.14c provides a brief description of the business relationship for clarification.

Requirements Analysis

The overall purpose of **requirements analysis** is to determine user, stakeholder, and organizational needs. For an accounts payable application, the stakeholders could include suppliers and members of the purchasing department. Questions that should be asked during requirements analysis include the following:

* Are these stakeholders satisfied with the current accounts payable application?
* What improvements could be made to satisfy suppliers and help the purchasing department?

requirements analysis
The determination of user, stakeholder, and organizational needs.

Asking Directly

One the most basic techniques used in requirements analysis is asking directly. **Asking directly** is an approach that asks users, stakeholders, and other managers about what they want and expect from the new or modified system. This approach works best for stable systems in which stakeholders and users clearly understand the system's functions. The role of the systems analyst during the analysis phase is to critically and creatively evaluate needs and define them clearly so that the systems can best meet them.

asking directly
An approach to gather data that asks users, stakeholders, and other managers about what they want and expect from the new or modified system.

Critical Success Factors

Another approach uses critical success factors (CSFs). As discussed earlier, managers and decision makers are asked to list only the factors that are critical to the success of their area of the organization.[57] A CSF for a production manager might be adequate raw materials from suppliers; a CSF for a sales representative could be a list of customers currently buying a certain type of product. Starting from these CSFs, the system inputs, outputs, performance, and other specific requirements can be determined.

The IS Plan

As we have seen, the IS plan translates strategic and organizational goals into systems development initiatives. The IS planning process often generates strategic planning documents that can be used to define system requirements. Working from these documents ensures that requirements analysis will address the goals set by top-level managers and decision makers. See Figure 8.15. There are unique benefits to applying the IS plan to define systems requirements. Because the IS plan takes a long-range approach to using information technology within the organization, the requirements for a system analyzed in terms of the IS plan are more likely to be compatible with future systems development initiatives.

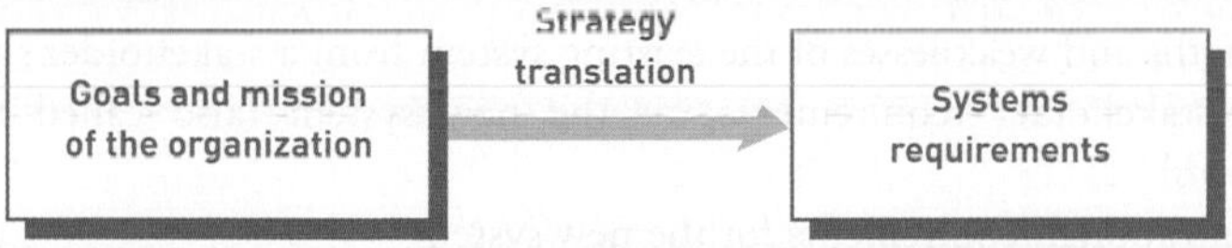

Figure 8.15

Converting Organizational Goals into Systems Requirements

Requirements Analysis Tools

A number of tools can be used to document requirements analysis, including CASE tools. As requirements are developed and agreed on, entity-relationship diagrams, data-flow diagrams, screen and report layout forms, and other types of documentation are stored in the CASE repository. These requirements might also be used later as a reference during the rest of systems development or for a different systems development project.

Object-Oriented Systems Analysis

The object-oriented approach can also be used during systems analysis. Like traditional analysis, problems or potential opportunities are identified during object-oriented analysis. Identifying key participants and collecting data are still performed. But instead of analyzing the existing system using data-flow diagrams and flowcharts, an object-oriented approach is used.

The section "Object-Oriented Systems Investigation" introduced a kayak rental example. A more detailed analysis of that business reveals that there are two classes of kayaks: single kayaks for one person and tandem kayaks that can accommodate two people. With the OO approach, a class is used to describe different types of objects, such as single and tandem kayaks. The classes of kayaks can be shown in a generalization/specialization hierarchy diagram. See Figure 8.16. KayakItem is an object that will store the kayak identification number (ID) and the date the kayak was purchased (datePurchased).

Figure 8.16

Generalization/Specialization Hierarchy Diagram for Single and Tandem Kayak Classes

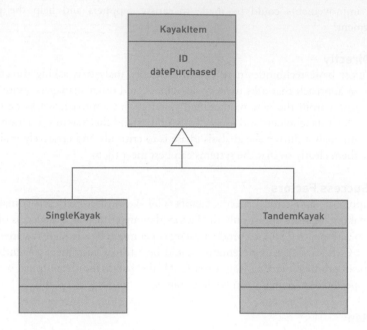

Of course, there could be subclasses of customers, life vests, paddles, and other items in the system. For example, price discounts for kayak rentals could be given to seniors (people over 65 years) and students. Thus, the Customer class could be divided into regular, senior, and student customer subclasses.

The Systems Analysis Report

Systems analysis concludes with a formal systems analysis report. It should cover the following elements:

- The strengths and weaknesses of the existing system from a stakeholder's perspective
- The user/stakeholder requirements for the new system (also called the *functional requirements*)
- The organizational requirements for the new system
- A description of what the new information system should do to solve the problem

Suppose analysis reveals that a marketing manager thinks a weakness of the existing system is its inability to provide accurate reports on product availability. These requirements and a preliminary list of the corporate objectives for the new system will be in the systems analysis report. Particular attention is placed on areas of the existing system that could be improved to meet user requirements. The table of contents for a typical report is shown in Figure 8.17.

Johnson & Florin, Inc.
Systems Analysis Report

CONTENTS

BACKGROUND INFORMATION
PROBLEM or NEED STATEMENT
DATA COLLECTION
DATA and REQUIREMENTS ANALYSIS
RECOMMENDATIONS
APPENDIXES of DOCUMENTS, TABLES, and CHARTS
GLOSSARY of TERMS

The systems analysis report gives managers a good understanding of the problems and strengths of the existing system. If the existing system is operating better than expected or the necessary changes are too expensive relative to the benefits of a new or modified system, the systems development process can be stopped at this stage. If the report shows that changes to another part of the system might be the best solution, the development process might start over, beginning again with systems investigation. Or, if the systems analysis report shows that it will be beneficial to develop one or more new systems or to make changes to existing ones, systems design, which is discussed next begins.

SYSTEMS DESIGN

The purpose of **systems design** is to answer the question "How will the information system solve a problem?" The primary result of the systems design phase is a technical design that details system outputs, inputs, and user interfaces; specifies hardware, software, databases, telecommunications, personnel, and procedures; and shows how these components are related. The new or modified system should take advantage of the latest developments in technology. Many companies, for example, are using cloud computing, where applications are run on the Internet instead of being developed and run within the company or organization.[58] One advantage of using cloud computing is easier management of information systems because everything is in one place on the Internet. General Electric, for example, is developing a "private" cloud that can only be accessed and used by General Electric's employees and managers.[59] According to General Electric's CTO, "You get efficiencies when you start to manage that as a single entity. You can flex that capacity across applications, back it up, monitor it, and manage it as one entity." There are, however, potential disadvantages of a systems development effort that relies on cloud computing.[60] One important risk is security. Organizations may not know who is able to hack into the Internet and get access to their sensitive and critical data.[61] Another risk is availability. If the Internet site that is providing the cloud computing application is unavailable or having technical problems, critical applications may not be available. Some cloud computing users have also complained about slow access and execution times for applications that are run using a cloud computing environment on the Internet.[62]

systems design
The stage of systems development that answers the question "How will the information system do what is necessary to solve a problem?"

Systems design is typically accomplished using the tools and techniques discussed earlier in this chapter. Depending on the specific application, these methods can be used to support and document all aspects of systems design. Two key aspects of systems design are logical and physical design.

Logical and Physical Design

logical design
A description of the functional requirements of a system.

Design has two dimensions: logical and physical. The **logical design** refers to what the system will do. It describes the functional requirements of a system. Today, for example, many stock exchanges, large hedge funds, and institutional stock investors include speed as a critical logical design element for new computer trading systems.[63] The objective is to increase profits by being faster in placing electronic trades than traditional computerized trading systems. Without logical design, the technical details of the system (such as which hardware devices should be acquired) often obscure the best solution. Logical design involves planning the purpose of each system element, independent of hardware and software considerations. The logical design specifications that are determined and documented include output, input, process, file and database, telecommunications, procedures, controls and security, and personnel and job requirements.

Security is always an important logical design issue for corporations and governments. In one survey, CIOs estimated that their organizations lost more than $4 billion annually in intellectual property theft as a result of inadequate security measures. [64] One Internet security firm estimated that losses due to security breaches were approximately $1 trillion in 2008. The United States is investigating a new military command position to coordinate cyber and Internet security for all military information systems.[65] According to the CIO of the United States, "We've got to be able to abstract the infrastructure from the applications. For example, when you look at security, it's easier to secure when you concentrate things than when you distribute them across the government."[66] Rules published in September 2005, for example, require that federal agencies incorporate security procedures in the design of new or modified systems. In addition, the Federal Information Security Management Act, enacted in 2002, requires federal agencies to make sure that security protection measures are incorporated into systems provided by outside vendors and contractors.

physical design
The specification of the characteristics of the system components necessary to put the logical design into action.

The **physical design** refers to how the tasks are accomplished, including how the components work together and what each component does. Physical design specifies the characteristics of the system components necessary to put the logical design into action. In this phase, the characteristics of the hardware, software, database, telecommunications, personnel, and procedure and control specifications must be detailed. These physical design components were discussed in Part 2 on technology. The New York Stock Exchange, for example has developed a physical design to build a large facility the size of several football fields to house super fast trading systems that can be used by large hedge funds and institutional investors to get the speed they specified in their logical designs.[67]

Object-Oriented Design

Logical and physical design can be accomplished using either the traditional approach or the object-oriented approach to systems development. Both approaches use a variety of design models to document the new system's features and the development team's understandings and agreements. Many organizations today are turning to OO development because of its increased flexibility. This section outlines a few OO design considerations and diagrams.

Using the OO approach, you can design key objects and classes of objects in the new or updated system.[68] This process includes considering the problem domain, the operating environment, and the user interface. The problem domain involves the classes of objects related to solving a problem or realizing an opportunity. In our Maui, Hawaii, kayak rental shop example discussed earlier in the chapter, and referring back to the generalization/specialization hierarchy showing classes we presented there, KayakItem in Figure 8.16 is an example of a problem domain object that will store information on kayaks in the rental program. The operating environment for the rental shop's system includes objects that interact with printers, system software, and other software and hardware devices. The user interface for the system includes objects that users interact with, such as buttons and scroll bars in a Windows program.

During the design phase, you also need to consider the sequence of events that must happen for the system to function correctly. For example, you might want to design the sequence of events for adding a new kayak to the rental program. The event sequence is often called a *scenario*, and it can be diagrammed in a sequence diagram. See Figure 8.18.

Figure 8.18

A Sequence Diagram to Add a New KayakItem Scenario

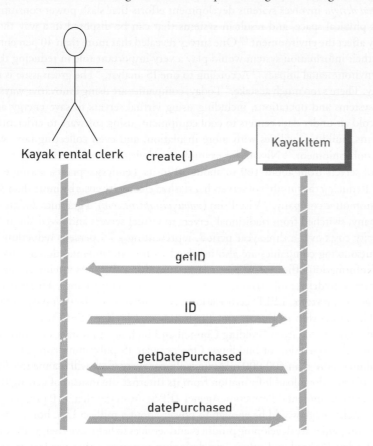

You read a sequence diagram starting at the top and moving down.

1. The Create arrow at the top is a message from the kayak rental clerk to the KayakItem object to create information on a new kayak to be placed into the rental program.
2. The KayakItem object knows that it needs the ID for the kayak and sends a message to the clerk requesting the information. See the getID arrow.
3. The clerk then types the ID into the computer. This is shown with the ID arrow. The data is stored in the KayakItem object.
4. Next, KayakItem requests the purchase date. This is shown in the getDatePurchased arrow.
5. Finally, the clerk types the purchase date into the computer. The data is also transferred to KayakItem object. This is shown in the datePurchased arrow at the bottom of Figure 8.18.

This scenario is only one example of a sequence of events. Other scenarios might include entering information about life jackets, paddles, suntan lotion, and other accessories. The same types of use case and generalization/specialization hierarchy diagrams can be created for each event, and additional sequence diagrams will also be needed.

ENVIRONMENTAL DESIGN CONSIDERATIONS

environmental design
Also called *green design*, it involves systems development efforts that slash power consumption, require less physical space, and result in systems that can be disposed in a way that doesn't negatively affect the environment.

Developing new systems and modifying existing ones in an environmentally sensitive way is becoming increasingly important for many IS departments. **Environmental design**, also called *green design,* involves systems development efforts that slash power consumption, require less physical space, and result in systems that can be disposed in a way that doesn't negatively affect the environment.[69] One survey revealed that more than 40 percent of CEOs said that their information system would play a very important role in reducing the organization's environmental impact.[70] According to one IS analyst, "The green issue is not going to go away. There's too much at stake." Today, companies are using innovative ways to design efficient systems and operations, including using virtual servers to save energy and space, pushing cold air under data centers to cool equipment, using software to efficiently control cooling fans, building facilities with more insulation, and even collecting rain water from roofs to cool equipment.[71] Nissan, for example, used virtual server technology to reduce its number of servers from about 160 to about 30 in its Tennessee plants, saving energy and money.[72] Reducing the number of servers has slashed electricity costs by more than 30 percent for the automotive company.[73] VistaPrint (*www.vistaprint.com*), a graphics design and printing company, switched from traditional servers to virtual servers and saved about $500,000 in electricity costs over a three-year period, representing a 75 percent reduction in energy usage. Outsourcing companies are also involved with environmental design. According to the chief information officer of Wipro, a large consulting and outsourcing company, "Our main priority is to deliver infrastructure services to drive energy savings. This includes building management systems, LEED certifications, consolidation of data centers, and virtualization."[74] The Leadership in Energy and Environmental Design (LEED) rating system was developed by the U.S. Green Building Council and includes a number of standards for the construction and operation of buildings. One popular IS publishing company that prints and distributes more than 400,000 magazines almost every week will plant a tree for the first 5,000 people that download information from its Internet site instead of getting the printed copy.[75] The Environmental Protection Agency (EPA) estimates that a 10 percent cut in data center electricity usage would be enough to power about a million U.S. homes every year.

Many companies are developing products and services to help save energy. PC companies, such as Hewlett-Packard and others, are designing computers that use less power and are made from recycled materials.[76] Some PCs consume energy even when they are turned off. Cell phone manufacturers are also starting to manufacture phones that consume less electricity.[77] EMC has developed new disk drives that use substantially less energy. Voltaic Generator (*www.voltaicsystems.com/bag*) has developed a solar PC case that charges batteries from sunlight or other light sources.[78] The solar-powered bag can power computers, cell phones, and other electronic devices. Environmental design also involves developing software and systems that help organizations reduce power consumption for other aspects of their operations. Carbonetworks and Optimum Energy, for example, have developed software products to help companies reduce energy costs by helping them determine when and how to use electricity. UPS developed its own software to reduce the miles its trucks and other vehicles drive by routing them more efficiently. The new software helped UPS cut million of miles per year, slash fuel costs, and reduce carbon emissions. Hewlett-Packard, Dell Computer, and others have developed procedures and machines to dispose of old computers and computer equipment in environmentally friendly ways.[79] VenJuvo (*www.venjuvo.com*) and other companies also recycle old electronics equipment and offer cash in some cases, depending on the age and type of equipment.[80] Old computers and computer equipment are fed into machines that shred them into small pieces and sort them into materials that can be reused. The process is often called *green death.* The U.S. government is also involved in environmental design. It has a plan to require federal agencies to purchase energy-efficient computer systems and equipment. The plan would require federal agencies to use the *Electronic Product Environmental Assessment Tool (EPEAT)* to analyze the energy usage of new systems.[81] Some estimate that the savings could be about $1 billion in five years.[82]

The U.S. Department of Energy rates products with the *Energy Star* designation to help people select products that save energy.[83] Today, utility companies are providing their corporate and individual customers with "smart meters" and specialized software that can help them reduce their power consumption and electric bills.[84]

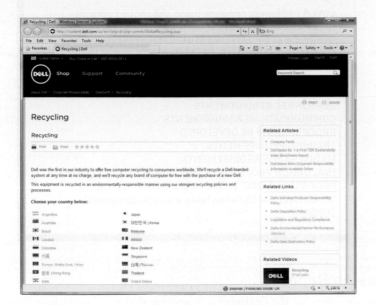

Companies such as Hewlett-Packard and Dell Computer dispose of old computers and computer equipment in environmentally friendly ways.

(Source: *www.dell.com/recycle*.)

Generating Systems Design Alternatives

Generating systems design alternatives often involves getting the involvement of single vendor or multiple vendors.[85] If the new system is complex, the original development team might want to involve other personnel in generating alternative designs. In addition, if new hardware and software are to be acquired from an outside vendor, a formal request for proposal (RFP) can be made.

Request for Proposals

The **request for proposal (RFP)** is a document that specifies in detail required resources such as hardware and software. The RFP is an important document for many organizations involved with large, complex systems development efforts. Smaller, less-complex systems often do not require an RFP. A company that is purchasing an inexpensive piece of software that will run on existing hardware, for example, might not need to go through a formal RFP process.

In some cases, separate RFPs are developed for different needs. For example, a company might develop separate RFPs for hardware, software, and database systems. The RFP also communicates these needs to one or more vendors, and it provides a way to evaluate whether the vendor has delivered what was expected. In some cases, the RFP is part of the vendor contract. The Table of Contents for a typical RFP is shown in Figure 8.19.

request for proposal (RFP)
A document that specifies in detail required resources such as hardware and software.

Evaluating and Selecting a Systems Design

Evaluating and selecting the best design involves achieving a balance of system objectives that will best support organizational goals. Normally, evaluation and selection involves both a preliminary and a final evaluation before a design is selected.

A **preliminary evaluation** begins after all proposals have been submitted. The purpose of this evaluation is to dismiss unwanted proposals. Several vendors can usually be eliminated by investigating their proposals and comparing them with the original criteria. The **final evaluation** begins with a detailed investigation of the proposals offered by the remaining vendors. The vendors should be asked to make a final presentation and to fully demonstrate the system. The demonstration should be as close to actual operating conditions as possible.

preliminary evaluation
An initial assessment whose purpose is to dismiss the unwanted proposals; begins after all proposals have been submitted.

final evaluation
A detailed investigation of the proposals offered by the vendors remaining after the preliminary evaluation.

Figure 8.19

A Typical Table of Contents for a
Request for Proposal

> Johnson & Florin, Inc.
> Systems Investigation Report
>
> **Contents**
>
> COVER PAGE (with company name and contact person)
> BRIEF DESCRIPTION of the COMPANY
> OVERVIEW of the EXISTING COMPUTER SYSTEM
> SUMMARY of COMPUTER-RELATED NEEDS and/or PROBLEMS
> OBJECTIVES of the PROJECT
> DESCRIPTION of WHAT IS NEEDED
> HARDWARE REQUIREMENTS
> PERSONNEL REQUIREMENTS
> COMMUNICATIONS REQUIREMENTS
> PROCEDURES to BE DEVELOPED
> TRAINING REQUIREMENTS
> MAINTENANCE REQUIREMENTS
> EVALUATION PROCEDURES (how vendors will be judged)
> PROPOSAL FORMAT (how vendors should respond)
> IMPORTANT DATES (when tasks are to be completed)
> SUMMARY

The Design Report

System specifications are the final results of systems design. They include a technical description that details system outputs, inputs, and user interfaces as well as all hardware, software, databases, telecommunications, personnel, and procedure components and the way these components are related. The specifications are contained in a **design report**, which is the primary result of systems design. The design report reflects the decisions made for systems design and prepares the way for systems implementation. The contents of the design report are summarized in Figure 8.20.

design report
The primary result of systems design, reflecting the decisions made and preparing the way for systems implementation.

Figure 8.20

A Typical Table of Contents for a
Systems Design Report

> Johnson & Florin, Inc.
> Systems Design Report
>
> **Contents**
>
> PREFACE
> EXECUTIVE SUMMARY of SYSTEMS DESIGN
> REVIEW of SYSTEMS ANALYSIS
> MAJOR DESIGN RECOMMENDATIONS
>   Hardware design
>   Software design
>   Personnel design
>   Communications design
>   Database design
>   Procedures design
>   Training design
>   Maintenance design
> SUMMARY of DESIGN DECISIONS
> APPENDICES
> GLOSSARY of TERMS
> INDEX

SYSTEMS IMPLEMENTATION

After the information system has been designed, a number of tasks must be completed before the system is installed and ready to operate. This process, called systems implementation, includes hardware acquisition, programming and software acquisition or development, user preparation, hiring and training of personnel, site and data preparation, installation, testing, start-up, and user acceptance. The typical sequence of systems implementation activities is shown in Figure 8.21.

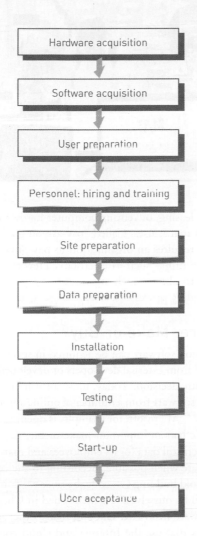

Figure 8.21

Typical Steps in Systems Implementation

Acquiring Hardware from an IS Vendor

To obtain the components for an information system, organizations can purchase, lease, or rent computer hardware and other resources from an IS vendor. An *IS vendor* is a company that offers hardware, software, telecommunications systems, databases, IS personnel, or other computer-related resources. Types of IS vendors include general computer manufacturers (such as IBM and Hewlett-Packard), small computer manufacturers (such as Dell and Sony), peripheral equipment manufacturers (such as Epson and SanDisk), computer dealers and distributors (such as The Shack and Best Buy), and chip makers such as Intel and AMD. Some of the most successful vendors include IBM (hardware and other services), Oracle (databases), Apple (personal computers), Microsoft (software), Accenture (IS consulting), and many others.[86] In addition, many new hardware vendors provide specialized equipment and services.[87] Venture capital firms have invested hundreds of millions of dollars in new hardware vendors. Hardware vendors can provide very small or very large systems. Many companies have multiple hardware vendors, but managing them can be difficult.[88] Different vendors must compete against each other to get an outsourcing contract with the organization. Then, the selected vendors must work together to develop an effective information system at a good price. Open communications among the outsourcing vendors is critical. Each vendor's work and pricing should be transparent and available to all the other outsourcing vendors. Over time, it is best to have a set of trusted, reliable outsourcing vendors that can be used in future systems development projects.

Computer dealers, such as Best Buy, manufacture built-to-order computer systems and sell computers and supplies from other vendors.

(Source: Daniel Acker/Bloomberg via Getty Images.)

In addition to buying, leasing, or renting computer hardware, companies can pay only for the computing services that it uses. Called "pay-as-you-go," "on-demand," or "utility" computing, this approach requires an organization to pay only for the computer power it uses, as it would pay for a utility such as electricity. Hewlett-Packard offers its clients a "capacity-on-demand" approach, in which organizations pay according to the computer resources actually used, including processors, storage devices, and network facilities.

Acquiring Software: Make or Buy?

As with hardware, application software can be acquired in several ways. As previously mentioned, it can be purchased from external developers or developed in-house. This decision is often called the **make-or-buy decision**. Today, most software is purchased. Individuals and organizations can purchase software from a number of online application stores or retail stores located around the country.[89] Microsoft, for example, is opening retail stores in Arizona and California. General Electric had an important make-or-buy decision for an Internet application that would provide critical data for its employees and managers.[90] The company tries to buy the necessary software when possible. According to the company's chief technology officer, "We're trying to avoid building GE code, because we don't have the energy to figure out how to do this right all by ourselves." As mentioned in Chapter 2, companies can also purchase open-source software from Red Hat and many other open-source software companies, including programs that use the Internet and cloud computing approaches.[91] Research has shown that many programmers and systems developers freely join groups to develop and refine open-source software.[92]

As mentioned in Chapter 2, *Software as a Service (SaaS)* allows businesses to subscribe to Web-delivered application software by paying a monthly service charge or a per-use fee.[93] Instead of acquiring software externally from a traditional software vendor, SaaS allows individuals and organizations to access needed software applications over the Internet. The Humane Society of the United States, for example, used a SaaS product called QualysGuard by Qualys (*www.qualys.com*) to obtain and process credit-card contributions from donors. Companies such as Google are using the cloud computing approach to deliver word processing, spreadsheet programs, and other software over the Internet.

make-or-buy decision
The decision regarding whether to obtain the necessary software from internal or external sources.

Medical Center Moves Patient Records to Cloud

Beth Israel Deaconess Medical Center (BID) is a large teaching hospital associated with Harvard Medical School that serves Boston and surrounding communities. BID is one of the top four recipients of biomedical research funding from the National Institutes for Health, winning nearly $200 million annually.

Moving to an electronic health record system (EHRS) has been a priority for BID for years. In 2008, long before the federal government offered stimulus funds for EHRS, BID rolled out its own EHRS. BID was motivated to move to an EHRS by requirements imposed by insurance companies. So-called pay-for-performance insurance plans require highly detailed and real-time documentation that is possible only through EHRS.

BID knew its 300 physicians would approve an EHRS, but earning the approval of the 900 other physicians associated with BID, but not employed by BID, was a bigger challenge. Many of those 900 physicians own small practices with limited resources. Along with their staff, they work 10-hour days, processing 40 patients a day, and have little time to invest in learning new technology. BID decided to invest in an EHRS for its own physicians, gambling that its associated physicians would eventually witness the value of an EHRS and buy in.

BID evaluated many EHRSs, weighing their benefits against their drawbacks. The systems analysts designing the EHRS were at a disadvantage because they didn't know exactly how many physicians and medical records the system would need to accommodate. They needed a system that could be scaled to match the amount of data. For this reason, they decided on a SaaS cloud computing system.

The solution came from a company named eClinicalWorks, which licenses its software to BID and provides the hosted service. The hosted service runs on virtual servers that make it easy to add more storage and processing resources as they are needed, without any interruption to service. For example, BID recently updated its security to a stronger form of encryption. The upgrade placed too much strain on the servers, so BID spent $20,000 to add more virtual resources to the service provider's hosted servers. The same scenario without virtualization would have cost BID $325,000 for additional hardware. This is a convincing argument in support of both virtualization and SaaS cloud computing services.

The medical industry has unique information security requirements placed on it by local and federal governments designed to provide patient privacy. Because of this, the move to EHRSs has been slow. The system designed by eClinicalWorks uses a thin client device that connects to PCs to encrypt medical data as it leaves the PC and decrypt data as it arrives. Other than that, the system requires no complicated software or set up, and it runs in a standard Web browser. The system also has an offline mode, so that records can be accessed even when clients are not connected to the network. Each time computers connect to eClinicalWorks, the records are synchronized with records on the server.

Because eClinicalWorks is easy to set up and use, BID associates are more likely to join the system. There are many other incentives as well. Besides the obvious convenience of access to medical records anywhere anytime, the EHRS also provides record keeping automation. Medical practices often hire several employees just to process the paperwork required by Medicaid and private insurers. The EHRS automates those processes, dramatically reducing the need for staff. BID has made the EHRS a requirement for associates who want to take advantage of the medical center's administrative, clinical, and technical support.

BID associates have been quick to see the benefits of the eClinicalWorks EHRS and are migrating to the new system. No longer do BID patients have to fill out forms in triplicate every time they visit a new physician or specialist. No longer do BID physicians have to dig through file cabinets for patient records—they have access to them anywhere, anytime on their tablet computers. No longer does staff have to work full time filling out insurance claims. The eClinicalWorks EHRS provides all of these services, and because it is hosted in the cloud, maintenance is not a concern for BID information systems staff.

Discussion Questions

1. What motivated BID to invest in an EHRS?
2. Why did BID decide on the eClinicalWorks SaaS as its EHRS solution?

Critical Thinking Questions

1. What special considerations does the medical industry face when implementing information systems?
2. What benefits are provided to medical centers, physicians, and patients by an EHRS?

SOURCES: Fogarty, Kevin, "Cloud servers help hospital with digital record," *Computerworld*, *www.computerworld.com/s/article/9160918/ Cloud_servers_help_hospital_with_digital_records ? taxonomyId=154&pageNumber=1*, February 23, 2010; Beth Israel Deaconess Medical Center Web site, *www.bidmc.org*, accessed March 5, 2010; eClinical-Works Web site, *www.eclinicalworks.com/products.php*, accessed March 5, 2010.

Acquiring Database and Telecommunications Systems

Because databases are a blend of hardware and software, many of the approaches discussed earlier for acquiring hardware and software also apply to database systems including open-source databases. *Virtual databases* and *database as a service (DaaS)* are popular ways to acquire database capabilities.[94] Sirius XM Radio, Bank of America, and Southwest Airlines, for example, use the DaaS approach to manage many of their database operations from the Internet.[95] In another case, a brokerage company was able to reduce storage capacity by 50 percent by using database virtualization. Wal-Mart gives its customers more information on how its databases are acquired and used.[96] According to a Wal-Mart's chief privacy officer, "We want to provide customers with more control over their own data, which is a big topic today for relationships with customers and their privacy."

With the increased use of e-commerce, the Internet, intranets, and extranets, telecommunications is one of the fastest-growing applications for today's organizations. Like database systems, telecommunications systems require a blend of hardware and software. For personal computer systems, the primary piece of hardware is a modem. For client/server and mainframe systems, the hardware can include multiplexers, concentrators, communications processors, and a variety of network equipment. Communications software will also have to be acquired from a software company or developed in-house. Again, the earlier discussion on acquiring hardware and software also applies to the acquisition of telecommunications hardware and software. As discussed earlier in this chapter and previous chapters, individuals and organizations are increasingly using the Internet and cloud computing to implement many new systems development efforts.[97] Systems analysts and programmers are also starting to use the Internet to develop applications.[98] Scott Schroeder and his company, Rabble +Rouser, developed an application that contained over 200 recipes for Weber Grills for iPhone users.[99] The $4.99 application was one of the best-selling applications at Apple's application store.

User Preparation

user preparation
The process of readying managers, decision makers, employees, other users, and stakeholders for new systems.

User preparation is the process of readying managers, decision makers, employees, other users, and stakeholders for the new systems. This activity is an important but often ignored area of systems implementation. When a new operating system or application software package is implemented, user training is essential. In some cases, companies decide not to install the latest software because the amount of time and money needed to train employees is too much. Because user training is so important, some companies provide training for their clients, including in-house, software, video, Internet, and other training approaches.

Providing users with proper training can help ensure that the information system is used correctly, efficiently, and effectively.

(Source: © Konstantin Chagin, 2010. Used under license from Shutterstock.com.)

IS Personnel: Hiring and Training

Depending on the size of the new system, an organization might have to hire and, in some cases, train new IS personnel. An IS manager, systems analysts, computer programmers, data-entry operators, and similar personnel might be needed for the new or modified system.

Site Preparation

The location of the new system needs to be prepared, a process called **site preparation**. For a small system, site preparation can be as simple as rearranging the furniture in an office to make room for a computer. With a larger system, this process is not so easy because it can require special wiring and air conditioning. A special floor, for example, might have to be built, under which the cables connecting the various computer components are placed, and a new security system might be needed to protect the equipment. Today, developing IS sites that are energy efficient is important for most systems development implementations. Security is also important for site preparation.[100] One company, for example, installed special security kiosks that let visitors log on and request a meeting with a company employee. The employee can see the visitor on his or her computer screen and accept or reject the visitor. If the visitor is accepted, the kiosk prints a visitor pass.

site preparation
Preparation of the location of a new system.

Data Preparation

Data preparation, or **data conversion**, involves making sure that all files and databases are ready to be used with new computer software and systems. If an organization is installing a new payroll program, the old employee-payroll data might have to be converted into a format that can be used by the new computer software or system. After the data has been prepared or converted, the computerized database system or other software will then be used to maintain and update the computer files.

data preparation, or data conversion
Making sure all files and databases are ready to be used with new computer software and systems.

Installation

Installation is the process of physically placing the computer equipment on the site and making it operational. Although normally the manufacturer is responsible for installing computer equipment, someone from the organization (usually the IS manager) should oversee the process, making sure that all equipment specified in the contract is installed at the proper location. After the system is installed, the manufacturer performs several tests to ensure that the equipment is operating as it should.

installation
The process of physically placing the computer equipment on the site and making it operational.

Testing

Good testing procedures are essential to make sure that the new or modified information system operates as intended.[101] Inadequate testing can result in mistakes and problems. Problems with a project to consolidate data center servers, for example, resulted in more than 160,000 Internet sites being shut down. The company that was trying to consolidate its database servers was hosting the Internet sites. Some Internet sites were down for more than six days. Better testing may have prevented these types of problems. Several forms of testing should be used, including testing each program (**unit testing**), testing the entire system of programs (**system testing**), testing the application with a large amount of data (**volume testing**), and testing all related systems together (**integration testing**), as well as conducting any tests required by the user (**acceptance testing**).

unit testing
Testing of individual programs.

system testing
Testing the entire system of programs.

volume testing
Testing the application with a large amount of data.

integration testing
Testing all related systems together.

acceptance testing
Conducting any tests required by the user.

Start-Up

Start-up, also called *cutover*, begins with the final tested information system. When start-up is finished, the system is fully operational. Start-up can be critical to the success of the organization. If not done properly, the results can be disastrous. In one case, a small manufacturing company decided to stop an accounting service used to send out bills on the same day they were going to start their own program to send out bills to customers. The manufacturing company wanted to save money by using their own billing program developed by an em-

start-up
The process of making the final tested information system fully operational.

ployee. The new program didn't work, the accounting service wouldn't help because they were upset about being terminated, and the manufacturing company wasn't able to send out any bills to customers for more than three months. The manufacturing company almost went bankrupt.

Various start-up approaches are available. See Figure 8.22. **Direct conversion** (also called *plunge* or *direct cutover*) involves stopping the old system and starting the new system on a given date. Direct conversion is usually the least desirable approach because of the potential for problems and errors when the old system is shut off and the new system is turned on at the same instant.

direct conversion (also called *plunge* or *direct cutover*)
Stopping the old system and starting the new system on a given date.

Figure 8.22

Start-Up Approaches

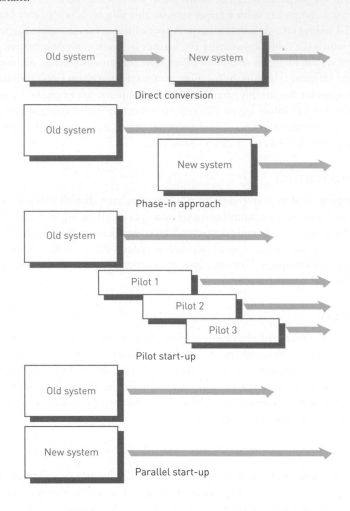

phase-in approach
Slowly replacing components of the old system with those of the new one. This process is repeated for each application until the new system is running every application and performing as expected; also called a *piecemeal approach*.

pilot start-up
Running the new system for one group of users rather than all users.

The **phase-in approach** is a popular technique preferred by many organizations. In this approach, sometimes called a *piecemeal approach*, components of the new system are slowly phased in while components of the old one are slowly phased out. When everyone is confident that the new system is performing as expected, the old system is completely phased out. This gradual replacement is repeated for each application until the new system is running every application. In some cases, the phase-in approach can take months or years.

Pilot start-up involves running the new system for one group of users rather than all users. For example, a manufacturing company with many retail outlets throughout the country could use the pilot start-up approach and install a new inventory control system at one of the retail outlets. When this pilot retail outlet runs without problems, the new inventory control system can be implemented at other retail outlets. The U.S. Army used a pilot approach to test new information systems and technology.[102] This lets the U.S. explore new technologies that can provide new and advanced capabilities. According to the Army's director of advanced technologies, "We're eager to explore technologies when they can deliver a capability that we don't have."

Parallel start-up involves running both the old and new systems for a period of time. The output of the new system is compared closely with the output of the old system, and any differences are reconciled. When users are comfortable that the new system is working correctly, the old system is eliminated.

User Acceptance

Most mainframe computer manufacturers use a formal **user acceptance document**—a formal agreement the user signs stating that a phase of the installation or the complete system is approved. This is a legal document that usually removes or reduces the IS vendor's liability for problems that occur after the user acceptance document has been signed. Because this document is so important, many companies get legal assistance before they sign the acceptance document. Stakeholders can also be involved in acceptance testing to make sure that the benefits to them are indeed realized.

parallel start-up
Running both the old and new systems for a period of time and comparing the output of the new system closely with the output of the old system; any differences are reconciled. When users are comfortable that the new system is working correctly, the old system is eliminated.

user acceptance document
A formal agreement signed by the user that states that a phase of the installation or the complete system is approved.

SYSTEMS OPERATION AND MAINTENANCE

Systems operation involves all aspects of using the new or modified system in all kinds of operating conditions. Getting the most out of a new or modified system during its operation is the most important aspect of systems operations for many organizations. Throughout this book, we have seen many examples of information systems operating in a variety of settings and industries. Thus, we will not cover the operation of an information system in detail in this section. To provide adequate support, many companies use a formal help desk. A *help desk* consists of people with technical expertise, computer systems, manuals, and other resources needed to solve problems and give accurate answers to questions. If you are having trouble with your PC and call a toll-free number for assistance, you might reach a help desk in India, China, or another country.

Systems maintenance involves checking, changing, and enhancing the system to make it more useful in achieving user and organizational goals. Organizations can perform systems maintenance in house or they can hire outside companies to perform maintenance for them.[103] Many companies that use database systems from Oracle or SAP, for example, often hire these companies to maintain their database systems. Systems maintenance is important for individuals, groups, and organizations. Individuals, for example, can use the Internet, computer vendors, and independent maintenance companies, including YourTechOnline.com (*www.yourtechonline.com*), Geek Squad (*www.geeksquad.com*), PC Pinpoint (*www.pcpinpoint.com*), and others. Organizations often have personnel dedicated to maintenance.

Software maintenance for purchased software can be 20 percent or more of the purchase price of the software annually.[104] While some CIOs complain about the high cost of software maintenance, others believe it is worth the cost. According to one CIO, "We've never viewed maintenance as a black hole you put your money into. Maintenance is part of the game." The maintenance process can be especially difficult for older software. A *legacy system* is an old system that might have been patched or modified repeatedly over time. An old payroll program in COBOL developed decades ago and frequently changed is an example of a legacy system. Legacy systems can be very expensive to maintain, and it can be difficult to add new features to some legacy systems. With about 11 million lines of older computer code, a large railroad company wasn't able to add the new features customers wanted.[105] At some point, it becomes less expensive to switch to new programs and applications than to repair and maintain the legacy system. Maintenance costs for older legacy systems can be 50 percent of total operating costs in some cases.

systems operation
Use of a new or modified system.

systems maintenance
A stage of systems development that involves checking, changing, and enhancing the system to make it more useful in achieving user and organizational goals.

SYSTEMS REVIEW

systems review
The final step of systems development, involving the analysis of systems to make sure that they are operating as intended.

Systems review, the final step of systems development, is the process of analyzing systems to make sure that they are operating as intended. The systems review process often compares the performance and benefits of the system as it was designed with the actual performance and benefits of the system in operation. The Transportation Security Agency (TSA), for example, used an approach called the *Idea Factory,* to review current information systems and recommend new ones or changes to existing systems.[106] According to the systems development director for TSA, "Often, the people on the front lines have the best ideas because they're the ones who interact with passengers." The Idea Factory was featured on the White House Web site.[107] In some cases, a formal audit of the application can be performed, using internal and external auditors.[108] Systems review can be performed during systems development, resulting in halting the new systems while they are being built because of problems.

System Performance Measurement

system performance measurement
Monitoring the system—the number of errors encountered, the amount of memory required, the amount of processing or CPU time needed, and other problems.

Systems review often involves monitoring the system, called **system performance measurement.** The number of errors encountered, the amount of memory required, the amount of processing or CPU time needed, and other problems should be closely observed. If a particular system is not performing as expected, it should be modified, or a new system should be developed or acquired. Comcast, the large cable provider, used Twitter to get user feedback on the performance of its information systems and all of its operations.[109] Some Comcast executives believe that using Twitter is like an "early-warning-system" that alerts Comcast to potential problems before they become serious and hurt system performance.

system performance products
Software that measures all components of the computer-based information system, including hardware, software, database, telecommunications, and network systems.

System performance products have been developed to measure all components of the information system, including hardware, software, database, telecommunications, and network systems. IBM Tivoli OMEGAMON can monitor system performance in real time.[110] Precise Software Solutions has system performance products that provide around-the-clock performance monitoring for ERP systems, Oracle database applications, and other programs.[111] HP also offers a software tool called Business Technology Optimization (BTO) software to help companies analyze the performance of their computer systems, diagnose potential problems, and take corrective action if needed.[112] When properly used, system performance products can quickly and efficiently locate actual or potential problems.

Measuring a system is, in effect, the final task of systems development. The results of this process can bring the development team back to the beginning of the development life cycle, where the process begins again.

SUMMARY

Principle

Effective systems development requires a team effort of stakeholders, users, managers, systems development specialists, and various support personnel, and it starts with careful planning.

The systems development team consists of stakeholders, users, managers, systems development specialists, and various support personnel. The development team is responsible for determining the objectives of the information system and delivering to the organization a system that meets its objectives.

A systems analyst is a professional who specializes in analyzing and designing business systems. The programmer is responsible for modifying or developing programs to satisfy user requirements. Other support personnel on the development team include technical specialists, either IS department employees or outside consultants. Depending on the magnitude of the systems development project and the number of IS systems development specialists on the team, the team may also include one or more IS managers.

An individual systems developer is a person that performs all of the systems development roles, including systems analyst, programmer, technical specialist, and other roles described in the above section. While individual systems developers can create applications for a group or entire organization, many specialize in developing applications for individuals. Individual users acquire applications for both personal and professional use. It is also possible for one person to be both an individual developer and user. The term end-user systems development describes any systems development project in which business managers and users assume the primary effort.

Information systems planning refers to the translation of strategic and organizational goals into systems development initiatives. Benefits of IS planning include a long-range view of information technology use and better use of IS resources. Planning requires developing overall IS objectives; identifying IS projects; setting priorities and selecting projects; analyzing resource requirements; setting schedules, milestones, and deadlines; and developing the IS planning document.

Principle

Systems development often uses different approaches and tools such as traditional development, prototyping, rapid application development, end-user development, computer-aided software engineering, and object-oriented development to select, implement, and monitor projects.

The five phases of the traditional SDLC are investigation, analysis, design, implementation, and maintenance and review. Systems investigation involves identifying potential problems and opportunities and considering them in light of organizational goals. Systems analysis seeks a general understanding of the solution required to solve the problem; the existing system is studied in detail and weaknesses are identified. Systems design involves creating new or modified system requirements. Systems implementation encompasses programming, testing, training, conversion, and operation of the system. Systems operation involves running the system once it is implemented. Systems maintenance and review entails monitoring the system and performing enhancements or repairs.

Prototyping is an iterative development approach that involves defining the problem, building the initial version, having users utilize and evaluate the initial version, providing feedback, and incorporating suggestions into the second version. Rapid application development (RAD) uses tools and techniques designed to speed application development. Its use reduces paper-based documentation, automates program source code generation, and facilitates user participation in development activities. An agile, or extreme programming, approach allows systems to change as they are being developed. RAD makes extensive use of the joint application development (JAD) process to gather data and perform requirements analysis. JAD involves group meetings in which users, stakeholders, and IS professionals work together to analyze existing systems, propose possible solutions, and define the requirements for a new or modified system.

The use of automated tools enables detailed development, tracking, and control of the project schedule. Effective use of these tools enables a project manager to deliver a high-quality system and to make intelligent trade-offs among cost, schedule, and quality. CASE tools can automate many of the systems development tasks, thus reducing the time and effort required to complete them while ensuring good documentation. With the object-oriented systems development (OOSD) approach, a project can be broken down into a group of objects that interact. Instead of requiring thousands or millions of lines of detailed computer instructions or code, the systems development project might require a few dozen or maybe a hundred objects.

Principle

Systems development starts with investigation and analysis of existing systems.

In most organizations, a systems request form initiates the investigation process. This form typically includes the problems in or opportunities for the system, objectives of systems investigation, overview of the proposed system, and expected costs and benefits of the proposed system. The systems investigation is designed to assess the feasibility of implementing solutions for business problems. An investigation team follows up on the request and performs a feasibility analysis that addresses technical, economic, legal, operational, and schedule feasibility. Object-oriented systems investigation is being used to a greater extent today. As a final step in the investigation process, a systems investigation report should be prepared to document relevant findings.

Systems analysis is the examination of existing systems, which begins once approval for further study is received from management. Additional study of a selected system allows those involved to further understand the system's weaknesses and potential improvement areas. An analysis team is assembled to collect and analyze data on the existing system.

Data collection methods include observation, interviews, and questionnaires. Data analysis manipulates the collected data to provide information. Data modeling is used to model organizational objects and associations using text and graphical diagrams. It is most often accomplished through the use of entity-relationship (ER) diagrams. Activity modeling is often accomplished through the use of data-flow diagrams (DFDs), which model objects, associations, and activities by describing how data can flow between and around various objects. DFDs use symbols for data flows, processing, entities, and data stores. The overall purpose of requirements analysis is to determine user and organizational needs. Object-oriented systems analysis also involves diagramming techniques, such as a generalization/specialization hierarchy diagram.

Principle

Designing new systems or modifying existing ones should always be aimed at helping an organization achieve its goals.

The purpose of systems design is to prepare the detailed design needs for a new system or modifications to an existing system. Logical systems design refers to the way the various components of an information system will work together. Physical systems design refers to the specification of the actual physical components.

If new hardware or software will be purchased from a vendor, a formal request for proposal (RFP) is needed. The RFP outlines the company's needs; in response, the vendor provides a written reply. Organizations have three alternatives for acquiring computer systems: purchase, lease, or rent. RFPs from various vendors are reviewed and narrowed down to the few most likely candidates. Near the end of the design stage, an organization prohibits further changes in the design of the system. The design specifications are then said to be frozen. After the vendor is chosen, contract negotiations can

begin. One of the most important steps in systems design is to develop a good contract if new computer facilities are being acquired. The final step is to develop a design report that details the outputs, inputs, and user interfaces. It also specifies hardware, software, databases, telecommunications, personnel, and procedure components and the way these components are related.

Environmental design, also called green design, involves systems development efforts that slash power consumption, take less physical space, and result in systems that can be disposed in a way that doesn't negatively affect the environment. A number of companies are developing products and services to help save energy. Environmental design also deals with how companies are developing systems to dispose of old equipment. The U.S. government is also involved in environmental design. It has a plan to require federal agencies to purchase energy-efficient computer systems and equipment. The plan would require federal agencies to use the Electronic Product Environmental Assessment Tool (EPEAT) to analyze the energy usage of new systems. The U.S. Department of Energy rates products with the Energy Star designation to help people select products that save energy and are friendly to the environment.

Principle

The primary emphasis of systems implementation is to make sure that the right information is delivered to the right person in the right format at the right time.

The purpose of systems implementation is to install a system and make everything, including users, ready for its operation. Systems implementation includes hardware acquisition, software acquisition or development, user preparation, hiring and training of IS personnel, site and data preparation, installation, testing, start-up, and user acceptance. Hardware acquisition requires purchasing, leasing, or renting computer resources from a vendor. Increasingly, companies are using service providers to acquire software, Internet access, and other IS resources.

Software can be purchased from external vendors or developed in house—a decision termed the *make-or-buy decision*. Implementation must also address database and telecommunications systems, user preparation, and IS personnel requirements. User preparation involves readying managers, employees, and other users for the new system. New IS personnel may need to be hired, and users must be well trained in the system's functions. The physical site of the system must be prepared, and any existing data to be used in the new system must be converted to the new format. Hardware is installed during the implementation step. Testing includes program (unit) testing, systems testing, volume testing, integration testing, and acceptance testing.

Start-up begins with the final tested information system. When start-up is finished, the system is fully operational. There are a number of different start-up approaches. Direct conversion (also called *plunge* or *direct cutover*) involves stopping the old system and starting the new system on a given date. With the phase-in approach, sometimes called a *piecemeal approach*, components of the new system are slowly phased in while components of the old one are slowly phased out. When everyone is confident that the new system is performing as expected, the old system is completely phased out. Pilot start-up involves running the new system for one group of users rather than all users. Parallel start-up involves running both the old and new systems for a period of time. The output of the new system is compared closely with the output of the old system, and any differences are reconciled. When users are comfortable that the new system is working correctly, the old system is eliminated. The final step of implementation is user acceptance.

Principle

Maintenance and review add to the useful life of a system but can consume large amounts of resources,

so they benefit from the same rigorous methods and project management techniques applied to systems development.

Systems operation is the use of a new or modified system. Systems maintenance involves checking, changing, and enhancing the system to make it more useful in obtaining user and organizational goals. Maintenance is critical for the continued smooth operation of the system. Some major reasons for maintenance are changes in business processes; new requests from stakeholders, users, and managers; bugs or errors in the program; technical and hardware problems; corporate mergers and acquisitions; government regulations; change in the operating system or hardware; and unexpected events, such as terrorist attacks.

Systems review is the process of analyzing systems to make sure that they are operating as intended. It involves monitoring systems to be sure they are operating as designed.

CHAPTER 8: SELF-ASSESSMENT TEST

Effective systems development requires a team effort of stakeholders, users, managers, systems development specialists, and various support personnel, and it starts with careful planning.

1. _____ is the activity of creating or modifying existing business systems. It refers to all aspects of the process—from identifying problems to be solved or opportunities to be exploited to the implementation and refinement of the chosen solution.

2. Which of the following individuals ultimately benefit from a systems development project?
   a. Computer programmers
   b. Systems analysts
   c. Stakeholders
   d. Senior-level managers

3. Like a contractor constructing a new building or renovating an existing one, chief information officer (CIO) takes the plans from the systems analyst and builds or modifies the necessary software. True or False?

Systems development often uses different approaches and tools such as traditional development, prototyping, rapid application development, end-user development, computer-aided software engineering, and object-oriented development to select, implement, and monitor projects.

4. Agile development allows systems to change as they are being developed. True or False?

5. _____ takes an iterative approach to the systems development process. During each iteration, requirements and alternative solutions to the problem are identified and analyzed, new solutions are designed, and a portion of the system is implemented.

Systems development starts with investigation and analysis of existing systems.

6. Feasibility analysis is typically done during which systems development stage?
   a. Investigation
   b. Analysis
   c. Design
   d. Implementation

7. Rapid application development (RAD) employs tools, techniques, and methodologies designed to speed application development. True or False?

Designing new systems or modifying existing ones should always be aimed at helping an organization achieve its goals.

8. Scenarios and sequence diagrams are used with _____.
   a. object-oriented design
   b. point evaluation
   c. incremental design
   d. nominal evaluation

9. _____ involves systems development efforts that slash power consumption and require less physical space.

10. The design report is the final result of system design that technical and detailed descriptions of the new system. True or False?

The primary emphasis of systems implementation is to make sure that the right information is delivered to the right person in the right format at the right time.

11. Software as a Service (SaaS) allows an organization to subscribe to Web-based applications and pay for the software and services actually used. True or False?

12. The phase-in approach to conversion involves running both the old system and the newsystem for a three months or longer. True or False?

Maintenance and review add to the useful life of a system but can consume large amounts of resources, so they benefit from

the same rigorous methods and project management techniques applied to systems development.

13. Reviewing and monitoring the number of errors encountered, the amount of memory required, the amount of processing or CPU time needed, and other problems should be closely observed. This is called
    a. Object review
    b. Structured review
    c. systems performance measurement
    d. Critical factors measurement

14. Monitoring a system after it has been implemented to make it more useful in achieving user and organizational goals is called _____.

CHAPTER 8: SELF-ASSESSMENT TEST ANSWERS

(1) Systems development, (2) c (3) False (4) True (5) Prototyping (6) a (7) True (8) a (9) environmental design (10) True (11) True (12) False (13) c (14) systems maintenance

REVIEW QUESTIONS

1. What is an information system stakeholder?
2. What is the goal of information systems planning? What steps are involved in IS planning?
3. What are the steps of the traditional systems development life cycle?
4. What is the difference between systems investigation and systems analysis? Why is it important to identify and remove errors early in the systems development life cycle?
5. What is end-user systems development? What are the advantages and disadvantages of end-user systems development?
6. List the different types of feasibility.
7. What is the result or outcome of systems analysis? What happens next?
8. How does the JAD technique support the RAD systems development life cycle?
9. What is prototyping?
10. What are the steps of object-oriented systems development?
11. What is an RFP? What is typically included in one? How is it used?
12. What is systems operation?
13. What activities go on during the user preparation phase of systems implementation?
14. Give three examples of a computer system vendor.
15. What are the financial options of acquiring hardware?
16. How can SaaS be used in software acquisition?
17. What are some of the reasons for program maintenance?
18. Describe how you back up the files you use at school.

DISCUSSION QUESTIONS

1. Why is it important for business managers to have a basic understanding of the systems development process?
2. Briefly describe the role of a system user in the systems investigation and systems analysis stages of a project.
3. You have decided to become an IS entrepreneur and develop applications for the iPhone and other PDAs.

Describe what applications you would develop and how you would do it.
4. Imagine that your firm has never developed an information systems plan. What sort of issues between the business functions and IS organization might exist?

5. You have been hired by your university to find an outsourcing company to perform the universities payroll function. What are your recommendations? Describe the advantages and disadvantages of the outsourcing approach for this application.

6. Briefly describe when you would use the object-oriented approach to systems development instead of the traditional systems development life cycle.

7. How important are communications skills to IS personnel? Consider this statement: "IS personnel need a combination of skills—one-third technical skills, one-third business skills, and one-third communications skills." Do you think this is true? How would this affect the training of IS personnel?

8. You have been hired to perform systems investigation for a French restaurant owner in a large metropolitan area. She is thinking of opening a new restaurant with a state-of-the-art computer system that would allow customers to place orders on the Internet or at Kiosks at restaurant tables. Describe how you would determine the technical, economic, legal, operational, and schedule feasibility for the restaurant and its new computer system.

9. Identify some of the advantages and disadvantages of purchasing a database package versus the DaaS approach.

10. You have been hired to design a computer system for a small business. Describe how you could use environmental design to reduce energy usage and the computer's impact on the environment.

11. Assume that you want to start a new video-rental business for students at your college or university. Go through logical design for a new information system to help you keep track of the videos in your inventory.

12. Identify some of the advantages and disadvantages of purchasing versus leasing hardware.

13. Identify some of the advantages and disadvantages of purchasing versus developing software.

14. Identify the various forms of testing used. Why are there so many different types of tests?

15. You have been hired to purchase a new billing and accounting system for a medium-sized business. Describe how you would start up the new system and place it into operation.

16. How would you go about evaluating a software vendor?

17. Assume that you have a personal computer that is several years old. Describe the steps you would use to perform a systems review to determine whether you should acquire a new PC.

18. Describe how you would select the best admissions software for your college or university. What features would be most important for school administrators? What features would be most important for students?

PROBLEM-SOLVING EXERCISES

1. You are developing a new information system for The Fitness Center, a company that has five fitness centers in your metropolitan area, with about 650 members and 30 employees in each location. This system will be used by both members and fitness consultants to track participation in various fitness activities, such as free weights, volleyball, swimming, stair climbers, and aerobic and yoga classes. One of the performance objectives of the system is that it must help members plan a fitness program to meet their particular needs. The primary purpose of this system, as envisioned by the director of marketing, is to assist The Fitness Center in obtaining a competitive advantage over other fitness clubs. Use a graphics program to develop a flowchart or a grid chart to show the major components of your information system and how the components are tied together.

2. For a business of your choice, use a graphics program to develop one or more Use Case diagrams and one or more Generalized/Specialized Hierarchy diagrams for your new business using the object-oriented approach.

3. You have been hired to develop a payroll program for a medium sized company. At a minimum, the application should have an hours-worked table that contains how many hours each employee worked and an employee table that contains information about each employee, including hourly pay rate. Design and develop these tables that could be used in a database for the payroll program.

TEAM ACTIVITIES

1. Your team should interview people involved in systems development in a local business or at your college or university. Describe the process used. Identify the users, analysts, and stakeholders for a systems development project that has been completed or is currently under development.

2. Your team has been hired to determine the requirements of a new medium-cost coffee bar to compete with higher-priced coffee shops like Starbucks. The new coffee bar will offer computer kiosks for customers to surf the Internet or order coffee and other products from the coffee bar. Using RAD and JAD techniques with your team, develop requirements for this new coffee bar and its computer system.

3. Your team should perform a systems review of a computer application being used at your college or university. Include the strengths and weaknesses of the computer application and describe how it could be improved from a student perspective.

WEB EXERCISES

1. Use the Internet to find two different systems development projects that failed to meet cost or performance objectives. Summarize the problems and what should have been done. You might be asked to develop a report or send an e-mail message to your instructor about what you found.
2. Cloud computing, whereby applications like word processing and spreadsheet analysis are delivered over the Internet, is becoming more popular. You have been hired to analyze the potential of a cloud computing application that performs payroll and invoicing over the Internet from a large Internet company. Describe the systems development steps and procedures you would use to analyze the feasibility of this approach.
3. Using the Web, search for information on the advantages and disadvantages of environmental design. Write a report on what you found

CAREER EXERCISES

1. Pick a career that you are considering. What type of information system would help you on the job? Perform technical, economic, legal, operational, and schedule feasibility for an information system you would like developed for you.
2. Research possible careers in developing applications for iPhones, other smartphones, and PDAs. Write a report that describes these opportunities. Include in your report applications that aren't currently available that you would find useful.

CASE STUDIES

Case One

Hotwire.com Gains Control of Hotel Partner Information

Hotwire.com advertises four-star hotels at two-star prices throughout North America and Europe. By building businesses partnerships with more than 7,000 hotels across North America, Hotwire assists its hotel partners in filling otherwise empty rooms and helps its customers by finding them rates discounted by as much as 60 percent.

Hotwire maintains a wealth of information about each of its 7,000 hotel partners. Besides basics such as location, amenities, customer reviews, and prices, it also maintains information about the number of available rooms. Much of the information Hotwire tracks changes frequently. Maintaining accurate and up-to-date information is a key to success in this highly competitive business.

Until recently, Hotwire maintained its hotel information on shared Excel spreadsheets. Hotwire employees kept the information up to date by accessing the shared spreadsheets over the corporate network. Various applications drew information from the spreadsheets to provide Hotwire the information it needed to provide its customers with deals. Unfortunately, the system was prone to errors and extremely difficult to maintain.

Hotwire systems analysts considered the problem and possible solutions. The data that the system manipulated was well defined, as were the processes that were applied to the data. The problem lay in the manner in which the data was stored and accessed.

It was clear to Hotwire analysts that a centralized database-driven system would allow Hotwire to gain better control over its data. Hotwire analysts considered costs and benefits of designing its own database and DBMS compared to outsourcing the service to a provider. Upon evaluation of the requirements for the system, they discovered that the data Hotwire was managing could be easily managed by a Customer Relationship Management (CRM) system, even though its partner hotels were not exactly customers.

Hotwire systems analysts evaluated a variety of CRM solutions and found the one from Salesforce to be most effective and feasible in terms of cost and technology. The Salesforce CRM allows Hotwire employees to access hotel data through a user-friendly Web-based interface from any Internet-connected computer. Hotwire data is securely stored and maintained on Salesforce servers, relieving Hotwire of that costly responsibility. Hotwire system engineers customize the CRM to deliver the information and reports needed by Hotwire managers and systems.

The new system has improved data reliability and accuracy, allowing better management of the information, happier customers, and improved relationships with Hotwire partners.

Discussion Questions

1. How do you think system engineers determined that it was a good time to invest in this information systems development project?
2. What benefits did Salesforce provide Hotwire.com over its previous system?

Critical Thinking Questions

1. What considerations are required in deciding whether to host your own system or outsource to a vendor like Salesforce?
2. How can Hotwire gauge the success of this information systems development project?

SOURCES: "Hotwire Uses Salesforce CRM to Keep Tabs on More Than 7,000 Hotel Partners Across North America," Salesforce success stories, www.salesforce.com, accessed May 25, 2010; Salesforce CRM Web site, www.salesforce.com/crm/products.jsp, accessed May 25, 2010; About Hotwire, www.hotwire.com/about-hotwire/press-room/factSheet.jsp, accessed May 25, 2010.

Case Two
Russian Sporting Goods Chain Scales Up Budgeting System

The Sportmaster Group is the largest sporting goods chain in Eastern Europe. It handles Sportlandia and Columbia brands and more than 200 other trademarks, including its own.

When Sportmaster was established, its CIO decided to build a budget planning system in house. The system was built with Excel spreadsheets and was well suited for a young start-up business. But in recent years, with stores spread across Eastern Europe, the executives at Sportmaster Group wanted to gain better control of their budgets and use their financial information to make wise strategic decisions. Due to the large amount of data involved, using the current Excel-based system had become burdensome, and the information provided was too limited to support corporate needs.

In the process of system maintenance and review, Sportmaster realized that to move to the next stage of growth, it needed to move to information systems used by large global corporations. In researching packages from a variety of vendors, the company settled on software from Cognos, an IBM company. To customize the Cognos software for its own needs, Sportmaster hired a company named IBS. Sportmaster chose IBS because it was a certified Cognos vendor and had experience working with Russian companies.

IBS consultants met with Sportmaster executives and information systems staff to discuss the expectations for the new budget analysis system. They worked on site so they could test prototypes on actual corporate data and have their progress reviewed by Sportmaster to confirm that they were on target.

The resulting system met the following goals defined by Sportmaster. According to the Cognos case study, the system could perform the following tasks:

- Create a basic gross profit budget
- Create budgets for investment activity, including opening new stores and capital investments
- Create an operating expenses budget for all corporate divisions, including more than "500 centers of responsibility"
- Create specialized budgets, including a consolidated Revenue and Expenditure Budget, a Cash-Flow Budget, and a Balance Sheet of Payables and Receivables
- Create a Revenue and Expenditure Budget for the divisions
- Integrate with external accounting systems

Sportmaster used a parallel start-up method, introducing the new budget planning and accounting system, keeping the old system available as a backup. After six months of successful use and a few tweaks to perfect the system, the company now fully depends on the new system and is enjoying its benefits. Sportmaster can access highly detailed budget reports that assist in making strategic decisions. The process for creating budget reports has been simplified, and the duration of the budget cycle is shortened. Operations that used to take days are now accomplished in near real time. Most important is that the reliability of the data is improved so that budget errors are minimized.

Discussion Questions

1. What motivated Sportmaster to start an IS project to build a new budget planning and accounting system?
2. What steps did the development team take to make sure that the project was completed in minimum time while meeting the company's needs?

Critical Thinking Questions

1. How can the level of detail of the information provided by a budget planning and accounting system affect a company's decision-making capability?
2. Why do you think Sportmaster decided to outsource the systems development project rather than work in house?

SOURCES: Cognos Staff, "Sportmaster Group," Cognos Case Study, www-01.ibm.com/software/success/cssdb.nsf/CS/ABRR-7WEBZ3?OpenDocument&Site=corp&cty=en_us, accessed March 3, 2010.

Questions for Web Case

See the Web site for this book to read about the Altitude Online case for this chapter. Following are questions concerning this Web case.

Altitude Online: Systems Investigation and Analysis Considerations

Discussion Questions

1. What important activities did Jon's team engage in during the Systems Investigation stage of the Systems Development Life Cycle?
2. Why are all forms of feasibility considerations especially important for an ERP development project?

Critical Thinking Questions

1. Why is the quality of the systems analysis report crucial to the successful continuation of the project?
2. Why do you think Jon felt the need to travel to communicate with Altitude Online colleagues rather than using e-mail or phone conferencing? What benefit does face-to-face communication provide in this scenario?

Altitude Online: Systems Design, Implementation, Maintenance, and Review Considerations

Discussion Questions

1. How did Jon's team coordinate with the vendor in the implementation stage of the systems development project?
2. What did Jon's team do in advance of contacting SAP that made the design and implementation systems proceed as smoothly as possible?

Critical Thinking Questions

1. What risks were involved in the systems development project?
2. What benefits were gained from this systems development project? Was it worth the risks?

NOTES

Sources for the opening vignette: "LEGO creates model business success with SAP and IBM," IBM Success stories, May 19, 2010, www-01.ibm.com/software/success/cssdb.nsf/CS/STRD-85KGS6?OpenDocument&Site=bladecenter&cty=en_us; "LEGO—about us," accessed May 23, 2010, www.lego.com/eng/info.

1 Kane, Y., "Seeking Fame in Apple's Sea of Apps," *The Wall Street Journal*, July 15, 2009, p. B1.
2 Anderson, Howard, "Project Triage: Skimpy Must Die," *Information Week*, March 16, 2009, p. 14.
3 Soat, John, "IT leaders are wrestling with how to bring informal collaboration into rigorous processes," *Information Week*, July 20, 2009, p. 17.
4 Rosato, Donna, "The 50 Best Jobs in America," *Money*, November 2009, p. 88.
5 Kane, Yukare Iwatani, "Apple Woos Developers with New iPhone," *The Wall Street Journal*, March 18, 2009, p. B6.
6 Kane, Y., "Apple Woos Developers with New iPhone Tools," *The Wall Street Journal*, March 18, 2009, p. B6.
7 Woolsey, Matt, "New Ball Game," *Forbes*, March 2, 2009, p. 50.
8 O'Brien, Jeffrey, "The Wizards of Apps," *Fortune*, May 25, 2009, p. 29.
9 Reena, J. and Burrows, P., "An All-Out Online Assault on the iPhone," *BusinessWeek*, April 6, 2009, p. 74.
10 Albro, Edward, "Google's Wave," *PC World*, August 2009, p. 14.
11 Weier, Mary H., "Google Tries to Make App Engine Practical," *Information Week*, April 13, 2009, p. 13.

12 MacMillan, D., et al, "The App Economy," *BusinessWeek*, November 2, 2009, p. 45.
13 Perez, Marin, "The iPhone as IT Security Tool," *Information Week*, November 30, 2009, p. 18.
14 Boudreau, John, "iPhone Has Musical Hit," *The Tampa Tribune*, March 30, 2009, p. 8.
15 Wildstorm, Stephen, "The Unstoppable iPhone," *BusinessWeek*, June 29, 2009, p. 63.
16 Mossberg W. and Bhehret K., "A Shopping Trip to the App Store for Your iPhone," *The Wall Street Journal*, July 23, 2009, p. D1.
17 Pressman, Aaron, "Figuring Your Finances," *BusinessWeek*, December 15, 2008, p. 78.
18 Wildstrom, Stephan, "A Stroll Through iPhone App Store," *BusinessWeek*, July 28, 2009, p. 74.
19 Kane, Y., "Breaking Apple's Grip on the iPhone," *The Wall Street Journal*, March 6, 2009, p. B1.
20 Aedo, I., et al, "End User Oriented Strategies," *Information Processing & Management*, January 2010, p. 11.
21 Cordoba, J., "Critical Reflection in Planning Information Systems," *Information Systems Journal*, March 2009, p. 123.
22 Preston, D. and Karahanna, E., "Antecedents of IS Strategic Alignment," *Information Systems Research*, June 2009, p. 159.
23 Colburn, M., "CIO Profiles," *Information Week*, March 16, 2009, p. 16.
24 Evans, Bob, "Stop Aligning IT with the Business," *Information Week*, January 19, 2009, p. 68.
25 Web, Warren, "Prototyping Kit Shortens Embedded-Systems-Development Schedule," *EDN*, April 9, 2009, p. 8.

26 Goodhue, D., et al, "Addressing Business Agility Challenges with Enterprise Systems," *MIS Quarterly Executive,* June 2009, p. 73.

27 Babcock, Charles, "IBM Adds Social Networking to Rational Team Development," *Information Week,* June 2, 2009, p. 24.

28 Vidgen, R., et al, "Coevolving Systems and the Organization of Agile Software Development," *Information Systems Research,* September 2009, p. 355.

29 Tubbs, Jerry, "Team Building Goes Viral," *Information Week,* February 22, 2010, p. 47.

30 Sarker, Saonee and Sarker, Suprateek, "Exploring Agility in Distributed Information Systems Development Teams," *Information Systems Research,* September 2009, p. 440.

31 Erickson, J., "Agile Development," *Information Week,* April 27, 2009, p. 31.

32 Conboy, Kieran, "Agility from First Principles," *Information Systems Research,* September, 2009, p. 329.

33 Tolfo, C., et al, "The Influence of Organizational Culture on the Adoption of Extreme Programming," *The Journal of Systems and Software,* November 2008, p. 1955.

34 Balijepally, V., et al, "Are Two Heads Better Than One for Software Development?" *MIS Quarterly,* March 2009, p. 91.

35 West, Dave, "Agile Processes Go Lean," *Information Week,* April 27, 2009, p. 32.

36 "The Rational Unified Process," *www-306.ibm.com/software/awdtools/rup/support...,* accessed June 2, 2008.

37 "Rational Case Studies," *www-01.ibm.com/software/success/cssdb.nsf/softwareL2VW?OpenView&Start=1&Count=30&RestrictToCategory=ratioanal_RationalUnifiedProcess,* accessed June 2, 2008.

38 Kettunen, Petri, "Adopting Key Lesson from Agile Manufacturing to Software Product Development," *Technovation,* June 2009, p. 408.

39 Weier, M., "How GM's CIO Looks at IT Restructuring," *Information Week,* June 8, 2009, p. 20.

40 McGregor, Jena, "The Chore Goes Offshore," *BusinessWeek,* March 23, 2009, p. 50.

41 Capell, Kerry, "Vodafone: Embracing Open Source with Open Arms," *BusinessWeek,* April 20, 2009, p. 52.

42 McGee, M., "Pay for Performance," *Information Week,* March 23, 2009, p. 32.

43 IBM Web site, *www.ibm.com,* accessed June 2, 2008.

44 EDS Web site, *www.eds.com,* accessed June 2, 2008.

45 Scheck, J. and Worthen, B., "Hewlett Packard Takes Aim at IBM," *The Wall Street Journal,* May 4, 2008, p. B1.

46 Accenture Web site, *www.accenture.com,* accessed June 2, 2008.

47 Sheth, N., "Wipro Sets Outsourcing Sights on Emerging Markets," *The Wall Street Journal,* May 13, 2009, p. B4B.

48 Dolinski, Catherine, "Medicaid Glitch Continues," *The Tampa Tribune,* February 6, 2009, p. 4.

49 Tiwana, A., "Governance-Knowledge Fit in Systems Development Projects," *Information Systems Research,* June 2009, p. 180.

50 Pereira, Rudy, "Embrace a Culture of Continuous Improvement," *Credit Union Magazine,* October 2009, p. 62.

51 Betts, Mitch, "The No. 1 Cause of IT Failure: Complexity," *Computerworld,* December 21, 2009, p. 4.

52 Capability Maturity Model for Software home page, *www.sei.cmu.edu,* accessed June 2, 2008.

53 Shang, S., et al, "Understanding the Effectiveness of the Capability Maturity Model," *Total Quality Management & Business Excellence,* Vol. 20, 2009, p. 219.

54 Staff, "An Object-Oriented Graphical Modeling for Power System Analysis," *International Journal of Modeling & Simulation,* Vol. 29, 2009, p. 71.

55 Ovide, S. and Vascellaro, J., "UAL Story Blame is Placed on Computer," *The Wall Street Journal,* September 10, 2008, p. B3.

56 Ioffe, Julia, "Tech RX for Health Care," *Fortune,* March 16, 2009, p. 36.

57 Sebora, T., et al, "Critical Success Factors for E-Commerce Entrepreneurship," *Small Business Economics,* March, 2009, p. 303.

58 Babcock, Charles, "Hybrid Clouds," *Information Week,* September 7, 2009, p. 15.

59 Hoover, Nicholas, "GE Puts the Cloud Model to the Test," *Information Week,* April 13, 2009, p. 32.

60 Wildstrom, Stephen, "What to Entrust to The Cloud," *BusinessWeek,* April 6, 2009, p. 89.

61 Greenberg, Andy, "No Phishing Zone," *Forbes,* April 27, 2009.

62 Healey, M., "Beat the Slow Commotion," *Information Week,* March 23, 2009, p. 40.

63 Patterson S. and Ng, S., "NYSE's Fast-Trade Hub," *The Wall Street Journal,* July 30, 2009, p. C1.

64 Erickson, J., "Information Security," *Information Week,* June 22, 2009, p. 45.

65 Gorman, S. and Dreazen, Y., "New Military Command to Focus on Cybersecurity," *The Wall Street Journal,* April 22, 2009, p. A2.

66 Hoover, Nicholas, "Fed CIO Scrutinizes Spending, Eyes Cloud," *Information Week,* March 16, 2009, p. 19.

67 Patterson, S. and Ng, S., "NYSE's Fast-Trade Hub," *The Wall Street Journal,* July 30, 2009, p. C1.

68 Marew, T., et al, "Tactics Based Approach for Integrating Non-Functional Requirements in Object-Oriented Analysis and Design," *Journal of Systems and Software,* October 2009, p. 1642.

69 Bustillo, Miguel, "Wal-Mart to Assign New Green Ratings," *The Wall Street Journal,* July 16, 2009, p. B1.

70 Pratt, Mary, "Slow Growing Green," *Computerworld,* January 1, 2009, p. 13.

71 Campbell, S. and Jeronimo, M., "Virtual Machines, Real Productivity," *Information Week,* January 26, 2009, p. 40.

72 Babcock, Charles, "Nissan Assembly Lines Roll With Fewer Servers," *Information Week,* July 6, 2009, p. 17.

73 Pratt, Mary, "Birth of an Energy Star," *Computerworld,* August 31, 2009, p. 24.

74 Badiga, L., "CIO Profiles," *Information Week,* March 2, 2009, p. 16.

75 Preston, Rob, "Get Serious About Going Green," *Information Week,* April 27, 2009, p. 60.

76 Carlton, Jim, "The PC Goes on an Energy Diet," *The Wall Street Journal,* September 8, 2009, p. R8.

77 Mies, Ginny, "Green Phones," *PC World,* September 2009, p. 28.

78 Voltaic Generator, *www.voltaicsystems.com/bag_generator.shtml,* accessed August 17, 2009.

79 Randall, David, "Be Green and Make A Buck," *Forbes,* March 2, 2009, p. 40.

80 Mossberg, Walter, "Where Old Gadgets Go to Breathe New Life," *The Wall Street Journal,* August 13, 2009, p. D8.

81 Staff, "Obama Orders Federal IT to Get Greener," *TechWeb,* October 6, 2009.

82 Staff, "Going Green Could Save Government $1 Billion in Five Years," *TechWeb,* January 22, 2008.

83 *www.energystar.gov/index.cfm?fuseaction=find_a_product,* accessed August 1, 2009.

84 Staff, "Ohm Economics," *Forbes,* February 2, 2009, p. 34.

85 Worthen, Ben, "Oracle Targets a New Rival: IBM," *The Wall Street Journal,* October 15, 2009, p. B4.

86 Staff, "The Infotech 100," *BusinessWeek,* June 1, 2009, p. 39.

87 Conry-Murray, A., "Engines of Innovation," *Information Week,* April 20, 2009, p. 30.

88 Poston, R., et al, "Managing the Vendor Set," *MIS Quarterly Executive,* June 2009, p. 45.

89 Wingfield, Nick, "Microsoft Seeks to Take A Bite Out Of Apple With New Stores," *The Wall Street Journal,* October 15, 2009, p. B1.

90 Hoover, Nicholas, "GE Puts the Cloud Model to the Test," *Information Week,* April 13, 2009, p. 32.

91 Babcock, Charles, "Red Hat to Certify Cloud-Ready Applications," *Information Week,* July 6, 2009, p. 14.

92 Hahn, J., et al, "Emergence of New Project Teams from Open Source Software Developer Networks," *Information Systems Research,* September 2008, p. 369.

93 Crosman, Penny, "SaaS Gains Street Traction," *Wall Street & Technology,* September 1, 2008, p. 37.

94 IT Redux Web site, *http://itredux.com/office-20/database/?family=Database,* accessed August 1, 2009.

95 Lai, Eric, "Cloud database vendors: What, us worry about Microsoft?" *Computerworld*, March 12, 2008, *www.computerworld.com/action/article.do?command=viewArticleBasic&articleId=9067979&pageNumber=1* .

96 Weier, Mary, "Wal-Mart Change Hints at Data-Driven Marketing," *Information Week,* July 20, 2009, p. 10.

97 Foley, John, "Gold in the Clouds," *Information Week,* September 28, 2009, p. 24.

98 Babcock, "Charles, "Platform As A Service," *Information Week,* October 5, 2009, p. 18.

99 Vuong, Andy, "Denver Developers Hatch iPhone Apps," *The Summit Daily News,* September 12, 2009, p. A19.

100 Hoover, N., "Tough Call," *Information Week,* February 23, 2009, p. 21.

101 Morrison, Scott, "Co-Op Field-Tests Software," *The Wall Street Journal,* October 28, 2009, p. B5A.

102 Hoover, N. and Foley, J., "Feds on the Edge," *Information Week,* July 6, 2009, p. 19.

103 Hodgson, J., "Rethinking Software Support," *The Wall Street Journal,* March 12, 2009, p. B8.

104 Weier, M., "Numbers Crunch," *Information Week,* January 26, 2009, p. 25.

105 Hoffman, Thomas, "Railroad Crossing," *Computerworld,* December 21, 2009, p. 22.

106 Hoover, N. and Foley, J., "Feds on the Edge," *Information Week,* July 6, 2009, p. 19.

107 The Idea factory, *www.whitehouse.gov/open/innovations/IdeaFactory,* accessed August 21, 2009.

108 Kelly, C., "Getting an F and Turning It Into Fun," *Computerworld,* May 26, 2008, p. 32.

109 Weier, Mary, "Comcast Team Tweets to Track, Douse Flames," *Information Week,* March 30, 2009, p. 20.

110 *www.ibm.com/software/tivoli/products,* accessed August 1, 2009.

111 *www.precise.com,* accessed August 1, 2009.

112 *https://h10078.www1.hp.com/cda/hpms/display/main/hpms_home.jsp?zn=bto&cp=1_4011_100__,* accessed August 1, 2009.

Information Systems in Business and Society

Chapter 9 The Personal and Social Impact of Computers

CHAPTER
· 9 ·

The Personal and Social Impact of Computers

PRINCIPLES	LEARNING OBJECTIVES
▪ Policies and procedures must be established to avoid waste and mistakes associated with computer usage.	▪ Describe some examples of waste and mistakes in an IS environment, their causes, and possible solutions. ▪ Identify policies and procedures useful in eliminating waste and mistakes. ▪ Discuss the principles and limits of an individual's right to privacy.
▪ Computer crime is a serious and rapidly growing area of concern requiring management attention.	▪ Explain the types of computer crime and their effects. ▪ Identify specific measures to prevent computer crime.
▪ Jobs, equipment, and working conditions must be designed to avoid negative health effects from computers.	▪ List the important negative effects of computers on the work environment. ▪ Identify specific actions that must be taken to ensure the health and safety of employees.
▪ Practitioners in many professions subscribe to a code of ethics that states the principles and core values that are essential to their work.	▪ Outline criteria for the ethical use of information systems.

Information Systems in the Global Economy
Facebook, United States
Balancing Profits and Privacy

Facebook has grown to be one of the most popular and influential online businesses. The enormously successful online social network has roughly half a billion members and more than a billion dollars in annual revenue, earned primarily from advertisements. Facebook's slogan is "Giving people the power to share and make the world more open and connected." Facebook's mission, combined with its popularity and influence, has often put the company in the hot seat when it comes to privacy and security concerns. Any lapses in ethical judgment on Facebook's part could negatively affect a great number of people and at the same time threaten the company's future.

Facebook walks a fine line between its mission—making the world more open and connected—and its obligation to protect the privacy of its members. Facebook's mission is largely influenced by the financial benefits it enjoys when it can share its users' information with the companies that advertise on its site. The more "open" Facebook users are with their information, the more money Facebook makes. The personal information that users store on Facebook is a gold mine for marketers and advertisers. Facebook users provide a detailed profile including their interests, likes and dislikes, opinions, day-to-day activities and thoughts, photos, lists of friends, and product purchase information. Combined, this information can provide businesses with valuable insight into the hearts and minds of Facebook members, which allows companies to pitch their products more effectively.

In 2010, Facebook rolled out new technology called Open Graph, which provides businesses with the ability to include Facebook "Like" links on Web pages of their products. When a Facebook user clicks the "Like" link on a product page, a message about the product is posted to the user's Facebook news feed, and the connection is stored in the user's Facebook profile. Open Graph provides Facebook users with the ability to conveniently express their interests and product loyalties online, share that information with friends, and build communities around products. However, privacy advocates view the technology as a serious threat to user privacy, because participating businesses can use the technology to more easily access users' profile information.

In early 2010, Facebook leaned too far toward sharing information at the risk of upsetting its users over privacy infringements. At the height of the Open Graph backlash, more than half of Facebook users polled indicated that they were considering deleting their Facebook accounts. Facebook responded by providing privacy controls that made it easier for users to indicate to Facebook to keep their information private. Still, privacy groups are not satisfied and have continuously appealed to Facebook to make all user information private unless the user "opts in" to sharing. Facebook has refused these appeals, arguing that agreeing would reduce the user data that it has access to and deeply cut into its profits. Thus, Facebook continuously teeters between privacy and profits, working to keep members content and to keep its profits increasing.

Facebook is not alone in its balancing act. All online businesses that profit from advertising are in the same situation, including Google, Yahoo!, MySpace, and many others. In fact, all businesses that collect member or customer private information have ethical and legal responsibilities for keeping that information safe, secure, and private. For example, Facebook must have a user's consent to share information with third parties. This is granted when a user creates a Facebook account and agrees to the terms of service. Facebook must also take measures to ensure that the user information it stores is safe from hackers. Because it stores so much valuable information, Facebook has become a target of all types of hacker attacks and schemes. Microblogging has similar concerns and filters all URLs referenced by its users to make sure the referenced sites are safe.

Facebook serves as the ultimate example of the pressures that businesses experience regarding the information they collect. These pressures come from a variety of sources. There is pressure from users or customers when they threaten to leave due to dissatisfaction with business practices. There is also pressure from local, state, and federal laws. For international businesses, this extends to the laws of all of the countries in which they do business. There is also the pressure of ethical standards to which a business holds itself and the ethical demands of special interest groups such as privacy advocates. Some of these pressures could be greatly reduced through the application of ethical standards to which a business could subscribe. Movements such as the Data Portability Project (*www.portabilitypolicy.org*) are pushing for standards to embrace what a business can and cannot do with the information it collects from its customers.

The issues addressed in this chapter are of growing importance to businesses and individuals. The valuable information that businesses store in Internet-connected databases is continuously targeted by hackers and swindlers. Businesses must maintain high ethical standards in the manner in which they treat that information and wisdom in the manner in which they secure that information in order to gain the trust of their customers.

As you read this chapter, consider the following:

- What are the primary concerns of corporations regarding security, privacy, and ethics?
- What strategies can assist a company with issues of security and privacy, and at what cost?

Why Learn About the Personal and Social Impact of the Internet?

Both opportunities and threats surround a wide range of nontechnical issues associated with the use of information systems and the Internet. The issues span the full spectrum—from preventing computer waste and mistakes, to avoiding violations of privacy, to complying with laws on collecting data about customers, to monitoring employees. If you become a member of a human resources, information systems, or legal department within an organization, you will likely be charged with leading the organization in dealing with these and other issues covered in this chapter. Also, as a user of information systems and the Internet, it is in your own self-interest to become well versed on these issues. You need to know about the topics in this chapter to help avoid becoming a victim of crime, fraud, privacy invasion, and other potential problems. This chapter begins with a discussion of preventing computer waste and mistakes.

Earlier chapters detailed the significant benefits of computer-based information systems in business, including increased profits, superior goods and services, and higher quality of work life. Computers have become such valuable tools that today's businesspeople have difficulty imagining work without them. Yet the information age has also brought the following potential problems for workers, companies, and society in general:

- Computer waste and mistakes
- Computer crime
- Privacy issues
- Work environment problems
- Ethical issues

This chapter discusses some of the social and ethical issues as a reminder of these important considerations underlying the design, building, and use of computer-based information systems. No business organization, and, hence, no information system, operates in a vacuum. All IS professionals, business managers, and users have a responsibility to see that the potential consequences of IS use are fully considered. Even entrepreneurs, especially those who use computers and the Internet, must be aware of the potential personal and social impact of computers.

Managers and users at all levels play a major role in helping organizations achieve the positive benefits of IS. These people must also take the lead in helping to minimize or eliminate the negative consequences of poorly designed and improperly utilized information systems. For managers and users to have such an influence, they must be properly educated. Many of the issues presented in this chapter, for example, should cause you to think back to some of the systems design and systems control issues discussed previously. They should also help you look forward to how these issues and your choices might affect your future use of information systems.

COMPUTER WASTE AND MISTAKES

Computer-related waste and mistakes are major causes of computer problems, contributing as they do to unnecessarily high costs and lost profits. Computer waste involves the inappropriate use of computer technology and resources. Computer-related mistakes refer to errors, failures, and other computer problems that make computer output incorrect or not useful; most of these are caused by human error. This section explores the damage that can be done as a result of computer waste and mistakes.

Computer Waste

The U.S. government is the largest single user of information systems in the world. It should come as no surprise, then, that the U.S. government also generates more waste than other organizations. For example, poorly designed information systems prevent the government from making sure recipients qualify before they receive a government payment; therefore, the government is creating a "do not pay list" to ensure that government checks are no longer sent to people who do not qualify. During 2009, almost $110 billion in federal funds were paid to the wrong person or for the wrong reason, including payments to 20,000 deceased people. Federal agencies will check the new list to ensure that the payees are in "good standing" (and alive!) before issuing a check.[1] Most observers wonder, "Why haven't we been doing this all along?"[2]

The government is not unique in this regard—the same type of waste and misuse found in the public sector also exists in the private sector. Some companies discard old software and computer systems when they still have value. Others waste corporate resources to build and maintain complex systems that are never used to their fullest extent.

A less dramatic, yet still relevant, example of waste is the amount of company time and money employees can waste playing computer games, sending unimportant e-mail, or browsing the Internet. Junk e-mail, or *spam*, also causes waste. People receive hundreds of e-mail messages advertising products and services not wanted or requested. Not only does this waste time, but it also wastes paper and computer resources. Worse yet, spam messages often carry attached files with embedded viruses that can cause networks and computers to crash or allow hackers to gain unauthorized access to systems and data.

Spam is considered a serious enough problem that the U.S. Department of Justice has prosecuted prolific spammers for violation of the CAN-SPAM Act. This law allows the sending of most e-mail spam as long as it adheres to three basic forms of compliance related to how receivers can unsubscribe, content of the e-mail, and the sending behavior. Since the act was passed in 2003, several people have been convicted and sentenced to up to six years in jail or up to a $1 million fine.[3]

A spam filter is software that attempts to block unwanted e-mail. One approach to filtering spam involves building lists of acceptable and unacceptable e-mail addresses. The lists can be created manually or automatically based on how the users keep or discard their e-mail. Another approach is automatic rejection of e-mail based on the content of the message or the appearance of keywords in the message. Rejected e-mail automatically goes to the spam

THE COMPUTER AS A TOOL TO COMMIT CRIME

A computer can be used as a tool to gain access to valuable information and as the means to steal thousands or millions of dollars. It is, perhaps, a question of motivation—many people who commit computer-related crime claim they do it for the challenge, not for the money. Credit card fraud—whereby a criminal illegally gains access to another's line of credit with stolen credit card numbers—is a major concern for today's banks and financial institutions. In general, criminals need two capabilities to commit most computer crimes. First, the criminal needs to know how to gain access to the computer system. Sometimes, obtaining access requires knowledge of an identification number and a password. Second, the criminal must know how to manipulate the system to produce the desired result. Frequently, a critical computer password has been talked out of a person, a practice called **social engineering**. Or, the attackers simply go through the garbage—**dumpster diving**—for important pieces of information that can help crack the computers or convince someone at the company to give them more access. In addition, more than 2,000 Web sites offer the digital tools—often without charge—that will let people snoop, crash computers, hijack control of a machine, or retrieve a copy of every keystroke. While some of the tools were intended for legitimate use to provide remote technical support or monitor computer usage, hackers take advantage of them to gain unauthorized access to computers or data.

Chris Nickerson is a master at employing social engineering techniques with the nerve to attempt exploits criminals might avoid. Nickerson has a legitimate job serving as a security expert and helping organizations test their security defenses. Businesses hire him to evaluate their vulnerability to social engineering crimes. Two of his jobs involved using social engineering techniques to gain access to an expensive jewelry store and car dealership. He was able to "acquire" expensive merchandise, including an exotic sports car.[20]

Cyberterrorism

Cyberterrorism has been a concern for countries and companies around the globe. The U.S. government considered the potential threat of cyberterrorism serious enough that in February 1998 it established the National Infrastructure Protection Center. This function was later transferred to the Homeland Security Department's Information Analysis and Infrastructure Protection Directorate to serve as a focal point for threat assessment, warning, investigation, and response for threats or attacks against the country's critical infrastructure, which provides telecommunications, energy, banking and finance, water systems, government operations, and emergency services. Successful cyberattacks against the facilities that provide these services could cause widespread and massive disruptions to the normal function of American society.

A **cyberterrorist** is someone who intimidates or coerces a government or organization to advance his political or social objectives by launching computer-based attacks against computers, networks, and the information stored on them.

International Multilateral Partnership Against Cyber Terrorism (IMPACT) is a global initiative against cyberterrorism. Its mission is to "bring together governments, academia, and cybersecurity experts to enhance the global community's capacity to prevent, defend against, and respond to cyber threats." It is headquartered in Malaysia's new planned township of science—a city named Cyberjaya.[21]

CIA Director Leon Panetta states that the U.S. must thwart thousands of attacks each day that come from China, Iran, Russia, and other unaffiliated attackers. He believes that "the next Pearl Harbor is likely to be a cyber attacker going after our grid ... and that can literally cripple this country."[22] According to FBI Director Robert Mueller, cyberterrorists "have executed numerous denial-of-service attacks. And they have defaced numerous Web sites, including Congress' Web site following President Obama's State of the Union speech."[23] Cyberterrorists from China, Russia, and other countries are believed to have invaded the U.S. power grid and planted software that could interrupt the system during a time of crisis or war.[24]

social engineering
Using social skills to get computer users to provide information that allows a hacker to access an information system or its data.

dumpster diving
Going through the trash cans of an organization to find secret or confidential information, including information needed to access an information system or its data.

cyberterrorist
Someone who intimidates or coerces a government or organization to advance his political or social objectives by launching computer-based attacks against computers, networks, and the information stored on them.

Identity Theft

Identity theft is a crime in which an imposter obtains key pieces of personal identification information, such as Social Security or driver's license numbers, to impersonate someone else. The information is then used to obtain credit, merchandise, and/or services in the name of the victim or to provide the thief with false credentials. The perpetrators of these crimes employ such an extensive range of methods that investigating them is difficult.

In some cases, the identity thief uses personal information to open new credit accounts, establish cellular phone service, or open a new checking account to obtain blank checks. In other cases, the identity thief uses personal information to gain access to the person's existing accounts. Typically, the thief changes the mailing address on an account and runs up a huge bill before the person whose identity has been stolen realizes there is a problem. The Internet has made it easier for an identity thief to use the stolen information because transactions can be made without any personal interaction.

More than 6 million customers of online brokerage firm TD Ameritrade were involved in a class action lawsuit resulting from a data theft that revealed customers' Social Security numbers, e-mail addresses, and account numbers. Many of the customers then received unsolicited e-mail ads that attempted to manipulate the prices of certain thinly traded stocks.[25]

Ben Bernanke, chairman of the Federal Reserve Board, and his wife, Anna, were one of hundreds of victims of an identity theft ring that used a combination of high-tech fraud and low-tech thievery to steal more than $2 million from the bank accounts of its unsuspecting victims. The thieves stole Anna Bernanke's purse containing the couple's joint checkbook and cashed several checks on the account, and then used their identities to make fraudulent deposits.[26]

Internet Gambling

Many people enjoy Internet gambling as a recreational and leisure activity. Baccarat, bingo, blackjack, pachinko, poker, roulette, and sports betting are all readily available online. The size of the online gambling market is not known, but one estimate is that Americans wager tens of billions of dollars each year in online games run by offshore operators.[27]

The revenues generated by Internet gambling represent a major untapped source of income for the state and federal governments. Representative Jim McDermott, who has offered a bill to levy federal and state taxes on Internet gambling, estimates the bill would generate $42 billion in new tax revenue over the next 10 years. However, many believe that such taxes are regressive in that they fall heaviest on those least able to pay them. This position is based on a study that showed that while people of all income levels played state lottery games, those people with an annual income of less than $10,000 spent nearly three times as much on gambling as those with annual income more than $50,000.[28]

THE COMPUTER AS A TOOL TO FIGHT CRIME

The computer is also used as a tool to fight computer crime. The computer and information systems are used to help in the recovery of stolen property, monitoring of sex offenders, and to better understand and diminish crime risks.

Recovery of Stolen Property

The Leads Online Web-based service system is used by law enforcement to recover stolen property. The system contains more than 250 million records in its database. Data is entered into the system from pawn brokers, secondhand dealers, and salvage yards. In some areas, state or local laws require that all such businesses register (with no charge for business owners) with LeadsOnline. The system allows law enforcement officers to search the database by item serial number or by individual. The system even has a partnership with eBay that makes it possible to locate possible stolen merchandise that has been listed for sale or sold online.[29]

The city of Rockford, Illinois uses the Leads Online system; one officer states: "We have made several arrests based on the information that has been provided to us and actually are on our way to solving some burglaries in the Wisconsin area" [as a result of using this system].[30]

Monitoring Sex Offenders

Offender Watch is a Web-based system used to track registered sex offenders. It stores the registered offender's address, physical description, and vehicle information. The public can access the information at *www.communitynotification.com*.[31] The information available varies depending on the county and state. For example, in Hamilton County, Ohio, the data is provided by the sheriff's department and allows the user to search for registered sex offenders by township, school district, zip code, or within one mile of an entered address. The information displayed includes a photo of all registered sex offenders, their description, and current addresses. Law enforcement agencies can search the database based on full or partial license plate number or vehicle description.

GPS tracking devices and special software are also used to monitor the movement of registered sex offenders. Clicking a name on the screen displays that individual's recent movements. Unfortunately, due to staffing limitations, law enforcement agents cannot continuously track offenders as they move about. The time required for continuous monitoring is too great, even though people have done terrible things while under surveillance—either cutting off the bracelet or doing crimes knowing they could be connected to the crime scene.[32]

Use of Geographic Information Systems

The ready availability of personal computers, coupled with the development of mapping and analysis software, has led law enforcement agencies to use crime-related data, powerful analysis techniques, and Geographic Information Systems (GIS) to better understand and even diminish crime risks. The use of such software enables law enforcement agencies to gain a quick overview of crime risk at a given address or in a given locale, as shown in Figure 9.1.

Figure 9.1

Mapping Crime Risk

CAP Index CRIMECAST reports provide a quick and thorough overview of the crime risk at any given address. A detailed map and spreadsheet of risk scores isolate and identify crime-related issues around a specific site (in this example is 545 Washington Blvd., Jersey City, N.J.).

(Source: Shillingford, David and Groussman, Jon D., "Using GIS to Fight Crime," *ISO Review*, March 2010.)

Current Scores (2009)	National	State	County
CAP Index	331	306	145
Homicide	319	316	190
Rape	83	148	173
Robbery	384	320	141
Aggravated assault	402	379	161
Crimes against persons	382	357	158
Burglary	48	114	130
Larceny	836	1248	697
Motor vehicle theft	123	150	89
Crimes against property	638	1046	613

CAP Index	National	State	County
Past – 2000	246	208	105
Present – 2009	331	306	145
Projected – 2014	319	322	155

CRIMECAST scores range from 0 to 2000 and indicate the risk of crime at a site compared to an average of 100. A score of 400 means that the risk is four times the average, and a score of 50 means the risk is half the average.

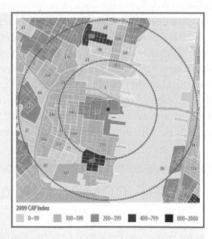

With GIS tools, law enforcement agencies can analyze crime data relative to other factors, including the locations of common crime scenes (such as convenience stores and gas stations) and certain demographic data (such as age and income distribution). Common GIS systems include the following:[33]

- The National Equipment Registry maps mobile equipment thefts in areas where peak equipment thefts have occurred so police and equipment owners can take appropriate action.
- The CompStat program uses GIS software to map crime and identify problem precincts. The program reduced crime in New York City in the 1990s and continues to be used in Baltimore, Los Angeles, and Philadelphia.
- CargoNet helps law enforcement and the transportation industry map cargo crimes and identify patterns.

THE COMPUTER AS THE OBJECT OF CRIME

A computer can also be the object of the crime, rather than the tool for committing it. Tens of millions of dollars worth of computer time and resources are stolen every year. Each time system access is illegally obtained, data or computer equipment is stolen or destroyed, or software is illegally copied, the computer becomes the object of crime. These crimes fall into several categories: illegal access and use, data alteration and destruction, information and equipment theft, software and Internet piracy, computer-related scams, and international computer crime.

Table 9.1

Common Methods Used to Commit Computer Crimes

Methods	Examples
Add, delete, or change inputs to the computer system.	Delete records of absences from class in a student's school records.
Modify or develop computer programs that commit the crime.	Change a bank's program for calculating interest so it deposits rounded amounts in the criminal's account.
Alter or modify the data files used by the computer system.	Change a student's grade from C to A.
Operate the computer system in such a way as to commit computer crime.	Access a restricted government computer system.
Divert or misuse valid output from the computer system.	Steal discarded printouts of customer records from a company trash bin.
Steal computer resources, including hardware, software, and time on computer equipment.	Make illegal copies of a software program without paying for its use.
Offer worthless products for sale over the Internet.	Send e-mails requesting money for worthless hair growth product.
Blackmail executives to prevent release of harmful information.	Eavesdrop on an organization's wireless network to capture competitive data or scandalous information.
Blackmail company to prevent loss of computer-based information.	Plant a logic bomb and send a letter threatening to set it off unless paid a considerable sum.

Illegal Access and Use

Crimes involving illegal system access and use of computer services are a concern to both government and business. Since the outset of information technology, computers have been

hacker
A person who enjoys computer technology and spends time learning and using computer systems.

criminal hacker (cracker)
A computer-savvy person who attempts to gain unauthorized or illegal access to computer systems to steal passwords, corrupt files and programs, or even transfer money.

script bunny
A cracker with little technical savvy who downloads programs called scripts, which automate the job of breaking into computers.

insider
An employee, disgruntled or otherwise, working solo or in concert with outsiders to compromise corporate systems.

plagued by criminal hackers. Originally, a **hacker** was a person who enjoyed computer technology and spent time learning and using computer systems. A **criminal hacker**, also called a **cracker**, is a computer-savvy person who attempts to gain unauthorized or illegal access to computer systems to steal passwords, corrupt files and programs, or even transfer money. In many cases, criminal hackers are people who are looking for excitement—the challenge of beating the system. Today, many people use the term hacker and cracker interchangeably. **Script bunnies** admire crackers but have little technical savvy. They are crackers who download programs called *scripts* that automate the job of breaking into computers. **Insiders** are employees, disgruntled or otherwise, working solo or in concert with outsiders to compromise corporate systems. The biggest threat for many companies is not external hackers, but their own employees. Insiders have extra knowledge that makes them especially dangerous—they know logon IDs, passwords, and company procedures that help them evade detection.

Catching and convicting criminal hackers remains a difficult task. Although the method behind these crimes is often hard to determine, even if the method is known, tracking down the criminals can take a lot of time. It took five years for the FBI to arrest and convict computer hacker Albert Gonzalez for leading a group of cybercriminals who hacked into the computer systems of major retailers and stole more than 90 million credit and debit card numbers. The loss to the various companies, banks, and insurers was estimated to be almost $200 million. Gonzalez was sentenced to 20 years in prison.[34]

Data and information are valuable corporate assets. The intentional use of illegal and destructive programs to alter or destroy data is as much a crime as destroying tangible goods. The most common of these programs are viruses and worms, which are software programs that, when loaded into a computer system, will destroy, interrupt, or cause errors in processing. Such programs are also called *malware*, and the growth rate for such programs is epidemic. It is estimated that hundreds of previously unknown viruses and worms emerge each day. Table 9.2 describes the most common types of malware.

Table 9.2

Common Types of Computer Malware

Type of malware	Description
Logic bomb	A type of Trojan horse that executes when specific conditions occur. Triggers for logic bombs can include a change in a file by a particular series of keystrokes or at a specific time or date.
Rootkit	A set of programs that enables its user to gain administrator level access to a computer or network. Once installed, the attacker can gain full control of the system and even obscure the presence of the rootkit from legitimate system administrators.
Trojan horse	A malicious program that disguises itself as a useful application or game and purposefully does something the user does not expect.
Variant	A modified version of a virus that is produced by the virus's author or another person by amending the original virus code.
Virus	Computer program file capable of attaching to disks or other files and replicating itself repeatedly, typically without the user's knowledge or permission.
Worm	Parasitic computer program that replicates, but unlike viruses, does not infect other computer program files. A worm can send the copies to other computers via a network.

In some cases, a virus or a worm can completely halt the operation of a computer system or network for days until the problem is found and repaired. In other cases, a virus or a worm can destroy important data and programs. If backups are inadequate, the data and programs might never be fully functional again. The costs include the effort required to identify and neutralize the virus or worm and to restore computer files and data, as well as the value of business lost because of unscheduled computer downtime.

Criminal hackers used a clever Trojan horse program that, when installed on a victim's computer, altered the text displayed by HTML code to either erase evidence of a money transfer transaction entirely or alter the amount of the money transfer and bank balances. The gang used this Trojan horse to steal more than €300,000 over three weeks from online customers of a German bank.[35]

Smartphones such as the iPhone and Google's Android phone that can run applications are also susceptible to malware.[36] Cybercriminals are distributing applications for smartphones that contain malware programmed to make calls to premium rate phone numbers around the globe and run up large phone bills without the owner's knowledge.[37]

Spyware

Spyware is software installed on a personal computer to intercept or take partial control over the user's interaction with the computer without the knowledge or permission of the user. Some forms of spyware secretly log keystrokes so that user names and passwords may be captured. Other forms of spyware record information about the user's Internet surfing habits and sites that have been visited. Still other forms of spyware change personal computer settings so that the user experiences slow connection speeds or is redirected to Web pages other than those expected. Spyware is similar to a Trojan horse in that users unknowingly install it when they download freeware or shareware from the Internet.

The IE Antivirus spyware is a fake spyware remover that scares unwary computer users into buying it. It uses pop-up windows and fake system notifications to tell the user his computer is infected and that he needs to buy IE Antivirus. If the user inquires further, the computer is scanned without the program asking permission. The results of the scan are faked to show an alarming number of security risks. The user is then directed to a Web site that explains how to order the bogus antivirus software.

Government officials in India suspect China of infecting several computers with spyware in a section of the Indian Ministry of External Affairs that deals with sensitive Pakistani affairs. The spyware caused e-mails to be secretly sent to a third-party e-mail account. India claims the attacks are part of an attempt to map India's network infrastructure in preparation of future attacks designed to disable or disrupt it during a conflict.[38]

Information and Equipment Theft

Data and information are assets or goods that can also be stolen. People who illegally access systems often do so to steal data and information. To obtain illegal access, criminal hackers require identification numbers and passwords. Some criminals try various identification numbers and passwords until they find ones that work. Using password sniffers is another approach. A **password sniffer** is a small program hidden in a network or a computer system that records identification numbers and passwords. In a few days, a password sniffer can record hundreds or thousands of identification numbers and passwords. Using a password sniffer, a criminal hacker can gain access to computers and networks to steal data and information, invade privacy, plant viruses, and disrupt computer operations.

In addition to theft of data and software, all types of computer systems and equipment have been stolen from offices. Portable computers such as laptops and portable storage devices (and the data and information stored in them) are especially easy for thieves to take. For example, a laptop computer containing a CD-ROM with the names, addresses, and Social Security numbers of more than 207,000 U.S. Army reservists was stolen from the offices of a government contractor.[39]

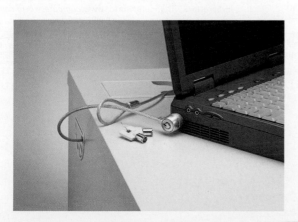

spyware
Software that is installed on a personal computer to intercept or take partial control over the user's interaction with the computer without knowledge or permission of the user.

password sniffer
A small program hidden in a network or a computer system that records identification numbers and passwords.

To fight computer crime, many companies use devices that disable the disk drive or lock the computer to the desk.

(Source: Courtesy of Kensington Computer Products Group.)

Data Theft: An Ongoing Concern for Businesses

Educational Credit Management Corp. (ECMC) is a nonprofit organization that assists with managing student loans for borrowers that enter bankruptcy. In March of 2010, the organization discovered that the tapes on which they archive their data had been stolen. The data that was on those tapes included the names, addresses, birth dates, and Social Security numbers of 3.3 million borrowers—a wealth of information that would allow the thieves to assume the borrowers' identities for considerable financial gain. ECMC notified the affected individuals and provided them with free credit monitoring to ensure that their credentials were not used illegally.

While the ECMC theft was impressive in its scope, it wasn't the largest data theft ever made. In 2006, a laptop and hard drive were stolen from the U.S. Department of Veterans Affairs containing the personal information of 26.5 million military veterans and their spouses. In 2008, the largest data theft occurred when 130 million credit card numbers were stolen from Heartland Payment System's servers. Heartland processes credit and debit card transactions for Visa, American Express, and other businesses. The data was stolen by hackers who infiltrated Heartland networks to gain access to the servers on which the data was stored.

As previously mentioned, computer hacker Albert Gonzalez was arrested for participating in the cybercrime ring that attacked Heartland and many others. The judge solicited valuable information about international cybercrime from Gonzalez prior to convicting him to 20 years in prison. Gonzalez informed the judge that international cybercrime rings have progressed from attacking individual businesses to attacking banks and organizations that handle large amounts of financial transaction data, such as Heartland.

Besides stealing the physical media on which data is stored and hacking into networks, data thieves also use collections of compromised computers called botnet armies to do their dirty work. For example, the Mariposa botnet, which was eventually dismantled in 2010, used 12.7 million infected PCs to steal credit card and bank account information. The botnet was shut down through a series of arrests in Spain. The Kneber botnet runs on infected computers spread across 126 countries and is designed to steal logon credentials for corporations, financial systems, and popular social networking and e-mail sites.

Often times, the data stolen in one attack is used to launch another attack or scam. For example, insurance company Aetna contacted 65,000 current and former employees letting them know that their Social Security numbers, e-mail addresses, and other private information may have been stolen from an external vendor's system. The company became aware of the theft when employees began receiving e-mails that referenced the stolen data.

Businesses aren't the only targets of data theft. Many colleges have been victimized as well. In 2009, the University of Florida-Gainesville suffered three attacks within three months. The final attack exposed the data of more than 97,000 students, faculty, and staff.

The Internet has become a major platform for criminal activity. Financial account information and personal information used for identity theft are often the goal of attacks, allowing hackers to rake in millions of dollars on the underground market. Hackers have many methods of attack, and no Internet-connected server can be considered 100 percent safe. It remains the responsibility of businesses to stay abreast of the latest security holes and patches, and to stay vigilant and watchful to react quickly when a data breach does occur. The best reaction to a data breach is immediate contact with those affected along with an offer to protect them from financial hardship.

Discussion Questions

1. What methods do hackers and thieves use to illegally access valuable information?
2. What is the best response for a business that has discovered private customer information stolen?

Critical Thinking Questions

1. What policies can financial institutions and governments put in place to protect consumers from data theft?
2. What practices can consumers use to help protect their own private information?

SOURCES: Kirk, Jeremy, "Company says 3.3M student loan records stolen," Computerworld, *www.computerworld.com*, March 29, 2010; Weil Nancy, "Gonzalez sentenced to 20 years for Heartland break-in," Computerworld, *www.computerworld.com*, March 26, 2010; Bright, Peter, "Spanish arrests mark the end of dangerous botnet," *Ars Technica, www.arstechnica.com*, March 20, 2010; Vijayan, Jaikumar, "Over 75,000 systems compromised in cyberattack," Computerworld, *www.computerworld.com*, February 18, 2010; Kirk, Jeremy, "Aetna warns 65,000 about Web site data breach," Computerworld, *www.computerworld.com*, May 28, 2009; Vijayan, Jaikumar, "Three months, three breaches at the Univ. of Florida-Gainesville," Computerworld, *www.computerworld.com*, February 22, 2009.

In many cases, the data and information stored in these systems are more valuable than the equipment. Vulnerable data can be used in identity theft. In addition, the organization responsible receives a tremendous amount of negative publicity that can cause it to lose existing and potential future customers. Often, the responsible organization offers to pay for credit monitoring services for those people affected in an attempt to restore customer goodwill and avoid law suits.

Safe Disposal of Personal Computers

Many companies donate personal computers they no longer need to schools, churches, or other organizations. Some sell them at a deep discount to their employees or put them up for sale on Internet auction sites such as eBay. However, care must be taken to ensure that all traces of any personal or company confidential data is completely removed. Simply deleting files and emptying the Recycle Bin does not make it impossible for determined individuals to view the data. Be sure to use disk-wiping software utilities that overwrite all sectors of your disk drive, making all data unrecoverable. For example, Darik's Boot and Nuke (DBAN) is free and can be downloaded from the SourceForge Web site.

Acer, Apple, Asus, Dell, Fujitsu, HP, Lenovo, Sony, and Toshiba all offer some sort of recycling program. The program may exchange old computers for credits toward the purchase of a new computer. Other programs may provide a simple pick-up for no or a very small fee.

Patent and Copyright Violations

Works of the mind, such as art, books, films, formulas, inventions, music, and processes that are distinct and "owned" or created by a single person or group, are called intellectual property. Copyright law protects authored works such as art, books, film, and music. Patent laws protect processes, machines, objects made by humans or machines, compositions of matter, and new uses of these items. Software is considered intellectual property and may be protected by copyright or patent law.

Software piracy is the act of unauthorized copying, downloading, sharing, selling, or installing of software. When you purchase software, you are purchasing a license to use it; you do not own the actual software. The license states how many times you can install the software. If you make more copies of the software than the license permits, you are pirating.

software piracy
The act of unauthorized copying or distribution of copyrighted software.

The Business Software Alliance (BSA) has become a prominent software antipiracy organization. Software companies, including Apple, Adobe, Hewlett-Packard, IBM, and Microsoft, contribute to the BSA. The BSA estimates that the 2009 global software piracy rate was 43 percent and amounted to $51 billion in lost sales.[40] Georgia, Zimbabwe, and Moldova have piracy rates exceeding 90 percent, while the U.S. has the lowest software piracy rate at just 20 percent.[41]

Penalties for software piracy can be severe. If the copyright owner brings a civil action against someone, the owner can seek to stop the person from using its software immediately and can also request monetary damages. The copyright owner can then choose between compensation for actual damages—which includes the amount it has lost because of the person's infringement, as well as any profits attributable to the infringement—and statutory damages, which can be as much as $150,000 for each program copied. In addition, the government can prosecute software pirates in criminal court for copyright infringement. If convicted, they could be fined up to $250,000 or sentenced to jail for up to five years, or both.

Another major issue in regards to copyright infringement is the downloading of music that is copyright protected. Estimates vary widely as to how much music piracy is costing the recording industry. One estimate from the Institute for Policy Innovation (an economic public policy organization that is a global partner of the Recording Industry Association of America) is that the global recording industry loses about $12.5 billion in revenue from music piracy every year.[42] However, independent research firm Jupiter Research believes that the losses are much lower.[43]

LimeWire was found guilty of copyright infringement in a case brought against it by the Recording Industry Association of America (RIAA). LimeWire boasts of 50 million unique monthly customers and downloads in excess of hundreds of thousands per day. At a maximum of $150,000 per copyright violation, LimeWire faces fines in excess of $1 billion.[44, 45]

In what represents the largest copyright infringement case in history, a class action lawsuit was filed on behalf of hundreds of authors against Google for copying and distributing their work without permission. Google has digitized millions of books to create a digital library and plans to provide Internet access to this vast storehouse of books. In setting up this storehouse, Google has asked that authors "opt out" of the program to allow access to their works rather than "opt in."[46]

Patent infringement is also a major problem for computer software and hardware manufacturers. It occurs when someone makes unauthorized use of another's patent. If a court determines that a patent infringement is intentional, it can award up to three times the amount of damages claimed by the patent holder. It is not unusual to see patent infringement awards in excess of $10 million.

To obtain a patent or to determine if a patent exists in an area a company seeks to exploit requires a search by the U.S. Patent Office; these can last longer than 25 months. Indeed, the patent process is so controversial that manufacturing firms, the financial community, consumer and public interest groups, and government leaders are demanding patent reform. One area that needs to be addressed is patent infringement. As Red Hat vice president and assistant general counsel Rob Tiller says, "The cost of software patent litigation since 1994 has vastly exceeded the profit generated by software patent licensing."[47]

HTC Corporation, maker of powerful mobile handsets, filed a complaint with the International Trade Commission (an independent U.S. federal agency that rules on cases involving imports that allegedly infringe intellectual property rights) against its competitor Apple. HTC claimed Apple violated five of its patents concerning cell phone directory hardware and software plus power management technology in portable devices. This came just a month after Apple filed patent infringement suits against HTC.[48]

Computer-Related Scams

People have lost hundreds of thousands of dollars on real estate, travel, stock, and other business scams. Today, many of these scams are being perpetrated with computers. Using the Internet, scam artists offer get-rich-quick schemes involving bogus real estate deals, tout "free" vacations with huge hidden costs, commit bank fraud, offer fake telephone lotteries, sell worthless penny stocks, and promote illegal tax-avoidance schemes.

Over the past few years, credit card customers of various banks have been targeted by scam artists trying to get personal information needed to use their credit cards. The scam works by sending customers an e-mail including a link that seems to direct users to their bank's Web site. At the site, they are greeted with a pop-up box asking them for their full debit card numbers, their personal identification numbers, and their credit card expiration dates. The problem is that the Web site is fake, operated by someone trying to gain access to customers' private information, a form of scam called *phishing*.

Traveling situations seem to be a popular setting for phishers. A writer received an e-mail from a reader he had communicated with two years ago. "I don't mean to inconvenience you right now," the e-mail stated, "but I am stuck in London and need $940 to get home." Another ploy phishers use frequently is to claim that they lost their luggage at the airport and need some cash quickly.[49]

Vishing is similar to phishing. However, instead of using the victim's computer, it uses the victim's phone. The victim is typically sent a notice or message to call to verify account information. If the victim returns the message, the caller asks for personal information such as a credit card account number or name and address. The information gained can be used in identity theft to acquire and use credit cards in the victim's name. Vishing criminals can even spoof the caller ID that appears with the message to make it appear as if it came from a legitimate source.[50] The Spoof Card, sold online for less than $5 dollars for 25 calls, causes phones to display a caller ID number specified by the caller rather than the actual number of the caller.[51]

International Computer Crime

Computer crime becomes more complex when it crosses borders. Money laundering is the practice of disguising illegally gained funds so that they seem legal. With the increase in electronic cash and funds transfer, some are concerned that terrorists, international drug dealers, and other criminals are using information systems to launder illegally obtained funds. An Australian arrested in Las Vegas was indicted for laundering more than $500 million in proceeds of U.S. gamblers and Internet gambling Web sites. It is charged that he processed the gambling transactions in the U.S. through the Automated Clearing House to offshore accounts held by the gambling companies.[52]

PREVENTING COMPUTER-RELATED CRIME

Because of increased computer use today, greater emphasis is placed on the prevention and detection of computer crime. Although all states have passed computer crime legislation, some believe that these laws are not effective because companies do not always actively detect and pursue computer crime, security is inadequate, and convicted criminals are not severely punished. However, all over the United States, private users, companies, employees, and public officials are making individual and group efforts to curb computer crime, and recent efforts have met with some success.

Crime Prevention by State and Federal Agencies

State and federal agencies have begun aggressive attacks on computer criminals, including criminal hackers of all ages. In 1986, Congress enacted the Computer Fraud and Abuse Act, which mandates punishment based on the victim's dollar loss.

The Department of Defense also supports the Computer Emergency Response Team (CERT), which responds to network security breaches and monitors systems for emerging threats. Law enforcement agencies are also increasing their efforts to stop criminal hackers, and many states are now passing new, comprehensive bills to help eliminate computer crimes. A complete listing of computer-related legislation by state can be found at *www.onlinesecurity.com/forum/article46.php*. Recent court cases and police reports involving computer crime show that lawmakers are ready to introduce newer and tougher computer crime legislation.

Crime Prevention by Corporations

Companies are also taking crime-fighting efforts seriously. Many businesses have designed procedures and specialized hardware and software to protect their corporate data and systems. Specialized hardware and software, such as encryption devices, can be used to encode data and information to help prevent unauthorized use. As discussed in Chapter 5, encryption is the process of converting an original electronic message into a form that can be understood only by the intended recipients. A key is a variable value that is applied using an algorithm to a string or block of unencrypted text to produce encrypted text or to decrypt encrypted text. Encryption methods rely on the limitations of computing power for their effectiveness—if breaking a code requires too much computing power, even the most determined code crackers will not be successful. The length of the key used to encode and decode messages determines the strength of the encryption algorithm.

As employees move from one position to another at a company, they can build up access to multiple systems if inadequate security procedures fail to revoke access privileges. It is clearly not appropriate for people who have changed positions and responsibilities to still have access to systems they no longer use. To avoid this problem, many organizations create role-based system access lists so that only people filling a particular role (e.g., invoice approver) can access a specific system.

Fingerprint authentication devices provide security in the PC environment by using fingerprint recognition instead of passwords. Laptop computers from Lenovo, Toshiba, and others have built-in fingerprint readers used to log on and gain access to the computer system and its data. The 2 GB Fingerprint Biometric USB Flash Memory Stick Drive requires users to swipe their fingerprints and match them to one of up to 10 trusted users to access the data. The data on the flash drive can also be encrypted for further protection.[53]

Crime-fighting procedures usually require additional controls on the information system. Before designing and implementing controls, organizations must consider the types of computer-related crime that might occur, the consequences of these crimes, and the cost and complexity of needed controls. In most cases, organizations conclude that the trade-off between crime and the additional cost and complexity weighs in favor of better system controls. Having knowledge of some of the methods used to commit crime is also helpful in preventing, detecting, and developing systems resistant to computer crime. Some companies actually hire former criminals to thwart other criminals.

The following list provides a set of useful guidelines to protect corporate computers from criminal hackers:

- Install strong user authentication and encryption capabilities on the corporate firewall.
- Install the latest security patches, which are often available at the vendor's Internet site.
- Disable guest accounts and null user accounts that let intruders access the network without a password.
- Do not provide overfriendly logon procedures for remote users (e.g., an organization that used the word "welcome" on their initial logon screen found they had difficulty prosecuting a criminal hacker).
- Restrict physical access to the server and configure it so that breaking into one server won't compromise the whole network.
- Dedicate one server to each application (e-mail, File Transfer Protocol, and domain name server). Turn audit trails on.
- Consider installing caller ID.
- Install a corporate firewall between your corporate network and the Internet.
- Install antivirus software on all computers and regularly download vendor updates.
- Conduct regular IS security audits.
- Verify and exercise frequent data backups for critical data.

Using Intrusion Detection Software

An **intrusion detection system** (IDS) monitors system and network resources and notifies network security personnel when it senses a possible intrusion. Examples of suspicious activities include repeated failed logon attempts, attempts to download a program to a server, and access to a system at unusual hours. Such activities generate alarms that are captured on log files. When they detect an apparent attack, intrusion detection systems send an alarm, often by e-mail or pager, to network security personnel. Unfortunately, many IDSs frequently provide false alarms that result in wasted effort. If the attack is real, network security personnel must make a decision about what to do to resist the attack. Any delay in response increases the probability of damage. Use of an IDS provides another layer of protection in case an intruder gets past the outer security layers—passwords, security procedures, and corporate firewall.

Security Dashboard

Many organizations use **security dashboard** software to provide a comprehensive display on a single computer screen of all the vital data related to an organization's security defenses, including threats, exposures, policy compliance, and incident alerts. The goal is to reduce the effort required for monitoring and to identify threats earlier. Data comes from a variety of sources, including firewalls, applications, servers, and other software and hardware devices. See Figure 9.2.

Some USB flash drives have built-in fingerprint readers to protect the data on the device.

(Source: Courtesy of Kanguru Solutions, *www.kanguru.com*.)

---

intrusion detection system (IDS)

Software that monitors system and network resources and notifies network security personnel when it senses a possible intrusion.

---

security dashboard

Software that provides a comprehensive display on a single computer screen of all the vital data related to an organization's security defenses, including threats, exposures, policy compliance and incident alerts.

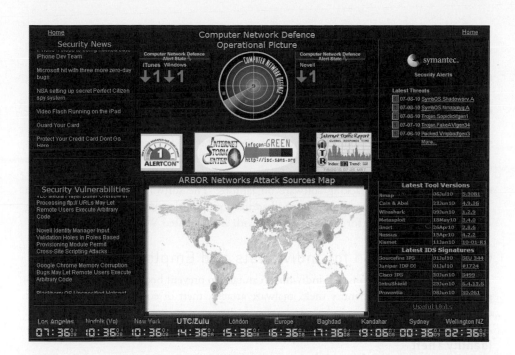

Figure 9.2

The Computer Network Defence Internet Operational Picture

The Computer Network Defence Internet Operational Picture, a security dashboard designed for the United Kingdom government and military networks, displays near real-time information on new and emerging cyber threats.

Del Monte Foods, a leading manufacturer of quality food and pet products, employs more than 5,400 people. The firm needed to provide an information infrastructure that would enable its employees to collaborate without risking the loss of proprietary company information. To that end, the company upgraded to software that provided protection against all forms of malware and could detect potential intrusions by unauthorized users. The firm also implemented a security dashboard that provides a snapshot of the data needed to identify and respond to security incidents immediately.[54]

Using Managed Security Service Providers (MSSPs)

Keeping up with computer criminals—and with new regulations—can be daunting for organizations. Criminal hackers are constantly poking and prodding, trying to breach the security defenses of companies. Also, such recent legislation as HIPAA, Sarbanes-Oxley, and the USA Patriot Act requires businesses to prove that they are securing their data. For most small and mid-sized organizations, the level of in-house network security expertise needed to protect their business operations can be quite costly to acquire and maintain. As a result, many are outsourcing their network security operations to managed security service providers (MSSPs) such as Counterpane, Guardent, IBM, Riptech, and Symantec. MSSPs monitor, manage, and maintain network security for both hardware and software. These companies provide a valuable service for IS departments drowning in reams of alerts and false alarms coming from virtual private networks (VPNs); antivirus, firewall, and intrusion detection systems; and other security monitoring systems. In addition, some provide vulnerability scanning and Web blocking/filtering capabilities.

IBM offers managed protection services that provide expert monitoring, management, and incident protection around the clock. The protection service is available for networks, servers, and desktops. When an Internet attack is detected, the service automatically blocks it without requiring human intervention. Taking the manual intervention step out of the process enables a faster response and minimizes damage from a criminal hacker. To encourage customers to adopt its service, IBM guarantees up to $50,000 in cash if the prevention service fails.[55]

Guarding Against Theft of Equipment and Data

Organizations need to take strong measures to guard against the theft of computer hardware and the data stored on it. Here are a few measures to be considered:

- Set clear guidelines on what kind of data (and how much of it) can be stored on vulnerable laptops. In many cases, private data or company confidential data may not be downloaded to laptops that leave the office.
- Require that data stored on laptops be encrypted and do spot checks to ensure that this policy is followed.
- Require that all laptops be secured using a lock and chain device so that they cannot be easily removed from an office area.
- Provide training to employees and contractors on the need for safe handling of laptops and their data. For example, laptops should never be left in a position where they can be viewed by the public, such as on the front seat of an automobile.
- Consider installing tracking software on laptops. The software sends messages via a wireless network to the specified e-mail address, pinpointing its location and including a picture of the thief (for those computers with an integrated Web cam).[56]

Crime Prevention for Individuals and Employees

This section outlines actions that individuals can take to prevent becoming a victim of computer crime, including identity theft, malware attacks, theft of equipment and data, and computer scams.

Identity Theft

Consumers can protect themselves from identity theft by regularly checking their credit reports with major credit bureaus, following up with creditors if their bills do not arrive on time, not revealing any personal information in response to unsolicited e-mail or phone calls (especially Social Security numbers and credit card account numbers), and shredding bills and other documents that contain sensitive information.

Some consumers contract with a service company that provides fraud monitoring services, helps you file required reports, and disputes unauthorized transactions in your accounts. Some services even offer identity theft guarantees of up to $1 million. Some of the more popular services include Trusted ID, Life Lock, Protect My ID, ID Watchdog, and Identity Guard. These services cost between $7 and $18 per month.

The U.S. Congress passed the Identity Theft and Assumption Deterrence Act of 1998 to fight identity theft. Under this act, the Federal Trade Commission (FTC) is assigned responsibility to help victims restore their credit and erase the impact of the imposter. It also makes identity theft a federal felony punishable by a prison term ranging from 3 to 25 years.

Malware Attacks

The number of personal computers infected with malware (viruses, worms, spyware, etc.) has reached epidemic proportions. As a result of the increasing threat of malware, most computer users and organizations have installed **antivirus programs** on their computers. Such software runs in the background to protect your computer from dangers lurking on the Internet and other possible sources of infected files. The latest virus definitions are downloaded automatically when you connect to the Internet, ensuring that your PC's protection is current. To safeguard your PC and prevent it from spreading malware to your friends and coworkers, some antivirus software scans and cleans both incoming and outgoing e-mail messages. The top four rated antivirus software for 2010 includes G-Data AntiVirus 2010, Symantec Norton Antivirus 2010, Kaspersky Lab Anti-Virus 2010, and Bit Defender Antivirus, which all cost in the range of $15 to $60.[57] See Figure 9.3.

Proper use of antivirus software requires the following steps:

1. *Install antivirus software and run it often.* Many of these programs automatically check for viruses each time you boot up your computer or insert a disk or CD, and some even monitor all e-mail, file transmissions, and copying operations.
2. *Update antivirus software often.* New viruses are created all the time, and antivirus software suppliers are constantly updating their software to detect and take action against these new viruses.

antivirus program
Software that runs in the background to protect your computer from dangers lurking on the Internet and other possible sources of infected files.

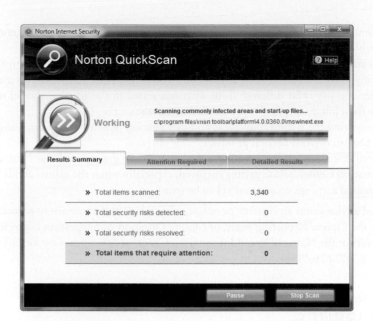

Figure 9.3

Antivirus Software

Antivirus software should be used and updated often.

3. *Scan all removable media, including CDs, before copying or running programs from them.* Hiding on disks or CDs, viruses often move between systems. If you carry document or program files on removable media between computers at school or work and your home system, always scan them.

4. *Install software only from a sealed package or secure Web site of a known software company.* Even software publishers can unknowingly distribute viruses on their program disks or software downloads. Most scan their own systems, but viruses might still remain.

5. *Follow careful downloading practices.* If you download software from the Internet or a bulletin board, check your computer for viruses immediately after completing the transmission.

6. *If you detect a virus, take immediate action.* Early detection often allows you to remove a virus before it does any serious damage.

Many e-mail services and ISP providers offer free antivirus protection. For example, AOL and MWEB (one of South Africa's leading ISPs) offer free antivirus software from McAfee.

Computer Scams
The following is a list of tips to help you avoid becoming a victim of a computer scam:

- Don't agree to anything in a high-pressure meeting or seminar. Insist on having time to think it over and to discuss your decision with someone you trust. If a company won't give you the time you need to check out an offer and think things over, you don't want to do business with them. A good deal now will be a good deal tomorrow; the only reason for rushing you is if the company has something to hide.

- Don't judge a company based on appearances. Flashy Web sites can be created and published in a matter of days. After a few weeks of taking money, a site can vanish without a trace in just a few minutes. You might find that the perfect money-making opportunity offered on a Web site was a money maker for the crook and a money loser for you.

- Avoid any plan that pays commissions simply for recruiting additional distributors. Your primary source of income should be your own product sales. If the earnings are not made primarily by sales of goods or services to consumers or sales by distributors under you, you might be dealing with an illegal pyramid.

- Beware of shills—people paid by a company to lie about how much they've earned and how easy the plan was to operate. Check with an independent source to make sure that the company and its offers are valid.

- Beware of a company's claim that it can set you up in a profitable home-based business but that you must first pay up front to attend a seminar and buy expensive materials.

Frequently, seminars are high-pressure sales pitches, and the material is so general that it is worthless.

- If you are interested in starting a home-based business, get a complete description of the work involved before you send any money. You might find that what you are asked to do after you pay is far different from what was stated in the ad. You should never have to pay for a job description or for needed materials.
- Get in writing the refund, buy-back, and cancellation policies of any company you deal with. Do not depend on oral promises.
- Do your homework. Check with your state attorney general and the National Fraud Information Center before getting involved, especially when the claims about a product or potential earnings seem too good to be true.

If you need advice about an Internet or online solicitation, or if you want to report a possible scam, use the Online Reporting Form or Online Question & Suggestion Form features on the Web site for the National Fraud Information Center at *http://fraud.org*, or call the NFIC hotline at 1-800-876-7060.

PRIVACY ISSUES

Another important social issue in information systems involves privacy. In 1890, U.S. Supreme Court Justice Louis Brandeis stated that the "right to be left alone" is one of the most "comprehensive of rights and the most valued by civilized man." Basically, the issue of privacy deals with this right to be left alone or to be withdrawn from public view. With information systems, privacy deals with the collection and use or misuse of data. Data is constantly being collected and stored on each of us. This data is often distributed over easily accessed networks and without our knowledge or consent. Concerns of privacy regarding this data must be addressed. Today many businesses have to handle many requests from law enforcement agencies for information about its employees, customers, and suppliers. Indeed, some phone and Internet companies have full-time employees whose role it is to deal with information requests from local, state, and federal law enforcement agencies.

With today's computers, the right to privacy is an especially challenging problem. More data and information are produced and used today than ever before. When someone is born, takes certain high school exams, starts a job, enrolls in a college course, applies for a driver's license, purchases a car, serves in the military, gets married, buys insurance, gets a library card, applies for a charge card or loan, buys a house, or merely purchases certain products, data is collected and stored somewhere in computer databases. A difficult question to answer is, "Who owns this information and knowledge?" If a public or private organization spends time and resources to obtain data on you, does the organization own the data, and can it use the data in any way it desires? Government legislation answers these questions to some extent for federal agencies, but the questions remain unanswered for private organizations.

Privacy and the Federal Government

The federal government has implemented a number of laws addressing personal privacy that are discussed in this section. With more than 6,300 retail stores, CVS Caremark is the largest pharmacy chain in the U.S. It also operates online and has mail order businesses. The Federal Trade Commission (FTC) began an investigation of CVS Caremark based on media reports that its pharmacies were throwing trash into open dumpsters that contained personal information about patients, their physicians, medications, credit card information, and driver's license data. CVS Caremark agreed to settle the FTC charges that it failed to take reasonable and appropriate security measures to protect the sensitive financial and medical information of its customers and employees, a violation of federal law. "The FTC order requires CVS Caremark to establish, implement, and maintain a comprehensive information security program designed to protect the security, confidentiality, and integrity of the personal information it collects from consumers and employees. It also requires the company to obtain, every two years for the next 20 years, an audit from a qualified, independent, third-party professional to ensure that its security program meets the standards of the order."[58] Concurrent with the FTC investigation, the Department of Health and Human Services (HHS) opened its own investigation. In a separate but related agreement, the company's pharmacy chain agreed to pay $2.25 million to resolve HHS allegations that it inappropriately disposed of health information protected by Health Insurance Portability and Accountability Act (HIPAA).[59]

The European Union has a data-protection directive that requires firms transporting data across national boundaries to have certain privacy procedures in place. This directive affects virtually any company doing business in Europe, and it is driving much of the attention being given to privacy in the United States.

Privacy at Work

The right to privacy at work is also an important issue. Employers are using technology and corporate policies to manage worker productivity and protect the use of IS resources. Employers are mostly concerned about inappropriate Web surfing, with 76 percent of employers monitoring the Web activity of their employees. Organizations also monitor employees' e-mail, with more than over half retaining and reviewing messages.[60] Statistics such as these have raised employee privacy concerns. In many cases, workers claim their right to privacy trumps their companies' rights to monitor employee use of IS resources.

E-Mail Privacy

E-mail also raises some interesting issues about work privacy. Federal law permits employers to monitor e-mail sent and received by employees. Furthermore, e-mail messages that have been erased from hard disks can be retrieved and used in lawsuits because the laws of discovery demand that companies produce all relevant business documents. On the other hand, the use of e-mail among public officials might violate "open meeting" laws. These laws, which apply to many local, state, and federal agencies, prevent public officials from meeting in private about matters that affect the state or local area.

E-mail has changed how workers and managers communicate in the same building or around the world. E-mail, however, can be monitored and intercepted. As with other services such as cell phones, the convenience of e-mail must be balanced with the potential of privacy invasion.

(Source: © Daniel Laflor/ iStockphoto.)

A 14-year employee was fired from SunTrust Mortgage for sending a racially insensitive chain e-mail she received at work.[61]

Instant Messaging Privacy

Using instant messaging (IM) to send and receive messages, files, and images introduces the same privacy issues associated with e-mail. As with e-mail, federal law permits employers to monitor instant messages sent and received by employees. Employers' major concern involves IMs sent by employees over their employer's IM network or using employer-provided phones.

To protect your privacy and your employer's property, do not send personal or private IMs at work. Here are a few other tips:

- Choose a nonrevealing, nongender-specific, unprovocative IM screen name (Sweet Sixteen, 2hot4u, UCLAMBA, all fail this test).
- Don't send messages you would be embarrassed to have your family members, colleagues, or friends read.
- Do not open files or click links in messages from people you do not know.
- Never send sensitive personal data such as credit card numbers, bank account numbers, or passwords via IM.

CNN International fired a 20-year veteran and senior editor for publishing a Twitter message that lamented the death of a Lebanese Shi'ite cleric who was an early mentor and spiritual leader of the militant group Hezbollah.[62]

Privacy and Personal Sensing Devices

RFID tags, essentially microchips with antenna, are embedded in many of the products we buy, from medicine containers, clothing, and library books to computer printers, car keys, and tires. RFID tags generate radio transmissions that, if appropriate measures are not taken, can lead to potential privacy concerns. Once these tags are associated with the individual who purchased the item, someone can potentially track individuals by the unique identifier associated with the RFID chip.

A handful of states, including Wisconsin, California, North Dakota, Missouri, Virginia, and Georgia, have reacted to the potential for abuse of RFID tags by passing legislation prohibiting the implantation of RFID chips under people's skin without their approval.[63] Advocates for RFID chip implantation argue their potential value in tracking children or criminals and their value in carrying an individual's medical records.

Privacy and the Internet

Some people assume that there is no privacy on the Internet and that you use it at your own risk. Others believe that companies with Web sites should have strict privacy procedures and be accountable for privacy invasion. Regardless of your view, the potential for privacy invasion on the Internet is huge. People wanting to invade your privacy could be anyone from criminal hackers to marketing companies to corporate bosses. Your personal and professional information can be seized on the Internet without your knowledge or consent. E-mail is a prime target, as discussed previously. Sending an e-mail message is like having an open conversation in a large room—people can listen to your messages. When you visit a Web site on the Internet, information about you and your computer can be captured. When this information is combined with other information, companies can find out what you read, what products you buy, and what your interests are.

Most people who buy products on the Web say it's very important for a site to have a policy explaining how personal information is used, and the policy statement must make people feel comfortable and be extremely clear about what information is collected and what will and will not be done with it. However, many Web sites still do not prominently display their privacy policy or implement practices completely consistent with that policy. The real issue that Internet users need to be concerned with is—what do content providers want to do with their personal information? If a site requests that you provide your name and address, you have every right to know why and what will be done with it. If you buy something and provide a shipping address, will it be sold to other retailers? Will your e-mail address be sold on a list of active Internet shoppers? And if so, you should realize that this e-mail list is no different from the lists compiled from the orders you place with catalog retailers. You have the right to be taken off any mailing list.

A potential solution to some consumer privacy concerns is the screening technology called the **Platform for Privacy Preferences** (P3P) being proposed to shield users from sites that don't provide the level of privacy protection they desire. Instead of forcing users to find and read through the privacy policy for each site they visit, P3P software in a computer's browser will download the privacy policy from each site, scan it, and notify the user if the policy does not match his preferences. (Of course, unethical marketers can post a privacy policy that does not accurately reflect the manner in which the data is treated.) The World Wide Web Consortium, an international industry group whose members include Apple, Commerce One, Ericsson, and Microsoft, is supporting the development of P3P. Version 1.1 of the P3P was released in February 2006 and can be found at *www.w3.org/TR/2006/WD-P3P11-20060210/ Overview.html.*

The Children's Online Privacy Protection Act (COPPA) was passed by Congress in October 1998. This act was directed at Web sites catering to children, requiring site owners to post comprehensive privacy policies and to obtain parental consent before they collect any personal information from children under 13 years of age. Web site operators who violate the rule could be liable for civil penalties of up to $11,000 per violation. COPPA has made an impact in the design and operations of Web sites that cater to children. For example, Lions Gate Entertainment, the operator of the *www.thebratzfilm.com* Web site, had to modify its site after the Council of Better Business Bureaus determined the site failed to meet the COPPA requirements. The Web site requested personally identifiable information to register for the Bratz Newsletter and register for a chance to win a trip to the premiere of *The Bratz Movie* without first obtaining verifiable parental consent.

A social network service employs the Web and software to connect people for whatever purpose. There are thousands of such networks, which have become popular among teenagers. Some of the more popular social networking Web sites include Bebo, Classmates, Facebook, Hi5, Imbee, MySpace, Namesdatabase, Tagged, and XuQa. Most of these Web sites allow you to easily create a user profile that provides personal details, photos, and even videos that can be viewed by other visitors to the Web site. Some of the Web sites have age restrictions or require that a parent register their preteen by providing a credit card to validate the parent's identity. Teens can provide information about where they live, go to school, their favorite music, and interests in hopes of meeting new friends. Unfortunately, they can also meet ill-intentioned strangers at these sites. Many documented encounters involve adults masquerading as teens attempting to meet young people for illicit purposes. Parents are advised to discuss potential dangers, check their children's profiles, and monitor their activities at such Web sites.

Internet Libel Concerns

Libel involves publishing an intentionally false written statement that is damaging to a person's or organization's reputation. Examples of Internet libel include an ex-husband posting lies about his wife on a blog, a disgruntled former employee posting lies about a company on a message board, and a jilted girlfriend posting false statements to her former boyfriend's Facebook account.

Individuals can post information to the Internet using anonymous e-mail accounts or screen names. This makes it more difficult, but not impossible, to identify the libeler. The offended party can file what is known as a "John Doe" lawsuit and use the subpoena power it grants to force the ISP to provide whatever information they have about the anonymous poster, including IP address, name, and street address. (Under Section 230 of the Communications Decency Act, ISPs are not usually held accountable for the bad behavior of their subscribers). A judge ordered an ISP to turn over identifying information on 178 anonymous posters who made allegedly defamatory comments about two individuals involved in a sexual assault case.[64]

Individuals, too, must be careful what they post on the Internet to avoid libel charges. In many cases, disgruntled former employees are being sued by their former employers for material posted on the Internet.

Catherine Crier, a former Dallas County judge, a TV news personality at CNN, ABC News, Fox News Channel, and an author of several nonfiction books, filed a John Doe lawsuit

Platform for Privacy Preferences (P3P)

A screening technology in Web browsers that shields users from Web sites that don't provide the level of privacy protection they desire.

against an unknown individual who posted untrue statements about her on her Wikipedia page.[65]

Filtering and Classifying Internet Content

To help parents control what their children see on the Internet, some companies provide *filtering software* to help screen Internet content. Many of these screening programs also prevent children from sending personal information over e-mail or through chat groups. This stops children from broadcasting their name, address, phone number, or other personal information over the Internet. The two approaches used are filtering, which blocks certain Web sites, and rating, which places a rating on Web sites. According to the 2010 Internet Filter Review, four of the top-rated filtering software packages costing less than $50.00 include: Net Nanny 6.5, CyberPatrol Parental Controls 7.7, PC Pandora 6.0, Safe Eyes 6.0, and OnLine Family Norton.[66]

Organizations also implement filtering software to prevent employees from visiting Web sites not related to work, particularly those involving gambling or those containing pornographic or other offensive material. Before implementing Web site blocking, the users must be informed about the company's policies and why they exist. It is best if the organization's Internet users, management, and IS organization work together to define the policy to be implemented. The policy should be clear about the repercussions to employees who attempt to circumvent the blocking measures.

The Internet Content Rating Association (ICRA) is a nonprofit organization whose members include Internet industry leaders such as America Online, Bell South, British Telecom, IBM, Microsoft, UUNet, and Verizon. Its specific goals are to protect children from potentially harmful material while safeguarding free speech on the Internet. Using the ICRA rating system, Web authors fill out an online questionnaire describing the content of their site—what is and isn't present. The broad topics covered include chat capabilities, the language used, nudity and sexual content, violence depicted, and other areas such as alcohol, drugs, gambling, and suicide. Based on the authors' responses, ICRA then generates a content label (a short piece of computer code) that the authors add to their site. Internet users (and parents) can set their browser to allow or disallow access to Web sites based on the objective rating information declared in the content label and their subjective preferences. Reliance on Web site authors to rate their own sites has its weaknesses, though. Web site authors can lie when completing the ICRA questionnaire so that their site receives a content label that doesn't accurately reflect the site's content. In addition, many hate groups and sexually explicit sites don't have an ICRA rating, so they will not be blocked unless a browser is set to block all unrated sites. Unfortunately, this option would block out so many acceptable sites that it could make Web surfing useless. For these reasons, site labeling is currently at best a complement to other filtering techniques.

The U.S. Congress has made several attempts to limit children's exposure to online pornography, including the Communications Decency Act (enacted 1996) and the Child Online Protection Act (enacted 1998). Within two years of their being enacted, the U.S. Supreme Court found that both these acts violated the First Amendment (freedom of speech) and ruled them to be unconstitutional. The Children's Internet Protection Act (CIPA) was signed into law in 2000 and later upheld by the Supreme Court in 2003. Under CIPA, schools and libraries subject to CIPA do not receive the discounts offered by the "E-Rate" program unless they certify that they have certain Internet safety measures in place to block or filter "visual depictions that are obscene, child pornography, or are harmful to minors." (The E-Rate program provides many schools and libraries support to purchase Internet access and computers).

Fairness in Information Use

Selling information to other companies can be so lucrative that many companies will continue to store and sell the data they collect on customers, employees, and others. When is this information storage and use fair and reasonable to the people whose data is stored and sold? Do people have a right to know about data stored about them and to decide what data is stored and used? As shown in Table 9.3, these questions can be broken down into four issues that should be addressed: knowledge, control, notice, and consent.

In the past few decades, significant laws have been passed regarding a person's right to privacy. Others relate to business privacy rights and the fair use of data and information.

Fairness Issues	Database Storage	Database Usage
The right to know	Knowledge	Notice
The ability to decide	Control	Consent

Knowledge. Should people know what data is stored about them? In some cases, people are informed that information about them is stored in a corporate database. In others, they do not know that their personal information is stored in corporate databases.

Control. Should people be able to correct errors in corporate database systems? This is possible with most organizations, although it can be difficult in some cases.

Notice. Should an organization that uses personal data for a purpose other than the original purpose notify individuals in advance? Most companies don't do this.

Consent. If information on people is to be used for other purposes, should these people be asked to give their consent before data on them is used? Many companies do not give people the ability to decide if information on them will be sold or used for other purposes.

Table 9.3

The Right to Know and the Ability to Decide Federal Privacy Laws and Regulations

The Privacy Act of 1974

The major piece of legislation on privacy is the Privacy Act of 1974 (PA74). PA74 applies only to certain federal agencies. The act, which is about 15 pages long, is straightforward and easy to understand. Its purpose is to provide certain safeguards for people against an invasion of personal privacy by requiring federal agencies (except as otherwise provided by law) to do the following:

- Permit people to determine what records pertaining to them are collected, maintained, used, or disseminated by such agencies
- Permit people to prevent records pertaining to them from being used or made available for another purpose without their consent
- Permit people to gain access to information pertaining to them in federal agency records, to have a copy of all or any portion thereof, and to correct or amend such records
- Ensure that they collect, maintain, use, or disseminate any record of identifiable personal information in a manner that ensures that such action is for a necessary and lawful purpose, that the information is current and accurate for its intended use, and that adequate safeguards are provided to prevent misuse of such information
- Permit exemptions from this act only in cases of an important public need for such exemption, as determined by specific law-making authority
- Be subject to civil suit for any damages that occur as a result of willful or intentional action that violates anyone's rights under this act

PA74, which applies to all federal agencies except the CIA and law enforcement agencies, established a Privacy Study Commission to study existing databases and to recommend rules and legislation for consideration by Congress. PA74 also requires training for all federal employees who interact with a "system of records" under the act. Most of the training is conducted by the Civil Service Commission and the Department of Defense. Another interesting aspect of PA74 concerns the use of Social Security numbers—federal, state, and local governments and agencies cannot discriminate against people for not disclosing or reporting their Social Security number.

Electronic Communications Privacy Act

This law was enacted in 1986 and deals with three main issues: 1) the protection of communications while in transit from sender to receiver, 2) the protection of communications held in electronic storage, and 3) the prohibition of devices to record dialing, routing, addressing, and signaling information without a search warrant. Under Title I of this law, the government is prohibited from intercepting electronic messages unless it obtains a court order based on probable cause. Title II prohibits access to wire and electronic communications for stored communications not readily accessible to the general public.

Gramm-Leach-Bliley Act

This act was passed in 1999 and requires all financial institutions to protect and secure customers' nonpublic data from unauthorized access or use. Under terms of this act, it was assumed that all customers approve of the financial institutions' collecting and storing their personal information. The institutions were required to contact their customers and inform them of this fact. Customers were required to write separate letters to each of their financial institutions and state in writing that they wanted to opt out of the data collection and storage process. Most people were overwhelmed with the mass mailings they received from their financial institutions and simply discarded them without ever understanding their importance.

USA Patriot Act

As discussed previously, the 2001 Uniting and Strengthening America by Providing Appropriate Tools Required to Intercept and Obstruct Terrorism Act (USA Patriot Act) was passed in response to the September 11 terrorism acts. Proponents argue that it gives necessary new powers to both domestic law enforcement and international intelligence agencies. Critics argue that the law removes many of the checks and balances that previously allowed the courts to ensure that law enforcement agencies did not abuse their powers. For example, under this act, Internet service providers and telephone companies must turn over customer information, including numbers called, without a court order if the FBI claims that the records are relevant to a terrorism investigation. Also, the company is forbidden to disclose that the FBI is conducting an investigation.

Table 9.4 lists additional laws related to privacy.

Corporate Privacy Policies

Even though privacy laws for private organizations are not very restrictive, most organizations are sensitive to privacy issues and fairness. They realize that invasions of privacy can hurt their business, turn away customers, and dramatically reduce revenues and profits. Consider a major international credit card company. If the company sold confidential financial information on millions of customers to other companies, the results could be disastrous. In a matter of days, the firm's business and revenues could be reduced dramatically. Therefore, most organizations maintain privacy policies, even though they are not required by law. Some companies even have a privacy bill of rights that specifies how the privacy of employees, clients, and customers will be protected. Corporate privacy policies should address a customer's knowledge, control, notice, and consent over the storage and use of information. They can also cover who has access to private data and when it can be used.

Multinational companies face an extremely difficult challenge in implementing data-collection and dissemination processes and policies because of the multitude of differing country or regional statutes. For example, Australia requires companies to destroy customer data (including backup files) or make it anonymous after it's no longer needed. Firms that transfer customer and personnel data out of Europe must comply with European privacy laws that allow customers and employees to access data about them and let them determine how that information can be used.

Web sites for a few corporate privacy policies are shown in Table 9.5.

A good database design practice is to assign a single unique identifier to each customer so each has a single record describing all relationships with the company across all its business units. That way, the organization can apply customer privacy preferences consistently throughout all databases. Failure to do so can expose the organization to legal risks—aside from upsetting customers who opted out of some collection practices. Again, the 1999 Gramm-Leach-Bliley Financial Services Modernization Act required all financial service institutions to communicate their data privacy rules and honor customer preferences.

Health Concerns

Organizations can increase employee effectiveness by payin[...]
in today's work environment. For some people, working w[...]
tional stress. Anxieties about job insecurity, loss of control, [...]
just a few of the fears workers might experience. In some cas[...]
that workers might sabotage computer systems and equipm[...]
can alert companies to potential problems. Training and c[...]
ployee and deter problems.

Heavy computer use can affect one's physical health as v[...]
a desk and using a computer for many hours a day qualifi[...]
can double the risk of seated immobility thromboembolisr[...]
clots in the legs or lungs. People leading a sedentary lifestyl[...]
undesirable weight gain, which can lead to increased fatigue a[...]
heart problems, and other serious ailments.

Other work-related health hazards involve emissions fr[...]
used equipment. Some studies show that poorly maintained[...]
into the air; others dispute the claim. Numerous studies o[...]
display screens have also resulted in conflicting theories. Alt[...]
believe that long-term exposure can cause cancer, studies ar[...]
any case, many organizations are developing conservative an[...]

Most computer manufacturers publish technical informat[...]
their computer monitors, and many companies pay close att[...]
Francisco was one of the first cities to propose a video displa[...]
requires companies with 15 or more employees who spend a[...]
with computer screens to give 15-minute breaks every two[...]
chairs and workstations are required if employees request th[...]

In addition to the possible health risks from radio frequen[...]
raised a safety issue—an increased risk of traffic accidents as[...]
tracted while driving by talking on their cell phones (or ope[...]
car navigation systems, or other computer devices). As a re[...]
illegal to operate a cell phone while driving.

From time to time, concern has been raised about heavy cel[...]
chance for brain cancer. A study by Interphone, an internatio[...]
was no such risk except for perhaps a small percentage of the [...]
Even here, the findings were inconclusive. [67]

Carpal tunnel syndrome (CTS) is an aggravation of the pat[...]
through the wrist (carpal tunnel). CTS involves wrist pain, a fee[...]
and difficulty grasping and holding objects. In the late 1990s[...]
claims were filed by people whose job required them to work[...]
day. However, a 2001 study by the Mayo Clinic found that[...]
seven hours per day) had the same rate of carpal tunnel as the [...]
that CTS is caused by factors other than the repetitive motion[...]

Law	Provisions
Fair Credit Reporting Act of 1970 (FCRA)	Regulates operations of credit-reporting bureaus, including how they collect, store, and use credit information
Tax Reform Act of 1976	Restricts collection and use of certain information by the Internal Revenue Service
Electronic Funds Transfer Act of 1979	Outlines the responsibilities of companies that use electronic funds transfer systems, including consumer rights and liability for bank debit cards
Right to Financial Privacy Act of 1978	Restricts government access to certain records held by financial institutions
Family Education Privacy Act 1974	Restricts collection and use of data by federally funded educational institutions, including specifications for the type of data collected, access by parents and students to the data, and limitations on disclosure
Electronic Communications Privacy Act of 1986	Defines provisions for the access, use, disclosure, interception, and privacy protections of electronic communications
Computer Matching and Privacy Act of 1988	Regulates cross-references between federal agencies' computer files (e.g. to verify eligibility for federal programs)
Video Privacy Act of 1988	Prevents retail stores from disclosing video rental records without a court order
Telephone Consumer Protection Act of 1991	Limits telemarketers' practices
Cable Act of 1992	Regulates companies and organizations that provide wireless communications services, including cellular phones
Computer Abuse Amendments Act of 1994	Prohibits transmissions of harmful computer programs and code, including viruses
Gramm-Leach-Bliley Act of 1999	Requires all financial institutions to protect and secure customers' nonpublic data from unauthorized access or use
USA Patriot Act of 2001	Requires Internet service providers and telephone companies to turn over customer information, including numbers called, without a court order, if the FBI claims that the records are relevant to a terrorism investigation
E-Government Act of 2002	Requires federal agencies to post machine-readable privacy policies on their Web sites and to perform privacy impact assessments on all new collections of data of ten or more people
Fair and Accurate Credit Transactions Act of 2003	Designed to combat the growing crime of identity theft; allows consumers to get free credit reports from each of the three major consumer credit reporting agencies every 12 months and to place alerts on their credit histories under certain circumstances

Table 9.4

Federal Privacy Laws and Their Provisions

Company	URL
Intel	www.intel.com/sites/sitewide/en_US/privacy/privacy.htm
Starwood Hotels & Resorts	www.starwoodhotels.com/corporate/privacy_policy.html
TransUnion	www.transunion.com/corporate/privacyPolicy.page
United Parcel Service	www.ups.com/content/corp/privacy_policy.html
Visa	www.corporate.visa.com/ut/privacy.jsp
Walt Disney Internet Group	http://disney.go.com/corporate/privacy/pp_wdig.html

Table 9.5

Corporate Privacy Policies

Individual E

Although numero
protect individual
many people are ta
you can take to pr

- *Find out what*
  a copy of your
  www.freecredit
  in the last
  www.equifax.c
  (888-397-374.
  request to a fee

- *Be careful whee*
  absolutely nece
  888, or 900 ca
  or financial ins
  consent.

- *Be proactive to*
  phone compan
  change your ac
  you can notify t
  copies of your c
  in the garbage.
  system. You ca
  the Direct Marl
  look under Cor

- *Take extra care*
  your credit card
  a site unless you
  of approval fron
  When you open
  data, make sure
  icon appears in
  without reviewii
  use credit card i
  is destroyed afte:

THE WORK EN

The use of computer
Jobs that require IS
inated. Corporate pr
them the concern th
integrated within the

However, the gro
up numerous avenue
telecommunications
markets in industries
aided by computers,
people with disabiliti
IS components drop
increased productivity
productivity and effic

---

at or just below eye level. Your wrists and hands should be in line with your forearms, with your elbows close to your body and supported. Your lower back needs to be well supported. Your feet should be flat on the floor. Take an occasional break to get away from the keyboard and screen. Stand up and stretch while at your workplace. Do not ignore pain or discomfort. Many workers ignore early signs of RSI, and as a result, the problem becomes much worse and more difficult to treat.

It is never too soon to stop unhealthy computer work habits. Prolonged computer use under poor working conditions can lead to carpal tunnel syndrome, bursitis, headaches, and permanent eye damage. Strain and poor office conditions cannot be left unchecked. Unfortunately, at times we are all distracted by pressing issues such as the organization's need to raise productivity, improve quality, meet deadlines, and cut costs. We become complacent and fail to pay attention to the importance of healthy working conditions.[68]

ETHICAL ISSUES IN INFORMATION SYSTEMS

code of ethics
A code that states the principles and core values that are essential to a set of people and that therefore govern their behavior.

As you've seen throughout this book in the "Ethical and Societal Issues" boxes, ethical issues deal with what is generally considered right or wrong. Laws do not provide a complete guide to ethical behavior. Just because an activity is defined as legal does not mean that it is ethical. As a result, practitioners in many professions subscribe to a **code of ethics** that states the principles and core values that are essential to their work and, therefore, govern their behavior. The code can become a reference point for weighing what is legal and what is ethical. For example, doctors adhere to varying versions of the 2000-year-old Hippocratic Oath, which medical schools offer as an affirmation to their graduating classes.

Some IS professionals believe that their field offers many opportunities for unethical behavior. They also believe that unethical behavior can be reduced by top-level managers developing, discussing, and enforcing codes of ethics. Various IS-related organizations and associations promote ethically responsible use of information systems and have developed useful codes of ethics. Founded in 1947, the Association for Computing Machinery (ACM) is the oldest computing society and boasts more than 80,000 members in more than 100 countries. The ACM has a code of ethics and professional conduct that includes eight general moral imperatives that can be used to help guide the actions of IS professionals. These guidelines can also be used for those who employ or hire IS professionals to monitor and guide their work. These imperatives are outlined in the following list: As an ACM member I will …

1. Contribute to society and human well-being.
2. Avoid harm to others.
3. Be honest and trustworthy.
4. Be fair and take action not to discriminate.
5. Honor property rights including copyrights and patents.
6. Give proper credit for intellectual property.
7. Respect the privacy of others.
8. Honor confidentiality.

(Source: ACM Code of Ethics and Professional Conduct, *http://www.acm.org/about/code-of-ethics?searchterm=code+of+ethics* accessed June 15, 2010.)

The mishandling of the social issues discussed in this chapter—including waste and mistakes, crime, privacy, health, and ethics—can devastate an organization. The prevention of these problems and recovery from them are important aspects of managing information and information systems as critical corporate assets. More organizations are recognizing that people are the most important component of a computer-based information system and that long-term competitive advantage can be found in a well-trained, motivated, and knowledgeable workforce that adheres to a set of principles and core values that help guide their actions.

SUMMARY

Principle

Policies and procedures must be established to avoid waste and mistakes associated with computer usage.

Computer waste is the inappropriate use of computer technology and resources in both the public and private sectors. Computer mistakes relate to errors, failures, and other problems that result in output that is incorrect and without value. At the corporate level, computer waste and mistakes impose unnecessarily high costs for an information system and drag down profits. Waste often results from poor integration of IS components, leading to duplication of efforts and overcapacity. Inefficient procedures also waste IS resources, as do thoughtless disposal of useful resources and misuse of computer time for games and personal use. Inappropriate processing instructions, inaccurate data entry, mishandling of IS output, and poor systems design all cause computer mistakes.

Preventing waste and mistakes involves establishing, implementing, monitoring, and reviewing effective policies and procedures. Companies should develop manuals and training programs to avoid waste and mistakes.

Principle

Computer crime is a serious and rapidly growing area of concern requiring management attention.

Some crimes use computers as tools. For example, a criminal can use a computer to manipulate records, counterfeit money and documents, commit fraud via telecommunications links, and make unauthorized electronic transfers of money.

A cyberterrorist is someone who intimidates or coerces a government or organization to advance his political or social objectives by launching computer-based attacks against computers, networks, and the information stored on them.

Identity theft is a crime in which an imposter obtains key pieces of personal identification information to impersonate someone else. The information is then used to obtain credit, merchandise, and services in the name of the victim, or to provide the thief with false credentials.

Although Internet gambling is popular, its legality is questionable within the United States.

The computer is also used as a tool to fight crime. The Leads Online Web-based system helps law enforcement officers recover stolen property. Offender Watch tracks registered sex offenders. Law enforcement agencies use GPS tracking devices and software to monitor the movement of registered sex offenders. Law enforcement agencies use crime-related data and powerful analysis techniques, coupled with GIS systems, to better understand and even diminish crime risks.

A criminal hacker, also called a *cracker*, is a computer-savvy person who attempts to gain unauthorized or illegal access to computer systems to steal passwords, corrupt files and programs, and even transfer money. Script bunnies are crackers with little technical savvy. Insiders are employees, disgruntled or otherwise, working solo or in concert with outsiders to compromise corporate systems. The greatest fear of many organizations is the potential harm that can be done by insiders who know system logon IDs, passwords, and company procedures.

Computer crimes target computer systems and include illegal access to computer systems by criminal hackers, alteration and destruction of data and programs by viruses, and simple theft of computer resources.

Malware is a general term for software that is harmful or destructive. There are many forms of malware, including viruses, variants, worms, Trojan horse attacks, logic bombs, and rootkits. Spyware is software installed on a personal computer to intercept or take partial control over the user's interactions with the computer without knowledge or permission of the user. A password sniffer is a small program hidden in a network or computer system that records identification numbers and passwords.

Computer managers and law enforcement agencies are emphasizing prevention and detection of computer crime. People can use antivirus software to detect the presence of all sorts of malware. Use of an intrusion detection system (IDS) provides another layer of protection in the event that an intruder gets past the outer security layers—passwords, security procedures, and corporate firewall. An IDS monitors system and network resources and notifies network security personnel when it senses a possible intrusion. Many small and mid-sized organizations are outsourcing their network security operations to managed security service providers (MSSPs), which monitor, manage, and maintain network security hardware and software. When discarding a computer, people should use disk wiping utilities to avoid the loss of personal or confidential data even after deleting files and emptying the Recycle Bin (or Trash).

Software piracy might represent the most common computer crime. It is estimated that the software industry lost $51 billion in revenue in 2009 to software piracy. The global recording industry loses as much as $12.5 billion in revenue from music piracy each year. Patent infringement is also a major problem for computer software and hardware manufacturers.

Computer-related scams, including phishing and vishing, have cost people and companies thousands of dollars. Computer crime is an international issue.

Security measures, such as using passwords, identification numbers, and data encryption, help to guard against illegal computer access, especially when supported by effective control procedures. Virus-scanning software identifies and removes damaging computer programs. Organizations can use a security dashboard to provide a comprehensive display of vital data related to its security defenses and threats.

Privacy issues are a concern with government agencies, e-mail use, corporations, and the Internet. The Children's Internet Protection Act protects minors using the Internet. The Privacy Act of 1974, with the support of other federal laws, establishes straightforward and easily understandable requirements for data collection, use, and distribution by federal agencies; federal law also serves as a nationwide moral guideline for privacy rights and activities by private organizations. The USA Patriot Act, passed only five weeks after the September 11 terrorist attacks, requires Internet service providers and telephone companies to turn over customer information, including numbers called, without a court order, if the FBI claims that the records are relevant to a terrorism investigation. Also, the company is forbidden to disclose that the FBI is conducting an investigation. Only time will tell how this act will be applied in the future.

The Gramm-Leach-Bliley Act requires all financial institutions to protect and secure customers' nonpublic data from unauthorized access or use. Under terms of this act, it is assumed that all customers approve of the financial institutions collecting and storing their personal information.

A business should develop a clear and thorough policy about privacy rights for customers, including database access. That policy should also address the rights of employees, including electronic monitoring systems and e-mail. Fairness in information use for privacy rights emphasizes knowledge, control, notice, and consent for people profiled in databases. People should know about the data that is stored about them and be able to correct errors in corporate database systems. If information on people is to be used for other purposes, they should be asked to give their consent beforehand. Each person has the right to know and the ability to decide.

Principle

Jobs, equipment, and working conditions must be designed to avoid negative health effects from computers.

Jobs that involve heavy use of computers contribute to a sedentary lifestyle, which increases the risk of health problems. Some critics blame computer systems for emissions of ozone and electromagnetic radiation. Use of cell phones while driving has been linked to increased car accidents.

The study of designing and positioning computer equipment, called ergonomics, has suggested some approaches to reducing these health problems. Ergonomic design principles help to reduce harmful effects and increase the efficiency of an information system. RSI prevention includes keeping good posture, not ignoring pain or problems, performing stretching and strengthening exercises, and seeking proper treatment. Although they can cause negative health consequences, information systems can also be used to provide a wealth of information on health topics through the Internet and other sources.

Principle

Practitioners in many professions subscribe to a code of ethics that states the principles and core values that are essential to their work.

A code of ethics states the principles and core values that are essential to the members of a profession or organization. Ethical computer users define acceptable practices more strictly than just refraining from committing crimes; they also consider the effects of their IS activities, including Internet usage, on other people and organizations. The Association for Computing Machinery developed guidelines and a code of ethics. Many IS professionals join computer-related associations and agree to abide by detailed ethical codes.

CHAPTER 9: SELF-ASSESSMENT TEST

Policies and procedures must be established to avoid waste and mistakes associated with computer usage.

1. Business managers and end users must work with IS professionals to implement and follow proper IS usage policies to ensure effective use of company resources. True or False?

2. Computer-related waste and mistakes are major causes of computer problems, contributing to unnecessarily high _____ and lost _____.

3. Preventing waste and mistakes involves establishing, implementing, _____, and reviewing effective policies and procedures.

Computer crime is a serious and rapidly growing area of concern requiring management attention.

4. According to the 2009 FBI Internet Crime Report, the dollar amount of Internet crime reported exceeded $560 million, an increase of more than 100 percent from 2008. True or False?

5. Convincing someone to give out a critical password is an example of _____.

6. Someone who intimidates or coerces a government or organization to advance his or her political or social objectives by launching computer-based attacks against computers, networks, and the information stored on them is called a(n) _____.

7. The _____ is the federal organization assigned to help identity theft victims restore their credit and erase the impact of the imposter.

8. Law enforcement agencies have combined crime-related data, powerful analysis techniques, and _____ to better understand and even diminish crime risks.

9. A rootkit is a type of Trojan horse that executes when specific conditions occur. True or False?

10. Malware that is capable of spreading itself from one computer to another is called a _____.
    a. logic bomb
    b. Trojan horse
    c. virus
    d. worm

11. A(n) _____ is a modified version of a virus that is produced by the virus's author or another person amending the original virus code.

12. Deleting files and emptying the Recycle Bin ensures that others cannot view the personal data on your recycled computer. True or False?

13. _____ is software that provides a comprehensive display on a single computer screen of all the vital data related to an organization's security defenses, including threats, exposures, policy compliance, and incident alerts.

Jobs, equipment, and working conditions must be designed to avoid negative health effects from computers.

14. A job that requires sitting at a desk and using a computer for many hours a day can double the risk of seated immobility thromboembolism, which is the _____.

Practitioners in many professions subscribe to a code of ethics that states the principles and core values that are essential to their work.

15. Just because an activity is defined as legal, it does not mean that it is ethical. True or False?

CHAPTER 9: SELF-ASSESSMENT TEST ANSWERS

(1) True (2) costs, profits (3) monitoring (4) False (5) social engineering (6) cyberterrorist (7) Federal Trade Commission (8) Geographic Information Systems (9) False (10) d (11) variant (12) False (13) Security dashboard (14) formation of blood clots in the legs or lungs (15) True

REVIEW QUESTIONS

1. What does the CAN-SPAM Act specifically allow?
2. What is a potential danger of using spam filters?
3. According to the 2009 FBI Internet Crime Report, how much did computer crime increase between 2008 and 2009?
4. What is social engineering?
5. Briefly discuss the seriousness with which the U.S. federal government views cyberterrorism.
6. How do you distinguish between a hacker and a criminal hacker?
7. Why are insiders one of the biggest threats for company computer systems?
8. What is a virus? What is a worm? How are they different?
9. What is vishing? What actions can you take to reduce the likelihood that you will be a victim of this crime?
10. What is filtering software? Why would organizations use such software?
11. What does intrusion detection software do? What are some of the issues with the use of this software?
12. What is the difference between a patent and a copyright? What copyright issues come into play when downloading software or music from a Web site?
13. What is a John Doe lawsuit? How would such a lawsuit arise?
14. What is ergonomics? How can it be applied to office workers?
15. What specific actions can you take to avoid spyware?
16. What is a code of ethics? Give an example.

DISCUSSION QUESTIONS

1. Discuss how policies and procedures can help prevent computer waste and mistakes.
2. Discuss at least three examples of the computer being used as a tool to fight crime.
3. Identify the risks associated with the disposal of obsolete computers. Discuss the steps that one must take to safely dispose of personal computers.
4. Identify five strong measures organizations can take to safeguard against the theft of computer hardware and the data stored on it.
5. Briefly discuss software piracy. What is it, how widespread is it, and who is harmed by it?
6. Identify and briefly discuss three acts of Congress that were implemented to limit children's exposure to online pornography.
7. Imagine that your friend regularly downloads copies of newly released, full-length motion pictures for free from the Internet and makes copies for others for a small fee. Do you think that this is ethical? Is it legal? Would you express any concerns to him?
8. Outline an approach, including specific techniques (e.g., dumpster diving, phishing, social engineering), that you could employ to gain personal data about the members of your class.
9. Your 12-year-old niece shows you a dozen or so photos of herself and a brief biography, including address and cell phone number that she plans to post on Facebook. What advice might you offer her about posting personal information and photos?
10. Imagine that you are a hacker and have developed a Trojan horse program. What tactics might you use to get unsuspecting victims to load the program onto their computer?
11. Briefly discuss the potential for cyberterrorism to cause a major disruption in our daily life. What are some likely targets of a cyberterrorist? What sort of action could a cyberterrorist take against these targets?
12. Do you believe that the National Security Agency should be able to collect the telephone call records of U.S. citizens without the use of search warrants? Why or why not?
13. Using information presented in this chapter on federal privacy legislation, identify which federal law regulates the following areas and situations: cross-checking IRS and Social Security files to verify the accuracy of information, customer liability for debit cards, your right to access data contained in federal agency files, the IRS obtaining personal information, the government obtaining financial records, and employers' access to university transcripts.
14. Briefly discuss the differences between acting morally and acting legally. Give an example of acting legally and yet immorally.

PROBLEM-SOLVING EXERCISES

1. Access the Web sites for the Recording Industry Association of America (RIAA), Motion Picture Association of America (MPAA), and Business Software Alliance (BSA) to get estimates of the amount of piracy worldwide for at least five years. Use a graphics package to develop a bar chart to show the amount of music, motion picture, and software piracy over a five-year time period.
2. Using spreadsheet software and appropriate forecasting routines, develop a forecast for the amount of piracy for next year. Document any assumptions you make in developing your forecast.
3. Do research on recent changes to the USA Patriot Act. Use presentation software to document these changes and present your position on the impact and legitimacy of these changes.

TEAM ACTIVITIES

1. The Digital Due Process Coalition includes the Internet's largest online service providers, such as Google, Microsoft, Facebook, AOL, and eBay, as well as social organizations such as the ACLU and the Bill of Rights Defense Committee. The Coalition is pushing for updates in the nation's electronic privacy law, the 1986 Electronic

Communications Protection Act. Do research to find out more about this organization and the four principles it wants to see implemented in new or revised privacy legislation. Prepare to debate either side of this statement: the Digital Due Process Coalition is doing worthwhile work to improve the privacy of electronic data.

2. Have each member of your team access ten different Web sites and summarize their findings in terms of the existence of data privacy policy statements. Did the site have such a policy? Was it easy to find? Was it complete and easy to understand? Did you find any sites using the P3P standard or ICRA rating method?

WEB EXERCISES

1. Do research on the Web and find information about efforts being taken to speed up and improve the patent application process. What are the forces that are driving these changes? Write a brief report summarizing your findings.
2. Do research on the Web to find the latest information about Google's ongoing battle with China over the filtering of Internet search results. Write a brief report documenting your findings. Do you think that Google has acted in an ethical manner in regards to its response to China's request to filter search results?
3. The Computer Emergency Response Team Coordination Center (CERT/CC) is located at the Software Engineering Institute (SEI), a federally funded research and development center at Carnegie Mellon University in Pittsburgh, Pennsylvania. Do research on the center and write a brief report summarizing its activities.

CAREER EXERCISES

1. You are a senior member of a marketing organization for a manufacturer of children's toys. A recommendation has been made to develop a Web site to promote and sell your firm's products as well as learn more about what parents and their children are looking for in new toys. Develop a list of laws and regulations that will affect the design of the Web site. Describe how these will limit the operation of your new Web site.
2. You have just begun a new position in customer relations for a retail organization that sells its products both online and in brick-and-mortar stores. Within your first week on the job, several customers have expressed concern about potential theft of customer data from the store's computer databases and identity theft. Who would you talk to within your organization to develop a satisfactory response to address your customers' concerns? What key points would you need to verify with the store's technical people?

CASE STUDIES

Case One
The 'My SHC Community' Privacy Disaster

Sears Holdings Corporation is the third largest retailer in the U.S., owning both Sears Roebuck and Kmart. Like all retailers, Sears is heavily invested in marketing research. Discovering customers' interests and purchasing habits is essential to successful marketing and sales campaigns.

Gathering data on customer interests can be difficult. The Internet helps considerably by allowing businesses to analyze the actions of visitors on their sites. Though helpful, Web analytics provide only a small glimpse into the online activities of customers; businesses cannot see what users do elsewhere on the Web. Some businesses have tried, typically with disastrous effects, to track users around the Web. For example, online advertiser DoubleClick was charged with violations of privacy laws when it placed ads on many Web sites to track users' movements around the Web. More recently, Facebook has been scrutinized for its Open Graph technology that allows it and its partners to track users' movements around the Web.

Sears Holdings Corporation (SHC) took its stab at tracking customers' online activities through a program it called My SHC Community. Customers of Sears and Kmart were encouraged to "become part of something new, something different." My SHC was framed as a "dynamic and highly interactive online community" where "your voice is heard and your opinion matters, and what you want and need counts!" Members were asked to install "research software" on their computers that would allow SHC to track their online browsing. To further entice customers to participate, SHC paid them $10 each.

After operating for several months, technology analysts discovered that the SHC "research" software collected a lot more information than users were aware of. It not only tracked URLs of Web pages visited, but also information typed into secure online forms. Such information included user names and passwords, credit card numbers, online shopping cart contents, essentially everything the user typed and every page the user visited. When Harvard Business School Assistant Professor Ben Edelman learned of the situation, he accused SHC of distributing spyware. Shortly thereafter, the Federal Trade Commission (FTC) took up the case.

The FTC found that Sears Holdings did not sufficiently notify its customers about what information was being gathered by the software. The only notice of what the software actually did was buried on page 10 of the software license that users may have glanced at before installing the software.

The FTC ordered Sears Holdings to discontinue distribution of the software, destroy all data gained from the experiment, and to stop collecting data from copies of the software still running on customer computers. Additionally, should Sears Holdings decide to attempt a similar experiment in the future, it must "clearly and prominently disclose the types of data the software will monitor, record, or transmit. This disclosure must be made prior to installation and separate from any user license agreement. Sears must also disclose whether any of the data will be used by a third party."

Sears Holdings has stated that while the software did collect sensitive user information, the company made "commercially viable efforts automatically to filter out confidential personally identifiable information such as user ID, password, credit card numbers, and account numbers, and made commercially viable efforts to purge our database of any such information if it was collected inadvertently."

Discussion Questions

1. What plan did Sears Holdings implement to gain more insight into customer's online shopping behavior?
2. Why was the plan considered by a Harvard Law Professor and the FTC to be unethical and unlawful?

Critical Thinking Questions

1. If Sears Holdings had implemented its plan in accordance with FTC guidelines, do you think any customers would have been willing to sacrifice their privacy for $10?
2. What other methods might Sears Holdings and other companies consider implementing to gather information about customers' online habits and interests?

SOURCES: Anderson, Nate, "FTC forces Sears, Kmart out of the spyware business," *Ars Technica*, www.arstechnica.com, September 13, 2009; McMillan, Robert, "Researcher Accuses Sears of Spreading Spyware," PCWorld, www.pcworld.com, January 2, 2008; My SHC Community Web page, www.myshccommunity.com, accessed June 26, 2010.

Case Two

U.S. Fights Fraud with Personal Certificates

One of the goals of the Obama administration has been to improve national information security to better protect the nation's infrastructure. Toward this goal, the President created a new post in his administration called the Cybersecurity Coordinator, appointing Howard Schmidt to the position.

One of Schmidt's first assignments was to develop a plan to effectively fight Internet fraud. Internet fraud is the most common type of Internet crime, covering a range of scams in which a buyer or seller assumes a false identity to trick someone out of money or merchandise. According to the FBI's Internet Crime Complaint Center, Internet fraud reports increased by 22.3 percent in 2009, with losses totaling close to $560 million. Among the most popular forms of fraud in 2009 were e-mail scams that used the FBI name to gain information (16.6 percent), undelivered merchandise or payment (11.9 percent), and advanced fee fraud (in which targets are asked to pay upfront) (9.8 percent). The one fact that is consistent across all cases of fraud is the use of a false identity. If real identities were used, the fraudsters would be caught.

Cybersecurity Coordinator Howard Schmidt has proposed a plan by which those involved in online transactions can be positively identified. Because credit cards and bank account numbers can be easily stolen and used by fraudsters, Schmidt's plan requires the possession of a smart identity card or digital certificate to buy and sell online.

The notion of national ID cards has been struck down by privacy advocates in the past, so Schmidt made his proposal a voluntary one. According to a draft plan, those who wish to authenticate their transactions would be able to acquire a secure identifier from a variety of service providers. Schwartz emphasizes that the system will balance efforts to maintain privacy while still collecting enough information about the card owner to ensure his identity.

Security vendors believe that the technology is already available for Schmidt's plan. For example, Google and Microsoft already have "single sign-on" systems for which people can use their credentials to sign on to partner sites. In this scenario, Google or Microsoft maintains information about the user, while the partner sites are not provided access to that private information. A similar system could be used for online transaction authentication. What is needed is a national push for standards so that businesses and consumers can invest in a technology with some assurance that it will be long lasting.

Security experts have expressed skepticism that a voluntary system will be effective. Some argue that unless all Internet users are required to use the system, many transactions will remain vulnerable. The government must walk a fine line between privacy and ensuring secure transactions. Only time will tell what type of authentication system will be regarded as acceptable to government, consumers, and businesses.

Discussion Questions

1. Why does the U.S. government think it's necessary to intervene in online transactions?
2. Why might the use of digital certificates or smart identity cards to fight Internet fraud be controversial?

Critical Thinking Questions

1. How might a digital certificate or smart identity card system be implemented so that individuals' privacy is maintained and yet transactions are made more secure?
2. What ways might a smart card system backfire or present other security or privacy concerns?

SOURCES: 2009 IC3 Annual Report, IC3 Web site, *www.ic3.gov/media/annual-report/2009_IC3Report.pdf*, accessed July 15, 2010; Baldor, Lolita, "White House Unveils Cybersecurity Plans," NewsFactor Network, *www.newsfactor.com*, June 29, 2010; Stokes, Jon, "White House wants to help you 'blog Anonymously'," *Ars Technica, www.arstechnica.com*, June 29, 2010; Markoff, John, "Taking the Mystery Out of Web Anonymity," *New York Times*, *www.nytimes.com*, July 2, 2010.

Questions for Web Case

See the Web site for this book to read about the Altitude Online case for this chapter. Following are questions concerning this Web case.

Altitude Online: The Personal and Social Impact of Computers

Discussion Questions

1. Why do you think extending access to a corporate network beyond the business's walls dramatically elevates the risk to information security?
2. What tools and policies can be used to minimize that risk?

Critical Thinking Questions

1. Why does information security usually come at the cost of user convenience?
2. How do proper security measures help ensure information privacy?

NOTES

Sources for the opening vignette: Rebbapragada, Narasu "What is your Facebook data worth?" Computerworld, *www.computerworld.com*, June 21, 2010; Bryant, Martin, "Ignore Facebook Open Graph at your peril – this is Web 3.0," The Next Web, *http://thenextweb.com*, April 21, 2010; Helft, Miguel, *Principles of Information Systems 8-66 CWR/lr* "Zuckerberg on the Hot Seat," *New York Times, www.nytimes.com*, June 2, 2010.; Levine, Barry, "Facebook Refuses Changes Urged by Privacy Groups," News Factor, *www.newsfactor.com*, June 18, 2010; Gaudin, Sharon, "More than half of Facebook users may quit site, poll finds," Computerworld, *www.computerworld.com*, May 21, 2010; Kincaid, Jason, "WSJ: Facebook, MySpace & Others Share Identifying User Data With Advertisers," TechCrunch, *www.techcrunch.com*, May 20, 2010.

1   O'Keefe, Ed, "Obama to Order Federal Agencies to Compile 'Do Not Pay List'," *Washington Post*, June 18, 2010.
2   Bewley, Elizabeth, "Fraud Fight Uses 'Do Not Pay' List," *Cincinnati Enquirer*, June 19, 2010, page A12.
3   Arthur, Charles, "Will Convicting Five Major Spammers Put an End to Spam?" *The Guardian*, June 30, 2009, *www.guardian.co.uk/technology/2009/jun/24/spam-newly-asked-questions/print*.
4   Staff, "Spam Filter Review for 2010," Top Ten Reviews, *http://spam-filter-review.toptenreviews.com*, accessed June 1, 2010.
5   Hananel, Sam, "Officials: Computer Error Affected Mine Scrutiny," *www.philly.com*, April 13, 2010.
6   Baxter, Elsa, "Computer Error Causes UAE Cash Machine Shutdown," *www.arabianbusiness.com/582494-computer-error-causes-uae-cash-machine-shutdown*, February 28, 2010.
7   Staff, "Computer Error Signals Erroneous Report of Earthquake in Northern California," Fox News, April 20, 2010.
8   Puzzanghera, Jim, "Regulators Still Working to Pinpoint Cause of Stock Market's Record Plunge," *Chicago Tribune*, May 11, 2010.

9   Barlow, Daniel, "State to Block X-Rated Web Sites," *Time Argus*, *timesargus.com/article/20100413/NEWS01/4130343/classifieds*, April 13, 2010
10   Hamilton, Tyler, "Dozens Fired at Bruce Power Over Web, E-Mail Use," *www.thestar.com/printarticle/693525*, September 10, 2009.
11   Klee, Andy, "SAP End User Training· An Interview with Shannon Hicks, South Carolina Enterprise Information Systems," IT Toolbox blog, *http://it.toolbox.com/blogs/saptraining*, posted January 6, 2010.
12   Doland, Angela, "Trial Opens for Accused French Rogue Trader," *Real Clear Markets, www.realclearmarkets.com/news/ap/finance_business/2010/Jun/08/trial_opens_for_accused_french_rogue_trader.html*, accessed June 8, 2010.
13   "Success Story: Quality High on the Menu for Specialty Food Smoker," University of Plymouth, *www.plymouth.ac.uk/files*, accessed June 8, 2010.
14   Staff, "Computer Crime Reports Increase 22% in 2009," *Crime News – Crime Prevention, http://crimeinamerica.net/2010/03/16/computer-crime-reports-increase*, March 16, 2010.
15   "Cyber Hackers Breached Jet Fighter Program," *Air Force Times*, April 23, 2009.
16   "U.S. Indicts Ohio Man and Two Foreign Residents in Alleged Ukraine-Based 'Scareware' Fraud Scheme That Caused $100 Million in Losses to Internet Victims Worldwide," Department of Justice press release, *http://chicago.fbi.gov/dojpressrel/pressrel10/cg052710.htm*, May 27, 2010.
17   "Chico Man Pleads Guilty to Embezzling $693,000 from Charity," Department of Justice press release, *www.justice.gov/criminal/cybercrime/randlePlea.pdf*, May 14, 2010.

18 "Indian National Sentenced to 81 Months in Prison for Role in International Online Brokerage 'Hack, Pump, and Dump' Scheme," Department of Justice Press Release, *www.cybercrime.gov/marimuthuSent.pdf*, April 26, 2010.

19 "Computer Crime Reports Increase 22% in 2009," *Crime News – Crime Prevention*, *http://crimeinamerica.net/2010/03/16/computer-crime-reports-increase*, March 16, 2010.

20 Stasiukonis, Steve, "Security's Top 4 Social Engineers of All Time," dark reading blog, *www.darkreading.com/blog/archives/2010/05/securitys_top_4.html*, posted May 26, 2010.

21 IMPACT Web site, "Mission & Vision: About Us," *www.impact-alliance.org/about_us.html*, accessed June 10, 2010.

22 Mann, Jack, "Panetta Warns Cyber Attack Could be Next Pearl Harbor," *The New New Internet*, April 21, 2010.

23 Cheek, Michael, "Threat of Cyber Terrorism 'Real and Expanding' Says FBI Director Mueller," *The New New Internet*, March 5, 2010.

24 Gorman, Siobhan, "Electricity Grid in U.S. Penetrated by Spies," *The Wall Street Journal*," April 8, 2009.

25 Funk, Josh, "TD Ameritrade Data Theft Settlement Gets Court OK, *Technology*, May 11, 2009.

26 Isikoff, Michael, "Bernanke Victimized by Identity Fraud Ring," *Newsweek*, August 25, 2009.

27 Editorial staff, "Tax Revenue to be Gained from Internet Gambling Isn't Worth Social Costs," *The Daily News*, June 6, 2010.

28 Editorial staff, "Tax Revenue to be Gained from Internet Gambling Isn't Worth Social Costs," *The Daily News*, June 6, 2010.

29 Leads Online Web site, *www.leadsonline.com/main/default.aspx*, accessed June 19, 2010.

30 Williams, Matt, "New Web System for Rockford Pawnshops Leads to Arrests," *Rockford Register Star*, April 9, 2010.

31 Community Notification Web site, *http://communitynotification.com*, accessed June 19, 2010.

32 Stetz, Michael, "Even GPS as a Tool to Fight Crime Has Limits," *Sign On Diego*, March 4, 2010.

33 Shullingford, David and Groussman, Jon D., "Using GIS to Fight Crime," *ISO Review*, March 2010.

34 Zetter, Kim, "TJX Hacker Gets 20 Years in Prison," *Wired*, March 25, 2010.

35 Zetter, Kim, "New Malware Rewrites Online Bank Statements to Cover Fraud," *Wired*, September 30, 2009.

36 "Kaspersky Predicts More iPhone, Android Attacks in 2010," *ABS CBN News*, *www.abs-cbnnews.com*, January 2, 2010.

37 Mills, Elinor, "Malware Found Lurking in Apps for Windows Mobile," *CNet News*, June 4, 2010.

38 Graham Cluley's blog, "Indian Government Computers Hit by Chinese Spyware Attack?" *www.sophos.com/blogs/gc/g/2009/02/16*, accessed June 20, 2010.

39 Staff, "Stolen Laptop Exposes Personal Data on 207,000 Army Reservists," *Krebs on Security*, *http://krebsonsecurity.com/2010/05/stolen-laptop-exposes-personal-data*, May 13, 2010.

40 Duffy, Jim, "Software Piracy Rate Up 2% in 2009, Study Finds," *Network World*, May 11, 2010.

41 Staff, "Crime Rate: Software Piracy Rate (most recent) by Country," *www.nationmaster.com/graph/cri_sof_pir_rat-crime-software-piracy-rate*, accessed June 11, 2010.

42 Staff, "Piracy: Online and on the Street," *www.riaa.com/physicalpiracy.php*, accessed June 12, 2010.

43 Staff, "Music Piracy Not That Bad, Industry Says," *CyberLaw Blog*, *http://cyberlaw.org.uk/2009/01/18/music-piracy-not-that-bad-industry-says*, January 18, 2009.

44 Kravets, David, "Recording Industry Says LimeWire on Hook for $1 Billion," *Wired*, June 8, 2010.

45 Kravets, David, "LimeWire Crushed in RIAA Infringement Lawsuit," *Wired*, May 12, 2010.

46 Craig, Brenda, "Google in Historic Lawsuit," Lawyers and Settlements Web site, *www.lawyersandsettlements.com/articles/14118/copyright-infringement-news-5.html*, May 12, 2010.

47 Paul, Ryan, "Red Hat Faces Another Patent Infringement Lawsuit over JBoss," *Ars Technica*, *http://arstechnica/com/open-source/news*, March 4, 2009.

48 Staff, "ITC to Probe Apple for HTC Patent Infringement," *Yahoo! News*, June 11, 2010.

49 Elliott, Christopher, "Travelers Vulnerable to 'Phishing' Scams," *CNN.com*, June 14, 2010.

50 Ledford, Jerri, "Vishing," *http://idtheft.about.com/od/glossary/g/Vishing.htm*, accessed June 18, 2010.

51 Staff, "SpoofCard," *www.spoofcard.com*, accessed June 18, 2010.

52 Hearns, Michael, "Australian in Las Vegas Suspected of Over 500 Million Laundered via Internet Gambling Web Sites," Money Laundering, *http://laundering money.blogspot.com/2010/australian-in-las-vegas-suspected of.html*, April 25, 2010.

53 Staff, "2GB Fingerprint Biometric USB Flash Memory Stick Drive AFU-082," *www.dinodirect.com/usb-flash-memory-stick-drive-finger-print-2gb-afu-082/AFFID-19.html*, accessed June 15, 2010.

54 Staff, "Microsoft Case Studies: Del Monte Foods Premium-Quality Food Producer Drives Growth with Security-Enhanced Collaboration Solutions," *www.microsoft.com/casestudies/Case_Study_Detail.aspx?casestudyid=4000007154*, May 5, 2010.

55 Staff, "Managed Protection Services," *www-935.ibm.com/services/us/index.wss/offering/iss/a1026962*, accessed June 20, 2010.

56 Srikanth, AD, "How to: Remotely Track Your Stolen Laptop?" Technology Bites, *www.teknobites.com/2010/04/06/how-to-remotely-track-your-stolen-laptop*, April 6, 2010.

57 Staff, "Top Standalone Antivirus Software for 2010," *PC World*, November 23, 2009.

58 "CVS Caremark Settles FTC Charges: Failed to Protect Medical and Financial Privacy of Customers and Employees," Federal Trade Commission Web site, *www.ftc.gov/opa/2009/02/cvs.shtm*, accessed July 13, 2010.

59 "CVS Caremark Settles FTC Charges: Failed to Protect Medical and Financial Privacy of Customers and Employees," Federal Trade Commission Web site, *www.ftc.gov/opa/2009/02/cvs.shtm*, accessed July 13, 2010.

60 "Survey: Many Companies Monitoring, Recording, Videotaping Employees," *Business Wire*, February 6, 2009.

61 Young, Chris I., "NAACP: Sun Trust Mortgage Worker Fired Over E-Mail," *Times-Dispatch*, March 18, 2010.

62 "CNN Fires Journalist for Tweeting Her Praise for Islamic Cleric," Greenslade Blog, *www.guardian.co.uk/media/greenslade/2010/jul/08/cnn-twitter/print*, accessed July 13, 2010.

63 Tencer, Daniel, "Virginia Delegates Pass Bill Banning Chip Implants as 'Mark of the Beast'," *Raw Story*, *http://rawstory.com/2010/02/virginia-passes-law-banning-chip-implants-mark-beast*, February 10, 2010.

64 Vijayan, Jaikumar, "Online Libel Case Stirs Up Free Speech Debate," *Computerworld*, October 15, 2009.

65 Wilonsky, Robert, "Catherine Crier Takes a Dallas 'John Doe' to Court Over Her Wikipedia Page," *Dallas Observer*, May 21, 2009.

66 Rubenking, Neil J., "Keep Your Child Safe Online," *PC Magazine*, March 1, 2010.

67 "NCI Statement: International Study Shows No Increased Risk of Brain Tumors from Cell Phone Use," National Cancer Institute Web page, *www.cancer.gov/newscenter/pressreleases/Interphone2010Results*, May 17, 2020.

68 "Creating an Injury-Free Workplace – How to Avoid Corporate Complacency," *http://businessinabullet.com/human-resources/cr...*, June 27, 2010.

acceptance testing Conducting any tests required by the user.

accounting MIS An information system that provides aggregate information on accounts payable, accounts receivable, payroll, and many other applications.

ad hoc DSS A DSS concerned with situations or decisions that come up only a few times during the life of the organization.

antivirus program Software that runs in the background to protect your computer from dangers lurking on the Internet and other possible sources of infected files.

application program interface (API) An interface that allows applications to make use of the operating system.

application service provider (ASP) A company that provides software, support, and the computer hardware on which to run the software from the user's facilities over a network.

arithmetic/logic unit (ALU) The part of the CPU that performs mathematical calculations and makes logical comparisons.

ARPANET A project started by the U.S. Department of Defense (DoD) in 1969 as both an experiment in reliable networking and a means to link the DoD and military research contractors, including many universities doing military-funded research.

artificial intelligence (AI) A field in which the computer system takes on the characteristics of human intelligence.

artificial intelligence systems People, procedures, hardware, software, data, and knowledge needed to develop computer systems and machines that demonstrate the characteristics of intelligence.

asking directly An approach to gather data that asks users, stakeholders, and other managers about what they want and expect from the new or modified system.

attribute A characteristic of an entity.

auditing Analyzing the financial condition of an organization and determining whether financial statements and reports produced by the financial MIS are accurate.

batch processing system A form of data processing where business transactions are accumulated over a period of time and prepared for processing as a single unit or batch.

blade server A server that houses many individual computer motherboards that include one or more processors, computer memory, computer storage, and computer network connections.

Bluetooth A wireless communications specification that describes how cell phones, computers, faxes, personal digital assistants, printers, and other electronic devices can be interconnected over distances of 10–30 feet at a rate of about 2 Mbps.

brainstorming A decision-making approach that consists of members offering ideas "off the top of their heads."

broadband communications A relative term but generally means a telecommunications system that can exchange data very quickly.

business intelligence (BI) The process of gathering enough of the right information in a timely manner and usable form and analyzing it to have a positive impact on business strategy, tactics, or operations.

business-to-business (B2B) e-commerce A subset of e-commerce where all the participants are organizations.

business-to-consumer (B2C) e-commerce A form of e-commerce in which customers deal directly with an organization and avoid intermediaries.

byte (B) Eight bits that together represent a single character of data.

Cascading Style Sheet (CSS) A markup language for defining the visual design of a Web page or group of pages.

central processing unit (CPU) The part of the computer that consists of three associated elements: the arithmetic/logic unit, the control unit, and the register areas.

centralized processing An approach to processing wherein all processing occurs in a single location or facility.

certificate authority (CA) A trusted third-party organization or company that issues digital certificates.

certification A process for testing skills and knowledge, which results in a statement by the certifying authority that confirms an individual is capable of performing a particular kind of job.

channel bandwidth The rate at which data is exchanged, usually measured in bits per second (bps).

character A basic building block of most information, consisting of uppercase letters, lowercase letters, numeric digits, or special symbols.

chief knowledge officer (CKO) A top-level executive who helps the organization use a KMS to create, store, and use knowledge to achieve organizational goals.

choice stage The third stage of decision making, which requires selecting a course of action.

client/server architecture An approach to computing wherein multiple computer platforms are dedicated to special functions, such as database management, printing, communications, and program execution.

clock speed A series of electronic pulses produced at a predetermined rate that affects machine cycle time.

code of ethics A code that states the principles and core values that are essential to a set of people and that therefore govern their behavior.

command-based user interface A user interface that requires you to give text commands to the computer to perform basic activities.

compact disc read-only memory (CD-ROM) A common form of optical disc on which data cannot be modified once it has been recorded.

competitive advantage A significant and (ideally) long-term benefit to a company over its competition.

competitive intelligence One aspect of business intelligence limited to information about competitors and the ways that knowledge affects strategy, tactics, and operations.

computer network The communications media, devices, and software needed to connect two or more computer systems or devices.

computer programs Sequences of instructions for the computer.

computer-aided software engineering (CASE) Tools that automate many of the tasks required in a systems development effort and encourage adherence to the SDLC.

computer-assisted manufacturing (CAM) A system that directly controls manufacturing equipment.

computer-based information system (CBIS) A single set of hardware, software, databases, telecommunications, people, and procedures that are configured to collect, manipulate, store, and process data into information.

computer-integrated manufacturing (CIM) Using computers to link the components of the production process into an effective system.

concurrency control A method of dealing with a situation in which two or more users or applications need to access the same record at the same time.

consumer-to-consumer (C2C) e-commerce A subset of e-commerce that involves consumers selling directly to other consumers.

content streaming A method for transferring large media files over the Internet so that the data stream of voice and pictures plays more or less continuously as the file is being downloaded.

control unit The part of the CPU that sequentially accesses program instructions, decodes them, and coordinates the flow of data in and out of the ALU, the registers, the primary storage, and even secondary storage and various output devices.

cost center A division within a company that does not directly generate revenue.

counterintelligence The steps an organization takes to protect information sought by "hostile" intelligence gatherers.

criminal hacker (cracker) A computer-savvy person who attempts to gain unauthorized or illegal access to computer systems to steal passwords, corrupt files and programs, or even transfer money.

critical path Activities that, if delayed, would delay the entire project.

culture A set of major understandings and assumptions shared by a group.

cybermall A single Web site that offers many products and services at one Internet location.

cyberterrorist Someone who intimidates or coerces a government or organization to advance his political or social objectives by launching computer-based attacks against computers, networks, and the information stored on them.

data Raw facts, such as an employee number, total hours worked in a week, inventory part numbers, or sales orders.

data administrator A nontechnical position responsible for defining and implementing consistent principles for a variety of data issues.

data analysis The manipulation of collected data so that the development team members who are participating in systems analysis can use the data.

data center A climate-controlled building or set of buildings that house database servers and the systems that deliver mission-critical information and services.

data collection Capturing and gathering all data necessary to complete the processing of transactions.

data correction The process of reentering data that was not typed or scanned properly.

data definition language (DDL) A collection of instructions and commands used to define and describe data and relationships in a specific database.

data dictionary A detailed description of all the data used in the database.

data editing The process of checking data for validity and completeness.

data item The specific value of an attribute.

data loss prevention (DLP) Systems designed to lock down—to identify, monitor, and protect—data within an organization.

data manipulation The process of performing calculations and other data transformations related to business transactions.

data manipulation language (DML) A specific language, provided with a DBMS, which allows users to access and modify the data, to make queries, and to generate reports.

data mart A subset of a data warehouse, used by small and medium-sized businesses and departments within large companies to support decision making.

data mining An information-analysis tool that involves the automated discovery of patterns and relationships in a data warehouse.

data model A diagram of data entities and their relationships.

data preparation, or data conversion Making sure all files and databases are ready to be used with new computer software and systems.

data storage The process of updating one or more databases with new transactions.

data store Representation of a storage location for data.

data warehouse A large database that collects business information from many sources in the enterprise, covering all aspects of the company's processes, products, and customers, in support of management decision making.

database An organized collection of facts and information.

database administrator (DBA) A skilled IS professional who directs all activities related to an organization's database.

database approach to data management An approach to data management whereby a pool of related data is shared by multiple information systems.

database management system (DBMS) A group of programs that manipulate the database and provide an interface between the database and the user of the database and other application programs.

data-flow diagram (DFD) A model of objects, associations, and activities that describes how data can flow between and around various objects.

data-flow line Arrows that show the direction of data element movement.

decentralized processing An approach to processing wherein processing devices are placed at various remote locations.

decision room A room that supports decision making, with the decision makers in the same building, combining face-to-face verbal interaction with technology to make the meeting more effective and efficient.

decision support system (DSS) An organized collection of people, procedures, software, databases, and devices used to support problem-specific decision making.

decision-making phase The first part of problem solving, including three stages: intelligence, design, and choice.

delphi approach A decision-making approach in which group decision makers are geographically dispersed; this approach encourages diversity among group members and fosters creativity and original thinking in decision making.

demand report A report developed to give certain information at someone's request rather than on a schedule.

design report The primary result of systems design, reflecting the decisions made and preparing the way for systems implementation.

design stage The second stage of decision making, in which you develop alternative solutions to the problem and evaluate their feasibility.

desktop computer A relatively small, inexpensive, single-user computer that is highly versatile.

dialogue manager A user interface that allows decision makers to easily access and manipulate the DSS and to use common business terms and phrases.

digital audio player A device that can store, organize, and play digital music files.

digital camera An input device used with a PC to record and store images and video in digital form.

digital certificate An attachment to an e-mail message or data embedded in a Web site that verifies the identity of a sender or Web site.

digital video disc (DVD) A storage medium used to store software, video games, and movies.

direct access A retrieval method in which data can be retrieved without the need to read and discard other data.

direct access storage device (DASD) A device used for direct access of secondary storage data.

direct conversion (also called *plunge* or *direct cutover*) Stopping the old system and starting the new system on a given date.

direct observation Directly observing the existing system in action by one or more members of the analysis team.

distributed database A database in which the data can be spread across several smaller databases connected via telecommunications devices.

distributed processing An approach to processing wherein processing devices are placed at remote locations but are connected to each other via a network.

document production The process of generating output records and reports.

domain The allowable values for data attributes.

domain expert The person or group with the expertise or knowledge the expert system is trying to capture (domain).

drill-down report A report providing increasingly detailed data about a situation.

dumpster diving Going through the trash cans of an organization to find secret or confidential information, including information needed to access an information system or its data.

e-commerce Any business transaction executed electronically between companies (business-to-business), companies and consumers (business-to-consumer), consumers and other consumers (consumer-to-consumer), business and the public sector, and consumers and the public sector.

economic feasibility The determination of whether the project makes financial sense and whether predicted benefits offset the cost and time needed to obtain them.

economic order quantity (EOQ) The quantity that should be reordered to minimize total inventory costs.

e-Government The use of information and communications technology to simplify the sharing of information, speed formerly paper-based processes, and improve the relationship between citizens and government.

electronic business (e-business) Using information systems and the Internet to perform all business-related tasks and functions.

electronic cash An amount of money that is computerized, stored, and used as cash for e-commerce transactions.

electronic commerce Conducting business activities (e.g., distribution, buying, selling, marketing, and servicing of products or services) electronically over computer networks.

electronic exchange An electronic forum where manufacturers, suppliers, and competitors buy and sell goods, trade market information, and run back-office operations.

electronic retailing (e-tailing) The direct sale of products or services by businesses to consumers through electronic storefronts, typically designed around an electronic catalog and shopping cart model.

end-user systems development Any systems development project in which the primary effort is undertaken by a combination of business managers and users.

enterprise data modeling Data modeling done at the level of the entire enterprise.

enterprise resource planning (ERP) system A set of integrated programs capable of managing a company's vital business operations for an entire multisite, global organization.

enterprise system A system central to the organization that ensures information can be shared across all business functions and all levels of management to support the running and managing of a business.

entity A general class of people, places, or things for which data is collected, stored, and maintained.

entity symbol Representation of either a source or destination of a data element.

entity-relationship (ER) diagrams Data models that use basic graphical symbols to show the organization of and relationships between data.

environmental design Also called *green design*, it involves systems development efforts that slash power consumption, require less physical space, and result in systems that can be disposed in a way that doesn't negatively affect the environment.

ergonomics The science of designing machines, products, and systems to maximize the safety, comfort, and efficiency of the people who use them.

exception report A report automatically produced when a situation is unusual or requires management action.

executive support system (ESS) Specialized DSS that includes all hardware, software, data, procedures, and people used to assist senior-level executives within the organization.

expert system A system that gives a computer the ability to make suggestions and function like an expert in a particular field.

explanation facility Component of an expert system that allows a user or decision maker to understand how the expert system arrived at certain conclusions or results.

Extensible Markup Language (XML) The markup language designed to transport and store data on the Web.

external auditing Auditing performed by an outside group.

extranet A network based on Web technologies that allows selected outsiders, such as business partners and customers, to access authorized resources of a company's intranet.

feasibility analysis Assessment of the technical, economic, legal, operational, and schedule feasibility of a project.

feedback Output that is used to make changes to input or processing activities.

field Typically a name, number, or combination of characters that describes an aspect of a business object or activity.

file A collection of related records.

final evaluation A detailed investigation of the proposals offered by the vendors remaining after the preliminary evaluation.

financial MIS An information system that provides financial information for executives and for a broader set of people who need to make better decisions on a daily basis.

five-forces model A widely accepted model that identifies five key factors that can lead to attainment of competitive advantage, including (1) the rivalry among existing competitors, (2) the threat of new entrants, (3) the threat of substitute products and services, (4) the bargaining power of buyers, and (5) the bargaining power of suppliers.

flexible manufacturing system (FMS) An approach that allows manufacturing facilities to rapidly and efficiently change from making one product to making another.

forecasting Predicting future events to avoid problems.

game theory The use of information systems to develop competitive strategies for people, organizations, or even countries.

Gantt chart A graphical tool used for planning, monitoring, and coordinating projects.

genetic algorithm An approach to solving large, complex problems in which many related operations or models change and evolve until the best one emerges.

geographic information system (GIS) A computer system capable of assembling, storing, manipulating, and displaying geographic information, that is, data identified according to its location.

graphical user interface (GUI) An interface that displays pictures (icons) and menus that people use to send commands to the computer system.

green computing A program concerned with the efficient and environmentally responsible design, manufacture, operation, and disposal of IS-related products.

grid computing The use of a collection of computers, often owned by multiple individuals or organizations, to work in a coordinated manner to solve a common problem.

group consensus approach A decision-making approach that forces members in the group to reach a unanimous decision.

group support system (GSS) Software application that consists of most elements in a DSS, plus software to provide effective support in group decision making; also called *group support system* or *compute-rized collaborative work system.*

hacker A person who enjoys computer technology and spends time learning and using computer systems.

handheld computer A single-user computer that provides ease of portability because of its small size.

hardware Computer equipment used to perform input, processing, and output activities.

heuristics "Rules of thumb," or commonly accepted guidelines or procedures that usually find a good solution.

hierarchy of data Bits, characters, fields, records, files, and databases.

highly structured problems Problems that are straightforward and require known facts and relationships.

HTML tags Codes that tell the Web browser how to format text—as a heading, as a list, or as body text—and whether images, sound, and other elements should be inserted.

human resource MIS (HRMIS) An information system that is concerned with activities related to employees and potential employees of an organization, also called a personnel MIS.

hyperlink Highlighted text or graphics in a Web document, that, when clicked, opens a new Web page containing related content.

Hypertext Markup Language (HTML) The standard page description language for Web pages.

identity theft A crime in which an imposter obtains key pieces of personal identification information, such as Social Security or driver's license numbers, to impersonate someone else.

IF-THEN statements Rules that suggest certain conclusions.

implementation stage A stage of problem solving in which a solution is put into effect.

inference engine Part of the expert system that seeks information and relationships from the knowledge base and provides answers, predictions, and suggestions similar to the way a human expert would.

informatics A specialized system that combines traditional disciplines, such as science and medicine, with computer systems and technology.

information A collection of facts organized in such a way that they have additional value beyond the value of the individual facts.

information center A support function that provides users with assistance, training, application development, documentation, equipment selection and setup, standards, technical assistance, and troubleshooting.

information service unit A miniature IS department.

information systems planning Translating strategic and organizational goals into systems development initiatives.

input The activity of gathering and capturing raw data.

insider An employee, disgruntled or otherwise, working solo or in concert with outsiders to compromise corporate systems.

installation The process of physically placing the computer equipment on the site and making it operational.

instant messaging A method that allows two or more people to communicate online in real time using the Internet.

institutional DSS A DSS that handles situations or decisions that occur more than once, usually several times per year or more. An institutional DSS is used repeatedly and refined over the years.

integration testing Testing all related systems together.

intelligence stage The first stage of decision making, in which you identify and define potential problems or opportunities.

intelligent agent Programs and a knowledge base used to perform a specific task for a person, a process, or another program; also called *intelligent robot* or *bot.*

intelligent behavior The ability to learn from experiences and apply knowledge acquired from experience, handle complex situations, solve problems when important information is missing, determine what is important, react quickly and correctly to a new situation, understand visual images, process and manipulate symbols, be creative and imaginative, and use heuristics.

internal auditing Auditing performed by individuals within the organization.

Internet The world's largest computer network, consisting of thousands of interconnected networks, all freely exchanging information.

Internet Protocol (IP) A communication standard that enables computers to route communications traffic from one network to another as needed.

Internet service provider (ISP) Any organization that provides Internet access to people.

intranet An internal network based on Web technologies that allows people within an organization to exchange information and work on projects.

intrusion detection system (IDS) Software that monitors system and network resources and notifies network security personnel when it senses a possible intrusion.

IP address A 64-bit number that identifies a computer on the Internet.

joining Manipulating data to combine two or more tables.

just-in-time (JIT) inventory An inventory management approach in which inventory and materials are delivered just before they are used in manufacturing a product.

key A field or set of fields in a record that is used to identify the record.

key-indicator report A summary of the previous day's critical activities, typically available at the beginning of each workday.

knowledge The awareness and understanding of a set of information and ways that information can be made useful to support a specific task or reach a decision.

knowledge acquisition facility Part of the expert system that provides a convenient and efficient means of capturing and storing all the components of the knowledge base.

knowledge base The collection of data, rules, procedures, and relationships that must be followed to achieve value or the proper outcome.

knowledge engineer A person who has training or experience in the design, development, implementation, and maintenance of an expert system.

knowledge user The person or group who uses and benefits from the expert system.

laptop computer A personal computer designed for use by mobile users; it is small and light enough to sit comfortably on a user's lap.

LCD display Flat display that uses liquid crystals—organic, oil-like material placed between two polarizers—to form characters and graphic images on a backlit screen.

learning systems A combination of software and hardware that allows the computer to change how it functions or reacts to situations based on feedback it receives.

legal feasibility The determination of whether laws or regulations may prevent or limit a systems development project.

linking Data manipulation that combines two or more tables using common data attributes to form a new table with only the unique data attributes.

local area network (LAN) A network that connects computer systems and devices within a small area, such as an office, home, or several floors in a building.

logical design A description of the functional requirements of a system.

magnetic disk A direct access storage device, with bits represented by magnetized areas.

magnetic tape A type of sequential secondary storage medium, now used primarily for storing backups of critical organizational data in the event of a disaster.

mainframe computer A large, powerful computer often shared by hundreds of concurrent users connected to the machine via terminals.

make-or-buy decision The decision regarding whether to obtain the necessary software from internal or external sources.

management information system (MIS) An organized collection of people, procedures, software, databases, and devices that provides routine information to managers and decision makers.

market segmentation The identification of specific markets to target them with advertising messages.

marketing MIS An information system that supports managerial activities in product development, distribution, pricing decisions, promotional effectiveness, and sales forecasting.

material requirements planning (MRP) A set of inventory-control techniques that help coordinate thousands of inventory items when the demand of one item is dependent on the demand for another.

metropolitan area network (MAN) A telecommunications network that connects users and their computers in a geographical area that spans a campus or city.

mobile commerce (m-commerce) Transactions conducted anywhere, anytime.

model base Part of a DSS that allows managers and decision makers to perform quantitative analysis on both internal and external data.

model management software (MMS) Software that coordinates the use of models in a DSS.

monitoring stage The final stage of the problem-solving process, in which decision makers evaluate the implementation.

multicore microprocessor A microprocessor that combines two or more independent processors into a single computer so they can share the workload and improve processing capacity.

multimedia Text, graphics, video, animation, audio, and other media that can be used to help an organization efficiently and effectively achieve its goals.

multiprocessing The simultaneous execution of two or more instructions at the same time.

natural language processing Processing that allows the computer to understand and react to statements and commands made in a "natural" language, such as English.

Near Field Communication (NFC) A very short-range wireless connectivity technology designed for cell phones and credit cards.

netbook computer The smallest, lightest, least expensive member of the laptop computer family.

nettop computer An inexpensive desktop computer designed to be smaller, lighter, and consume much less power than a traditional desktop computer.

network operating system (NOS) Systems software that controls the computer systems and devices on a network and allows them to communicate with each other.

network-management software Software that enables a manager on a networked desktop to monitor the use of individual computers and shared hardware (such as printers); scan for viruses; and ensure compliance with software licenses.

networks Computers and equipment that are connected in a building, around the country, or around the world to enable electronic communications.

neural network A computer system that can act like or simulate the functioning of a human brain.

nominal group technique A decision-making approach that encourages feedback from individual group members, and the final decision is made by voting, similar to the way public officials are elected.

nonprogrammed decision A decision that deals with unusual or exceptional situations.

notebook computer Smaller than a laptop computer, an extremely lightweight computer that weighs less than 6 pounds and can easily fit in a briefcase.

object-oriented database A database that stores both data and its processing instructions.

object-oriented database management system (OODBMS) A group of programs that manipulate an object-oriented database and provide a user interface and connections to other application programs.

object-oriented systems development (OOSD) An approach to systems development that combines the logic of the systems development life cycle with the power of object-oriented modeling and programming.

object-relational database management system (ORDBMS) A DBMS capable of manipulating audio, video, and graphical data.

off-the-shelf software Software mass-produced by software vendors to address needs that are common across businesses, organizations, or individuals.

online analytical processing (OLAP) Software that allows users to explore data from a number of perspectives.

online transaction processing (OLTP) A form of data processing where each transaction is processed immediately, without the delay of accumulating transactions into a batch.

operational feasibility The measure of whether the project can be put into action or operation.

optimization model A process to find the best solution, usually the one that will best help the organization meet its goals.

organic light-emitting diode (OLED) display Flat display that uses a layer of organic material sandwiched between two conductors, which, in turn, are sandwiched between a glass top plate and a glass bottom plate so that when electric current is applied to the two conductors, a bright, electro-luminescent light is produced directly from the organic material.

organization A formal collection of people and other resources established to accomplish a set of goals.

organizational change How for-profit and nonprofit organizations plan for, implement, and handle change.

organizational culture The major understandings and assumptions for a business, corporation, or other organization.

output Production of useful information, usually in the form of documents and reports.

parallel computing The simultaneous execution of the same task on multiple processors to obtain results faster.

parallel start-up Running both the old and new systems for a period of time and comparing the output of the new system closely with the output of the old system; any differences are reconciled. When users are comfortable that the new system is working correctly, the old system is eliminated.

password sniffer A small program hidden in a network or a computer system that records identification numbers and passwords.

p-card (procurement card or purchasing card) A credit card used to streamline the traditional purchase order and invoice payment processes.

perceptive system A system that approximates the way a person sees, hears, and feels objects.

personal area network (PAN) A network that supports the interconnection of information technology within a range of 33 feet or so.

phase-in approach Slowly replacing components of the old system with those of the new one. This process is repeated for each application until the new system is running every application and performing as expected; also called a *piecemeal approach*.

physical design The specification of the characteristics of the system components necessary to put the logical design into action.

pilot start-up Running the new system for one group of users rather than all users.

pixel A dot of color on a photo image or a point of light on a display screen.

planned data redundancy A way of organizing data in which the logical database design is altered so that certain data entities are combined, summary totals are carried in the data records rather than calculated from elemental data, and some data attributes are repeated in more than one data entity to improve database performance.

plasma display A type of display using thousands of smart cells (pixels) consisting of electrodes and neon and xenon gases that are electrically turned into plasma (electrically charged atoms and negatively charged particles) to emit light.

Platform for Privacy Preferences (P3P) A screening technology in Web browsers that shields users from Web sites that don't provide the level of privacy protection they desire.

portable computer A computer small enough to carry easily.

predictive analysis A form of data mining that combines historical data with assumptions about future conditions to predict outcomes of events, such as future product sales or the probability that a customer will default on a loan.

preliminary evaluation An initial assessment whose purpose is to dismiss the unwanted proposals; begins after all proposals have been submitted.

primary key A field or set of fields that uniquely identifies the record.

problem solving A process that goes beyond decision making to include the implementation stage.

procedures The strategies, policies, methods, and rules for using a CBIS.

process A set of logically related tasks performed to achieve a defined outcome.

process symbol Representation of a function that is performed.

processing Converting or transforming data into useful outputs.

productivity A measure of the output achieved divided by the input required.

profit center A department within an organization that focuses on generating profits.

Program Evaluation and Review Technique (PERT) A formalized approach for developing a project schedule.

programmed decision A decision made using a rule, procedure, or quantitative method.

programmer A specialist responsible for modifying or developing programs to satisfy user requirements.

programming languages Sets of keywords, symbols, and rules for constructing statements that people can use to communicate instructions to a computer.

project deadline The date the entire project is to be completed and operational.

project milestone A critical date for the completion of a major part of the project.

project schedule A detailed description of what is to be done.

projecting Manipulating data to eliminate columns in a table.

proprietary software One-of-a-kind software designed for a specific application and owned by the company, organization, or person that uses it.

prototyping An iterative approach to the systems development process in which at each iteration requirements and alternative solutions to a problem are identified and analyzed, new solutions are designed, and a portion of the system is implemented.

quality control A process that ensures that the finished product meets the customers' needs.

questionnaires A method of gathering data when the data sources are spread over a wide geographic area.

Radio Frequency Identification (RFID) A technology that employs a microchip with an antenna to broadcast its unique identifier and location to receivers.

random access memory (RAM) A form of memory in which instructions or data can be temporarily stored.

rapid application development (RAD) A systems development approach that employs tools, techniques, and methodologies designed to speed application development.

read-only memory (ROM) A nonvolatile form of memory.

record A collection of data fields all related to one object, activity, or individual.

redundant array of independent/inexpensive disks (RAID) A method of storing data that generates extra bits of data from existing data, allowing the system to create a "reconstruction map" so that if a hard drive fails, the system can rebuild lost data.

relational model A database model that describes data in which all data elements are placed in two-dimensional tables, called *relations*, which are the logical equivalent of files.

reorder point (ROP) A critical inventory quantity level that calls for more inventory to be ordered for an item when the inventory level drops to the reorder point or critical level.

replicated database A database that holds a duplicate set of frequently used data.

request for proposal (RFP) A document that specifies in detail required resources such as hardware and software.

requirements analysis The determination of user, stakeholder, and organizational needs.

return on investment (ROI) One measure of IS value that investigates the additional profits or benefits that are generated as a percentage of the investment in IS technology.

revenue center A division within a company that generates sales or revenues.

rich Internet application (RIA) Software that has the functionality and complexity of traditional application software, but does not require local installation and runs in a Web browser.

robotics The development of mechanical or computer devices that perform tasks requiring a high degree of precision or that are tedious or hazardous for humans.

rule A conditional statement that links conditions to actions or outcomes.

satisficing model A model that will find a good—but not necessarily the best—solution to a problem.

schedule feasibility The determination of whether the project can be completed in a reasonable amount of time.

scheduled report A report produced periodically, such as daily, weekly, or monthly.

schema A description of the entire database.

script bunny A cracker with little technical savvy who downloads programs called scripts, which automate the job of breaking into computers.

search engine A valuable tool that enables you to find information on the Web by specifying words that are key to a topic of interest, known as keywords.

Secure Sockets Layer (SSL) A communications protocol used to secure sensitive data during e-commerce.

security dashboard Software that provides a comprehensive display on a single computer screen of all the vital data related to an organization's security defenses, including threats, exposures, policy compliance and incident alerts.

selecting Manipulating data to eliminate rows according to certain criteria.

semistructured or unstructured problems More complex problems in which the relationships among the pieces of data are not always clear, the data might be in a variety of formats, and the data is often difficult to manipulate or obtain.

sequential access A retrieval method in which data must be accessed in the order in which it is stored.

sequential access storage device (SASD) A device used to sequentially access secondary storage data.

server A computer used by many users to perform a specific task, such as running network or Internet applications.

site preparation Preparation of the location of a new system.

smart card A credit card–sized device with an embedded microchip to provide electronic memory and processing capability.

smartphone A phone that combines the functionality of a mobile phone, personal digital assistant, camera, Web browser, e-mail tool, and other devices into a single handheld device.

social engineering Using social skills to get computer users to provide information that allows a hacker to access an information system or its data.

software The computer programs that govern the operation of the computer.

Software as a Service (SaaS) A service that allows businesses to subscribe to Web-delivered business application software by paying a monthly service charge or a per-use fee.

software piracy The act of unauthorized copying or distribution of copyrighted software.

software suite A collection of single programs packaged together in a bundle.

speech-recognition technology Input devices that recognize human speech.

spyware Software that is installed on a personal computer to intercept or take partial control over the user's interaction with the computer without knowledge or permission of the user.

stakeholders People who, either themselves or through the organization they represent, ultimately benefit from the systems development project.

start-up The process of making the final tested information system fully operational.

steering committee An advisory group consisting of senior management and users from the IS department and other functional areas.

storage area network (SAN) A special-purpose, high-speed network that provides high-speed connections among data-storage devices and computers over a network.

strategic alliance (strategic partnership) An agreement between two or more companies that involves the joint production and distribution of goods and services.

strategic planning Determining long-term objectives by analyzing the strengths and weaknesses of the organization, predicting future trends, and projecting the development of new product lines.

structured interview An interview where the questions are written in advance.

supercomputers The most powerful computer systems with the fastest processing speeds.

supply chain management (SCM) A system that includes planning, executing, and controlling all activities involved in raw material sourcing and procurement, converting raw materials to finished products, and warehousing and delivering finished product to customers.

system performance measurement Monitoring the system—the number of errors encountered, the amount of memory required, the amount of processing or CPU time needed, and other problems.

system performance products Software that measures all components of the computer-based information system, including hardware, software, database, telecommunications, and network systems.

system testing Testing the entire system of programs.

systems analysis The systems development phase involving the study of existing systems and work processes to identify strengths, weaknesses, and opportunities for improvement.

systems analyst A professional who specializes in analyzing and designing business systems.

systems design The systems development phase that defines how the information system will do what it must do to obtain the problem solution.

systems development The activity of creating or modifying existing business systems.

systems implementation The systems development phase involving the creation or acquisition of various system components detailed in the systems design, assembling them, and placing the new or modified system into operation.

systems investigation The systems development phase during which problems and opportunities are identified and considered in light of the goals of the business.

systems investigation report A summary of the results of the systems investigation and the process of feasibility analysis and recommendation of a course of action.

systems maintenance A stage of systems development that involves checking, changing, and enhancing the system to make it more useful in achieving user and organizational goals.

systems maintenance and review The systems development phase that ensures the system operates and modifies the system so that it continues to meet changing business needs.

systems operation Use of a new or modified system.

systems request form A document filled out by someone who wants the IS department to initiate systems investigation.

systems review The final step of systems development, involving the analysis of systems to make sure that they are operating as intended.

tablet computer A portable, lightweight computer with no keyboard that allows you to roam the office, home, or factory floor carrying the device like a clipboard.

technical feasibility Assessment of whether the hardware, software, and other system components can be acquired or developed to solve the problem.

technology acceptance model (TAM) A model that describes the factors leading to higher levels of acceptance and usage of technology.

technology diffusion A measure of how widely technology is spread throughout the organization.

technology infrastructure All the hardware, software, databases, telecommunications, people, and procedures that are configured to collect, manipulate, store, and process data into information.

technology infusion The extent to which technology is deeply integrated into an area or department.

telecommunications The electronic transmission of signals for communications; enables organizations to carry out their processes and tasks through effective computer networks.

telecommunications medium Any material substance that carries an electronic signal to support communications between a sending and receiving device.

thin client A low-cost, centrally managed computer with essential but limited capabilities and no extra drives (such as CD or DVD drives) or expansion slots.

total cost of ownership (TCO) The sum of all costs over the life of an information system, including the costs to acquire components such as the

technology, technical support, administrative costs, and end-user operations.

traditional approach to data management An approach to data management whereby each distinct operational system used data files dedicated to that system.

transaction Any business-related exchange, such as payments to employees, sales to customers, and payments to suppliers.

transaction processing cycle The process of data collection, data editing, data correction, data manipulation, data storage, and document production.

transaction processing system (TPS) An organized collection of people, procedures, software, databases, and devices used to record completed business transactions.

tunneling The process by which VPNs transfer information by encapsulating traffic in IP packets over the Internet.

ultra wideband (UWB) A form of short range communications that employs extremely short electromagnetic pulses lasting just 50 to 1,000 picoseconds that are transmitted across a broad range of radio frequencies of several gigahertz.

Uniform Resource Locator (URL) A Web address that specifies the exact location of a Web page using letters and words that map to an IP address and a location on the host.

unit testing Testing of individual programs.

unstructured interview An interview where the questions are not written in advance.

user acceptance document A formal agreement signed by the user that states that a phase of the installation or the complete system is approved.

user interface The element of the operating system that allows you to access and command the computer system.

user preparation The process of readying managers, decision makers, employees, other users, and stakeholders for new systems.

users People who will interact with the system regularly.

utility program Program that helps to perform maintenance or correct problems with a computer system.

value chain A series (chain) of activities that includes inbound logistics, warehouse and storage, production, finished product storage, outbound logistics, marketing and sales, and customer service.

virtual reality The simulation of a real or imagined environment that can be experienced visually in three dimensions.

virtual reality system A system that enables one or more users to move and react in a computer-simulated environment.

virtual tape A storage device for less frequently needed data so that it appears to be stored entirely on tape cartridges, although some parts of it might actually be located on faster hard disks.

virtual workgroups Teams of people located around the world working on common problems.

vision systems The hardware and software that permit computers to capture, store, and manipulate visual images.

volume testing Testing the application with a large amount of data.

Web Server and client software, the hypertext transfer protocol (http), standards, and mark-up languages that combine to deliver information and services over the Internet.

Web 2.0 The Web as a computing platform that supports software applications and the sharing of information among users.

Web browser Web client software such as Internet Explorer, Firefox, Chrome, Safari, and Opera used to view Web pages.

Web log (blog) A Web site that people can create and use to write about their observations, experiences, and opinions on a wide range of topics.

Web portal A Web page that combines useful information and links and acts as an entry point to the Web— they typically include a search engine, a subject directory, daily headlines, and other items of interest. Many people choose a Web portal as their browser's home page (the first page you open when you begin browsing the Web).

wide area network (WAN)   A telecommunications network that connects large geographic regions.

Wi-Fi   A medium-range wireless telecommunications technology brand owned by the Wi-Fi Alliance.

workstation   A more powerful personal computer used for mathematical computing, computer-aided design, and other high-end processing, but still small enough to fit on a desktop.

Worldwide Interoperability for Microwave Access (WiMAX)   The common name for a set of IEEE 802.16 wireless metropolitan area network standards that support various types of communications access.

Subject

Note: A boldface page number indicates a key term and the location of its definition in the text.

C

Company Names